# SUPERMACROPOROUS CRYOGELS

## Biomedical and Biotechnological Applications

# SUPERMACROPOROUS CRYOGELS

## Biomedical and Biotechnological Applications

Edited by
## ASHOK KUMAR

CRC Press
Taylor & Francis Group
Boca Raton London New York

CRC Press is an imprint of the
Taylor & Francis Group, an **informa** business

Cover photos:

·   Upper photograph on the cover is courtesy of Dr. Phil Salmon, Bruker microCT, Belgium, and illustrates microCT image of cryogel showing pore distribution.

·   Lower photograph on the cover is contributed by Ms. Jyoti Kumari and Prof. Ashok Kumar, Indian Institute of Technology Kanpur, India, confocal microscopic image of cell laden cryogel scaffold.

CRC Press
Taylor & Francis Group
6000 Broken Sound Parkway NW, Suite 300
Boca Raton, FL 33487-2742

First issued in paperback 2019

© 2016 by Taylor & Francis Group, LLC
CRC Press is an imprint of Taylor & Francis Group, an Informa business

No claim to original U.S. Government works

ISBN-13: 978-0-4822-2881-6 (hbk)
ISBN-13: 978-0-367-86947-2 (pbk)

This book contains information obtained from authentic and highly regarded sources. Reasonable efforts have been made to publish reliable data and information, but the author and publisher cannot assume responsibility for the validity of all materials or the consequences of their use. The authors and publishers have attempted to trace the copyright holders of all material reproduced in this publication and apologize to copyright holders if permission to publish in this form has not been obtained. If any copyright material has not been acknowledged please write and let us know so we may rectify in any future reprint.

Except as permitted under U.S. Copyright Law, no part of this book may be reprinted, reproduced, transmitted, or utilized in any form by any electronic, mechanical, or other means, now known or hereafter invented, including photocopying, microfilming, and recording, or in any information storage or retrieval system, without written permission from the publishers.

For permission to photocopy or use material electronically from this work, please access www.copyright.com (http://www.copyright.com/) or contact the Copyright Clearance Center, Inc. (CCC), 222 Rosewood Drive, Danvers, MA 01923, 978-750-8400. CCC is a not-for-profit organization that provides licenses and registration for a variety of users. For organizations that have been granted a photocopy license by the CCC, a separate system of payment has been arranged.

**Trademark Notice:** Product or corporate names may be trademarks or registered trademarks, and are used only for identification and explanation without intent to infringe.

---
**Library of Congress Cataloging-in-Publication Data**
---

Names: Kumar, Ashok, 1963- , editor.
Title: Supermacroporous cryogels : biomedical and biotechnological applications / editor, Ashok Kumar.
Description: Boca Raton : Taylor & Francis, 2016. | Includes bibliographical references and index.
Identifiers: LCCN 2016000474 | ISBN 9781482228816 (alk. paper)
Subjects: | MESH: Cryogels | Biocompatible Materials | Biomedical Technology
Classification: LCC R857.M3 | NLM QT 37.5.P7 | DDC 610.28--dc23
LC record available at http://lccn.loc.gov/2016000474

---

**Visit the Taylor & Francis Web site at**
http://www.taylorandfrancis.com

**and the CRC Press Web site at**
http://www.crcpress.com

# Contents

Preface ................................................................................................................. ix
Editor .................................................................................................................. xi
Contributors ..................................................................................................... xiii

## SECTION I  Synthesis and Characterization of Cryogel Matrices

**Chapter 1**   Cryogels and Related Research: A Glance over the Past Few Decades ........................................................................................ 3

   *Deepak Bushan Raina and Ashok Kumar*

**Chapter 2**   Synthesis and Characterization of Cryogels ..................................... 35

   *Apeksha Damania, Arun Kumar Teotia and Ashok Kumar*

**Chapter 3**   Production of Synthetic Cryogels and the Study of Porosity Thereof ... 91

   *Irina N. Savina and Igor Yu. Galaev*

**Chapter 4**   Fabrication and Characterization of Cryogel Beads and Composite Monoliths ................................................................ 111

   *Junxian Yun, Linhong Xu, Dong-Qiang Lin, Kejian Yao and Shan-Jing Yao*

## SECTION II  Application of Supermacroporous Cryogels in Biomedical Engineering

**Chapter 5**   Cryogel Tissue Phantoms with Uniform Elasticity for Medical Imaging ..................................................................................... 145

   *Azizeh-Mitra Yousefi, Corina S. Drapaca and Colin J. Kazina*

**Chapter 6**   Cryogels in Regenerative Medicine ................................................ 175

   *Irina N. Savina, Rostislav V. Shevchenko, Iain U. Allan, Matthew Illsley and Sergey V. Mikhalovsky*

**Chapter 7** Biocompatibility of Macroporous Cryogel Materials ..................... 195
Akhilesh Kumar Shakya and Ashok Kumar

**Chapter 8** Cryogel Biomaterials for Musculoskeletal Tissue Engineering ....... 215
Ruchi Mishra, Sumrita Bhat and Ashok Kumar

**Chapter 9** Cryogels for Neural Tissue Engineering ......................................... 251
Tanushree Vishnoi and Ashok Kumar

**Chapter 10** Supermacroporous Cryogels as Scaffolds for Pancreatic Islet
Transplantation ................................................................................ 277
Y. Petrenko, A. Petrenko, P. Vardi and K. Bloch

# SECTION III   Application of Supermacroporous Cryogels in Biotechnology

**Chapter 11** Enzymatic Biocatalysts Immobilized on/in the Cryogel-Type
Carriers ............................................................................................ 307
Elena N. Efremenko, Ilya V. Lyagin and Vladimir I. Lozinsky

**Chapter 12** Application of Cryogels in Water and Wastewater Treatment ......... 331
Linda Önnby

**Chapter 13** Cryogels in High Throughput Processes ......................................... 361
Joyita Sarkar, Ankur Gupta and Ashok Kumar

**Chapter 14** Cryogels: Applications in Extracorporeal Affinity Therapy ............ 387
Handan Yavuz, Nilay Bereli, Gözde Baydemir, Müge Andaç,
Deniz Türkmen and Adil Denizli

**Chapter 15** Cryogel Bioreactor for Therapeutic Protein Production .................. 417
Era Jain and Ashok Kumar

**Chapter 16** Supermacroporous Functional Cryogel Stationary Matrices
for Efficient Cell Separation ............................................................. 443

*Akshay Srivastava, Samarjeet Singh and Ashok Kumar*

**Index** ................................................................................................................ 463

# Preface

Macroporous hydrogels propound fascinating prospects in biomedicine and biotechnology, and among these 'cryogels' have emerged as a versatile class of hydrogels. Cryogels are supermacroporous polymeric gels that are fabricated by cryogelation technology, that is, at moderately frozen conditions, using ice as a porogen. Supermacroporous implies a pore size of 100 μm and above. Such large pore size engenders intriguing properties such as allowing easy passage of crude samples without clogging the pores and rendering enough space for cell growth without impeding the exchange of nutrients and waste. A large variety of monomers, polymers, cross-linkers and synthesis conditions can be used to produce these gels that provide scope to exploit various combinations to get the desired property of the matrix. Extensive studies have been conducted on the potential of these cryogels in various fields, which is reflected by the increase in the number of scientific publications as well as the number of research groups working on cryogels over the past 2–3 decades. The cryogels have been successfully used as a matrix for chromatography and separation. A wide range of analytes including proteins, nucleic acids, microbial and animal cells, small biomolecules such as endocrine disruptors, viruses and so on has been separated on cryogel matrices. In addition, using appropriate polymers and monomers makes the cryogel biocompatible, thus making it an excellent matrix for biomedical and tissue engineering applications. Easy scale-up and scale-down of these cryogels has also made them suitable as a matrix for a bioreactor for production and high throughput systems for analysis, respectively.

The diversity of the applications of the cryogel triggered exponential augmentation and advancement of research on cryogels. In fact, this miscellany and development of cryogel has ushered one of the main operations of the industry, Protista International AB, Sweden. The biotechnology division of this industry has mainly dealt with development, research and production of cryogels. At present, substantial progress has been made in terms of clinical trials and commercialization of the cryogel matrices.

In three sections this book thoroughly covers all the facets of cryogels starting from their synthesis to their applications in various fields. The first introductory chapter provides a quick and highlighted look into the history, synthesis, properties and application of cryogels and would help in understanding the concept of cryogels to a naive reader in this field. The first section of the book features the synthesis and characterization of cryogels. This section elaborates the synthesis of the matrices in different formats like beads, monoliths and so on from diverse precursors of various origins and different methods of inducing gel formation. It also focuses on the factors affecting the properties of cryogels; their morphological, physical and chemical characterization as well as their surface modification.

In the second section, the biomedical aspects of cryogels are discussed. It categorically describes the biocompatible nature of the cryogels and their recent usage in medical imaging by creating phantoms of various tissues. The performance of cryogels has been incredible in skeletal tissue engineering. A polymeric cryogel for repair of cartilage defect and an inorganic organic biocomposite cryogel for restoring a bone defect are undergoing clinical trials; both are discussed in detail in this section. The section also focuses on the application of cryogels in regenerative medicine and as a scaffold for neural and pancreatic tissue engineering. The third section deals with the application of cryogels in biotechnology. It covers a wide area providing a comprehensive view of the involvement of cryogels in biocatalysis, cell separation, wastewater treatment, high throughput processes, bioreactor and so on. A cryogel as a matrix for protein immobilization, bioreactor and cell separation has eminent applications in both research laboratories and industries. In fact, a cryogel-based leukocyte depletion filter is on the threshold of commercialization with the technology being transferred to an industry. Recently, cryogel is also evolving to play a crucial role in relieving the plight on the environment by showing propitious outcome in wastewater treatment and high throughout analysis of endocrine disruptors. All these applications have been proffered exhaustively in this section.

This is the first book that is dedicated to conferring the applications of supermacroporous cryogels meticulously. I have tried to give an account of the evolutions in cryogel research in the past decade by presenting our research developments and by collecting research contributions from researchers around the globe. Meanwhile, the previous research on the development of cryogels from the point of view of polymer chemistry has also been taken into consideration and has been discussed in depth. The book has been crafted with the notion to instill the concept of cryogels in an unacquainted mind and to update adept researchers about recent developments.

This book would not have been conceivable without all the research carried out on cryogels. Therefore, I express my gratitude toward all scientists and researchers around the world who have been exploring cryogels. I also appreciate the help that has been provided by all my past and present students to accomplish this book, especially Apeksha Damania and Joyita Sarkar. I also acknowledge the Department of Science and Technology and the Department of Biotechnology, Ministry of Science and Technology, Government of India, and my institute, IIT Kanpur, for financially assisting me in carrying out research on cryogels. Lastly, I thank CRC Press, Taylor & Francis Group, Boca Raton, Florida, for considering this proposal and publishing this book.

**Ashok Kumar**
*Kanpur, India*

# Editor

**Professor Ashok Kumar, PhD,** earned his doctorate in biotechnology from IIT Roorkee, India and has carried out postdoctoral research at Lund University, Sweden and Nagoya University, Japan. At present, Prof. Kumar is a professor of bioengineering and holds the Lalit Mohan Kapoor endowed chair professorship in the Department of Biological Sciences and Bioengineering at the Indian Institute of Technology Kanpur, India. He has carried out profound research on synthesis, properties and disparate applications of cryogels and smart polymers. His research interests are in biomaterials, tissue engineering, stem cell technology, regenerative medicine, bioprocess engineering, bioseparations and environmental biotechnology. He has published more than 125 peer-reviewed research articles, has several patents and has edited five books. He is an executive board member of the Asian Federation of Biotechnology and serves on the editorial boards and expert committees of many reputable journals and scientific organizations. His major achievement in biotechnology and biomedical research has been the fabrication of biocompatible cryogel biomaterials for the repair of bone and cartilage defects, stem cell separation technology, bioartificial liver support, bioreactor and matrices for air and water filtration, which are at the advanced stage of applications. Prof. Kumar has carried out numerous national and international scientific projects and has been awarded with GRO Samsung project award, Korea and the TATA Innovation Fellowship from DBT, Ministry of Science and Technology, Government of India for his research contributions.

# Contributors

**Iain U. Allan**
University of Brighton
Brighton, United Kingdom

**Müge Andaç**
Hacettepe University
Ankara, Turkey

**Gözde Baydemir**
Hacettepe University
Ankara, Turkey

**Nilay Bereli**
Hacettepe University
Ankara, Turkey

**Sumrita Bhat**
Indian Institute of Technology Kanpur
Kanpur, India

and

University of Calgary
Calgary, Canada

**K. Bloch**
Sackler Faculty of Medicine
Tel Aviv University
Tel Aviv, Israel

**Apeksha Damania**
Indian Institute of Technology Kanpur
Kanpur, India

**Adil Denizli**
Hacettepe University
Ankara, Turkey

**Corina S. Drapaca**
Pennsylvania State University
University Park, Pennsylvania

**Elena N. Efremenko**
M.V. Lomonosov Moscow State University
Moscow, Russia

**Igor Yu. Galaev**
DSM Biotechnology Center
Delft, Netherlands

**Ankur Gupta**
Indian Institute of Technology Kanpur
Kanpur, India

**Matthew Illsley**
University of Brighton
Brighton, United Kingdom

**Era Jain**
Saint Louis University
Saint Louis, Missouri

**Colin J. Kazina**
University of Manitoba
Manitoba, Canada

and

Winnipeg Children's Hospital
Winnipeg, Canada

**Ashok Kumar**
Indian Institute of Technology Kanpur
Kanpur, India

**Dong-Qiang Lin**
Zhejiang University
Hangzhou, China

**Vladimir I. Lozinsky**
A.N. Nesmeyanov Institute
 of Organoelement Compounds
Russian Academy of Sciences
Moscow, Russia

**Ilya V. Lyagin**
M.V. Lomonosov Moscow State
 University
Moscow, Russia

**Sergey V. Mikhalovsky**
University of Brighton
Brighton, United Kingdom

and

Nazarbayev University
Astana, Kazakhstan

**Ruchi Mishra**
The Ohio State University
Columbus, Ohio

**Linda Önnby**
Lund University
Lund, Sweden

**A. Petrenko**
Institute for Problems of Cryobiology
 and Cryomedicine
Kharkov, Ukraine

**Y. Petrenko**
Institute for Problems of Cryobiology
 and Cryomedicine
Kharkov, Ukraine

**Deepak Bushan Raina**
Indian Institute of Technology Kanpur
Kanpur, India

and

Lund University Hospital
Lund, Sweden

**Joyita Sarkar**
Indian Institute of Technology Kanpur
Kanpur, India

**Irina N. Savina**
University of Brighton
Brighton, United Kingdom

**Akhilesh Kumar Shakya**
Texas Tech University
Lubbock, Texas

**Rostislav V. Shevchenko**
Pharmidex Pharmaceutical Services Ltd.
London, United Kingdom

**Samarjeet Singh**
Indian Institute of Technology Kanpur
Kanpur, India

**Akshay Srivastava**
National University of Ireland
Galway, Ireland

**Arun Kumar Teotia**
Indian Institute of Technology Kanpur
Kanpur, India

**Deniz Türkmen**
Hacettepe University
Ankara, Turkey

**P. Vardi**
Sackler Faculty of Medicine
Tel Aviv University
Tel Aviv, Israel

**Tanushree Vishnoi**
Indian Institute of Technology Kanpur
Kanpur, India

**Linhong Xu**
China University of Geosciences (Wuhan)
Wuhan, China

**Kejian Yao**
Zhejiang University of Technology
Hangzhou, China

**Shan-Jing Yao**
Zhejiang University
Hangzhou, China

**Handan Yavuz**
Hacettepe University
Ankara, Turkey

**Azizeh-Mitra Yousefi**
Miami University
Oxford, Ohio

**Junxian Yun**
Zhejiang University of Technology
Hangzhou, China

# Section I

Synthesis and Characterization of Cryogel Matrices

# 1 Cryogels and Related Research

## *A Glance over the Past Few Decades*

*Deepak Bushan Raina and Ashok Kumar**

## CONTENTS

1.1 Cryogels: The Origin ..................................................................................4
1.2 Gels and Different Gelation Methods ........................................................5
1.3 Introduction to Cryogels ............................................................................6
    1.3.1 Process of Cryogelation ..................................................................7
    1.3.2 How Are Cryogels Different from Conventional Gels? ..................8
    1.3.3 The History and Statistics ...............................................................9
        1.3.3.1 Cryogels: From a Historical Perspective ..........................9
        1.3.3.2 Cryogels: From a Statistical Perspective ........................11
1.4 Cryogels: Synthesis and Properties ..........................................................15
    1.4.1 Cryogel Synthesis .........................................................................15
    1.4.2 Cryogel Properties ........................................................................16
        1.4.2.1 Supermacroporous Architecture with Interconnected Pores ...16
        1.4.2.2 Rapid Swelling Mechanism ............................................16
        1.4.2.3 Mechanical Stability and Viscoelasticity .......................17
        1.4.2.4 Chemical Stability and Modifiability .............................18
        1.4.2.5 Biocompatibility .............................................................18
        1.4.2.6 Ease of Fabrication, Economical Operations and Longer Shelf Life ...18
1.5 Cryogel Applications ................................................................................19
    1.5.1 Cryogels in Immobilization, Separation and Imprinting ..............19
    1.5.2 Cryogels in Tissue Engineering and Biomedical Applications ....21
    1.5.3 Cryogels in Environmental Biotechnology ..................................23
    1.5.4 Other Applications of Cryogels ....................................................26
Acknowledgements ..........................................................................................27
List of Abbreviations .......................................................................................27
References .......................................................................................................28

---

* Corresponding author: E-mail: ashokkum@iitk.ac.in; Tel: +91-512-2594051.

## ABSTRACT

This chapter describes how cryogels have evolved as promising matrices in various biotechnological and biomedical application areas. Cryogels date back nearly 40 years and here we describe the origin of these materials followed by characterization and their applications and also emphasize the differences in cryogelation and other related techniques. The main aim is to familiarize readers with the principles underlying the mechanism of cryogelation, how these matrices are characterized, provide historical and statistical data on their evolution over the past few decades and finally provide a brief idea of their applications in the areas of separation sciences, tissue engineering, environmental sciences, and so on.

## KEYWORDS

Cryogels, synthesis, characterization, applications, cryogel research overview

## 1.1 CRYOGELS: THE ORIGIN

The earliest reports of cryogel synthesis date back nearly 40 years; however, they gained importance in the early 1980s (Lozinsky et al., 1982, 2003; Lozinsky, 2002). Lozinsky and his co-workers at the Institute of Organo-Element Compounds, Russian Academy of Sciences, took up the research on studying the principles underlying the process of cryogelation. The initial studies focused on understanding the mechanism of cryogel synthesis and characterizing the macroporous architecture of cryogels. They experimented with how cross-linked polyacrylamide (pAAm) cryogels could be obtained by freezing the precursors and initiators at subzero temperatures to successfully achieve a highly interconnected, macroporous structure (Lozinsky et al., 1982; Belavtseva et al., 1984). Their initial studies also explained that these gels could be prepared by using different solvent systems, such as, for instance, replacing water by formamide or dimethylsulphoxide (DMSO) (Lozinsky et al., 1984a,b). These studies also explained the osmotic characteristics of synthesized cryogels along with the relation of precursor concentrations on important cryogel properties like osmotic characteristics, microstructure architecture and so on (Lozinsky et al., 1984a,b). One of the earliest available studies also provided an insight into the supermacroporous architecture by means of various microscopic tools such as light, scanning and transmission electron microscopy. This first of its kind study unfolded some mesmerizing results. The macroporous architecture was analysed for the first time using advanced imaging tools and the study concluded with the authenticity of the principle of cryostructurization by observing pores ranging from a few microns up to hundreds of microns and proved the heterogeneity of the system (Lozinsky et al., 1986a,b). The following experiments concluded with some very interesting results such as the effect of temperature regime at both the times of freezing and defreezing the polymeric structure, the effect of concentration of precursors on cross-linking and so on (Lozinsky et al., 1986a, 1989a). The rudimentary idea behind the first few experiments was to establish the worth of these macroporous structures in the widespread field of biotechnology.

Once the supermacroporous cryogels were characterized by the existing techniques from that time, further advancements started seeking out various applications of these matrices. From the available literature, the first few applications of cryogels were based on immobilization of bacterial cells and enzymes onto the cryogel matrices (Lozinsky et al., 1989b). Cryogels gained popularity in the 1990s when different groups around the world started exploiting the physicochemical properties of these supermacroporous materials. Since then, these macroporous gel matrices have been used extensively in various biotechnological and biomedical areas such as chromatography; carriers for enzymes, cells and other biomolecules and macromolecules; molecular imprinting; environmental biotechnology; tissue engineering and therapeutics; high throughput screening of drugs and hazardous molecules; and so on (Bölgen et al., 2007; Kumar and Srivastava, 2010; Asliyuce et al., 2012; Onnby et al., 2013).

The following sections of the chapter will mainly deal with a cursory glance over the method of cryogelation and cryogel characterization techniques, while a major focus will be on how these materials have come to the fore and served numerous biological applications and their importance as a material of the future.

## 1.2 GELS AND DIFFERENT GELATION METHODS

Most often gels are defined as polymer systems that are held together by bonds that are physical or chemical in nature and such systems do not flow over time. A gel can be thought of as a system where the solute is immobilized in the mobile solvent phase. An example could be a hydrogel formed by reducing the temperature of a solution of agarose below its gel point. Once the gelation of the system occurs, the polymeric chains form an entangled network of nonfluctuating, physical bonds resulting in immobilization of the solute polymer, while the mobile solvent fills in the gaps between the individual polymer chains (Clark et al., 1983; Lozinsky et al., 2003). Irrespective of a very high percentage of solvent existing in gels, these structures still possess varying strengths ranging from soft to very hard (Clark et al., 1983; DeRossi et al., 1991). This is mainly due to a three-dimensional (3D) network of bonds that is homogenously spread around the structure. While there are many factors that decide the strength of gels, the nature or type of bond formed plays a pivotal role. The structure of the polymer mainly contributes to the nature of the bonds and thus it is very important to choose polymers selectively for every end use. A variety of methods can be implemented for achieving gels with different physicochemical properties and Table 1.1 provides a list of different gelation methods along with some examples.

Since the main aim of this chapter is to provide the readers with a platform to understand the mechanism of cryogels and how these gels fit into various different biotechnological applications, we will now emphasize *cryogels* from here onward. As far as different gelation methods are concerned, cryogels can be fabricated from almost all methods mentioned in Table 1.1 with the exception of thermotropic gelation because it demands high reaction temperatures, contrary to the process of cryogel formation.

**TABLE 1.1**
**Different Gelation Methods**

| Gelation Method | Gelation Mechanism | Examples |
| --- | --- | --- |
| Chemotropic | A covalent network of chemical bonds between reacting molecules mostly caused by a reactive functional group and a cross-linker | pAAM, chitosan gels, etc. |
| Ionotropic | As the name suggests, ionic interactions lead to bond formation | Alginate-poly lysine |
| Chelatotropic | Chelation leads to co-ordination bonds giving rise to a stable structure | Calcium-alginate gels |
| Solvotropic | Changes in solvent composition leads to bond formation | Cellulose acetate fibers |
| Thermotropic | An increased heat in the system leads to the gelation of precursors | poly (*N*-isopropyl acrylamide) (pNiPAAm), hydroxyethyl cellulose |
| Psychotropic | A gel system achieved at low temperatures (nonfreezing, low temperatures) | Agarose, starch |
| Cryotropic | Freezing of the polymer system causes formation of highly stable bonds | pAAm cryogel, chitosan cryogel |

*Source:* Modified from Lozinsky, V. I., Galaev, I. Y., Plieva, F. M., Savina, I. N., Jungvid, H. and Mattiasson, B. (2003). *TRENDS in Biotechnology* 21(10): 445–451. With permission.

## 1.3 INTRODUCTION TO CRYOGELS

**Cryogels** are supermacroporous gel matrices fabricated at moderately freezing conditions (Lozinsky et al., 2003; Kumar et al., 2010). The word itself is the wedlock of two words—'cryo' and 'gel', with the former having been derived from the Greek word '*Kruos*' meaning frost, and the latter being coined by Scottish chemist Thomas Graham to refer to materials that are jelly-like, possessing varying stiffness and strength, and are held together by various chemical bonds that do not change with time. Cryogels are formed by the cross-linking, polymerization or physical entanglement or gelation of monomeric or polymeric precursors under subzero temperatures to form a network of polymeric chains distributed heterogeneously along the matrix (Lozinsky et al., 2001; Berillo et al., 2012). The frozen solvent acts as a porogen, which upon thawing unveils a highly macroporous structure with high interconnectivity of pores that allows free movement of gases, liquid and submicron particles without much difficulty (Vladimir, 2002; Kumar et al., 2006a,b; Jain and Kumar, 2009). The process of cryogelation can be seen as that of permafrost formation, where, due to freezing temperatures, different parts of land remain frozen most of the time, and upon thawing, they give rise to polygonal and interconnected pieces of land (Kumar et al., 2010).

## 1.3.1 PROCESS OF CRYOGELATION

The process begins with dissolving the precursors in an appropriate solvent, most commonly water (Sharma et al., 2013). While most monomers or polymers are soluble in water, some are partially soluble or insoluble and thus appropriate additives are used to achieve a homogenous solution of the solute in the solvent (Bhat et al., 2011). However, different solvent systems can be used based on the requirements of the fabrication process (Zhang and Chu, 2003). Once the solute is completely dissolved, precooling of the solution is recommended in order to ensure a homogenous reaction temperature as well as to reduce the rate of reaction (Jain and Kumar, 2013). In some cases, though, where precooling is unmanageable, it is possible to skip this step. An example could be a solution containing agarose, which cannot be cooled below 39°C as it would form a gel at room temperature. Once the solution is precooled, an appropriate cross-linker or initiator can be added followed by leaving the system undisturbed at subzero temperatures for 12–16 h. Upon thawing the gel matrices, the solvent melts, leading to pore formations. The cryogelation process is explained in detail in Figure 1.1.

As evident from the figure, the solution containing the precursors is allowed to freeze at temperatures generally 10 degrees below the solvent freezing point. As soon as ice crystal formation initiates, there is formation of an *unfrozen liquid microphase* (ULMP) (Lozinsky et al., 2003; Kumar et al., 2010), which represents a very high concentration of precursors in minimal volume of the solvent. Micro-CT analysis also shows changes in porosity of the frozen structure over time indicated by the increase in porosity and pore size of the system (Figure 1.1b). Polymer systems are generally comprised of water that can be classified as 'free water', meaning water that is free of solute molecules, or 'bound water', which indicates the portion that has maximum interaction with the solute (Wolfe et al., 2002). This bound water contains a very high concentration of reaction precursors, which is also referred to as 'cryo-concentration' (Lozinsky et al., 2001). Despite the subzero temperatures, it is observed that the phenomenon of cryo-concentration catalyses the reaction between the precursors and cross-linkers or initiators, leading to the formation of a highly interconnected polymer network. The phenomenon of cryo-concentration is majorly the reason of such a strong gel matrix formation at subzero temperatures and it is also reported that sometimes these reactions proceed faster than the same reaction carried out at room temperature. The free water, however, keeps forming ice crystals, which progressively grow until they meet the facet of another crystal (Kumar et al., 2010). The shape and size of these crystals depends on various factors with the most important being the concentration of the precursors and freezing temperature (Kumar et al., 2006a,b; Ivanov et al., 2007). Upon thawing the frozen system, the ice crystals melt, giving way to large interconnected pores while the ULMP forms the strong polymer network (Lozinsky et al., 2003). The pore size of cryogels varies from a few micrometers up to several hundred micrometers (Kumar et al., 2006a,b). Such a large pore size enables the free flow of gases, liquids and particulate matter and makes them important gel matrices in various biotechnological processes (Kumar and Bhardwaj, 2008; Jain and Kumar, 2009).

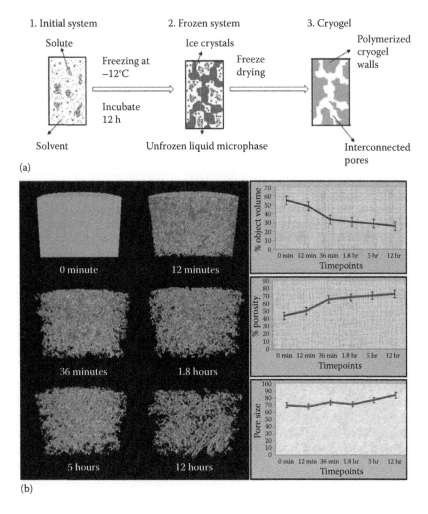

**FIGURE 1.1** Cryogelation process. (a) Schematic of the cryogelation process beginning from the initial system consisting of a solute and the solvent, a frozen system consisting of an unfrozen liquid microphase and finally the cryogel structure with supermacroporous architecture. (b) Microcomputed tomographic analysis of the cryogelation process indicating the synthesis mechanism and structural changes in the porous structure over time. The graphs in (b) show the quantification of volume and porosity change in the cryogel structure over time. (Reprinted from *Materials Today* 13(11), Kumar, A., Mishra, R., Reinwald, Y. and Bhat, S., Cryogels: Freezing unveiled by thawing, 42–44, Copyright 2010, with permission from Elsevier.)

### 1.3.2 How Are Cryogels Different from Conventional Gels?

For the readers to understand the differences that demarcate cryogels from gels synthesized by other methods, let us consider a typical hydrogel formed at room temperature. Once an appropriate stimulus (e.g., pH, temperature, cross-linker, initiator, etc.) is applied to the solution containing the polymer, the precursors start reacting to form physical or chemical bonds while the solvent molecules fill in the void between the polymeric chains. Once all reactive groups are exhausted, the reaction stops, leaving behind a connected network of polymers held together by different bonds. On the contrary, cryogelation takes place at subzero temperatures and thus allows the freezing of the solvent molecules that are free of solute, while the solute is concentrated in the ULMP where the reaction accelerates owing to the formation of what is called 'cryo-concentration', a very high concentration of precursors at subzero temperatures, and different types of bonds are formed depending on the structure of the polymers (Lozinsky et al., 2003). Upon completion of the reaction, the gel is then thawed, unleashing a very highly connected and supermacroporous 3D architecture. Ice crystals give way to pores while reactions in the UFLM maintain the structural integrity of the cryogels.

Structurally, one might find some resemblance between cryogels and gels synthesized by freeze-drying. Such gels also possess a similar architecture in many cases, wherein the solvent is frozen and later subjected to sublimation that leads to the formation of pores. However, a marked difference lies in the synthesis procedure and the gel formation does not take place due to the UFLM or the cryoconcentration, but instead takes place normally like in other systems described earlier. Since a majority of their structure contains water or another solvent, the system still can be frozen and sublimed to form a porous structure. However, to produce a continuous structure of any desired shape by this process is practically impossible and there is limited control on the morphology of the structure (Lozinsky et al., 2003). The process of solvent removal in cryogels is much simpler and does not require lyophilization every time, which introduces further simplicity to the system and imparts ease to the user during fabrication.

### 1.3.3 THE HISTORY AND STATISTICS

#### 1.3.3.1 Cryogels: From a Historical Perspective

Ever since the inception of cryogels as gel matrices of biotechnological interest, myriad different monomers/polymers has been used to fabricate cryogel matrices (Lozinsky, 2002). This section comprises information on how cryogels were developed and became the choice of biotechnological interest for many scientists across the globe and also provides some information on how different types of cryogels can be synthesized. As mentioned earlier, cryogels are more than four decades old now and owing to their physicochemical properties, they have widely been used in a variety of applications. The very early available reports suggest that Lozinski and co-workers fabricated and characterized the first few cryogel matrices with selective polymeric precursors. Poly-acrylamide (pAAm) and poly-vinyl alcohol (PVA) cryogels have been the oldest of cryogels matrices synthesized in that time (Lozinsky et al., 1984a,b, 1986a,b,c). The initial studies suggest that the research was more focused on understanding the principle of cryo-structurization of polymeric systems including PVA and pAAm. Thus, this research group was successfully able to synthesize the so-called 'cryogel' that would soon become a material offering a multitude of applications to researchers around the world. After fabrication of these gel matrices, further focus was to understand the basic mechanism underlying the whole process. Further research started explaining subtle but critical temperature differences that could alter the dynamics of cross-linked structure formation. Reports became available on how different freezing regimes could control critical parameters like pore morphology, osmotic characteristics and so on (Lozinsky et al., 1986a,b,c). A proof of the concept of cryo-structurization was also provided in the report 'Study of Cryo-Structurization of the Polymer Systems' wherein state-of-the-art imaging tools were used to explain the phenomenon of cryo-structurization (Belavtseva et al., 1984). It would be fair to say that the initial few reports on cryogels were dedicated to understanding the process of cryogelation along with studying critical parameters that could directly or indirectly alter their properties (Lozinsky et al., 1982, 1984a,b, 1986a,b,c). After successfully understanding the structure of cryogels, it was sought out to put these matrices to use in biotechnological processes and one of the first available reports indicated that cryogels were used in entrapping bacterial cells for the production of 3-flouro-L-tyrosine. *Citrobacterium intermedius* cells containing L-tyrosine phenol lyase can catalyse the production of 3-flouro-L-tyrosine (Lozinsky et al., 1989b). This is one of the first reports that emphasizes the very first biotechnological application of cryogels. Until then, hydrogels from different polymers were used to perform such tasks. However, they had some limitations, which were overcome by the use of cryogels. The first decade of cryogel inception saw fabrication and characterization during the first half of the decade while cryogels were also starting to be used for biotechnological applications toward the end of the first decade.

The same research group continued working on cryogels and their applications through the next decade as well. They continued producing biologically important molecules like amino acids, enzymes and so on. Lozinsky and co-workers immobilized different bacterial cells and enzymes to enhance the production of various biologically important molecules like L-lysine, L-proline, ethanol and so on (Velizarov et al., 1992; Gough et al., 1998) and worked on the development of cryogel-based biosensors. While the previous decade mainly involved the use of synthetic polymers for cryogel fabrication, in the next decade cryogels from naturally existing polymers were also introduced. In the later part of this decade, cryogels had gained international recognition and a few other groups started exploring the applications that cryogel monoliths could provide (Doretti et al., 1998).

In the year 2000 and onward, many groups around the world became interested in these macroporous structures and started using these matrices for various applications. During the first few years of the decade, the macroporous structure of cryogels was used in the area of chromatography. The first few applications included chromatography of microbial cells and enzymes based on ion exchange, affinity and immobilized metal affinity chromatography (IMAC), respectively (Arvidsson et al., 2002, 2003; Lozinsky et al., 2003). After these reports were published, a huge number of research groups adopted the cryogel-based separation of various molecules and cells. In the coming years, cryogels were widely used in the production and purification of monoclonal antibodies, chromatographic separation of large or small biomolecules, separation of bacterial and mammalian cells, enzymes, purification of proteins, cell organelles and so on (Kumar et al., 2003a,b, 2005; Dainiak et al., 2004). Over the course of many years, new cryogel materials were synthesized using synthetic or natural polymers or were often modified based on the type of application (Lozinsky, 2002). An important point worth noticing is that cryogel matrices have continually been characterized with newer and more state-of-the-art analytical tools. In the later part of the decade, cryogels were used in applications focused on tissue engineering, biomedical engineering and regenerative medicine. A number of groups fabricated cryogel scaffolds largely from natural polymers for tissue engineering applications. Owing to their large pore size and mechanical stability, cryogel scaffolds mimic native tissue in many ways as compared to other scaffolds and thereby become useful in regenerating various tissues such as bone, cartilage, skin, muscle and so forth.

### 1.3.3.2 Cryogels: From a Statistical Perspective

A statistical overview on the development of cryogel-based research is provided in this section. As mentioned previously, cryogels were studied by very few groups in the initial years until they were being used for various applications. The first two decades of cryogel inception saw slow growth in terms of being used widely among different research groups and also in the number of publications based on cryogels. However, at the beginning of the millennium, cryogel-based research grew exponentially mainly because of the variety of applications for which cryogel matrices could be used. Figure 1.2 presents information on how cryogel-based research appears to be increasing based on the number of publications per year.

As is evident from Figure 1.2, a very gradual increase in the number of publications can be seen in the first two decades. All the publications during that phase mainly deal with cryogel characterization and very limited applications as discussed earlier (Lozinsky et al., 1982, 1986a,b,c, 1996, 1997; Belavtseva et al., 1984; Gough et al., 1998; Lozinsky and Plieva, 1998). However, during the first decade of the millennium, cryogels were known internationally and many groups started exploiting the applications of cryogel matrices (Plieva et al., 2004a,b; Babac et al., 2006; Kumar et al., 2006; Van Vlierberghe et al., 2006; Yao et al., 2006; Wang et al., 2008). Figure 1.3 indicates the increasing acceptability of cryogels by different research groups across the world.

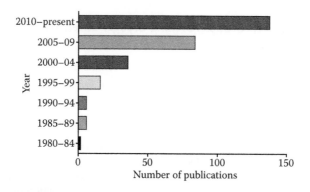

**FIGURE 1.2** Increasing trend of cryogel-based research publications.

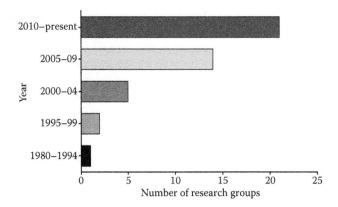

**FIGURE 1.3** Increasing trend in the number of research groups working with cryogels.

In the initial years of cryogel inception, cryogels were limited to specific research groups. However, in the following years, cryogel research was being conducted on many continents and currently there are more than 20 research groups worldwide that are extensively working on cryogel-based research in different areas. Some of the groups that have extensively contributed in the development of cryogel-based research are mentioned in Table 1.2.

With increasing numbers of research groups pursuing research on cryogels, the number of cryogel-based applications also started increasing and different groups used cryogels for different applications. Figure 1.4 indicates different areas of biotechnological applications of cryogels.

The most widely used application area of cryogel matrices is immobilization, separation and imprinting, majorly because of the macroporous architecture of cryogels that allows unhindered flow of liquid as well as solid micro- and nanoparticles (Kumar and Bhardwaj, 2008; Jain and Kumar, 2009). Apart from that, immobilization of various molecules of interest, cells or ligands is very convenient on cryogels (Efremenko et al., 2006; Plieva et al., 2006; Dainiak et al., 2007; Stanescu et al.,

**TABLE 1.2**
**A Few Research Groups Pursuing Cryogel-Based Research and Their Affiliations**

| Research Groups | Affiliation |
| --- | --- |
| Lozinski and co-workers | Institute of Organo-Element Compounds, Russia |
| Mattaisson and co-workers | Lund University, Sweden |
| Kumar and co-workers | Indian Institute of Technology Kanpur, India |
| Piskin and co-workers | Hacettepe University, Turkey |
| Denizli and co-workers | Hacettepe University, Turkey |
| Bloch and co-workers | Tel Aviv University, Israel |
| Yao and co-workers | Zhejiang University of Technology, Hangzhou, China |
| Mikhalovsky and co-workers | University of Brighton, UK |

# Cryogels and Related Research 13

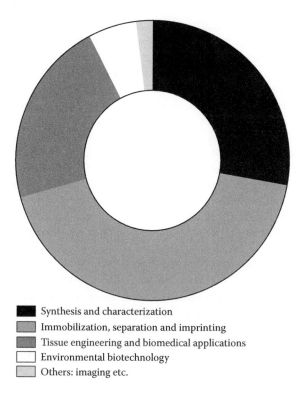

■ Synthesis and characterization
▨ Immobilization, separation and imprinting
▨ Tissue engineering and biomedical applications
□ Environmental biotechnology
▨ Others: imaging etc.

**FIGURE 1.4** Pie chart indicating various cryogel-based research areas and their respective shares.

2010; Jain and Kumar, 2013). More than 40% of cryogel-based research extensively explains different applications of cryogels in separation and molecular imprinting using cryogel matrices. This particular application of cryogels is exploited by a maximum number of groups working on cryogels than any other application area, thereby indicating the importance of such materials in the area of separation technology (Plieva et al., 2004a,b, 2007; Yun et al., 2007; Andac et al., 2008, 2013; Yan et al., 2008; Önnby et al., 2012; Srivastava et al., 2012; Dario Arrua et al., 2013). Other than separation science, cryogels have been used extensively in tissue engineering, regenerative medicine and biomedical applications. Tissue engineering is the second most widely explored application area of cryogels with more than 20% of total research focusing on such applications (Petrenko et al., 2011; Bhat et al., 2013; Vishnoi and Kumar, 2013a,b). Due to a high availability of natural polymers that mimic the extra cellular matrix and continual exchange of gases and liquid media between the porous structures, cryogels are well-suited matrices for tissue engineering. Moreover, delivery of bioactive molecules becomes relatively easier by using cryogels instead of other types of materials. A one-step fabrication approach also makes cryogels special and it is easier to fabricate such matrices under good manufacturing practices (GMP). This increases the possibility of using cryogel-based scaffolds as scaffolds of choice for regenerating human tissues.

It is interesting to observe that even though cryogels are more than four decades old, approximately 25% of cryogel-based research has been performed to synthesize and characterize newer cryogel matrices. Scientists around the world have been using the latest techniques to better understand these materials to enhance their use in the above-stated areas. The progress in areas like polymer chemistry also provides ease to researchers to tailor cryogel matrices based on specific applications. A lot of natural and synthetic polymers have been used thus far to fabricate cryogels and state-of-the-art techniques like scanning and transmission electron, atomic force and confocal microscopy, μ-CT, X-ray diffraction, liquid extrusion porosimetry, contact angle goniometry, energy dispersive X-rays and so on have been used for the characterization of these matrices. In recent years, cryogels have also been used in the area of environmental biotechnology with a significant degree of success (Önnby et al., 2012). A description of different applications of cryogels will be given in the coming sections and an explanation that is more detailed can be found in the forthcoming chapters. Apart from the above-mentioned applications, cryogels have also been used for the development of various biosensors, as imaging phantoms, particulate filters and so on.

As previously mentioned, the initial applications of cryogels were mostly in the area of separation sciences but with increasing research in other areas and better understanding of various biological and chemical systems, cryogels started being used in various other application areas as discussed earlier. Figure 1.5 indicates a paradigm shift in the cryogel-based applications since their inception.

Evidently, in the first phase between 1985 and 2005, cryogels were widely synthesized and used in areas of immobilization, separation and imprinting (Lozinsky and Plieva, 1998; Lozinsky et al., 2001; Babac et al., 2006). However, in the later phase

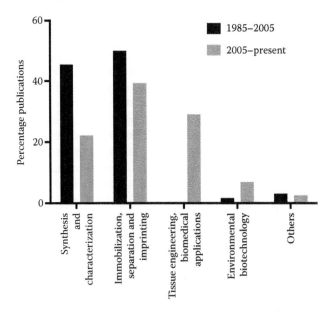

**FIGURE 1.5** Paradigm shift in various cryogel-based applications over the past few decades.

newer applications of cryogel-based research emerged. Noteworthy to observe is the area of tissue engineering and biomedical applications that appears to be a stand-alone emerging application of the later phase (Bölgen et al., 2007; Bhat et al., 2011). Apart from tissue engineering, environmental biotechnology has also been sought out for use in the later phase. The information in Figure 1.5 can be used to conclude that there is a visible paradigm shift in the type of applications cryogel matrices have been used for during the two phases. Scientists around the globe have been looking for various new applications of cryogels mainly because a better understanding of these systems is available due to the extensive research that is being carried out on cryogels.

## 1.4 CRYOGELS: SYNTHESIS AND PROPERTIES

### 1.4.1 Cryogel Synthesis

The nature of bond formation in cryogels depends on the structure of the polymeric precursors. Cryogels are mainly formed either by physical gelation or by cross-linking or polymerization of precursors (Lozinsky et al., 2001). Some of them consist of a stand-alone precursor while some also exist as a composite of different polymers depending on the type of application. Accordingly, a cross-linker or an initiator is chosen in order to achieve a stable gel matrix with nonfluctuating physical or chemical bonds that hold the system together with the passage of time. Physical bonding is the simplest way to achieve a cryogel structure; however, the stability of such bonds is far less when compared to chemical bonds. An example could be the PVA or agarose cryogel, wherein the network is held together by means of noncovalent interactions in the form of physical bonds (Clark et al., 1983). On the contrary, the addition of a cross-linker leads to the formation of a stronger mesh of polymeric chains due to the formation of covalent bonds. To understand this phenomenon, an example of pAAm cryogels can be considered, wherein free radical copolymerization takes place between the monomers of acrylamide and $N,N$-methylenebisacrylamide (MBAAm) in the presence of suitable initiators such as ammonium persulfate (APS) and tetramethylethylenediamine (TEMED). Another example of a cross-linked network can be a cryogel made from gelatin or chitosan wherein the amidogen links rapidly to an aldehyde or a carbodimide from two very common cross-linkers—glutaraldehyde and 1-ethyl-3-(3-dimethylaminopropyl) carbodimide (commonly known as EDC)—to form an imine, amide and ester, respectively (Sharma et al., 2013). Cryogel composites consisting of more than one polymer can also be fabricated by either using a single cross-linker that has the affinity to react with functional groups on both polymers or using different cross-linkers. Chitosan-gelatin cryogels use chitosan and gelatin as polymers and both possess a free amidogen on their structure and thus a common cross-linker such as glutaraldehyde or EDC can be used to achieve a semi-interpenetrating (semi-IPN) network of chitosan and gelatin chains (Kathuria et al., 2009). When there is an absence of a common reactive group in polymers of choice, different cross-linkers can be used to achieve a network of homogenously distributed polymers. Poly-acrylonitrile (PAN)-chitosan cryogels are formed by two separate mechanisms taking place in a single ULMP (Jain and Kumar, 2009).

Free radical copolymerization of acrylonitrile monomers with MBAAm and cross-linking of free amino groups on chitosan with aldehyde groups of glutaraldehyde takes place simultaneously to form a homogenously distributed semi-IPN of PAN-chitosan. Poly-hydroxyethylmethacrylate (pHEMA)-gelatin cryogels also exhibit a similar phenomenon (Singh et al., 2010).

Apart from a composite cryogel fabrication, several groups around the world have also been successful in covalently coupling various functional groups on a cryogel monolith. The coupled functional groups then act as anchors for ligands or as ligands themselves for immobilizing various chemical and biological molecules and macromolecules. One of the most commonly used cryogels in such applications is the poly-N,N-dimethylacrylamide (DMAAm) cryogel copolymerized with allylglycidyl ether (AGE) and MBAAm in the presence of APS and TEMED (Plieva et al., 2006). An epoxy group is added to immobilize protein A as a ligand, homogenously to ensure high affinity binding of the target molecules. The activation of an epoxy group is a two-step process and is done to avoid any steric hindrans that might interfere when the ligand binds to the target molecule. The cryogel is allowed to react with a solution of ethylenediamine followed by an interaction with glutaraldehyde. In this manner, a large spacer arm of seven carbon atoms is created which enhances the binding of target molecules (lymphocytes in this case) to the ligand (Kumar et al., 2003a). A similar approach has also been used for poly-acrylamide cryogels copolymerized with AGE to covalently couple copper (Cu II) ions on the cryogel monolith using iminodiacetic acid (IDA) as a chelating ligand. This particular study focuses on the capture of a clinically important enzyme urokinase continually produced by human fibrosarcoma cells HT 1080 in a bioreactor setup (Kumar et al., 2006a). Cryogels have been used in numerous applications in different areas and the details will be discussed in other chapters.

Table 1.3 provides a list of cryogels that have been used thus far along with information about appropriate initiators/cross-linkers.

### 1.4.2 Cryogel Properties

A brief description of a number of characteristic physicochemical properties of cryogels is provided in this section to present a brief idea about different properties of cryogels while detailed descriptions will be given in subsequent chapters.

#### 1.4.2.1 Supermacroporous Architecture with Interconnected Pores

One of the most characteristic properties of cryogel matrices is the supermacroporous architecture of these gels (Belavtseva et al., 1984). The pore size is tunable and can be modulated by different factors (Plieva et al., 2005). The pores range from a few microns to several hundred microns ensuring an unhindered flow of matter like transport of gases, fluids and even nano- and microsized molecules (Lozinsky et al., 2003; Tripathi and Kumar, 2011). Interconnected pores also impart a high flow rate of fluids or gases across these gels without any resistance or back flow.

#### 1.4.2.2 Rapid Swelling Mechanism

Owing to the large interconnected pores, cryogels exhibit a very rapid uptake of a majority of fluids. Most cryogels reach a maximum of their swelling capacity within

## TABLE 1.3
### List of Different Cryogels with Polymers, Cross-Linkers/Initiators and Mechanism

| Cryogel | Cross-Linker/Initiator | Mechanism | Refs. |
|---|---|---|---|
| PVA | –/glutaraldehyde | Physical gelation or chemical cross-linking | Lozinsky et al., 1997, Plieva et al., 2007 |
| PVA-tetraethylortho silicate (TEOS)-agarose | – | Physical gelation | Mishra and Kumar, 2011 |
| pAAm | MBAAm | Free radical polymerization | Yao et al., 2006 |
| PAN | MBAAm | Free radical polymerization | Jain et al., 2009 |
| pNIPAAm | MBAAm | Free radical polymerization | Srivastava et al., 2007 |
| Poly(4-vinyl pyridine-co-divinyl benzene) | PEGDA | Free radical polymerization | Gupta et al., 2013 |
| HEMA | MBAAm | Free radical polymerization | Andac et al., 2008 |
| HEMA-gelatin | PEGDA and glutaraldehyde | Free radical polymerization and chemical cross-linking | Singh et al., 2010 |
| pAAM-gelatin | MBAAm/glutaraldehyde or EDC-NHS | Free radical polymerization and chemical cross-linking | Jain and Kumar, 2013 |
| Chitosan | Glutaraldehyde/oxidized dextran | Chemical cross-linking | Lozinsky et al., 1982 |
| Gelatin | Glutaraldehyde/EDC-NHS | Chemical cross-linking | Fassina et al., 2010 |
| Gelatin-agarose | Glutaraldehyde/EDC-NHS | Chemical cross-linking | Tripathi et al., 2009 |
| Chitosan-agarose-gelatin | Glutaraldehyde/EDC-NHS | Chemical cross-linking | Bhat et al., 2011 |
| Gelatin-fibrinogen | Glutaraldehyde | Chemical cross-linking | Dainiak et al., 2010 |
| Collagen | Dialdehyde starch | Chemical cross-linking | Mu et al., 2010 |
| Gelatin-hyaluronic acid | EDC-NHS | Chemical cross-linking | Chang et al., 2013 |

a couple of minutes (Bhat et al., 2011). This property makes cryogels very feasible for applications that require rapid uptake of fluids.

### 1.4.2.3 Mechanical Stability and Viscoelasticity

Cryogels are highly stable structures, which is mainly because of cryo-structuration that causes a strong network of polymers throughout the structure. Another important feature that cryogels exhibit is tunable mechanical strength, which depends on several factors like concentration of precursors, temperature regime and so on.

Depending on the type of application, cryogels can range from very soft to very tough with their compressive strengths varying from a few kilopascals to several megapascals (Bhat et al., 2011; Mishra and Kumar, 2011). Various uni-axial or multi-axial compression analysers have been used around the globe to characterize cryogels based on their strength.

Visco-elasticity is another important and tunable feature of cryogels that makes them suitable for a wide range of applications (Van Vlierberghe et al., 2010). Most cryogels exhibit a nature that is more elastic while they are far less viscous, which makes them structurally stable and enhances their shelf life (Vishnoi and Kumar, 2013a,b).

### 1.4.2.4 Chemical Stability and Modifiability

Cryogels are chemically stable, and reactions occurring at subzero temperatures yield similar results as those carried out at room temperature. Fourier transform infrared (FT-IR), X-ray diffraction (XRD) and energy dispersive X-ray (EDX) are a few tools that are used to evaluate and confirm the chemical structures of these gel matrices (Mishra and Kumar, 2011; Sharma et al., 2013). One of the most important features of cryogels is the flexibility they provide in terms of adding functionalities. Various functionally active groups, biologically active molecules and so on, can be coupled to these monoliths to achieve the desired functionality.

### 1.4.2.5 Biocompatibility

Many studies that have been carried out in recent years indicate that cryogels are highly biocompatible and do not disturb the immune system (Shakya et al., 2013). This property of cryogels makes them highly demanding in studies that involve cell/tissue material interactions.

### 1.4.2.6 Ease of Fabrication, Economical Operations and Longer Shelf Life

Cryogelation can be considered a one-step process wherein precursors are allowed to react at subzero temperatures for a fixed period of time and are eventually thawed to achieve a macroporous structure. Unlike other process, it does not require any special intermittent steps and the fabrication is much more streamlined and up-scalable with limited watch of the fabricator. Cryogels of various sizes and shapes can be achieved with little degrees of variation. Minimal batch-to-batch variations are observed due to a single step of fabrication, which adds commercial importance to these matrices. Moreover, the procedure does not demand any special setup and adds economical ease to the end user.

Unlike many other materials, cryogels have a very long shelf life and they tend to maintain their structural integrity for infinite time. No special storage conditions need to be developed and they can be stored in moisture-free bags at room temperature for longer periods of time. A more detailed description of many other properties is offered in other chapters in this book. The focus here has mainly been on covering some major characteristic features of the matrices that would enable the reader to have a better understanding of the upcoming contents of this chapter.

## 1.5 CRYOGEL APPLICATIONS

The purpose of this review is to explain cryogels and their application areas to researchers from various domains; the other chapters in this book will discuss in detail the process of synthesizing and characterizing cryogel matrices and include a number of reviews that are dedicated to explaining various applications of cryogels based on the research that has been accomplished thus far. In this chapter, we have discussed in brief different types of cryogels, the synthesis and characterization, and then emphasized more on the emergence of cryogel matrices from a historical and statistical point of view. A more logical denouement to this chapter would also be to introduce various applications wherein cryogels have been put to use. Here we address a few applications from each area to ensure that readers gain information from different applications of cryogels. Each application area is described briefly in the following sections and a much more detailed description will be provided in later chapters.

### 1.5.1 CRYOGELS IN IMMOBILIZATION, SEPARATION AND IMPRINTING

The earliest applications of cryogel matrices that are available in the literature deal with the immobilization of *Citrobacterium intermedius* cells for the production of 3-flouro-L-tyrosine (Lozinsky et al., 1989b). The cells were immobilized on a 10% (w/v) PVA cryogel to enhance the production of the amino acid derivative that was commonly used in studying enzyme-substrate specificity and protein analysis and so on. The study resulted in only a slight increase in the production of the amino acid derivative; however, it laid a foundation for researchers to further explore the application of these matrices. In the coming years, a variety of bacterial, fungal and mammalian cells have been immobilized on cryogel matrices for enhanced production of various biochemically important molecules. Various proteins and enzymes have also been reported to be immobilized on cryogel matrices. Recently, concavlin A was immobilized on an epoxy-activated composite cryogel matrix of PVA to ensure the capture of horseradish peroxidase (HRP) (Hajizadeh et al., 2012). Hybridoma cells were immobilized on a semi-interpenetrating network of pAAm-chitosan in a bioreactor setup to continually enhance the production of monoclonal antibodies (mAb) and urokinase. The study showed that the efficiency of the cryogel-based setup was fourfold higher than what one would obtain using a tissue culture flask (Jain and Kumar, 2013). Many ligands have also been coupled to cryogel matrices to ensure capture of biological molecules and macromolecules and cells of different types. An example could be a protein A based affinity ligand coupled to an epoxy-activated cryogel column to separate antibody labeled cells from unlabeled cells. Due to the large and connected pore structure, the labeled cells would bind to the ligand via the Fc receptor while the unlabeled cells would pass through the matrix unhindered. The bound cells are then recovered with a high specificity and viability just by mechanically squeezing the cryogel monolith with a high percentage of recovery. This setup for affinity chromatography is very compact and does not require large amounts of processing time and results in a high viability of target cells (Kumar and Srivastava, 2010). A similar study carried out a few years earlier successfully separated T-lymphocytes and B-lymphocytes in the same fashion with high efficiency (Kumar et al., 2003a, 2005) as shown in Figure 1.6.

In another study, *N*-methacryloyl-(l)-histidine methyl ester (MAH) was immobilized as a chelating ligand on a pHEMA cryogel matrix to purify cytochrome c from rat liver homogenate (Tamahkar et al., 2011). Biocompatible and inert pHEMA cryogels were covalently coupled with a dye molecule—Cibacron blue F3GA—to reduce the levels of albumin. The study showed a 77% reduction of albumin from the serum sample after passing through the cryogel column (Andac et al., 2008, 2012). Yao and co-workers separated useful immunoglobulin (IgG) molecules from bovine milk whey by using pHEMA cryogel grafted with 2-(dimethylamino) ethyl methacrylate. The results indicate that they successfully isolated and purified approximately 95% of pure IgG molecules present in the bovine whey using cryogels

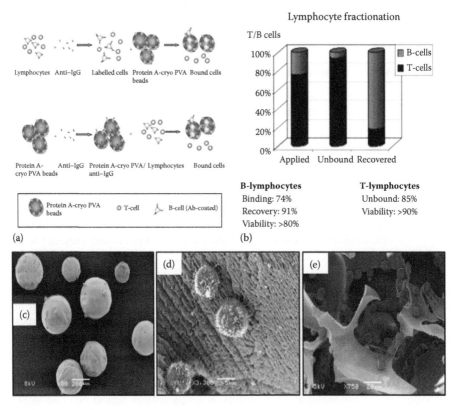

**FIGURE 1.6** Cell separation using cryogels. Panel (a) is a schematic of cryogel column preparation and cell separation using an affinity-based separation approach while panel (b) shows the efficiency of the system. Panel (c) indicates cryo-PVA beads (pseudocolored to green) used for cell separation. Scanning electron microscopy (SEM) images in panels (d) and (e) indicate bound lymphocytes (pseudocolored to red) to the cryogel column. (From Kumar, A., Rodríguez-Caballero, A., Plieva, F. M., Galaev, I. Y., Nandakumar, K. S., Kamihira, M., Holmdahl, R., Orfao, A. and Mattiasson, B.: Affinity binding of cells to cryogel adsorbents with immobilized specific ligands: Effect of ligand coupling and matrix architecture. *Journal of Molecular Recognition*. 2005. 18(1). 84–93. Copyright Wiley-VCH Verlag GmbH & Co. KGaA. Reproduced with permission.)

(Dong et al., 2013). The same cryogel matrix was incorporated with macroporous cellulose beads and used for purification of IgG molecules and albumin from human serum with purities of 83% and 98%, respectively (Ye et al., 2013). Recently, cryogels have also been fabricated by adding functional nanoparticles, either charged or neutral, to purify proteins with a high percentage of purity and greater ease as compared with other methods. Cryogels have also been very successful in the area of molecular imprinting because they offer a uniform polymer matrix to create the shape-memory effect.

In a recent study, Mattiasson and co-workers were successful in molecular imprinting of nano-sized particles in a cryogel matrix to successfully remove propanol from either aqueous solutions directly or from complex plasma samples without prior protein precipitation. The results indicate high selectivity and stability of the cryogel matrices in the given application (Hajizadeh et al., 2013). pHEMA-based molecularly imprinted cryogels were synthesized and used for selective binding experiments on human serum albumin in the presence of other competitive proteins like human transferrin and myoglobin. The study reported a very high depletion of human albumin with a high degree of reusability of the same cryogel column many times with a minimal reduction in the efficiency of depletion of albumin (Andac et al., 2013). Similar experiments were performed with hemoglobin with similar results and a fast protein liquid chromatography approach was used to purify antihepatitis B antibodies using molecularly imprinted cryogel membranes (Asliyuce et al., 2012). Other than the above applications, molecularly imprinted cryogels have also been used to isolate various other molecules like bilirubin, hemoglobin, interferon-alpha and several other biomolecules with high efficiency (Baydemir et al., 2009; Derazshamshir et al., 2010; Ertürk et al., 2013).

Cryogels have been used successfully in the area of separation sciences mainly due to the supermacroporous structure of these matrices that allows rapid flow of solid and liquid particles. Apart from that, the ease of coupling of various ligands, functional monomers and so on, to the surface of cryogels also makes it very convenient to target various biomolecules easily. In this section, we have seen that cryogels have made a significant contribution in the areas of immobilization, chromatography and molecular imprinting, thereby justifying their significance as important gel matrices of current times. Research that is more rigorous is being carried out to put these materials to further use and to exploit the advantages offered by these matrices. It has been seen in earlier sections that separation science is the most widely used application area of cryogels with a maximum number of groups working on this particular area of application.

### 1.5.2 CRYOGELS IN TISSUE ENGINEERING AND BIOMEDICAL APPLICATIONS

The area of tissue engineering has been in practice for more than five decades and cryogels were not introduced in this area until very recently. The tissue engineering approach has been widespread and there has been a constant urge to look out for newer materials that can mimic the extracellular matrix better in order to replace or regenerate a damaged tissue. As mentioned earlier, owing to a better understanding of polymer systems, polymer chemists have been continually coming up with various natural polymers that mimic the composition of the native tissue as closely as possible, thus making it easier to recreate a congenial three-dimensional environment for cells with minimal rejections. In the later part of the first decade of the millennium, cryogels were put to use in the field of tissue engineering. This area of application has emerged much faster than many other areas of applications mainly due to the rising demands in this area. From a statistical point of view the paradigm has completely shifted from other areas of applications of cryogels to the area of tissue engineering. Various research groups have been performing extensive research on using cryogels as scaffolds for tissue reconstruction and many of them have been successful enough to put these materials in preclinical and clinical trials. A cryogel-based tissue engineering approach has been successfully used in regenerating soft as well as hard tissues like neural, skin, cartilage and bone tissues, respectively (Sharma et al., 2013; Vishnoi and Kumar, 2013a,b; Gupta et al., 2014; Mishra et al., 2014).

Kumar and co-workers developed a biocomposite cryogel matrix with biphasic properties containing calcium and bioactive glass with silanol groups as the inorganic phase and PVA and alginate in the organic phase. This composite cryogel was aimed at healing critical sized cranial bone defects. The study concluded that the cryogel-treated groups exhibited a significantly higher amount of bone formation when compared to the nontreated group. The amount of bone regenerated was quantified using microcomputed tomography to make a quantitative analysis. The qualitative analysis confirmed the presence of terminally differentiated osteoblasts at the defect site using immunostaining and qRT-PCR and the results suggest a high amount of osteoblasts in the experimental group (Mishra et al., 2014). In another study, the researchers were also successful in fabricating a highly elastic and macroporous composite cryogel consisting of chitosan, agarose and gelatin for cartilage tissue engineering (Bhat et al., 2011). Chitosan and gelatin were used mainly due to their high biocompatibility and providing anchor sites to various cells. Agarose provides elasticity in the gel and at the defect site helps to regenerate the highly elastic cartilage tissue. The smart selection of polymers leads to the development of neo-cartilage *in vitro* (Bhat et al., 2013). The same material was then later used to heal a subchondral cartilage and bone defect in a rabbit model (Gupta et al., 2014). The cartilage in the treated group was regenerated well with close histological resemblance to native cartilage. Moreover, the cartilage tissue thus developed was articular in nature and showed no signs of fibrous cartilage. The study suggests that cryogels can be used to heal critical cartilage defects as well.

A conducting cryogel scaffold was also successfully developed by this research group to deliver alpha-ketoglutarate to the neural cells and to aid in the regeneration of the neural tissue (Vishnoi and Kumar, 2013a,b). Piskin and his group also synthesized novel pHEMA-lactate-dextran cryogels for regenerating critical-sized bone defects (Bölgen et al., 2008). A biphasic cryogel scaffold was designed which consisted of an organic gelatinous phase for cellular adhesion and an inorganic phase in the form of hydroxyapatite to mimic the native bone matrix. The matrix was supplied with vascular endothelial growth factor (vEGF) to enhance the bone regeneration process (Ozturk et al., 2013). They also studied the tissue responses when biodegradable cryogel scaffolds are implanted in animals to understand the feasibility of using such materials in animal models and translating such scaffolds into higher models (Bölgen et al., 2009).

In another study, the chondrocytes were transfected to express BMP-7 just to recreate an *in vivo* situation, and seeded transfected cells on cryogel scaffolds were implanted into cartilage defects in rabbits. The study was carried out for a period of four months, and significant differences were observed in the transfected cells compared to the cells that were not transfected. Both quantitative and qualitative differences were observed with a high statistical significance (Odabas et al., 2013). Vlierberghe and co-workers studied the differentiation process of human bone marrow stromal cells seeded in gelatin cryogel. The cryogels were modified by adding various bone proteins like osteocalcin, collagen type 1 and 3 and so on. The effect of scaffold pretreatment with various bioactive molecules on the cell fate was observed (Fassina et al., 2010). In later studies, a similar approach was used to provide electromagnetic stimulation to the scaffold that led to enhanced binding of cell ECM proteins to the scaffold and increased the proliferation of cells on the scaffolds as compared to controls with no stimulation (Saino et al., 2011). Apart from tissue regeneration, cryogel biomaterials were also used in a bioreactor setup to enhance the production of clinically important therapeutic molecules. The cryogels were used in either monolith or bead format to improve the production of urokinase and monoclonal antibodies. A first of this type of study had promising results and established another important application of cryogels—a simple and disposable setup to simplify the process and claim that the whole setup does not require a long time and the reactor can be run up to two months or more (Jain and Kumar, 2013).

Thus, it clearly establishes the importance of cryogel biomaterials in the area of tissue engineering and biomedical applications. The supermacroporous architecture of cryogels enables sufficient transfer of nutrients and gases to the inhabiting cells, thus providing an amiable dwelling required for a healthy cell proliferation and functioning. A large number of commercially approved natural polymers also make it possible to tailor various cryogel scaffolds as per specific tissue needs. Figure 1.7 provides a brief insight into the cryogel-based tissue engineering approach.

### 1.5.3 Cryogels in Environmental Biotechnology

We have seen how cryogels have been widely used in completely different sets of applications; however, the supermacroporous architecture of these gel matrices is a property that is commonly explored for all these application areas. We have also mentioned that cryogel matrices can be altered easily in terms of attaching specific ligands to the matrix, choice of polymer or functional co-monomers and so on. This is one of the properties of cryogel matrices that has continually been explored by various research groups that have been successful in their efforts to a great extent. Just like the targeted capture of various biomolecules, cryogels can also be efficient in environmental biotechnology applications. One of the important roles of these matrices in environmental biotechnology is the selective and highly efficient capture and removal of environmentally hazardous molecules. In one of the earliest studies, Lozinsky's group performed bioremediation of diesel-contaminated soil in a laboratory setup. The group used PVA cryogel with immobilized microorganisms

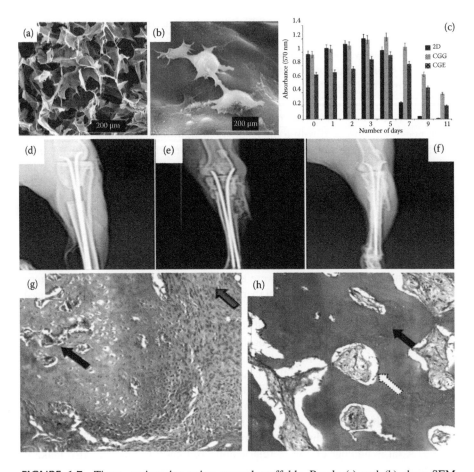

**FIGURE 1.7** Tissue engineering using cryogel scaffolds. Panels (a) and (b) show SEM images of carrageenan-gelatin cryogels with microstructure details and proliferation, attachment of Cos-7 fibroblast cells (pseudocolored to green), respectively. Panel (c) shows *in vitro* biocompatibility of Cos-7 fibroblasts on carrageenan-gelatin cryogels using MTT assay. Panels (d), (e) and (f) indicate X-ray radiographs of a critical-sized femoral defect, defect treated with gelatin-hydroxyapatite cryogel scaffold loaded with vascular endothelial growth factor (vEGF) 6 weeks and 12 weeks postsurgery, respectively. Panels (g) and (h) indicate histological evaluation of defect treated with cryogels 6 and 12 weeks postimplantations with constant bone remodeling. (From Sharma, A., Bhat, S., Vishnoi, T., Nayak, V. and Kumar, A. (2013). *BioMed Research International* 2013: 478279, 15. Copyright (2013), authors and Hindawi Publishing Corporation; Ozturk, B. Y., Inci, I., Egri, S., Ozturk, A. M., Yetkin, H., Goktas, G., Elmas, C., Piskin, E. and Erdogan, D. (2013). *European Journal of Orthopaedic Surgery & Traumatology* 23(7): 767–774. Copyright (2012), Springer-Verlag. With permission.)

to bio-augment cryogel scaffolds in eliminating diesel contamination. They demonstrated better results using cryogels with immobilized microorganisms as compared to commercially available remediation fluid (Cunningham et al., 2004) (Figure 1.8).

A fluidized-bed bioreactor setup with *Rhodococcal* cells in order to treat petroleum-contaminated water was also established. It was claimed to have achieved

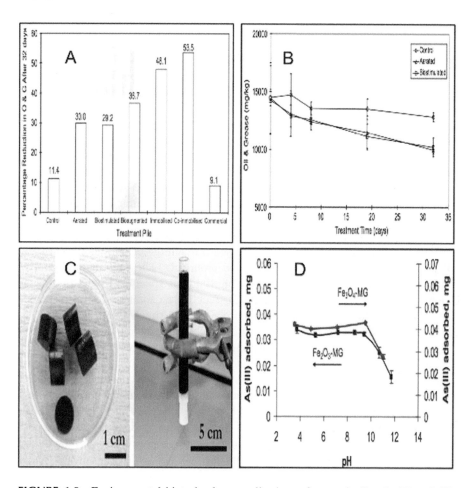

**FIGURE 1.8** Environmental biotechnology applications of cryogels. Panels (A) and (B) represent reduction of oil and grease levels in the soil samples passed through PVA cryogel matrix immobilized with bacteria. Panel (C) represents iron nanoparticle loaded cryogel cylinders and column setup while the image in panel (D) indicates the arsenic capture efficiency of the designed cryogel column. (Reprinted from *International biodeterioration & Biodegradation*, 54(2), Cunningham, C., Ivshina, I., Lozinsky, V., Kuyukina, M. and Philp, J., Bioremediation of diesel-contaminated soil by microorganisms immobilised in polyvinyl alcohol, 167–174, Copyright 2004, with permission from Elsevier; Reprinted from *Journal of Hazardous Materials*, 192(3), Savina, I. N., English, C. J., Whitby, R. L. D., Zheng, Y., Leistner, A., Mikhalovsky, S. V. and Cundy, A. B., High efficiency removal of dissolved As(III) using iron nanoparticle-embedded macroporous polymer composites, 1002–1008. Copyright 2011, with permission from Elsevier.)

70–100% success in removal of n-alkanes while poly-aromatic hydrocarbons (PAHs) were also removed with an efficiency of 66–70% (Kuyukina et al., 2009). Mattiasson and co-workers also used a similar approach to remove heavy metals from drinking water (Önnby et al., 2010). Endocrine disruptors and their catabolic effects on human health have been better understood in the last decade and thus emphasis was placed on the removal of such molecules from different sources. Thus, PVA cryogels were used to remove 17β-estradiol, 4-nonylphenol and atrazine from water. Quantity binding was achieved in the spiked samples for all analytes, making the use of cryogels very promising in the area of environmental biotechnology (Le Noir et al., 2009). Removal of benzo-a-pyrene, a highly carcinogenic PAH, was achieved using a hydrophobic column of poly(4-vinyl pyridine) co-polymerized with divinyl benzene. It successfully removed more than 70% of the targeted molecule in a high throughput platform using a 96-well plate for parallel processing the samples (Gupta et al., 2013). This particular study shows the applicability of the cryogel matrices in the removal of such hazardous molecules with high efficiency in terms of sample number and speed of operation. Other research groups have also been exploring cryogels in the area of environmental biotechnology. Iron-aluminum double oxides were incorporated within a cryogel column for the removal of arsenic, which is one of the most toxic elements and even at the lowest concentrations can be very harmful. Wastewater containing arsenic was passed through the column to evaluate the efficiency and the results were shown to be promising. The column successfully eliminated arsenic from water to permissible limits as specified by the European Union guidelines (Önnby et al., 2012, 2013). Tackling an additional pH adjustment step and a pre-oxidation step makes this process much simpler, and such columns can be efficiently used for removing arsenic from highly contaminated sources.

In a similar approach, an attempt to remove cadmium from water was tried using synthesized PVA cryogels with incorporated hydroxyapatite at variable concentrations. A fixed-bed column was used in the study, which showed that the height of the bed did not have any significant effect on the cadmium binding efficiency (Wang and Min, 2008). However, the concentration of hydroxyapatite did affect the overall efficiency of the column. Mikhalovsky and his group also made use of cryogel columns in the removal of arsenic by embedding iron nanoparticles within cryogel monoliths. The columns were capable of eliminating arsenic with high efficiency and at a wider range of pH (Savina et al., 2011) (Figure 1.8). Many other cryogel applications in the area of environmental biotechnology have been explored with very promising results. With more groups working on environmental applications of cryogels, it is speculated that this area of cryogel applications will also tend to grow with coming time. A brief view of a few applications has been provided in this section and the other contributions in this book give very detailed descriptions of a few such applications to enable our readers to understand the different approaches in detail.

### 1.5.4 Other Applications of Cryogels

Cryogel matrices have been used widely as a tool of biotechnological interest in very conventional biotechnological applications like the ones mentioned above. However, at the same time, cryogels have also been used in other areas of applications like, for example, development of biosensors and so on. Veronese and co-workers used poly(ethylene glycol) (PEG) and glucose oxidase in a PVA cryogel to detect the presence of glucose (Doretti et al., 1998). They evaluated the performance of the synthesized glucose sensor with varying conditions of pH and temperature. The functioning of the sensor was evaluated electrochemically and the study provided positive results.

A couple of years later, the same group developed a sensor for acetylcholine. In this study, acetyl cholinesterase was used in a PVA cryogel membrane. The membrane was also immobilized with PEG modified with an enzyme choline oxidase. This newly synthesized polymeric cryogel was then employed on a platinum electrode to electrochemically detect the presence of hydrogen peroxide. This approach was useful in accessing the efficiency of the sensor under varying conditions of temperature and pH (Doretti et al., 2000). Apart from being used in biosensor applications, cryogels have recently been reported to be used as phantoms for bio-imaging of structurally complex tissues. Magnetic resonance elastography (MRE) is one of the latest techniques used to study the mechanical properties of tissues such as liver and brain, and the existing imaging techniques are limited mainly due to the absence of appropriate phantom materials that can mimic the mechanical properties of such tissue. Because of this, the use of PVA cryogels as phantoms for such imaging techniques has been developed. This is due to a high degree of resemblance of PVA cryogels to soft tissue mentioned above in terms of mechanical strength and elasticity. In order to increase the similarities between the native tissue and the phantom, the researchers report that they have developed a unique freeze–thaw regime to ensure a homogenous polymer structure throughout the matrix (Minton et al., 2012; Iravani et al., 2014). An interesting invention was made by fabricating supermacroporous cryogel matrices as efficient filters for cigarette smoke (Kumar et al., 2010). These filters removed more than 95% of the tar content from cigarette smoke without affecting the nicotine content.

Therefore, in conclusion, in this chapter an overview was provided to explain how cryogels have come to be important matrices for biotechnological applications in the past few decades. The advances in polymer chemistry have led to a continual development of these materials while newer characterization techniques have been used to characterize them. Due to amicable properties, these matrices have established their importance in several biotechnological areas with increasing numbers of groups performing cryogel-based research.

## ACKNOWLEDGEMENTS

The authors acknowledge all the research and developments made in the area of cryogels and acknowledge all those researchers who have contributed in this area.

## LIST OF ABBREVIATIONS

| | |
|---|---|
| AGE | Allylglycidyl ether |
| APS | Ammonium persulfate |
| BMP | Bone morphogenic protein |
| DMSO | Dimethylsulphoxide |
| ECM | Extra cellular matrix |
| EDC | 1-Ethyl-3-(3-dimethylaminopropyl) carbodimide |
| EDX | Energy dispersive X-ray |
| FT-IR | Fourier transform infrared |
| GMP | Good manufacturing practices |
| HRP | Horseradish peroxidase |

| | |
|---|---|
| IDA | Iminodiacetic acid |
| IgG | Immunoglobulin |
| IMAC | Immobilized metal affinity chromatography |
| mAb | Monoclonal antibodies |
| MAH | *N*-Methacryloyl-(l)-histidine methyl ester |
| MBAAm | *N*,*N*-Methylenebisacrylamide |
| μ-CT | Micro computed tomography |
| MRE | Magnetic resonance elastography |
| pAAm | Poly-acrylamide |
| PAH | Poly-aromatic hydrocarbons |
| PAN | Poly-acrylonitrile |
| pDMAAm | Poly-*N*,*N*-dimethylacrylamide |
| PEG | Poly(ethylene glycol) |
| PEGDA | Poly(ethylene glycol) diacrylate |
| pHEMA | Poly-hydroxyethylmethacrylate |
| pNiPAAm | Poly(*N*-isopropyl acrylamide) |
| PVA | Poly-vinyl alcohol |
| TEMED | Tetramethylethylenediamine |
| TEOS | Tetraethylorthosilicate |
| ULMP | Unfrozen liquid microphase |
| vEGF | Vascular endothelial growth factor |
| XRD | X-ray diffraction |

## REFERENCES

Andac, M., Galaev, I. and Denizli, A. (2012). Dye attached poly(hydroxyethyl methacrylate) cryogel for albumin depletion from human serum. *Journal of Separation Science* **35**(9): 1173–1182.

Andac, M., Galaev, I. Y. and Denizli, A. (2013). Molecularly imprinted poly (hydroxyethyl methacrylate) based cryogel for albumin depletion from human serum. *Colloids and Surfaces B: Biointerfaces* **109**: 259–265.

Andac, M., Plieva, F. M., Denizli, A., Galaev, I. Y. and Mattiasson, B. (2008). Poly(hydroxyethyl methacrylate)-based macroporous hydrogels with disulfide cross-linker. *Macromolecular Chemistry and Physics* **209**(6): 577–584.

Arvidsson, P., Plieva, F. M., Lozinsky, V. I., Galaev, I. Y. and Mattiasson, B. (2003). Direct chromatographic capture of enzyme from crude homogenate using immobilized metal affinity chromatography on a continuous supermacroporous adsorbent. *Journal of Chromatography A* **986**(2): 275–290.

Arvidsson, P., Plieva, F. M., Savina, I. N., Lozinsky, V. I., Fexby, S., Bülow, L., Yu Galaev, I. and Mattiasson, B. (2002). Chromatography of microbial cells using continuous supermacroporous affinity and ion-exchange columns. *Journal of Chromatography A* **977**(1): 27–38.

Asliyuce, S., Uzun, L., Yousefi Rad, A., Unal, S., Say, R. and Denizli, A. (2012). Molecular imprinting based composite cryogel membranes for purification of anti-hepatitis B surface antibody by fast protein liquid chromatography. *Journal of Chromatography B* **889**: 95–102.

Babac, C., Yavuz, H., Galaev, I. Y., Pişkin, E. and Denizli, A. (2006). Binding of antibodies to concanavalin A-modified monolithic cryogel. *Reactive and Functional Polymers* **66**(11): 1263–1271.

Baydemir, G., Bereli, N., Andaç, M., Say, R., Galaev, I. Y. and Denizli, A. (2009). Supermacroporous poly(hydroxyethyl methacrylate) based cryogel with embedded bilirubin imprinted particles. *Reactive and Functional Polymers* **69**(1): 36–42.

Belavtseva, E., Titova, E., Lozinsky, V., Vainerman, E. and Rogozhin, S. (1984). Study of cryostructurization of polymer systems. *Colloid and Polymer Science* **262**(10): 775–779.

Berillo, D., Elowsson, L. and Kirsebom, H. (2012). Oxidized dextran as crosslinker for chitosan cryogel scaffolds and formation of polyelectrolyte complexes between chitosan and gelatin. *Macromolecular Bioscience* **12**(8): 1090–1099.

Bhat, S., Lidgren, L. and Kumar, A. (2013). *In vitro* neo-cartilage formation on a three-dimensional composite polymeric cryogel matrix. *Macromolecular Bioscience* **13**(7): 827–837.

Bhat, S., Tripathi, A. and Kumar, A. (2011). Supermacroprous chitosan–agarose–gelatin cryogels: *In vitro* characterization and *in vivo* assessment for cartilage tissue engineering. *Journal of the Royal Society Interface* **8**(57): 540–554.

Bölgen, N., Plieva, F., Galaev, I. Y., Mattiasson, B. and Pişkin, E. (2007). Cryogelation for preparation of novel biodegradable tissue-engineering scaffolds. *Journal of Biomaterials Science, Polymer Edition* **18**(9): 1165–1179.

Bölgen, N., Vargel, I., Korkusuz, P., Güzel, E., Plieva, F., Galaev, I., Mattiasson, B. and Pişkin, E. (2009). Tissue responses to novel tissue engineering biodegradable cryogel scaffolds: An animal model. *Journal of Biomedical Materials Research Part A* **91A**(1): 60–68.

Bölgen, N., Yang, Y., Korkusuz, P., Güzel, E., El Haj, A. J. and Pişkin, E. (2008). Three-dimensional ingrowth of bone cells within biodegradable cryogel scaffolds in bioreactors at different regimes. *Tissue Engineering Part A* **14**(10): 1743–1750.

Chang, K.-H., Liao, H.-T. and Chen, J.-P. (2013). Preparation and characterization of gelatin/hyaluronic acid cryogels for adipose tissue engineering: *In vitro* and *in vivo* studies. *Acta Biomaterialia* **9**(11): 9012–9026.

Clark, A. H., Richardson, R. K., Ross-Murphy, S. B. and Stubbs, J. M. (1983). Structural and mechanical properties of agar/gelatin co-gels. Small-deformation studies. *Macromolecules* **16**(8): 1367–1374.

Cunningham, C., Ivshina, I., Lozinsky, V., Kuyukina, M. and Philp, J. (2004). Bioremediation of diesel-contaminated soil by microorganisms immobilised in polyvinyl alcohol. *International Biodeterioration & Biodegradation* **54**(2): 167–174.

Dainiak, M. B., Allan, I. U., Savina, I. N., Cornelio, L., James, E. S., James, S. L., Mikhalovsky, S. V., Jungvid, H. and Galaev, I. Y. (2010). Gelatin–fibrinogen cryogel dermal matrices for wound repair: Preparation, optimisation and *in vitro* study. *Biomaterials* **31**(1): 67–76.

Dainiak, M. B., Galaev, I. Y., Kumar, A., Plieva, F. M. and Mattiasson, B. (2007). Chromatography of living cells using supermacroporous hydrogels, cryogels. *Cell Separation*, Springer: 101–127.

Dainiak, M. B., Kumar, A., Plieva, F. M., Galaev, I. Y. and Mattiasson, B. (2004). Integrated isolation of antibody fragments from microbial cell culture fluids using supermacroporous cryogels. *Journal of Chromatography A* **1045**(1): 93–98.

Dario Arrua, R., Nordborg, A., Haddad, P. R. and Hilder, E. F. (2013). Monolithic cryopolymers with embedded nanoparticles. I. Capillary liquid chromatography of proteins using neutral embedded nanoparticles. *Journal of Chromatography A* **1273**: 26–33.

Derazshamshir, A., Baydemir, G., Andac, M., Say, R., Galaev, I. Y. and Denizli, A. (2010). Molecularly imprinted PHEMA-based cryogel for depletion of hemoglobin from human blood. *Macromolecular Chemistry and Physics* **211**(6): 657–668.

DeRossi, D., Kajiwara, K., Osada, Y. and Yamauchi, A. (1991). In: *Polymer gels*, Springer.

Dong, S., Chen, L., Dai, B., Johnson, W., Ye, J., Shen, S., Yun, J., Yao, K., Lin, D. Q. and Yao, S. J. (2013). Isolation of immunoglobulin G from bovine milk whey by poly (hydroxyethyl methacrylate)-based anion-exchange cryogel. *Journal of Separation Science* **36**(15): 2387–2393.

Doretti, L., Ferrara, D., Gattolin, P., Lora, S., Schiavon, F. and Veronese, F. M. (1998). PEG-modified glucose oxidase immobilized on a PVA cryogel membrane for amperometric biosensor applications. *Talanta* **45**(5): 891–898.

Doretti, L., Ferrara, D., Lora, S., Schiavon, F. and Veronese, F. M. (2000). Acetylcholine biosensor involving entrapment of acetylcholinesterase and poly (ethylene glycol)-modified choline oxidase in a poly (vinyl alcohol) cryogel membrane. *Enzyme and Microbial Technology* **27**(3): 279–285.

Efremenko, E. N., Spiricheva, O. V., Veremeenko, D. V., Baibak, A. V. and Lozinsky, V. I. (2006). L (+)-Lactic acid production using poly (vinyl alcohol)-cryogel-entrapped Rhizopus oryzae fungal cells. *Journal of Chemical Technology and Biotechnology* **81**(4): 519–522.

Ertürk, G., Bereli, N., Tümer, M. A., Say, R. and Denizli, A. (2013). Molecularly imprinted cryogels for human interferon-alpha purification from human gingival fibroblast culture. *Journal of Molecular Recognition* **26**(12): 633–642.

Fassina, L., Saino, E., Visai, L., Avanzini, M. A., Cusella De Angelis, M., Benazzo, F., Van Vlierberghe, S., Dubruel, P. and Magenes, G. (2010). Use of a gelatin cryogel as biomaterial scaffold in the differentiation process of human bone marrow stromal cells. *Engineering in Medicine and Biology Society (EMBC),* 2010 Annual International Conference of the IEEE, IEEE.

Gough, S., Barron, N., Zubov, A., Lozinsky, V. and McHale, A. (1998). Production of ethanol from molasses at 45 C using Kluyveromyces marxianus IMB3 immobilized in calcium alginate gels and poly (vinyl alcohol) cryogel. *Bioprocess Engineering* **19**(2): 87–90.

Gupta, A., Bhat, S., Jagdale, P. R., Chaudhari, B. P., Lidgren, L., Gupta, K. C. and Kumar, A. (2014). Evaluation of three-dimensional chitosan-agarose-gelatin cryogel scaffold for the repair of subchondral cartilage defects: An *in vivo* study in a rabbit model. *Tissue Engineering* **20**(23–24): 3101–3111.

Gupta, A., Sarkar, J. and Kumar, A. (2013). High throughput analysis and capture of benzo [a] pyrene using supermacroporous poly (4-vinyl pyridine- *co*-divinyl benzene) cryogel matrix. *Journal of Chromatography A* **1278**: 16–21.

Hajizadeh, S., Kirsebom, H., Leistner, A. and Mattiasson, B. (2012). Composite cryogel with immobilized concanavalin A for affinity chromatography of glycoproteins. *Journal of Separation Science* **35**(21): 2978–2985.

Hajizadeh, S., Xu, C., Kirsebom, H., Ye, L. and Mattiasson, B. (2013). Cryogelation of molecularly imprinted nanoparticles: A macroporous structure as affinity chromatography column for removal of β-blockers from complex samples. *Journal of Chromatography A* **1274**: 6–12.

Iravani, A., Mueller, J. and Yousefi, A.-M. (2014). Producing homogeneous cryogel phantoms for medical imaging: A finite-element approach. *Journal of Biomaterials Science, Polymer Edition* **25**(2): 181–202.

Ivanov, R. V., Lozinsky, V. I., Noh, S. K., Han, S. S. and Lyoo, W. S. (2007). Preparation and characterization of polyacrylamide cryogels produced from a high-molecular-weight precursor. I. Influence of the reaction temperature and concentration of the crosslinking agent. *Journal of Applied Polymer Science* **106**(3): 1470–1475.

Jain, E. and Kumar, A. (2009). Designing supermacroporous cryogels based on polyacrylonitrile and a polyacrylamide–chitosan semi-interpenetrating network. *Journal of Biomaterials Science, Polymer Edition* **20**(7–8): 877–902.

Jain, E. and Kumar, A. (2013). Disposable polymeric cryogel bioreactor matrix for therapeutic protein production. *Nature Protocols* **8**(5): 821–835.

Jain, E., Srivastava, A. and Kumar, A. (2009). Macroporous interpenetrating cryogel network of poly(acrylonitrile) and gelatin for biomedical applications. *Journal of Materials Science: Materials in Medicine* **20**(1): 173–179.

Kathuria, N., Tripathi, A., Kar, K. K. and Kumar, A. (2009). Synthesis and characterization of elastic and macroporous chitosan–gelatin cryogels for tissue engineering. *Acta Biomaterialia* **5**(1): 406–418.

Kumar, A., Bansal, V., Andersson, J., Roychoudhury, P. K. and Mattiasson, B. (2006a). Supermacroporous cryogel matrix for integrated protein isolation: Immobilized metal affinity chromatographic purification of urokinase from cell culture broth of a human kidney cell line. *Journal of Chromatography A* **1103**(1): 35–42.

Kumar, A., Bansal, V., Nandakumar, K. S., Galaev, I. Y., Roychoudhury, P. K., Holmdahl, R. and Mattiasson, B. (2006b). Integrated bioprocess for the production and isolation of urokinase from animal cell culture using supermacroporous cryogel matrices. *Biotechnology and Bioengineering* **93**(4): 636–646.

Kumar, A. and Bhardwaj, A. (2008). Methods in cell separation for biomedical application: Cryogels as a new tool. *Biomedical Materials* **3**(3): 034008.

Kumar, A., Mishra, R., Reinwald, Y. and Bhat, S. (2010). Cryogels: Freezing unveiled by thawing. *Materials Today* **13**(11): 42–44.

Kumar, A., Plieva, F. M., Galaev, I. Y. and Mattiasson, B. (2003a). Affinity fractionation of lymphocytes using a monolithic cryogel. *Journal of Immunological Methods* **283**(1–2): 185–194.

Kumar, A., Rodríguez-Caballero, A., Plieva, F. M., Galaev, I. Y., Nandakumar, K. S., Kamihira, M., Holmdahl, R., Orfao, A. and Mattiasson, B. (2005). Affinity binding of cells to cryogel adsorbents with immobilized specific ligands: Effect of ligand coupling and matrix architecture. *Journal of Molecular Recognition* **18**(1): 84–93.

Kumar, A., Sami, H., Srivastava, A. and Ghatak, A. (2010). Cryotropic hydrogels and their use as filters. PCT/2010/050285.

Kumar, A. and Srivastava, A. (2010). Cell separation using cryogel-based affinity chromatography. *Nature Protocols* **5**(11): 1737–1747.

Kumar, A., Wahlund, P. O., Kepka, C., Galaev, I. Y. and Mattiasson, B. (2003b). Purification of histidine-tagged single-chain Fv-antibody fragments by metal chelate affinity precipitation using thermoresponsive copolymers. *Biotechnology and Bioengineering* **84**(4): 494–503.

Kuyukina, M. S., Ivshina, I. B., Serebrennikova, M. K., Krivorutchko, A. B., Podorozhko, E. A., Ivanov, R. V. and Lozinsky, V. I. (2009). Petroleum-contaminated water treatment in a fluidized-bed bioreactor with immobilized *Rhodococcus* cells. *International Biodeterioration & Biodegradation* **63**(4): 427–432.

Le Noir, M., Plieva, F. M. and Mattiasson, B. (2009). Removal of endocrine-disrupting compounds from water using macroporous molecularly imprinted cryogels in a moving-bed reactor. *Journal of Separation Science* **32**(9): 1471–1479.

Lozinsky, V. and Plieva, F. (1998). Poly (vinyl alcohol) cryogels employed as matrices for cell immobilization. 3. Overview of recent research and developments. *Enzyme and Microbial Technology* **23**(3): 227–242.

Lozinsky, V., Plieva, F., Galaev, I. and Mattiasson, B. (2001). The potential of polymeric cryogels in bioseparation. *Bioseparation* **10**(4–5): 163–188.

Lozinsky, V., Vainerman, E., Ivanova, S., Titova, E., Shtil'man, M., Belavtseva, E. and Rogozhin, S. (1986a). Study of cryostructurization of polymer systems. VI. The influence of the process temperature on the dynamics of formation and structure of cross-linked polyacrylamide cryogels. *Acta polymerica* **37**(3): 142–146.

Lozinsky, V., Vainerman, E. and Rogozhin, S. (1982). Study of cryostructurization of polymer systems. 2. The influence of freezing of a reacting mass on the properties of products in the preparation of covalently cross-linked gels. *Colloid and Polymer Science* **260**: 776–780.

Lozinsky, V. I. (2002). Cryogels on the basis of natural and synthetic polymers: Preparation, properties and application. *Russian Chemical Reviews* **71**(6): 489–511.

Lozinsky, V. I., Faleev, N. G., Zubov, A. L., Ruvinov, S. B., Antonova, T. V., Vainerman, E. S., Belikov, V. M. and Rogozhin, S. V. (1989b). Use of PVA-cryogel entrapped Citrobacter intermedius cells for continuous production of 3-fluoro-L-tyrosine. *Biotechnology Letters* **11**(1): 43–48.

Lozinsky, V. I., Galaev, I. Y., Plieva, F. M., Savina, I. N., Jungvid, H. and Mattiasson, B. (2003). Polymeric cryogels as promising materials of biotechnological interest. *TRENDS in Biotechnology* **21**(10): 445–451.

Lozinsky, V. I., Morozova, S. A., Vainerman, E. S., Titova, E. F., Shtil'Man, M. I., Belavtseva, E. M. and Rogozhin, S. V. (1989a). Study of cryostructurization of polymer systems. VIII. Characteristic features of the formation of crosslinked poly(acryl amide) cryogels under different thermal conditions. *Acta Polymerica* **40**(1): 8–15.

Lozinsky, V. I., Vainerman, E. S., Domotenko, L. V., Mamtsis, A. M., Titova, E. F., Belavtseva, E. M. and Rogozhin, S. V. (1986b). Study of cryostructurization of polymer systems VII. Structure formation under freezing of poly(vinyl alcohol) aqueous solutions. *Colloid and Polymer Science* **264**(1): 19–24.

Lozinsky, V. I., Vainerman, E. S., Ivanova, S. A., Titova, E. F., Shtil'man, M. I., Belavtseva, E. M. and Rogozhin, S. V. (1986c). Study of cryostructurization of polymer systems. VI. The influence of the process temperature on the dynamics of formation and structure of cross-linked polyacrylamide cryogels. *Acta Polymerica* **37**(3): 142–146.

Lozinsky, V. I., Vainerman, E. S., Korotaeva, G. F. and Rogozhin, S. V. (1984a). Study of cryostructurization of polymer systems. *Colloid and Polymer Science* **262**(8): 617–622.

Lozinsky, V. I., Vainerman, E. S., Titova, E. F., Belavtseva, E. M. and Rogozhin, S. V. (1984b). Study of cryostructurization of polymer systems. *Colloid and Polymer Science* **262**(10): 769–774.

Lozinsky, V. I., Zubov, A. L. and Titova, E. F. (1996). Swelling behavior of poly (vinyl alcohol) cryogels employed as matrices for cell immobilization. *Enzyme and Microbial Technology* **18**(8): 561–569.

Lozinsky, V. I., Zubov, A. L. and Titova, E. F. (1997). Poly (vinyl alcohol) cryogels employed as matrices for cell immobilization. 2. Entrapped cells resemble porous fillers in their effects on the properties of PVA-cryogel carrier. *Enzyme and Microbial Technology* **20**(3): 182–190.

Minton, J. A., Iravani, A. and Yousefi, A.-M. (2012). Improving the homogeneity of tissue-mimicking cryogel phantoms for medical imaging. *Medical Physics* **39**(11): 6796–6807.

Mishra, R., Goel, S. K., Gupta, K. C. and Kumar, A. (2014). Biocomposite cryogels as tissue-engineered biomaterials for regeneration of critical-sized cranial bone defects. *Tissue Engineering Part A* **20**(3–4): 751–762.

Mishra, R. and Kumar, A. (2011). Inorganic/organic biocomposite cryogels for regeneration of bony tissues. *Journal of Biomaterials Science, Polymer Edition* **22**(16): 2107–2126.

Mu, C., Liu, F., Cheng, Q., Li, H., Wu, B., Zhang, G. and Lin, W. (2010). Collagen cryogel cross-linked by dialdehyde starch. *Macromolecular Materials and Engineering* **295**(2): 100–107.

Odabas, S., Feichtinger, G., Korkusuz, P., Inci, I., Bilgic, E., Yar, A., Cavusoglu, T., Menevse, S., Vargel, I. and Piskin, E. (2013). Auricular cartilage repair using cryogel scaffolds loaded with BMP-7-expressing primary chondrocytes. *Journal of Tissue Engineering and Regenerative Medicine* **7**(10): 831–840.

Önnby, L., Giorgi, C., Plieva, F. M. and Mattiasson, B. (2010). Removal of heavy metals from water effluents using supermacroporous metal chelating cryogels. *Biotechnology Progress* **26**(5): 1295–1302.

Önnby, L., Pakade, V., Mattiasson, B. and Kirsebom, H. (2012). Polymer composite adsorbents using particles of molecularly imprinted polymers or aluminium oxide nanoparticles for treatment of arsenic contaminated waters. *Water Research* **46**(13): 4111–4120.

Önnby, L., Svensson, C., Mbundi, L., Busquets, R., Cundy, A. and Kirsebom, H. (2013). γ-Al2O3-based nanocomposite adsorbents for arsenic (V) removal: Assessing performance, toxicity and particle leakage. *The Science of the Total Environment* **473**: 207–214.

Ozturk, B. Y., Inci, I., Egri, S., Ozturk, A. M., Yetkin, H., Goktas, G., Elmas, C., Piskin, E. and Erdogan, D. (2013). The treatment of segmental bone defects in rabbit tibiae with vascular endothelial growth factor (vEGF)-loaded gelatin/hydroxyapatite 'cryogel' scaffold. *European Journal of Orthopaedic Surgery & Traumatology* **23**(7): 767–774.

Petrenko, Y. A., Ivanov, R. V., Petrenko, A. Y. and Lozinsky, V. I. (2011). Coupling of gelatin to inner surfaces of pore walls in spongy alginate-based scaffolds facilitates the adhesion, growth and differentiation of human bone marrow mesenchymal stromal cells. *Journal of Materials Science: Materials in Medicine* **22**(6): 1529–1540.

Plieva, F., Bober, B., Dainiak, M., Galaev, I. Y. and Mattiasson, B. (2006). Macroporous polyacrylamide monolithic gels with immobilized metal affinity ligands: The effect of porous structure and ligand coupling chemistry on protein binding. *Journal of Molecular Recognition* **19**(4): 305–312.

Plieva, F. M., Andersson, J., Galaev, I. Y. and Mattiasson, B. (2004a). Characterization of polyacrylamide based monolithic columns. *Journal of Separation Science* **27**(10–11): 828–836.

Plieva, F. M., Galaev, I. Y. and Mattiasson, B. (2007). Macroporous gels prepared at sub-zero temperatures as novel materials for chromatography of particulate-containing fluids and cell culture applications. *Journal of Separation Science* **30**(11): 1657–1671.

Plieva, F. M., Karlsson, M., Aguilar, M.-R., Gomez, D., Mikhalovsky, S. and Galaev, I. Y. (2005). Pore structure in supermacroporous polyacrylamide based cryogels. *Soft Matter* **1**(4): 303–309.

Plieva, F. M., Savina, I. N., Deraz, S., Andersson, J., Galaev, I. Y. and Mattiasson, B. (2004b). Characterization of supermacroporous monolithic polyacrylamide based matrices designed for chromatography of bioparticles. *Journal of Chromatography B* **807**(1): 129–137.

Saino, E., Fassina, L., Van Vlierberghe, S., Avanzini, M., Dubruel, P., Magenes, G., Visai, L. and Benazzo, F. (2011). Effects of electromagnetic stimulation on osteogenic differentiation of human mesenchymal stromal cells seeded onto gelatin cryogel. *International Journal of Immunopathology and Pharmacology* **24**(1 Suppl 2): 1–6.

Savina, I. N., English, C. J., Whitby, R. L. D., Zheng, Y., Leistner, A., Mikhalovsky, S. V. and Cundy, A. B. (2011). High efficiency removal of dissolved As(III) using iron nanoparticle-embedded macroporous polymer composites. *Journal of Hazardous Materials* **192**(3): 1002–1008.

Shakya, A. K., Holmdahl, R., Nandakumar, K. S. and Kumar, A. (2014). Polymeric cryogels are biocompatible, and their biodegradation is independent of oxidative radicals. *Journal of Biomedical Materials Research Part A* **102**(10): 3409–3418.

Sharma, A., Bhat, S., Vishnoi, T., Nayak, V. and Kumar, A. (2013). Three-dimensional supermacroporous carrageenan-gelatin cryogel matrix for tissue engineering applications. *BioMed Research International* **2013**: 478279, 15.

Singh, D., Nayak, V. and Kumar, A. (2010). Proliferation of myoblast skeletal cells on three-dimensional supermacroporous cryogels. *International Journal of Biological Sciences* **6**(4): 371–381.

Srivastava, A., Jain, E. and Kumar, A. (2007). The physical characterization of supermacroporous poly(N-isopropylacrylamide) cryogel: Mechanical strength and swelling/de-swelling kinetics. *Materials Science and Engineering: A* **464**(1–2): 93–100.

Srivastava, A., Shakya, A. K. and Kumar, A. (2012). Boronate affinity chromatography of cells and biomacromolecules using cryogel matrices. *Enzyme and Microbial Technology* **51**(6): 373–381.

Stanescu, M. D., Fogorasi, M., Shaskolskiy, B. L., Gavrilas, S. and Lozinsky, V. I. (2010). New potential biocatalysts by laccase immobilization in PVA cryogel type carrier. *Applied Biochemistry and Biotechnology* **160**(7): 1947–1954.

Tamahkar, E., Bereli, N., Say, R. and Denizli, A. (2011). Molecularly imprinted supermacroporous cryogels for cytochrome c recognition. *Journal of Separation Science* **34**(23): 3433–3440.

Tripathi, A. and Kumar, A. (2011). Multi-featured macroporous agarose–alginate cryogel: Synthesis and characterization for bioengineering applications. *Macromolecular Bioscience* **11**(1): 22–35.

Tripathi, A., Kathuria, N. and Kumar, A. (2009). Elastic and macroporous agarose–gelatin cryogels with isotropic and anisotropic porosity for tissue engineering. *Journal of Biomedical Materials Research Part A* **90**(3): 680–694.

Van Vlierberghe, S., Cnudde, V., Masschaele, B., Dubruel, P., De Paepe, I., Jacobs, P., Van Hoorebeke, L., Unger, R., Kirkpatrick, C. and Schacht, E. (2006). Porous gelatin cryogels as cell delivery tool in tissue engineering. *Journal of Controlled Release* **116**(2): e95–e98.

Van Vlierberghe, S., Dubruel, P. and Schacht, E. (2010). Effect of cryogenic treatment on the rheological properties of gelatin hydrogels. *Journal of Bioactive and Compatible Polymers* **25**(5): 498–512.

Velizarov, S. G., Rainina, E. I., Sinitsyn, A. P., Varfolomeyev, S. D., Lozinsky, V. I. and Zubov, A. L. (1992). Production of L-lysine by free and PVA-cryogel immobilized Corynebacterium glutamicum cells. *Biotechnology Letters* **14**(4): 291–296.

Vishnoi, T. and Kumar, A. (2013a). Comparative study of various delivery methods for the supply of alpha-ketoglutarate to the neural cells for tissue engineering. *BioMed Research International* 2013: 294679, 11.

Vishnoi, T. and Kumar, A. (2013b). Conducting cryogel scaffold as a potential biomaterial for cell stimulation and proliferation. *Journal of Materials Science: Materials in Medicine* **24**(2): 447–459.

Vladimir, I. L. (2002). Cryogels on the basis of natural and synthetic polymers: Preparation, properties and application. *Russian Chemical Reviews* **71**(6): 489.

Wang, L., Shen, S., He, X., Yun, J., Yao, K. and Yao, S.-J. (2008). Adsorption and elution behaviors of bovine serum albumin in metal-chelated affinity cryogel beds. *Biochemical Engineering Journal* **42**(3): 237–242.

Wang, X. and Min, B. G. (2008). Comparison of porous poly (vinyl alcohol)/hydroxyapatite composite cryogels and cryogels immobilized on poly (vinyl alcohol) and polyurethane foams for removal of cadmium. *Journal of Hazardous Materials* **156**(1): 381–386.

Wolfe, J., Bryant, G. and Koster, K. L. (2002). What is 'unfreezable water', how unfreezable is it and how much is there? *Cryo Letters* **23**(3): 157–166.

Yan, C., Shen, S., Yun, J., Wang, L., Yao, K. and Yao, S. J. (2008). Isolation of ATP from a yeast fermentation broth using a cryogel column at high flow velocities. *Journal of Separation Science* **31**(22): 3879–3883.

Yao, K., Shen, S., Yun, J., Wang, L., He, X. and Yu, X. (2006). Preparation of polyacrylamide-based supermacroporous monolithic cryogel beds under freezing-temperature variation conditions. *Chemical Engineering Science* **61**(20): 6701–6708.

Ye, J., Yun, J., Lin, D. Q., Xu, L., Kirsebom, H., Shen, S., Yang, G., Yao, K., Guan, Y. X. and Yao, S. J. (2013). Poly(hydroxyethyl methacrylate)-based composite cryogel with embedded macroporous cellulose beads for the separation of human serum immunoglobulin and albumin. *Journal of Separation Science* **36**(24): 3813–3820.

Yun, J., Shen, S., Chen, F. and Yao, K. (2007). One-step isolation of adenosine triphosphate from crude fermentation broth of Saccharomyces cerevisiae by anion-exchange chromatography using supermacroporous cryogel. *Journal of Chromatography B* **860**(1): 57–62.

Zhang, X.-Z. and Chu, C.-C. (2003). Synthesis of temperature sensitive PNIPAAm cryogels in organic solvent with improved properties. *Journal of Materials Chemistry* **13**(10): 2457–2464.

# 2 Synthesis and Characterization of Cryogels

*Apeksha Damania, Arun Kumar Teotia and Ashok Kumar**

## CONTENTS

| | | |
|---|---|---|
| 2.1 | Introduction | 36 |
| 2.2 | The Process of Gelation | 37 |
| 2.3 | Different Methods of Inducing Gel Formation | 38 |
| | 2.3.1 Physical Cross-Linking | 38 |
| | 2.3.2 Chemical Cross-Linking | 44 |
| | 2.3.3 Addition Polymerization | 44 |
| |     2.3.3.1 Cryogelation by Free Radical Polymerization | 44 |
| |     2.3.3.2 Mechanism of Free Radical Polymerization | 48 |
| |     2.3.3.3 Thermal Initiators | 51 |
| |     2.3.3.4 Redox Initiation | 52 |
| |     2.3.3.5 Photochemical Initiators | 52 |
| |     2.3.3.6 Radiation Initiation | 52 |
| 2.4 | Cryogelation Methods and Factors Affecting Cryogelation by Free Radical Polymerization | 53 |
| | 2.4.1 Cryogelation by Free Radical Polymerization Using UV Irradiation | 54 |
| | 2.4.2 Cryogelation by Functional Group Cross-Linking | 55 |
| |     2.4.2.1 Cross-Linking with Aldehydes | 55 |
| |     2.4.2.2 Cross-Linking with Adipic Acid Hydrazide | 55 |
| |     2.4.2.3 Cross-Linking with Ethyl(dimethylaminopropyl) Carbodiimide-*N*-Hydroxysuccinimide (EDC-NHS) | 56 |
| 2.5 | Physical Properties of Cryogels and Factors Affecting These Properties | 56 |
| | 2.5.1 Freezing Rate and Temperature | 57 |
| | 2.5.2 Effect of Porogen and Solvent | 57 |
| | 2.5.3 Effect of Monomer and Initiator Concentrations | 58 |
| 2.6 | Characterization of Cryogels | 59 |
| | 2.6.1 Morphological and Textural Features | 59 |
| | 2.6.2 Mechanical and Thermal Properties | 69 |
| |     2.6.2.1 Compression Test | 69 |

* Corresponding author: E-mail: ashokkum@iitk.ac.in; Tel: +91-512-2594051.

    2.6.2.2 Fatigue Test ........................................................................................ 70
    2.6.2.3 Rheology Studies .............................................................................. 71
  2.6.3 Chemical Properties ...................................................................................... 73
    2.6.3.1 Fourier Transform Infrared Spectroscopy (FTIR) .............. 73
    2.6.3.2 X-Ray Diffraction (XRD) .................................................................. 73
    2.6.3.3 XPS and Elemental Analysis .......................................................... 74
  2.6.4 Biological Properties ..................................................................................... 74
    2.6.4.1 Protein Adsorption Studies .............................................................. 74
    2.6.4.2 *In Vitro* Biocompatibility ................................................................. 75
2.7 Conclusion ..................................................................................................................... 76
Acknowledgements ................................................................................................................ 77
List of Abbreviations .............................................................................................................. 77
References ................................................................................................................................ 78

## ABSTRACT

Hydrogels are aqueous cross-linked polymeric network systems generated by cross-linking polymer solutions. Polymeric hydrogel systems have been synthesized in different formats, ranging from microporous hydrogels used in drug delivery to macroporous spongy gels used in tissue engineering, by employing different techniques depending on the intended application. One such process of synthesising macroporous hydrogels is *cryogelation*. In this technique, gelation takes place at subzero temperatures in a semifrozen state. On thawing, a macroporous gel is generated with a pore size distribution of 10 μm to 200 μm. These types of macroporous polymeric gels, referred to commonly as **cryogels**, can be generated from a large number of synthetic, semisynthetic and natural polymers. Cryogels are finding applications in diverse fields such as chromatographic separation, cell culture systems and tissue engineering. These spongy polymeric gels can be characterized for their various intrinsic properties such as interconnected porosity and better mechanical properties, using different analytical techniques such as scanning electron microscopy (SEM), thermogravimetric analysis/differential scanning calorimetry (TGA/DSC), X-ray diffraction, porosity and so on.

## KEYWORDS

Cryogel, interconnected pores, intrinsic properties, analytical techniques

## 2.1 INTRODUCTION

Over the years, macroporous hydrogels have found applications in various areas such as tissue engineering applications (Bölgen et al., 2007, 2009; Van Vlierberghe et al., 2011), as separation matrices (Kumar et al., 2003; Kumar and Srivastava, 2010) and in the form of bioreactors (Kumar et al., 2006; Jain and Kumar, 2013) as well as other industrial, medical and pharmaceutical applications (Gustavsson et al., 2009; Omidian and Park, 2009). Although there have been many strategies used for the formation of these macroporous hydrogels, one method that has come to be widely accepted as the best way to form macroporous hydrogels is the method of *cryogelation*. Simply put, cryogelation is the formation of hydrogels at subzero temperatures. The concept of cryogelation arose from the principle that a difference in solubility of solutes in liquid and solid water results in the concentration of the solute in the liquid or aqueous phase. This 'pushing out' effect, commonly termed cryoconcentration, with a gradual decrease in the freezing temperature, forms the basis of cryostructuration or cryotropic gelation (Gun'ko et al., 2013). Hence, contrary to conventional hydrogels, which are homophase systems, macroporous hydrogels, more commonly referred to as cryogels, are heterophase systems, with the solvent (water in most cases) found both within the interconnected pores as well as bound to the polymer network (Plieva et al., 2007). Typically, cryogels have a pore diameter $1 < d < 300$ μm. The life science classification classifies pore diameters $d < 0.1$ μm, $0.1 < d < 100$ μm and $d > 100$ μm as nanopores, micropores and macropores, respectively (Gun'ko et al., 2013). This system of classification has been used to term/classify cryogels as macroporous hydrogels.

A variety of factors determines the physical properties of cryogels. These include degree and type of cross-linking, concentration of monomers/polymers, temperature of gelation, cooling rate, pH of solvent and amount of solvent. All of these, in one way or another, affect the physical or mechanical properties of cryogels, such as the pore size, pore distribution, mechanical strength and elasticity of the cryogels.

A number of techniques have been implemented to study the overall structure and morphology of cryogels, to identify their characteristic features, which also plays an important role in determining the application of the cryogel. These techniques include but are not limited to microscopy, mercury porosimetry, nitrogen adsorption, thermogravimetry, water-uptake studies, X-ray diffraction studies, NMR cryoporometry and Fourier transform infrared spectroscopy.

This chapter reviews the different methods and mechanisms involved in cryogel synthesis, the factors affecting cryogelation and the overall characterization of cryogels for their use in different applications.

## 2.2 THE PROCESS OF GELATION

Gelation can take place in two different forms:

1. Cross-linking of a polymer without a solvent, that is, polymerization of polymer/monomers in powder state.
2. Cross-linking of a polymer homogeneously dissolved in a liquid medium.

The second system is the most commonly used system. In this system, the solution of monomers in appropriate solvent is gelated by incorporating an appropriate cross-linking agent such as an ionotropic agent or a chelatotropic agent. Gelation may also take place due to a change in the thermal state of the system (i.e., decreasing affinity of solvent toward monomer) or by partial crystallization of the polymer or colloid (in ceramics) leading to phase transition.

The initial state of a monomer or poly-disperse branched monomer is called 'sol' and the polymerization or cross-linking process results in the formation of an infinite chain length of molecules with decreasing solubility. This transition from a finite size monomer to an infinite chain polymer is called a 'sol–gel transition' and is also known as 'gelation'. The point at which gelation first occurs in the system is called the 'gel point' (Rubinstein and Colby, 2003).

In multicomponent systems, at moderate frozen conditions, microsolidification takes place where apart from the solid phase, there exists a liquid microphase between the solid solvent crystals where the reaction can proceed (Pincock and Kiovsky, 1965). It was noted that this liquid phase contains a high concentration of reactants and may even accelerate catalysis in water and ice (Bruice and Butler, 1964).

Depending on the mechanism used for gelation, cryogels can be categorized as follows (Lozinsky, 2002):

1. Chemotropic: induced by the formation of chemical bonds between polymer chains, resulting in a three-dimensional (3D) covalent network of gels
2. Ionotropic: induced by the formation of ionic bonds by ion exchange
3. Chelatotropic: induced by chelation
4. Thermotropic: induced by heating
5. Psychrotropic: induced by cooling without freezing
6. Cryotropic: induced by freezing
7. Solvotropic: induced by a change in the solvent properties

## 2.3 DIFFERENT METHODS OF INDUCING GEL FORMATION

### 2.3.1 Physical Cross-Linking

This type of gelation occurs in polymers such as agarose (Lozinsky et al., 2008), collagen (Podorozhko et al., 2000) and polyvinyl alcohol (PVA) (Lozinsky and Plieva, 1998; Hassan and Peppas, 2000; Lozinsky et al., 2007). Multiple freeze–thaw cycles increase the gel strength of cryogels. PVA provides a classic example for the study of physical gelation occurring in molecules due to a freeze–thaw treatment (Kobayashi et al., 1995; Lozinsky, 1998; Ricciardi et al., 2004; Auriemma et al., 2008; Lozinsky et al., 2008). Cryotropic gelation due to physical cross-linking via hydrogen bond formation among polymer chains has been studied in PVA, PVA-gelatin, PVA-casein systems (Stauffer and Peppast, 1992; Bajpai and Saini, 2005a,b; El Fray et al., 2007) and in agarose (Lozinsky et al., 2008).

The cryotropic treatment leads to the generation of cryogels with interconnected micro- and macroporous networks (Lozinsky, 1998). These structures have high modulus and show increased mechanical stability and elasticity over a large range of deformations due to their complex architecture (Watase and Nishinari, 1989; Lozinsky et al., 2007).

Surfactant incorporated (anionic sodium dodecyl sulphate and cationic dodecyl trimethyammonium bromide [DTAB]) gas-filled macroporous polyvinyl alcohol (PVA) cryogels were also synthesized by freeze–thaw cycles. These showed on average a large pore size of ~180 μm, which was attributed to the coalescence of air bubbles generated in the heterogeneous gel matrix (Lozinsky et al., 2005). Results showed that the nature of the surfactant incorporated in the PVA polymer mixture and the cryogenic treatment affects the architecture, structuring, distribution and size of the pores formed in PVA cryogel, thus influencing its mechanical and thermal properties (Lozinsky et al., 2005). Similar results were observed regarding the incorporation of γ-cyclodextrin on the cryogel of PVA and on its mechanical, rheological and other physiochemical properties (Hernández et al., 2004a,b).

Physically cross-linked cryogels of blends of PVA with different materials such as chitosan (Park and Kim, 2006; Yang et al., 2008), silk fibroin (Li et al., 2001) and polyamido amine (PAMAM) (Wu et al., 2004) have been synthesized by providing either just freeze–thaw cycles to the blend mixtures or freeze–thaw cycles in addition to γ-irradiation to the cryogel of the blend (Park and Kim, 2006). In another approach, homogeneous physically cross-linked cryogels of composites of PVA-cellulose (Millon and Wan, 2006), PVA-clay (Bandi and Schiraldi, 2006), PVA-resin (amberlite) (Savina et al., 2005) and PVA-tetramethoxysilane (TMOS) (Lozinsky et al., 2007) were synthesized. In this approach, a homogeneous colloidal suspension of PVA and of material to be incorporated is generated and this composite blend is then given freeze–thaw cycles to generate a composite cryogel. Incorporating active materials in a cryogel matrix by this approach gives rise to composite cryogels of different size, shape, materials, architecture and properties, which can be used for diverse applications.

Physically cross-linked cryogel monoliths or beads have been synthesized for application in separation science and chromatographic processes. These matrices are finding increased application in the above-mentioned fields as an alternative to classical packed bed columns (Yu et al., 2007). Cryogel matrices provide flexibility as a suitable/required property that can be incorporated into the matrix either during the fabrication process (cryogelation) or postfabrication (functionalization of cryogels). In this direction, PVA cryogels were synthesized in which a molecularly imprinted polymer (MIP) with 17β-estradiol (an endocrine disruptor) as a template was incorporated (Le Noir et al., 2007) for generating a solid phase extraction (SPE) system.

The properties of cryogels can be modified postfabrication of cryogels ('functionalization'). This approach provides a technique for modifying the surfaces of cryogel matrices. One such use is to make surfaces more biocompatible and to incorporate in them cell adhesion entities for application in biological sciences and tissue engineering. Using these approaches, PVA cryogel surfaces have been successfully functionalized with hydroxyapatite, which were used for culturing mouse fibroblasts (Kaneko et al., 2004). In another work, the PVA cryogel was successfully functionalized with protein-A onto which anti-CD34 antibody was attached against the CD34 antigen found on the cell surface. This type of cryogel was used for capturing human acute myeloid leukemia KG-1 cells (Kumar et al., 2005).

Polysaccharides are another class of polymeric materials that can cross-link physically and undergo gelation. Apart from physical gelation, these polymers can be cross-linked chemically via polyelectrolyte complexation (ionotropic linkage), for example, alginic acid. Polysaccharides like agarose, carrageenan and gellan undergo reversible thermal-gelation. During cryogelation, structured pores form due to the formation of ice crystals, which act as porogens leading to the formation of macropores in the cross-linked matrix. Gelation in polysaccharides occurs due to the formation of hydrogen bonds among the functional groups present in the polymer chains. In polysaccharide cryogel systems, polymerization can occur at two different stages in the process. First, cryogelation (polymerization) may take place in the chemically cross-linked system through covalent bond formation among molecules. Second, cryogelation may take place during the thawing step in a process referred to bond formation in the polymeric systems of amylopectin, amylose, maltodextrin, xanthan, locust bean gum (LBG), β-glucan, hyaluronan, carboxymethyl curdlan (CMc) and carboxymethyl cellulose (CMC) (Lozinsky et al., 2000a,b,c; Zhang et al., 2013). Physical gelation provides an advantage that the process is highly biocompatible because it does not involve the use of any toxic cross-linking agents that need to be extracted before use, or the use of damaging UV or γ-radiations. Hence, this method can be safely used with biological systems such as cells, enzymes and other biomolecules without causing any unwanted reactions or interactions with them (Coviello et al., 2007; Silva et al., 2010; Hennink and Van Nostrum, 2012).

Physically cross-linked polysaccharide cryogels are prepared by providing a freezing treatment at –18 to –20°C. It was observed that lowering the freezing temperature of the system near its glass transition temperature (Tg) leads to an increase in gel strength. However, if the cryogelation temperature is decreased below the transition temperature, it leads to cryogels with decreased stability. This may be due to a decreased mobility of macromolecules in the unfrozen microphase (Lozinsky, 2002; Lazaridou and Biliaderis, 2004; Doyle et al., 2006). The formation of a thermo-reversible β-glucan gelatinous physical gel from barley extract by freeze–thaw treatment (Morgan and Ofman, 1998) was analysed by Lazardou and Biladeris for the mechanism of cryogelation (Lazaridou and Biliaderis, 2004). They observed that the strength of these physically cross-linked cryogels increased with an increase in polymer concentration and a decrease in pore size. The strength of the cryogels further increased when they were given repetitive freeze–thaw cycles, which leads to the generation of cryo-structured gels. The gel structure depended on the β-glucan molecular size and the ratio of cellotriose to cellotetrose in β-glucan (Lazaridou and Biliaderis, 2004). Additionally, the effect of incorporating polyols in the gels was studied and it was observed that the addition of polyols leads to weaker gels due to retardation in cryo-structuration (Lazaridou et al., 2008). With increasing concentration of polyols, a higher number of cycles were required to generate cryogels. The samples with low molecular size β-glucan gave fibrous structures, whereas those with high molecular size gave gelatinous cryo-precipitates. Adding polyols in increasing concentrations leads to the transformation of generated cryogel from fibrilar architecture to that with smoother homogeneous microstructures with nodes in them (Lazaridou et al., 2008).

Curdlan, another polysaccharide and its carboxymethyl derivative, carboxymethyl curdlan (CMc), both form physically cross-linked hydrogels similar to carboxymethyl cellulose (CMC). The strength and stiffness of gels increased when higher polymer concentrations were used and freeze–thaw cycles implemented to the polymer system (Zhang et al., 2002; Cong et al., 2003; Jin et al., 2006; Nishinari et al., 2009; Wu et al., 2012).

Cryogels of xanthan gum, which is an anionic polysaccharide, were generated by freeze–thaw treatment and characterized for their physicochemical properties (Giannouli and Morris, 2003). In addition, LBG, which undergoes physical thermo-reversible gelation, was also used for the synthesis of macroporous gels by cryogelation at subzero temperatures. Gelation was attributed to the formation of hydrogen bonds between hydroxyl and carboxyl groups of the polysaccharide chain (Lozinsky et al., 2000a,b,c). Gelation was accelerated when LBG was annealed at high temperatures due to molecular conformation change in solid state (Hatakeyama et al., 2005).

The mechanical properties and thermal stability of physically cross-linked cryogels depend on the rate of thawing. A thawing rate of 0.03°C min$^{-1}$ was found optimal with respect to gelation yield and mechanical and thermal properties of the gel (Lozinsky et al., 2000a,b,c; Lozinsky, 2002; Vaikousi and Biliaderis, 2005; Doyle et al., 2006). When the rate of thawing was kept low, the gels obtained showed increased mechanical strength, elasticity and thermal properties compared to gels that were thawed at rapid rates. The effects of incorporating sugars such as sucrose, glucose, fructose and sorbitol on the physical characteristics of the gels were also studied (Doyle et al., 2006). At lower sugar concentrations, the cryogels obtained showed higher mechanical properties whereas at higher sugar concentrations the cryogel strength decreased. This effect was attributed to a higher concentration of polymers in an unfrozen liquid microphase because of a reduction of the water content with increased sugar concentration. At very high concentrations, the sugar molecules inhibit polymer–polymer interactions, as they tend to bind to the polymer chains, hindering hydrogen bond formation and subsequently decreasing gel strength (Doyle et al., 2006).

It has been noted that a prolonged incubation of frozen systems at temperatures above the glass transition temperature (~ −5 to 0°C) increases gel strength by favouring the interaction between polymeric chains because of increased polymeric chain mobility. A similar effect was observed by giving multiple freeze–thaw cycles to the polymer system. This happens because of the exclusion of polymers from the solvent system by growing ice crystals and the formation of a concentrated nonfrozen liquid microphase (NFLMP). This NFLMP generates a polymer-rich region in a uniform phase that favours polymer chain interactions and pushes the polymer chains closer and closer in subsequent freeze–thaw cycles (Lozinsky et al., 2000a; Lozinsky, 2002; Lazaridou and Biliaderis, 2004; Vaikousi and Biliaderis 2005; Doyle et al., 2006; Mao and Chen, 2006; Lazaridou et al., 2008).

Cryogels of starch have been synthesized by giving freeze–thaw cycles to its hydrocolloid. The low temperature treatment leads to the generation of spongy gels due to cryotropic gelation (Lozinsky et al., 2000a). The morphology and texture of these cryogels depends on the amylopectin/amylose ratio in the polymer, concentration of the polymer system and the condition of cryogenic treatment (Eliasson and Kim, 1992; Navarro et al., 1997; Lozinsky et al., 2000b,c).

Cryogenic treatment of an aqueous system of maltodextrin resulted in the formation of spongy cryogels whose properties depended on the polymer concentration in the system, freezing temperatures and the duration and rate of thawing (Lozinsky et al., 2002). Cryogels of agarose were prepared by physical gelation under cryogenic treatment. The process resulted in the synthesis of spongy cryogels. These macroporous cryogels were used for culturing rat pancreatic insulinoma cells (Bloch et al., 2005; Lozinsky et al., 2008). As PVA and polysaccharides both undergo physically cross-linked gelation, efforts are made to use both of these and generate composite cryogels that have unique properties of both the gels. Based on these concepts, a PVA-carboxymethyl cellulose composite cryogel was synthesized by a freeze–thaw technique (Xiao and Gao, 2008). In this system, an increase in the ratio of CMC to PVA results in a decrease in crystallinity and melting temperature of the gels. These cryogels show a unique property of pH responsive swelling behaviour, enabling them to be used as drug delivery vehicles (Xiao and Gao, 2008). Interactions between CMC and PVA generate intra- and intermolecular entanglements in the polymer chain as shown in the Figure 2.1. Physically cross-linked PVA cryogels require cross-linking with glutraldehyde for stability to be functional in certain applications (Plieva et al., 2006). For example, in the physically cross-linked system of gelatin-PVA, chemical cross-linking was required to have optimum stability for functional use (Bernhardt et al., 2009; Vrana et al., 2010).

PVA was used to generate composite cryogel with dextran and chitosan (Takamura et al., 1992; Cascone et al., 1999; Kim et al., 2008; Yang et al., 2008). Under freeze–thaw conditions, the PVA chains form highly ordered and structured microcrystalline zones. These zones act as junction points in the matrix network between other chains of the same polymer in a homogeneous system or other chains of different polymer molecules in a heterogeneous system. The hydrogen bond formation between the hydroxyl groups also leads to intermolecular links and gelation (Yokoyama et al., 1986; Lozinsky, 2002; Lozinsky et al., 2012, 2014).

Comparative studies on hyaluronic acid hydrogels (Okamoto and Miyoshi, 2002; Luan et al., 2012) and freeze-dried gels by cryotropic gelation in hyaluronic acid solutions were performed to evaluate the mechanism of cryogelation in physically cross-linked gelation systems (Luan et al., 2012). In hyaluronic acid hydrogels in addition to hydroxyl (–OH) groups (similar to those in PVA), the carboxyl (–COO$^-$) and methylamine (–NHCH$_3$) groups also are responsible for hydrogen bond formation. In addition to these, the hydrophobic interactions among molecules cooperatively are responsible for gelation in hydrogel systems (Okamoto and Miyoshi, 2002). The freeze-dried porous gels of hyaluronic acid were synthesized from an acidified solution (presence of excess H$^+$ ions) and a hyaluronic acid solution with Na$^+$ ions incorporated into the system. It was observed that in the presence of excess H$^+$ ions, the ionization of carboxyl groups in hyaluronic acid decreased and they were available for hydrogen bonding between hyaluronic acid chains.

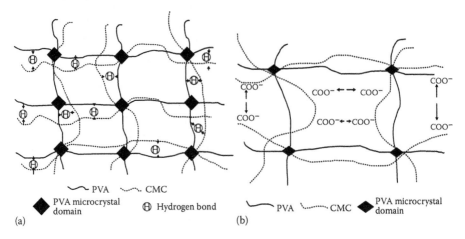

**FIGURE 2.1** Network models of CMC/PVA complex gels (a) in HCl (0.1 M, pH 1.2) and (b) in PBS (0.1 M, pH 7.4). (From Xiao, C. and Gao, Y.: Preparation and properties of physically cross-linked sodium carboxymethylcellulose/poly(vinyl alcohol) complex hydrogels. *Journal of Applied Polymer Science*. 2008. 107(3). 1568–1572. Copyright Wiley-VCH Verlag GmbH & Co. KGaA. Reproduced with permission.)

# Synthesis and Characterization of Cryogels

When $Na^+$ ions are present in the system they interact with carboxyl (–COO⁻) groups and hinder them from forming hydrogen bonds, thus decreasing overall gel strength. Similarly, the cryotropic gels of hyaluronic acid were synthesized by freezing the hyaluronic acid solution acidified and containing $Na^+$ ions. It was observed that there occurred no gelation in solutions containing $Na^+$ ions with shorter duration of freezing treatment (3 days). But weak gels were formed on prolonged freeze treatment (8 days) with strength similar to that of gels formed after 3 days of freeze treatment in hyaluronic acid solutions without any $Na^+$ ions (Luan et al., 2012). Similar results were also observed in hydrogel systems of polymers with hydrogen bonding induced gelation (Lozinsky et al., 1996, 2000a), indicating that NaCl interfered with the formation of hydrogen bonds and inhibiting gelation (Luan et al., 2012). Similar effects were observed in the presence of other chaotropic agents, for example, urea in the system (Luan et al., 2012) (Figure 2.2).

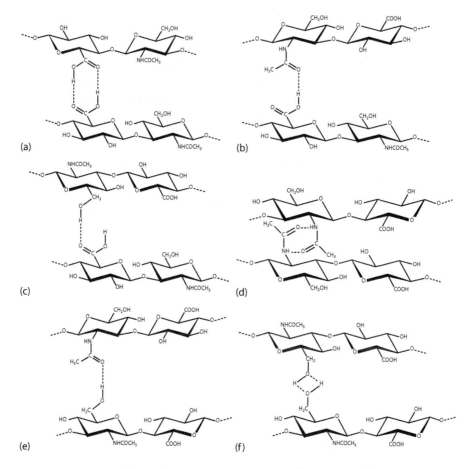

**FIGURE 2.2** The possible forms of hydrogen bonds between –COOH, –OH and –NHCOCH₃ groups in hyaluronic acid gels. (Reprinted from *Carbohydrate Polymers* 87(3), Luan, T., Wu, L., Zhang, H. and Wang, Y., A study on the nature of intermolecular links in the cryotropic weak gels of hyaluronan, 2076–2085. Copyright 2012, with permission from Elsevier.)

## 2.3.2 CHEMICAL CROSS-LINKING

Cryogel synthesis can take place by polymerization or cross-linking between polymer chains using different types of chemical cross-linking methods. Usually similar cross-linking systems or approaches can be employed in cryogelation, which are commonly employed in conventional polymeric hydrogel synthesis such as chemical cross-linking by free radical polymerization, UV irradiation (Bryant and Anseth, 2002, 2003; Park et al., 2003; Petrov et al., 2006, 2007), γ-radiation, or using redox initiated polymerization systems (Hong et al., 2007). Another approach is to induce polymerization by cross-linking functional groups in the polymer chains. This is a common technique used for water-soluble monomers and micromonomers such as proteins and polysaccharides. Common types of reactions employed in this technique include Schiff-base formation, Michael-type addition, peptide ligation and so on (Figures 2.3 and 2.4 and Table 2.1).

## 2.3.3 ADDITION POLYMERIZATION

### 2.3.3.1 Cryogelation by Free Radical Polymerization

Free radical polymerization reactions are rapid reactions consisting of chain reaction steps—initiation, propagation and termination. Free radical initiators are generated by hemolytic cleavage of covalent bonds.

Free radical reactions can be initiated by application of heat (thermal initiation), UV/visible light (photochemical initiation), radiation (ionizing), redox reagents, electricity (electrochemical) or any other method that can generate free radicals in the reaction mixture. One of the commonly used polymer classes for which free radical initiated polymerization is employed are monomers with vinyl groups. In these molecules, on suitable induction, for example, thermal induction, the heat treatment ruptures the pi (π) bonds, forming two-headed free radicals, which act as free radical initiators. The general bond dissociation energy is C–H > C–C > C–N > O–O.

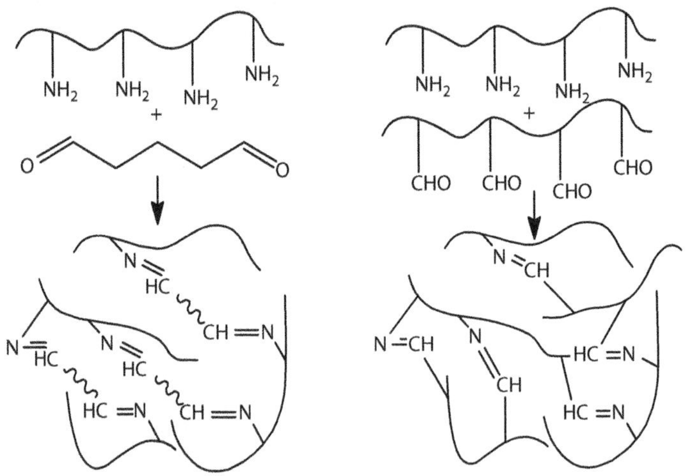

**FIGURE 2.3** Aldehyde-induced cross-linking.

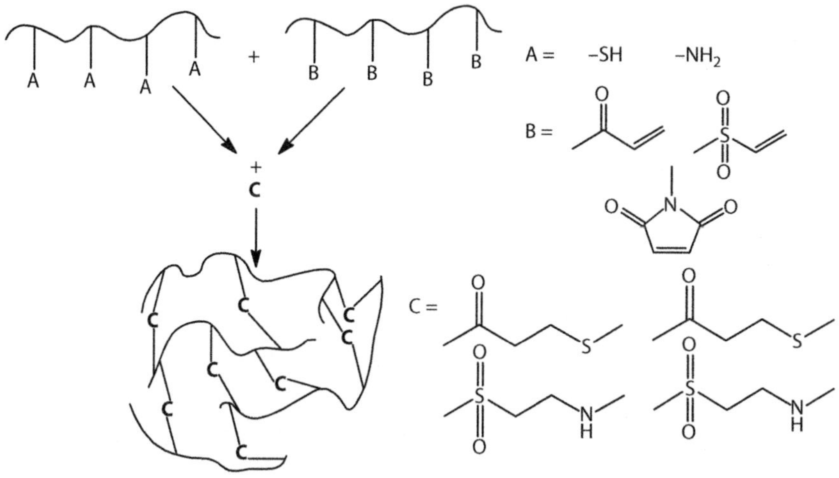

# Synthesis and Characterization of Cryogels

**FIGURE 2.4** Cross-linking by Michael addition.

**TABLE 2.1**
**Different Commonly Used Cross-Linking Agents**

| S. No. | Name | Structure |
|---|---|---|
| 1. | Glutraldehyde | 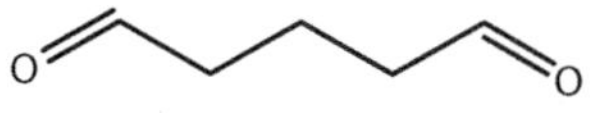 |
| 2. | EDC | 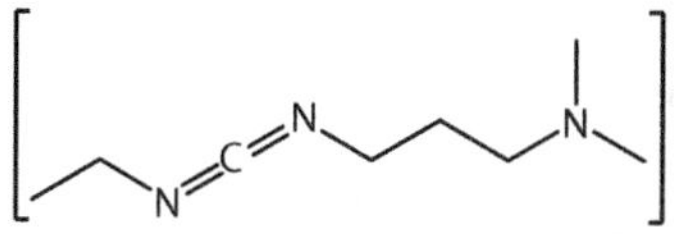 |
| 3. | Genipin | 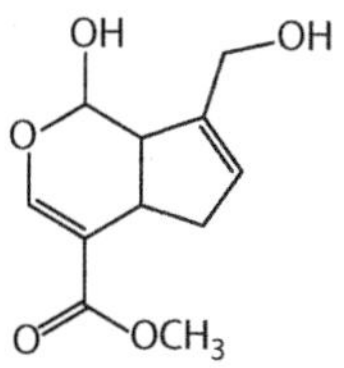 |
| 4. | Epichlorhydrin | 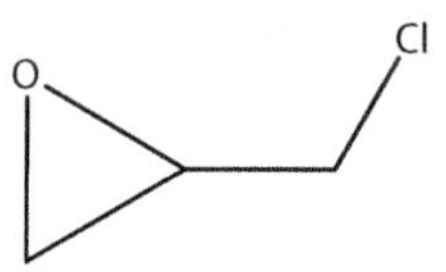 |

Commonly used initiators are ammonium peroxide, potassium peroxide, benzoyl peroxide and 2,2′-azo-*bis*-isobutyronitrile (AIBN). AIBN requires temperatures in the range of 70–80°C for initiation; benzoyl peroxide requires temperatures from 60°C to 70°C, whereas ammonium peroxide can initiate a reaction by generating free radicals from room temperature to subzero temperatures. Table 2.2 gives a list of some commonly used monomers for addition polymerization.

## TABLE 2.2
## Some Commonly Used Monomers for Addition Polymerization

| S. No. | Name | Structure |
|---|---|---|
| 1. | Poly(acrylonitrile) | |
| 2. | Poly(vinyl acetate) | |
| 3. | Poly(vinyl alcohol) | |
| 4. | Poly(vinyl butyral) | |
| 5. | Poly(vinyl chloride) | |
| 6. | Poly(vinylidene chloride) | |
| 7. | Polytetrafluoroethylene (PTFE) (Teflon) | |
| 8. | Polyethylene | |

*(Continued)*

## TABLE 2.2 (CONTINUED)
## Some Commonly Used Monomers for Addition Polymerization

| S. No. | Name | Structure |
|---|---|---|
| 9. | Poly isoprene | |
| 10. | Polystyrene | |
| 11. | Poly(methylmethacrylate) | |
| 12. | Vinyl acetate | |
| 13. | N-Vinyl pyrrolidone | |
| 14. | Acrylic acid | |
| 15. | Methacrylic acid | |
| 16. | Methylmethacrylate | |

*(Continued)*

## TABLE 2.2 (CONTINUED)
### Some Commonly Used Monomers for Addition Polymerization

| S. No. | Name | Structure |
|---|---|---|
| 17. | Acrylamide | (acrylamide structure) |
| 18. | N-isopropyl acrylamide | (structure; $R_1 = H$, $R_2 = $ isopropyl) |
| 19. | N,N-Methylene-bis-acrylamide | (structure) |
| 20. | (Hydroxyethyl)methacrylate | (structure) |

UV radiation can initiate polymerization (free radical) by disrupting selected bonds to generate free radicals. This photocatalysed reaction can be carried out even at room temperature, for example, diphenyl ketone on exposure to UV radiation breaks to generate two free radicals (Oda et al., 1978; Sato et al., 1979a):

$$Ph-CO-R \xrightarrow{UV/heat} Ph-\dot{C}=O + \dot{R}$$

Free radical polymerization is one of the most commonly used methods for polymerization reactions. There is the addition of monomer units to an activated center at the terminal of the existing chain in a chain growth addition polymerization. In general, for a free radical reaction to occur it is necessary to initiate the reaction using an initiator or activator. In certain cases, polymerization can occur even in the absence of initiators due to internal generation of free radicals. Table 2.3 lists some commonly used free radical initiators.

### 2.3.3.2 Mechanism of Free Radical Polymerization

The free radical chain growth polymerizations can be initiated in different ways. Based on the type of initiator that can be used, the different initiations are classified

## TABLE 2.3
## Some Commonly Used Free Radical Initiators

| S. No. | Thermal Free Radical Initiators | Solvent System | Structure |
|---|---|---|---|
| 1. | Ammonium persulphate | Aqueous | |
| 2. | Potassium persulphate | Aqueous | |
| 3. | 2,2'-azo-*bis*-isobutyronitrile (AIBN) | Organic/nonaqueous | |
| 4. | 1,1'-azo-*bis*(cyclohexane-carbonitrile) (ABCN) | Organic/nonaqueous | |
| 5. | Benzoyl peroxide | Organic/nonaqueous | |
| 6. | Tert-butyl peroxide | Organic/nonaqueous | |

as thermal initiation, redox initiation, photochemical initiation, ionizing radiations, electrolysis and so forth. All of these reactions involve identical steps:

1. Initiation: Radical generation from non-radical species
2. Propagation: Radical addition to a substituted alkene
3. Termination by disproportionation: Atom transfer and atom abstraction reactions
4. Radical-radical recombination reactions

$$\text{I} \longrightarrow \text{I}^{\cdot}$$
$$\text{I}^{\cdot} + \text{M–X} \longrightarrow \text{I–M–X}^{\cdot}$$
$$\text{I–M–X}^{\cdot} + \text{X}^{\cdot}\text{–M–I} \longrightarrow \text{I–M–X–M–X}$$
$$[\text{M}]_n\text{–X}^{\cdot} + \text{X}^{\cdot}\text{–}[\text{M}]_n \longrightarrow \text{--}[\text{M}]_n\text{–}[\text{M}]_n\text{--}$$

where I = initiator, M = monomer and $[\text{M}]_n$ = polymer.

The conversion of monomers to polymers depends on the kinetic and thermodynamic feasibility of the reaction. A polymerization reaction is feasible only if the free energy ($\Delta G$) of the reaction, that is, the free energy of the product (polymer) and the reactants (monomers), is negative; then the reaction will occur only under certain conditions (Braun et al., 2001; Bower, 2002; Odian, 2004). The feasibility of a polymerization reaction depends not only on thermodynamic feasibility but also on kinetic feasibility, that is, whether the reaction proceeds at a sufficient rate under specific conditions for proper initiation and propagation of polymerization. This implies that even if the polymerization reaction is thermodynamically feasible, the reaction has to be carried out at specific conditions at which it is also kinetically feasible in order to achieve a completely feasible polymerization (Odian, 2004).

Different kinds of initiators such as radical, cationic and anionic are used in chain polymerization reactions. These initiators cannot be used randomly to initiate a reaction because different classes of initiators do not work equally efficiently for all the monomers and solvent systems. The monomers containing vinyl, aldehyde and ketone groups preferentially undergo polymerization with anionic and cationic initiators. The vinyl group preferentially is initiated by free radical initiators whereas a carbonyl group is unaffected by free radical initiators.

The C–C double bond in the vinyl group is activated by both radical and ionic initiators; the vinyl group can undergo either hemolytic or heterolytic cleavage.

The type of reaction by which a vinyl monomer will polymerize depends on the type of substitution present on the molecule and its resonance or inductive properties on the monomer. If the substituent is an electropositive (electron donating) species (such as alkoxy, alkyl, alkenyl and phenyl), it increases electron density on a vinyl group facilitating action with a cationic species.

$$\text{CH}_2^{\delta-} = \text{CH}_2 \leftarrow \text{Y}^{\delta+}$$

The substituents stabilize the cationic species (carbocation) formed by the resonance effect (e.g., as in vinyl ether).

## Synthesis and Characterization of Cryogels

The electron withdrawing groups such as, alkenyl or phenyl groups decreases electron density on the reacting species, thus its resonance stabilizes the anionic propagating species in the reaction. The free radical initiators act on any carbon–carbon double bond. The radical species thus formed are neutral and do not stringently attack the pi ($\pi$) bond for reaction. The propagating radical species is stabilized by resonance stabilization (by delocalization).

The different radical species formed undergoes polymerization by head-to-head, head-to-tail or tail-to-tail polymerization depending on a predominance of radical stabilization and steric hindrance by substituent groups (Bovey et al., 1977; Cais and Kometani, 1984; Ovenall and Uschold, 1991; Bovey and Mirau, 1996; Guiot et al., 2002).

The various types of initiation mechanisms used for free radical polymerization reactions are thermal, redox, photochemical, ionization radiation and electrochemical methods (Moad et al., 2002; Denisov et al., 2003).

### 2.3.3.3 Thermal Initiators

The type of compound that can be used as a thermal initiator in polymerization reactions depends on its bond dissociation energy. Compounds with too low or high dissociation energy will dissociate too rapidly or too slowly, respectively. A few classes of compounds generally used in polymerization reactions have O–O, S–S or N–O bonds with bond energies in the desired range of ~100–170 KJmol$^{-1}$.

Azo compounds are also widely used thermal initiators. The most common of these is 2,2'-azo-*bis*-isobutyronitrile (AIBN). Other azo compounds include 2,2'-azo-*bis*(2,4-dimethylpentanenitrile), 4,4'-azo-*bis*(4-cyano valeric acid) and 1,1'-azo-*bis*(cyclohexane carbonitrile) (Sheppard and Mark, 1988). These compounds readily dissociate to give rise to a stable dinitrogen molecule and a free radical species (Oda et al., 1978; Sato et al., 1979b).

### 2.3.3.4 Redox Initiation

Oxidation-reduction reactions can lead to the generation of free radicals and thus can be used as initiators in polymerization reactions (Sarac, 1999). This class of initiators can generate radicals at a sufficient rate and in reasonable quantities over a wide range of temperatures. Even at subzero temperatures, these reactions can be initiated photolytically, thermally or through radiation. Most of these redox initiators can function in aqueous systems. However, they can also be used in nonaqueous systems using acyl peroxides as initiators and amines as reducing agents, for example, benzoyl peroxide and *N,N*-dialkylaniline (O'Driscoll et al., 1965; Morsi et al., 1977).

### 2.3.3.5 Photochemical Initiators

In photochemical initiated reactions, the radicals are generated by employing UV or visible light irradiation (Oster and Yang, 1968). The irradiation either leads to direct decomposition of monomers into free radicals or initiates a redox reaction by energy transfer to redox compound and generation of free radicals. The initiation rate in these reactions depends on the state of the system and can be controlled by source of light, radicals, light intensity, duration and temperature of the system. The selection of a monomer system depends on the structure of the monomer. For example, photolysis of a monomer will depend on whether it can absorb UV radiation. Photolysis will occur more efficiently in monomer systems having a conjugate double bond with other groups (e.g., styrene, methylmethacrylate). These systems provide an advantage as they are able to absorb UV radiation above vacuum UV region (200 nm), hence initiating these systems requires no special facility. Some of the systems in which photochemical initiator systems are used are acrylates, unsaturated polyester styrene, dithiol-diene (Decker, 1996) and so on.

### 2.3.3.6 Radiation Initiation

Polymerization can also be initiated using high-energy radiations such as electrons, neutrons, $\alpha$-particles ($He^{2+}$), $\gamma$-radiation and electromagnetic radiations. The energy from these radiations leads to molecular excitation and subsequently generates free radicals. This effect is more prominent due to the higher energy levels of these radiations than that of UV or visible light. The reaction mechanism is, however, similar to that of photochemical initiated systems. Radiation causes loss of $\pi$-electrons generating a radical cation ($C^{\cdot+}$).

$$C + radiation \rightarrow C^{\cdot+} + e^-$$

The radical cation, depending on reaction conditions, can propagate at the radical or cation center or dissociate to generate different radicals and cationic species depending on reaction conditions.

$$C^{\cdot+} \to A^{\cdot} + B^{\cdot}$$

At higher temperatures, radiation polymerizations are generally radical polymerizations because at higher temperatures unstable ionic species generate free radicals. The initiation rate depends on radiation intensity, type of radiation particles, as well as concentration of ionizable species. The depth of penetration of the radiations is much more than that of UV/visible radiation, thus providing more flexibility in terms of application and the formats in which polymerization can be carried out (El-Hadi, 2003; Goharian et al., 2007).

Free radical polymerization is employed in the synthesis of different types of polymers in different formats such as discs, beads, sheets and monolithic columns (El-Hadi, 2003; Goharian et al., 2007).

Polymerization at cryo-conditions (cryogelation) is a technique used for fabricating macroporous hydrogels (cryogels). In cryogelation, usually water-soluble polymers or monomers are used in an aqueous system. The gelation process is carried out at subzero temperatures and generally takes place in the semifrozen aqueous phase where the formed ice crystals act as porogens embedded in the cross-linked polymeric heterogeneous phase, which upon thawing forms a macroporous structure (Lozinsky, 2002). Cryogels from various types of monomers such as acrylamide and its derivatives (Plieva et al., 2005, 2006a) and methacrylic acid and its derivatives (Savina et al., 2007; Andac et al., 2008) have been synthesized. Cryogels from cellulose and its derivatives were prepared by UV initiated cross-linking (Petrov et al., 2006, 2007; Chang and Zhang, 2011).

The characteristics of cryogels formed by polymerization via a free radical mechanism at subzero temperatures depends on various factors such as composition of reaction mixture and monomers, cryogelation conditions, geometries and material of molds (Franks, 1982; Lozinsky, 2002; Zhan et al., 2013; Carvalho et al., 2014).

## 2.4 CRYOGELATION METHODS AND FACTORS AFFECTING CRYOGELATION BY FREE RADICAL POLYMERIZATION

Free radical induced cross-linking polymerization of a variety of monomers is used for synthesis of cryogels. The vinyl group containing acrylamide and its derivatives are some of the commonly used monomers. A mix of ammonium persulphate and N,N,N′,N′-tetramethyl ethylenediamine (TEMED) is used to initiate polymerization in this system (Plieva et al., 2005, 2006a,b,c; Dinu et al., 2007). The use of APS/TEMED as an initiator system provides an advantage that it is able to generate free radicals at an appropriate rate and in sufficient amounts to initiate and sustain polymerization even at subzero temperatures. This system has been used for generating cross-linked polymeric cryogels of systems such as polyacrylamide (pAAm), poly-dimethylaminoethyl-methacrylate (pDMAEMA), poly-dimethylacrylamide (pDMAAm), poly-hydroxyethyl-methacrylate (pHEMA), poly-ethyleneglycol (PEG), poly-N-isopropylacrylamide (pNIPAAm), dextran-methacrylate, allyl-agarose and HEMA-L-lactide-dextran macromers (Plieva et al., 2005, 2006a,c, 2007, 2008). Macroporous hydrogels of dextran were generated using a glycidyl methacrylate derivative of dextran and free radical polymerization at subzero temperatures (Plieva et al., 2006b).

### 2.4.1 Cryogelation by Free Radical Polymerization Using UV Irradiation

UV irradiation induces the initiation of a free radical polymerization reaction by generating a free radical species (Figure 2.5). After the generation of free radicals, polymerization proceeds onto the propagation step and is completed with the termination step. The UV-induced polymerization process is widely employed in the synthesis of polymers and in the generation of polymeric macroporous hydrogel systems in frozen aqueous systems. The advantages of using this form of free radical polymerization include low cost, biocompatibility (as there is no use of chemical cross-linking and/or initiating systems) and lastly polymerization can be carried out to form complex geometries of cryogels and hydrogels. Recently, cryogels of poly(ethyleneglycol) and cellulose derivatives hydroxy-ethylcellulose (HEC), hydroxylpropyl-methylcellulose (HPMC) and 2-hydroxy-ethylcellulose were synthesized using UV-induced polymerization in semifrozen aqueous systems (Doycheva et al., 2004; Petrov et al., 2006). In a similar study, *Saccharomyces cerevisiae* cells were entrapped in UV cross-linked hydroxyethylcellulose/poly(ethyleneglycol) core

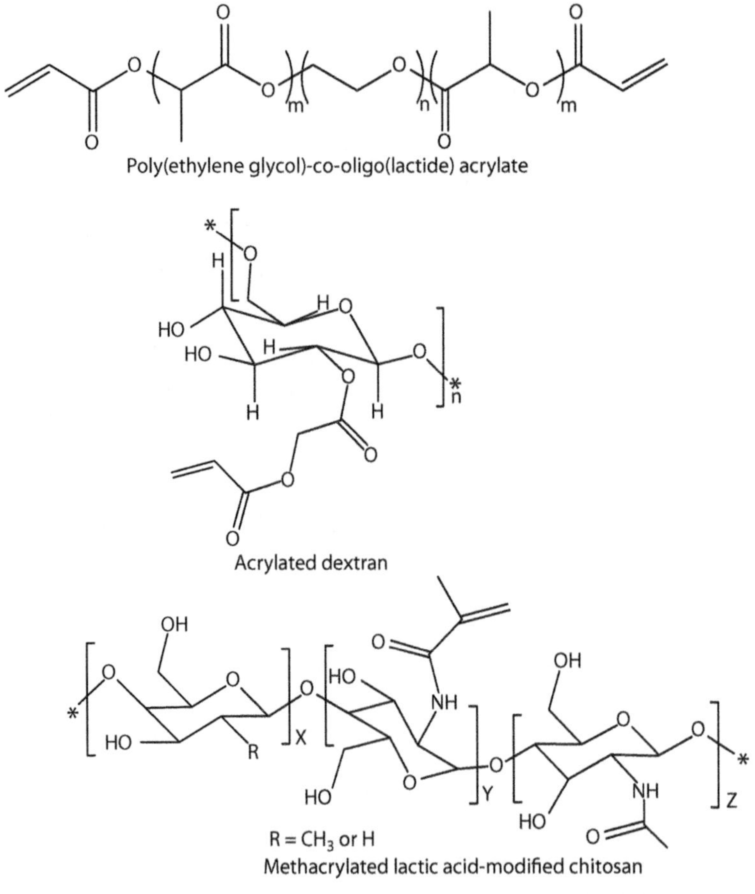

**FIGURE 2.5** Common monomers used in UV-induced gelation (polymerization).

Synthesis and Characterization of Cryogels 55

shell double layered gels composed of cryogel core made up of hydroxyethylcellulose and hydrogel shell of poly(ethylene oxide) (Velickova et al., 2010).

Simionescu and co-workers synthesized cryogels of atelocollagen:dimethyl silanediol-hyaluronic acid (DMSHA):poly(ε-caprolactone) diisocyanate for tissue engineering applications using UV-induced cross-linking in a semifrozen state. The generated material was evaluated for biocompatibility by *in vivo* implantation studies and showed quite promising results (Simionescu et al., 2013).

However, UV-induced polymerization also poses a problem for systems in which cells are incorporated initially in the matrix before gelation. The cells are exposed to UV at a high intensity and the heat evolved over longer durations may have detrimental effects on cell viability and metabolic activity (Bryant et al., 2000; Łukaszczyk et al., 2005; Hiemstra et al., 2007). Macroporous cryogels of gelatin were synthesized by UV treatment to form a polymeric gelatin–methacrylate system (Dubruel et al., 2007; Van Vlierberghe et al., 2007; Uygun et al., 2014).

### 2.4.2 Cryogelation by Functional Group Cross-Linking

#### 2.4.2.1 Cross-Linking with Aldehydes

For functional group cross-linking, different types of cross-linking chemistries are employed, for example, cross-linking by Schiff-base formation between aldehyde and amino groups in the polymer chain using an aldehyde like glutaraldehyde as a cross-linker. An amine group containing polymers can be cross-linked using glutaraldehyde under mild conditions as compared to their cross-linking under drastic conditions such as low pH, high temperatures and high pressures. Glutaraldehyde is used to generate hydrogels (Willmott et al., 1985; Tabata and Ikada, 1989; Jameela and Jayakrishnan, 1995) of proteins and polysaccharides (Bloch et al., 2005; Tripathi et al., 2009; Tripathi and Kumar, 2011). It undergoes an addition reaction via Schiff-base formation and is incorporated in the polymer chain during cross-linking. After gelation, the cross-linker has to be washed out of the gel before using it for any application. Glutaraldehyde has been shown to be very toxic to cells even at very low concentrations (Draye et al., 1998; Hennink and Van Nostrum, 2012), hence the need to wash away the residual groups. Even though glutaraldehyde at high concentrations is toxic, it is used extensively in the synthesis of cryogels of a large number of polymers and polysaccharides for application in tissue engineering, separation sciences and cell immobilization systems. Cryogels of different materials have been generated, using glutaraldehyde as a cross-linker with an adequate amount of washing, such as chitosan-gelatin (Vishnoi and Kumar, 2013), chitosan-agarose-gelatin (Bhat et al., 2011), carrageenan-gelatin (Sharma et al., 2013) and so on.

#### 2.4.2.2 Cross-Linking with Adipic Acid Hydrazide

Adipic acid is another cross-linking agent that is used to generate hydrogels and cryogels. Adipic acid hydrazide was used as a cross-linking agent to generate hydrogels at normal conditions (Lee et al., 2000). However, it is also employed in the synthesis of macroporous hydrogels by freeze–thaw treatment (cryogelation) (Gavrilaş et al., 2012; Iravani et al., 2014).

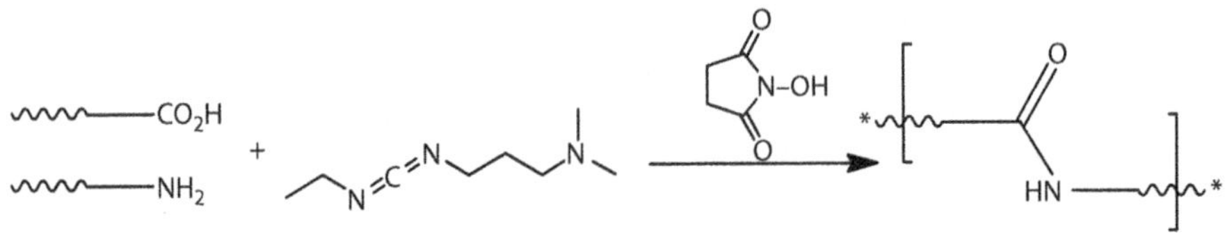

**FIGURE 2.6** Carbodiimide (EDC-NHS) based cross-linking.

### 2.4.2.3 Cross-Linking with Ethyl(dimethylaminopropyl) Carbodiimide-*N*-Hydroxysuccinimide (EDC-NHS)

EDC-NHS chemistry has been used to form hydrogels and cryogels of polymers with hydroxyl groups, amine groups or carboxyl groups or their derivatives (Figure 2.6). The advantage of using EDC-NHS is that it is a zero adducts cross-linker and no additional molecules with any undesired properties are added to the cross-linked molecules. It is quite commonly used in the synthesis of hydrogels of different types of polymers such as gelatin (Kuijpers et al., 2000), chitosan (Ye et al., 2005), dextran (Zhang et al., 2005), collagen-chitosan (Rafat et al., 2008) and alginate (Eiselt et al., 1999).

The EDC-NHS cross-linking chemistry was not extensively employed in the generation of supermacroporous matrials. Yet recently, more and more articles employing this technique are coming up as the technique is maturing. It has been employed in the synthesis of agarose-alginate cryogels (Tripathi and Kumar, 2011), agarose-gelatin cryogels (Tripathi et al., 2009; Vishnoi and Kumar, 2013), gelatin-hyaluronic acid cryogels (Tsung et al., 2011) and PEG-heparin cryogels (Welzel et al., 2012).

## 2.5 PHYSICAL PROPERTIES OF CRYOGELS AND FACTORS AFFECTING THESE PROPERTIES

Macroporous polymeric cryogels are generated by freezing a homogeneous aqueous polymer solution and initiating polymerization and a cross-linking process. The freezing process causes a phase separation in the system and generates two distinct phases—one is a crystalline solid phase generated by crystallization of the water molecules in the system and the other phase generated due to this phase separation is the cryo-concentrated nonfreezing liquid microphase (NFLMP) (Lozinsky, 2002). This cryoconcentration of monomers and cross-linkers in NFLMP increases the rate of polymerization within a certain range of negative temperature usually below the glass transition temperature (Tg) (Lozinsky, 2002). This cryogelation at certain conditions such as temperature, rate of cooling or concentration of monomers in initial solution affects the physical properties of cryogels such as pore size, pore distribution, pore architecture and interconnectivity, pore wall thickness and so on. These properties affect the applicability of cryogels in certain areas. For example, interconnectivity and pore size affect the mass transfer of fluid material through these macroporous materials.

Factors affecting different properties in the macroporous cryogels are as follows.

## 2.5.1 Freezing Rate and Temperature

Freezing rate is a crucial parameter that affects the rate of freezing of a liquid solvent and at the same time affects the rate of polymerization of the monomers. The monomer properties, molecular weight and concentration lower the freezing point of the solution (freezing point depression). It is observed that at slower freezing rates, larger pores with a high level of interconnectivity are obtained, whereas at lower temperatures (rapid freezing rates), weak cryogels with very low interconnectivity are obtained. Rapid freezing of the system may sometimes causes supercooling of the solvent without initiation of ice crystal formation, hence resulting in the formation of smaller and irregular pores (Chen and Lee, 1998; Lozinsky, 2002; Plieva et al., 2004; Ozmen and Okay, 2005; Plieva et al., 2006a,c; He et al., 2007). This can be explained based on nucleation rates and the creation of a viscous unfrozen liquid phase, which further slows down the rate of crystallization in the reaction mixture. At slightly higher freezing temperatures, not too many nuclei are formed and the crystal growth from the few nucleation points formed is uniform. Crystals of long-range order and directionality are formed giving rise to large size interconnected pores. At lower freezing temperatures, a large number of nucleation points are formed, which leads to smaller ice crystals being formed and results in fewer interconnected pores. These gels also showed lower mechanical strength with very low mass flow rates (Plieva et al., 2004, 2005, 2006a,c, 2007).

## 2.5.2 Effect of Porogen and Solvent

In the case of cryogels, the solvent, usually water or a mixture of water and another solvent, for example, dimethyl sulphoxide (DMSO) or dimethyl formamide (DMF), acts as the porogen. The solvent and solvent-porogen mixture can affect the porous properties of the matrix. Hence, this can be used to control the pore properties of cryogels without necessarily changing the polymer composition because the solvent-porogen combination may affect the solubility of the polymer/monomer in the solvent and hence affect the nature of solvent crystals generated (Viklund et al., 1996; Peters et al., 1999).

The solvent and porogen are selected based on the freezing point of the solvents. There are two types of solvents: the first is a solvent that is soluble with the polymeric network and the second is that which induces a phase separation in the homogeneous system. The solvent should freeze at temperatures lower than the glass transition temperature of the polymer to be able to act as a porogen. Some of the solvents used are formamide, 1,4-dioxane, DMSO and DMF. All of these have been evaluated and used as porogens in an aqueous system for the synthesis of cryogels (Ozmen and Okay, 2005; Plieva et al., 2006a,b,c; Perez et al., 2007; Zhang et al., 2013).

Formamide and 1,4-dioxane have been used as porogens in the aqueous phase polyacrylamide cryogelation system at −20°C (Plieva et al., 2006a). Both of these solvents provide different pore properties to the polyacrylamide-bisacrylamide porous polymeric network. This effect occurs because the polar molecules of formamide are associated with monomer molecules, whereas the less soluble formamide induces phase separation during polymerization (Plieva et al., 2006a,b) (Figure 2.7).

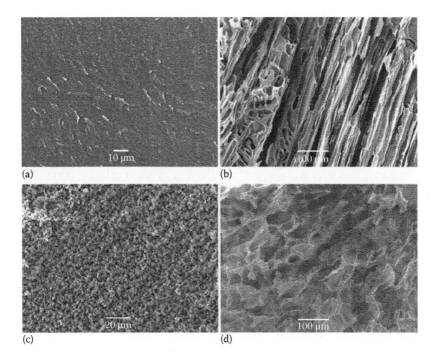

**FIGURE 2.7** SEM images showing pore architecture in pAAm gel prepared at 22°C (a); and pAAm cryogel prepared at –20°C (b) with 95% formamide and 0.5% initiating system (APS/TEMED); and conventional pAAm gel synthesized at 22°C (c); and pAAm cryogel synthesized at –20°C (d) with 95% dioxane and 1.4% initiating system (APS/TEMED). (From Plieva, F., Huiting, X., Galaev, I. Y., Bergenståhl, B. and Mattiasson, B., *Journal of Materials Chemistry* 16(41), 4065–4073, 2006. Reproduced by permission of The Royal Society of Chemistry.)

This difference was attributed to the association of AAm and MBAAm to formamide resulting in better dissolution of MBAAm (hydrophobic) and generating a more homogeneous system. There is less cryostructuring due to the lack of formation of high concentration MBAAm regions in water; therefore, the water-dioxane mixture resulted in the generation of microporous gels (Plieva et al., 2006a,b).

### 2.5.3 Effect of Monomer and Initiator Concentrations

The polymer/monomer concentration, molecular weight and composition ratio drastically affect the properties of macroporous gels formed (Ivanov et al., 2007; Perez et al., 2007). The variation in monomer composition ratio leads to variations in the morphologies of the porous structures formed. A higher concentration of monomers and cross-linkers in the solution leads to an early start of polymerization and phase separation, hence giving rise to a more cross-linked porous gel.

In the case of APS and TEMED initiated free radical reactions, the initial concentration of these activators in the polymerization reaction affects the rate of initiation of reaction and the length of the polymer chains formed. In systems where the reactions propagate slowly, large molecular weight polymer chains are formed as compared to systems with high reaction rates where shorter polymer chains are formed that lead to the generation of brittle and rigid gels. A similar effect is observed when a high amount of cross-linker is present in the system, with high degrees of cross-linking resulting in rigid and brittle gels.

In the freeze systems, the rate of initiation and therefore the amount of initiators present in the system are crucial because two situations can arise in these systems, First, if the rate of polymerization and cross-linking is slow, the freezing of the solution can occur before the gelation can take place and thus the polymerization reaction occurs in NFLMP (Lozinsky, 2002; Plieva et al., 2006). Second, if the polymerization reaction is at a rapid rate, then the polymerization of the reaction system will happen before the initiation of crystallization. The crystallization phenomenon and effect of initiators were studied in acrylamide systems (Plieva et al., 2006) and poly-*N*-isopropylacrylamide systems (Perez et al., 2007). The effect of monomer concentration on the nature of pores formed was studied in 1.2% w/v and 5.0% w/v initiation systems. In both systems, pores of similar size were observed. But in the high concentration initiator system, less flow through rates were observed. This was attributed to the formation of less interconnected pores due to a lack of formation of long range order interconnected ice crystals in the system (Lozinsky, 2002). As ice crystals generated are already entrapped in the polymer gel formed, these gels also showed less elasticity and mechanical strength (Ivanov et al., 2007; Kathuria et al., 2009; Tripathi et al., 2009; Van Vlierberghe et al., 2009).

## 2.6 CHARACTERIZATION OF CRYOGELS

Macroporous cryogels have found their use in various areas including but not limited to tissue engineering applications, as bioseparation units and as bioreactors. The practical application of a cryogel largely depends on its macromolecular structure, physical and chemical properties and in the case of tissue engineering applications, its biological characteristics. Cryogels may be broadly classified into the following:

1. Morphology and textural features
2. Mechanical and thermal properties
3. Chemical properties
4. Biological properties

### 2.6.1 MORPHOLOGICAL AND TEXTURAL FEATURES

One of the most commonly used methods to study the structure and morphology of cryogels is microscopy. Starting from simple optical/light microscopy to the more complex confocal laser scanning microscopy, various microscopic methods have been applied to study the characteristic features of cryogels.

Cryogels are often colored with a dye or a fluorescein probe for optical microscopy. Light microscopy images of cryogels reveal their macroporous features, with a regular distribution of macropores (seen as oval shapes) ranging in size from 10–30 μm for dry samples to 50–80 μm for swollen/hydrated samples (Belavtseva et al., 1984). A major disadvantage of using light microscopy to study cryogels is its lack of clarity in the images. In addition, optical microscopy provides information only in two dimensions (i.e., only surface features). For a three-dimensional study of cryogels, electron microscopy provides information that is more reliable.

For obtaining microscopic images of cryogels using electron microscopy, a freeze-drying or lyophilization procedure is adapted to obtain dry or dehydrated cryogel samples. This process may result in pore wall shrinkage and hence affect porosity, pore size and pore size distribution (Spiller et al., 2008). To overcome these problems, environmental scanning electron microscopy (ESEM) and cryo-SEM techniques have been used to study cryogel morphology (Henderson et al., 2013). Both of these methods allow for observation of cryogels in their natural hydrated state, with ESEM providing an additional benefit of being able to monitor the changes in the structural features of the cryogel as it undergoes slow dehydration (Plieva et al., 2005). The microscopic images are shown in Figure 2.8.

In cryo-SEM, the wet or hydrated cryogel is placed on a holder where it is flash-frozen using liquid nitrogen and later freeze-fractured to give a freshly cut slice of the cryogel sample. After sublimation of the frozen water, the sample is sputter-coated with gold/palladium and examined. The advantage of this technique, which is more time consuming than regular SEM or ESEM, is that it can be used to clearly image the cryogel in its frozen hydrated state following rapid vitrification (El Fray et al., 2007). Images obtained from cryo-SEM are clearer as compared to images obtained from ESEM (Henderson et al., 2013).

Savina and co-workers studied the morphology of HEMA-AGE cryogels using various microscopic techniques (Savina et al., 2011). The authors studied the cryogels in both hydrated as well as freeze-dried states, and observed that the images obtained for both were more or less similar and any diminution in the pore size or pore wall thickness was insignificant. The authors attribute the slight changes in these features to minor shrinkage of polymer walls as opposed to the earlier belief

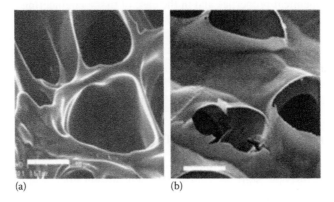

**FIGURE 2.8** Microscopic observation of hydrated polysaccharide-based cryogel scaffolds by ESEM (a); and dehydrated scaffolds by SEM (b). (Reprinted from *Acta Biomaterialia* 6(9), Autissier, A., Visage, C. L., Pouzet, C., Chaubet, F. and Letourneur, D., Fabrication of porous polysaccharide-based scaffolds using a combined freeze-drying/cross-linking process, 3640–3648, Copyright 2010, with permission from Elsevier.)

# Synthesis and Characterization of Cryogels

that the entire 3D structure was collapsing due to drying. Their study brings to fore the fact that most of the water in hydrated cryogels is found in the macropores and not in the walls of the macropores, which are relatively thin, hence, providing cryogels with mechanical strength, a sponge-like morphology and elasticity (Savina et al., 2011), features deemed important for scaffold engineering.

Textural characteristics of cryogels obtained using electron microscopy images can be further quantitatively confirmed using more advanced microscopic techniques like confocal laser scanning microscopy (CLSM) and multiphoton microscopy (MPM) (Savina et al., 2011). In both these methods, a very thin slice (approximately 1 μm) of the wet cryogel is cut and stained with a fluorescent stain like FITC. After incubation with the stain, the sample is washed with PBS and water to remove any unbound stain. The principle of working in both techniques is very similar. The cryogel is optically sectioned in its X-Y plane along the Z-axis a multiple number of times at a known thickness (1 μm). This produces a projection of images along the Z-axis known as Z-stacks. The Z-stack is further analysed using imaging software such as Image J or Fiji (Savina et al., 2011) to determine the textural features such as percentage porosity, surface area, pore diameter and wall thickness. It is important to note that the values obtained for these features using the imaging software of CLSM and MPM are comparable to the values obtained using SEM, ESEM and cryo-SEM (Figures 2.9 and 2.10).

Another method, closely related to CLSM in principle, that has been used of late to measure the textural features of cryogels is the micro-computed tomography (micro-CT

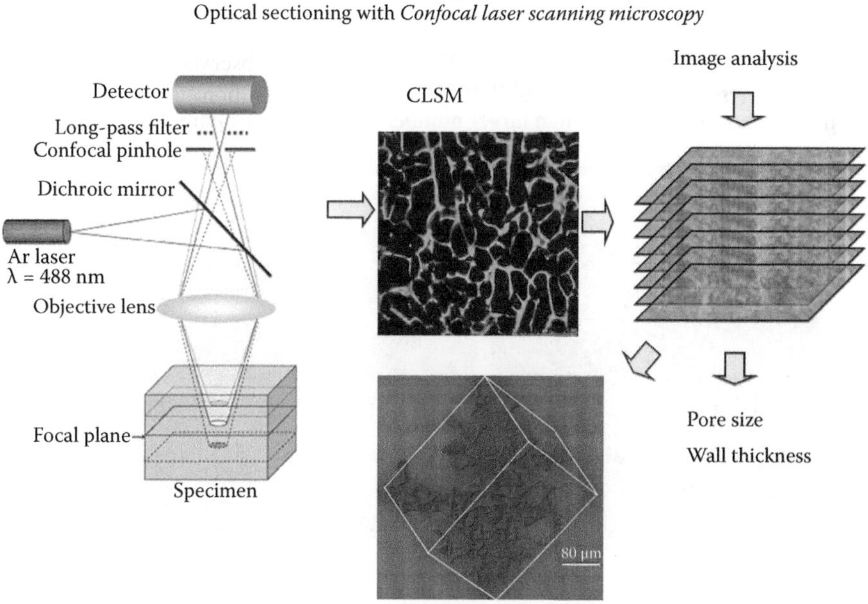

**FIGURE 2.9** Schematic representing the principle of confocal laser scanning microscopy (CLSM). (Reprinted from *Advances in Colloid and Interface Science* 187–188, Gun'ko, V. M., Savina, I. N. and Mikhalovsky, S. V., Cryogels: morphological, structural and adsorption characterization, 1–46, Copyright 2013, with permission from Elsevier.)

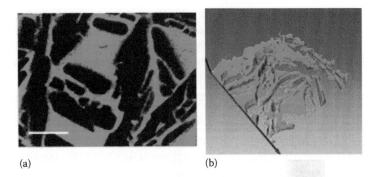

**FIGURE 2.10** 2D confocal laser scanning microscopy (CLSM) image of polysaccharide-based cryogel scaffold (a) and its 3D reconstruction using image software (b). (Reprinted from *Acta Biomaterialia* 6(9), Autissier, A., Visage, C. L., Pouzet, C., Chaubet, F. and Letourneur, D., Fabrication of porous polysaccharide-based scaffolds using a combined freeze-drying/cross-linking process, 3640–3648, Copyright 2010, with permission from Elsevier.)

or μ-CT) method. It is a tomographic technique that uses X-rays to create cross-sections of an object. The sections are further used to reconstruct a 3D virtual model. The term micro is used because the pixel sizes of the cross sections obtained are in the micron range. Like CLSM, μ-CT is used to measure the structural characteristics of hydrated cryogels using imaging software. Vlierberghe and co-workers synthesized porous gelatin cryogels as potential tissue engineering scaffolds to study the effect of concentration of gelatin, cooling rate and temperature gradients on the textural characteristics of cryogels (Van Vlierberghe et al., 2007). For this study, SEM and micro-CT were used to determine porosity, pore size and pore distribution in the gelatin cryogels. Higher gelatin concentrations resulted in a larger number of pores with smaller diameters, due

**FIGURE 2.11** Micro-CT image representation of gelatin cryogel (a) and SEM micrograph of the same cryogel (b). Scale bars represent 500 μm. (Reprinted with permission from Van Vlierberghe, S., Cnudde, V., Dubruel, P., Masschaele, B., Cosijns, A., De Paepe, I., Jacobs, P. J., Van Hoorebeke, L., Remon, J. P. and Schacht, E., *Biomacromolecules* 12(5), 1387–1408. Copyright 2007 American Chemical Society.)

# Synthesis and Characterization of Cryogels

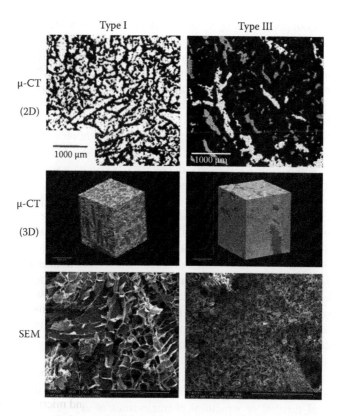

**FIGURE 2.12** Micro-CT and SEM images of 5% (w/v) (type I) and 15% (w/v) gelatin (type III) cryogels. The top row shows 2D μ-CT images, middle row 3D μ-CT (the pores are dark gray, and the material is light gray), last row SEM micrographs. The scale bars represent 1000 μm (μ-CT) and 500 μm (SEM). (Reprinted with permission from Van Vlierberghe, S., Cnudde, V., Dubruel, P., Masschaele, B., Cosijns, A., De Paepe, I., Jacobs, P. J., Van Hoorebeke, L., Remon, J. P. and Schacht, E., *Biomacromolecules* 12(5), 1387–1408. Copyright 2007 American Chemical Society.)

to higher nucleation rates (Figures 2.11 and 2.12). Also, slower cooling rates resulted in larger pores as compared to faster cooling rates (135 μm at cooling rate of 0.15°C/min versus 65 μm at cooling rate of 0.83°C/min) (Van Vlierberghe et al., 2007).

In brief, the cryogel samples to be scanned are first saturated with iron (III) chloride and incubated overnight. Later, they are washed with water and freeze-dried or lyophilized. The samples are then scanned at different known resolutions. The sample rotates around a stable vertical axis as the X-ray source and detector are fixed. Data obtained in the form of X-ray radiographs is then converted into 2D images using reconstruction software, which are further reconstructed to 3D images. These virtual 3D models are analysed using software like μCT analySIS (Van Vlierberghe et al., 2007) or Morpho+ (Savina et al., 2007) to determine structural features like pore size, pore wall thickness and pore size distribution.

In addition to providing quantitative information on the textural properties of cryogels, micro-CT has an added advantage over SEM by quantitatively determining the interconnectivity of pores in cryogels (Savina et al., 2007), a feature unique to cryogelation, and necessary for its application as tissue engineering scaffolds and separation matrices. Whereas in SEM, the interconnectivity of pores is observed qualitatively on the micrographic images, in micro-CT, interconnectivity of pores is expressed as the number of objects that can be discriminated within the cryogel sample (Figure 2.13). The higher the number, the lower the pore interconnectivity, that is, the number of objects denotes the number of pores or void volumes occupied in the material. Hence, for a perfectly interconnected porous material, the number of objects will be just one. It was observed that the micro-CT analysis of the PHEMA-MH cryogel returned only one object, indicating the presence of one large pore with a supposedly complex structure that ran throughout or across the sample (Savina et al., 2007).

Recently our group has shown the use of micro-CT as a technique to study the process of cryogelation, a phenomenon that not many available methodologies have been able to capture. Formation of polyacrylamide cryogels was studied over the 12 hours of cryogelation. It was observed that the percentage object volume decreased with time as polymerization proceeds due to the formation of polymeric walls that surround the pores (Kumar et al., 2010). This study helped to practically illustrate what was until now a hypothetically understood process. In yet another study, carrageenan-gelatin cryogel scaffolds were studied using micro-CT. The results showed the distribution of the porous structure throughout the cryogel scaffolds and details such as pore wall thickness, pore interconnectivity, as well as pore size were obtained (Figure 2.14).

Table 2.4 shows some of the significant differences between SEM and micro-CT as techniques used to study the textural or morphological features of cryogels.

The study of textural or structural characteristics of cryogels is not limited to microscopic studies only. Various other methods are available for determining the porosity of macroporous cryogels. Of these, the most common and simplest measure

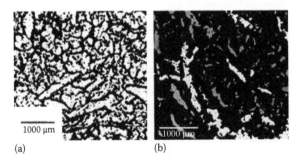

**FIGURE 2.13** 2D µ-CT images showing the difference in interconnectivity of gelatin cryogel with the difference in gelatin concentration. The 5% gelatin (a) was seen to have a more interconnecting network over the 15% gelatin scaffold (b). The 15% gelatin cryogel showed more numbers of individual pores whereas the 5% gelatin cryogel showed a complete pore network, as indicated by the white portion in both images. Scale bars represent 1000 µm. (Reprinted with permission from Van Vlierberghe, S., Cnudde, V., Dubruel, P., Masschaele, B., Cosijns, A., De Paepe, I., Jacobs, P. J., Van Hoorebeke, L., Remon, J. P. and Schacht, E., *Biomacromolecules* 12(5), 1387–1408. Copyright 2007 American Chemical Society.)

# Synthesis and Characterization of Cryogels

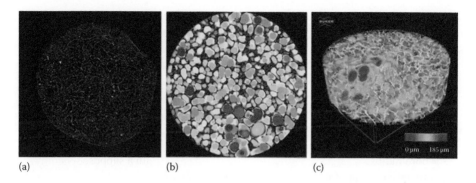

(a) (b) (c)

**FIGURE 2.14** Representation of cross section of carrageenan-gelatin cryogel scaffold using micro-CT (a). Colour-coded image of the pore's size distribution in three orthogonal planes using DataViewer (b). 3D volume rendering of the original dataset of sample carragenan-gelatin scaffold (white membranes), overlayed with 3D colour-coded pore size distribution (c).

**TABLE 2.4**
**Difference between SEM and Micro-CT**

| SEM | Micro-CT |
| --- | --- |
| Destructive technique | Nondestructive technique |
| Resolution 100 nm | Resolution 10 μm |
| 2D imaging | 3D imaging |

of determining the porosity of a cryogel is based on the Archimedes principle (Kathuria et al., 2009). According to this method, the percentage porosity of a cryogel can be calculated using the following equation:

$$\% porosity = \frac{(W_2 - W_3 - W_s)/\rho}{(W_1 - W_3)/\rho} * 100,$$

where
$W_1$ = weight of specific gravity bottle and ethanol
$W_2$ = weight of specific gravity bottle and ethanol and cryogel section
$W_3$ = weight of specific gravity bottle and ethanol after removing cryogel section
$W_s$ = ethanol saturated cryogel section
$\rho$ = density of ethanol

Another method of determining the porous nature of cryogels is using mercury intrusion porosimetry. In this method, a nonwetting liquid (in this case Hg) is intruded into the material at high pressures. This external pressure with which the mercury is being forced into the pores of the material will be opposed by the liquid's surface tension. The pressure that is required to force mercury into the pores of the material is inversely proportional to the size of the pore, that is, the higher the pressure required, the smaller the pore size.

The pore radius is calculated based on the Washburn equation (Rigby et al., 2003):

$$p \cdot r = -2\gamma \cos\theta,$$

where
 $p$ = applied pressure
 $r$ = pore radius
 $\gamma$ = surface tension of mercury (approximately 480 mN/m)
 $\theta$ = contact angle of mercury on solid surface (approximately 140°)

The total pore volume is equivalent to the total volume of mercury that is intruded at maximum pressure applied.

Total pore surface area is calculated using the equation by Rootare and Prenzlow (1967):

$$S = \frac{1}{\gamma \cos\theta} \int_0^{V_{tot}} p\, dV,$$

where size distribution is determined using the equation by Ritter and Drake (1945):

$$D(d) = \frac{p}{d} \bigg/ \frac{dV}{dp}.$$

Mercury intrusion porosimetry was used to study the porosity of chitosan-gelatin cryogels synthesized at various monomer concentrations by Kathuria et al. (2009). It was observed that mercury intrusion porosimetry was capable of identifying micropores (laying in the range 30–50 μm) due to the high pressures that force mercury into these smaller pores as well. It is important to note that these micropores are of significant importance in cryogels synthesized for use as tissue engineering scaffolds, as they aid in the unhindered cell migration and flow of low molecular weight solvents and gases like oxygen (Kathuria et al., 2009).

In another study of cryogels, mercury porosimetry is used to study pore size distribution, as a means to understand the interactions between carbon microparticles and the base PVA cryogel, to enable the practical application of such composite cryogels (Zheng et al., 2012). Data from mercury porosimetry showed that the PSD of the activated charcoal beads alone showed a significant contribution of broad nanopores and micropores and that these AC particles were able to maintain their integrity even at high mercury pressures. However, on addition to the PVA-glutaraldehyde cryogels, the PSD showed a significant loss in macroporosity, primarily due to the high mercury pressures and the soft nature of the material. The mercury porosimetry data indicated the loss in macroporosity of the ACs on interaction with the PVA-GA cryogels, a property important for the application of ACs as medicinal adsorbents (Zheng et al., 2012) (Figure 2.15). One of the major drawbacks of mercury porosimetry is that large pores with small openings are filled at high pressures indicating them as small pores, and hence leading to miscalculated pore size distributions.

Another technique very similar to mercury porosimetry, and most often used to confirm the porosimetry data, is the nitrogen adsorption method. It is based on the property of a gas (in this case nitrogen) to adsorb onto the surface of a material through weak Van der Waal's interactions. Measuring the amount of gas adsorbed/desorbed at different pressures and a constant temperature gives rise to adsorption/desorption isotherms, which can be used later to characterize pore volume, pore size distributions and so on of materials. Through this technique, two features—pore size distribution and specific surface area—can be determined, based on which the practical application of a cryogel may be established. However, the nitrogen adsorption method is a very time-consuming technique and has a determination range much lower than that of mercury porosimetry. The nitrogen adsorption method can detect pores in the range 3–200 nm, whereas mercury porosimetry can detect pores in the range 7 nm to 14 μm (Westermarck et al., 1998).

Last but not the least, another technique for determining porosity and pore size distribution is the use of NMR cryoporometry (Gun'ko et al., 2010). In this method, liquid is imbibed into a porous structure and allowed to freeze. Then, it is slowly warmed, causing the frozen liquid to melt. NMR is then used to measure the quantity of the liquid melted as a function of temperature. In NMR cryoporometry, small

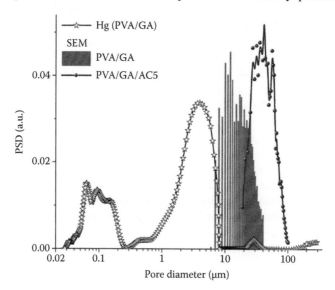

**FIGURE 2.15** Pore size distribution (PSD) based on mercury porosimetry and SEM analysis for PVA/GA and PVA/GA/AC5 cryogels. The PSD based on mercury porosimetry covers pores in the micropore range due to the ability of mercury to be forced into these small pore openings. (Reprinted with permission from Zheng, Y., Gun'ko, V. M., Howell, C. A., Sandeman, S. R., Phillips, G. J., Kozynchenko, O. P., Tennison, S. R., Ivanov, A. E. and Mikhalovsky, S. V., *ACS Applied Materials & Interfaces*, 4(11), 5936–5944. Copyright 2012 American Chemical Society.)

crystals of a liquid in a pore melt at a lower temperature as compared to the bulk liquid; hence, the pore size is inversely proportional to the depression in melting point. NMR cryoporometry was used as a tool to study the cryogelation process as well as study the progress of the polymerization reaction through changes in amounts of unfrozen water (Kirsebom et al., 2009). However, it was a qualitative study and no quantitative data were provided for the same.

There are many states of water that can be identified in both cryogels and hydrogels. First, there is the unbound water (UBW), which does not interact with the gel pore walls and is entirely bulk water. Second is the weakly bound water (WBW), which makes weak interactions with the gel pore walls as well as the surface of the gel. It freezes at subzero temperatures close to 0°C. Third is the strongly bound water (SBW), which strongly interacts with the surface of the cryogel either through hydrogen bonds or through electrostatic interactions. This state of water normally remains frozen at temperatures less than −15°C. The amount of water in these different states largely affects the interactions of cryogels with solutes, biomolecules and cells, especially within the pores and their pore walls. Hence, the need to study the amounts of different types of water in the cryogel, which in the end, helps in understanding the adsorption/desorption and diffusion processes within cryogels. NMR cryoporometry is an important tool used to evaluate these different states of water in cryogels (Gun'ko et al., 2013). The data on the amount of states of water can be used to derive the pore size distribution data for various kinds of cryogels. NMR cryoporometry revealed a range of pores from nano (< 2 nm) to broad nano (2–50 nm) to macro (200–250 μm) (Savina et al., 2011).

There have been many instances where cryogels have been used as bioreactor systems, either for the production of biomolecules (Kumar et al., 2006) or as separation units (Plieva et al., 2004). In each of these applications, the flow rate of solvent through the cryogel bioreactor plays a crucial role. This flow rate or resistance to flow in some cases is a good measure of the interconnectivity of pores in the cryogel. A decreased flow rate or high backpressure indicates collapsing of gel walls, and hence blocking of pores. In the case of a pure gelatin cryogel, the constricted pore morphology of the gel results in high resistance to flow. Meanwhile, a high flow rate or minimal flow resistance indicates continuously connected pores (Kathuria et al., 2009).

Another study that helps shed light on pore interconnectivity of cryogels is the swelling–deswelling kinetics analysis. This parameter indicates how much and how quickly cryogels absorb the solvent from their surrounding, indicating the interconnectivity of the gels (Srivastava et al., 2007).

The percentage water uptake ($Wu$) is given by the following equation:

$$Wu(\%) = \frac{W_t - W_g}{W_e} \cdot 100,$$

where
$W_t$ = weight at regular intervals
$W_g$ = weight of dry gel
$W_e$ = weight of water in swollen gels at equilibrium a particular temperature

# Synthesis and Characterization of Cryogels

The swelling ratio (*SR*) is given by the following equation:

$$SR = \frac{W_s - W_d}{W_d},$$

where
$W_s$ = weight of swollen gel
$W_d$ = weight of dry gel

In a study comparing swelling–deswelling kinetics of cryogels versus hydrogels (Srivastava et al., 2007), it was observed that cryogels attained swelling equilibrium within approximately 20 minutes whereas hydrogels took more than 2 days. This difference is a clear marker for the difference in pore morphology and pore wall thickness between hydrogels and cryogels. Whereas hydrogels have smaller pore sizes and thick pore walls, cryogels have large pore sizes and thin pore walls, a feature distinctly advantageous for the use of cryogels as tissue engineering scaffolds and separation units. A faster swelling–deswelling rate represents good pore interconnectivity as the solvent is able to move across the network freely by convection. However, slower swelling–deswelling rates as in hydrogels represent poor pore interconnectivity (Srivastava et al., 2007). The same applies for the swelling ratio as well. A higher swelling ratio indicates a highly porous system, allowing for a faster exchange of materials.

## 2.6.2 Mechanical and Thermal Properties

A number of mechanical and thermal properties are of significant importance. Characterizing the bulk tensile and compressive strength of cryogel is necessary to determine the application of the cryogel. Since local matrix stiffness affects cell proliferation and differentiation (Wells, 2008; Hadjipanayi et al., 2009; Bott et al., 2010), measurement of these mechanical properties of scaffolds becomes all the more important. Some of the methods used to characterize cryogels based on their mechanical properties include the following.

### 2.6.2.1 Compression Test

The tensile stress of a material, calculated by the following equation:

$$Stress(\sigma) = \frac{P}{A},$$

where
$P$ = load
$A$ = cross-sectional area

is a measure of the strength of a material when loaded under tension, that is, it is an indicator of the material's resistance to failure.

On the other hand, strain calculated as

$$Strain(\epsilon) = \frac{\delta}{L_o},$$

where
  $\delta$ = deformation
  $L_o$ = original length

is a measure of the stiffness of a material, that is, the load needed to induce a certain amount of deformation in the material.

The relationship between these two mechanical properties of a material is given by the following equation:

$$\sigma = E\epsilon,$$

where
  $E$ = constant of proportionality called the Young's modulus or modulus of elasticity

In tension/compression testing, the cryogel sample is stretched (tension) or compressed (compression) between two parallel plates at a maximum load and the modulus of elasticity is determined based on the slope of the stress–strain curve.

### 2.6.2.2 Fatigue Test

This is equivalent to a cyclic compression test, and is used to determine the point of failure in a material when subjected to a series of stress amplitudes. Fatigue strength is defined as the maximum alternating stress that a material will withstand for a given number of cycles.

It has been observed that cyclic mechanical strain plays an important role as a mechanical signal in the development of smooth muscle cells in vascularised tissue engineered constructs (Kim and Mooney, 2000; Ceccarelli et al., 2012). In addition, cyclic deformation has been observed to improve ECM production and mechanical strength of tissue-engineered chondrocyte constructs used for articular cartilage regeneration (Hunter et al., 2004; Kisiday et al., 2004). Hence the need to design scaffolds with enhanced mechanical properties. In this effect, Kathuria and co-workers performed extensive mechanical studies on chitosan-gelatin cryogels to study their potential as tissue engineering scaffolds (Kathuria et al., 2009). The cryogel from gelatin only showed very little stress on deformation, whereas the cryogel from chitosan only showed elastic behavior until 10% compression after which it cracked, indicating its brittle nature. The chitosan-gelatin cryogels at different monomer concentrations showed a slight increase in the Young's modulus on increasing concentration. However, a similar stress–strain pattern was observed and the difference in the Young's modulus was not significant. The gels were able to return to their original shape even after being compressed to 80% of their original length, showing their elastic nature.

Similarly, the chitosan-gelatin cryogels did not crack on being subjected to a fatigue test, and there was no sign of deformation even at high frequencies and at high strain.

The above results show the potential of the chitosan-gelatin cryogel to be used as a scaffold for tissue engineering of load bearing tissues like cartilage. In a similar study, it was observed that the addition of agarose to a chitosan-gelatin scaffold increased the elastic modulus significantly (39–44 kPa of chitosan-agarose-gelatin [CAG] [Bhat et al., 2011] as opposed to 36–39 kPa of chitosan-gelatin [CG] [Kathuria et al., 2009]). These CAG scaffolds also did not undergo any deformation when subjected to fatigue tests, showing elasticity and mechanical stability of the cryogels.

### 2.6.2.3 Rheology Studies

The use of rheological analysis to characterize cryogels is a very common technique to study the viscoelastic properties of cryogels (Hernández et al., 2004). In one study of PVA composite cryogels synthesized with different amounts of cellulose (Păduraru et al., 2012), the values of the storage modulus ($G'$) were significantly higher than the values for the loss modulus ($G''$). This is a very typical result for gels and solid-like materials (Păduraru et al., 2012) and indicates increasing elasticity of the gels with increasing cellulose content. A higher storage modulus for higher cellulose concentrations indicates that cellulose plays a role in improving the mechanical strength of the PVA cryogel. Similar rheological studies were performed on PVA cryogels synthesized in aqueous media as well as salt solutions (Patachia et al., 2009). Comparing values of $G'$ to $G''$ indicated higher values for $G'$ to those of $G''$ confirming the formation of a network structure during cryogelation. In addition, it was observed that the deformation of the cryogel (as indicated by increase or decrease in $G'$, with increasing strain amplitude) showed a dependence on the type of salt used for cryogel formation. Through this study, a lot can be deduced about the relationship between storage modulus and loss modulus of a cryogel, as well as the relationship between crystallinity/cross-linking in cryogel and a storage modulus. All of these features make it possible to decide the mechanical strength and elasticity of the cryogel, making it easier to classify as to what application the cryogel is most suited.

An important way of characterizing cryogels is based on their thermal properties. The most commonly used techniques to study the thermal properties of cryogels include thermogravimetric analysis (TGA) and differential scanning calorimetry (DSC). Both of these methods provide information about the crystallinity or order of cryogel, hence providing critical information on the stability of cryogels at various temperatures. In addition, TGA/DSC methods play an important role in evaluating the strength of interaction between copolymers in the case of chemical linking as well as in studying the effect of incorporation of biomolecules/ligands on the order or structure of cryogels.

DSC was used to study the crystallinity of poly(vinylalcohol)/polyester cryogels (PVA-PES) (Paranhos et al., 2007). Briefly, 5–10 mg of dried cryogel sample was placed in an aluminium pan and heated at 10°C/min from 40 to 250°C. A change in the $\Delta H_m$ on addition of PES to the PVA cryogel indicates a change in the degree of crystallinity of PVA cryogels. Addition of PES to the cryogel increases its amorphous content allowing more PVA to cross-link with PES, hence causing a decrease in crystallinity, which eventually causes an increase in the water uptake capacity of the cryogels. A similar observation was seen when PVA cryogels were loaded with dyes (Papancea et al., 2010). The loading of the dye significantly reduced the degree of crystallinity, as measured by the ratio of $\Delta H_m$ for cryogel loaded with dye to $\Delta H_m$ of polymer in its completely crystalline state.

The thermal stability of PVA-cellulose cryogels was studied using TGA/DSC (Varganici et al., 2013). Pure cellulose decomposes in two stages, with major degradation occurring in the second stage at temperatures not less than 300°C and maximum at 345°C. Meanwhile, pure PVA cryogel decomposes in four stages, with the third showing the highest amount of mass loss at temperature $T = 306°C$. The composite cryogel PVA/cellulose also decomposes in four stages. However, the temperature for decomposition in the second stage is lower than that of pure gels with a decrease in the temperature as the cellulose content increases. This is also shown by a decrease in the intensity of DTG curve peaks at the second stage. These results clearly show that there has been hydrogen bonding between the two components of the cryogel. The hydrogen bond between the –OH group of glucose ring and the –OH group of PVA results in enhanced thermal stability of the cryogel (Figure 2.16).

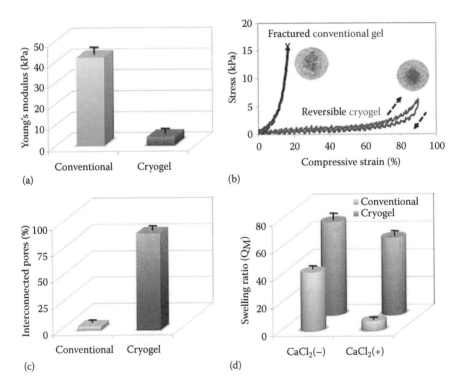

**FIGURE 2.16** A comparative study of the mechanical and structural properties of macroporous hydrogels (cryogels) and conventional hydrogels. Young's modulus (a); stress–strain curve (b); evaluation of pore interconnectivity (c); swelling ratio (d), for conventional versus macroporous hydrogels. (From Bencherif, S. A., Sands, R. W., Bhatta, D., Arany, P., Verbeke, C. S., Edwards, D. A. and Mooney, D. J., *Proceedings of the National Academy of Sciences*, 109(48), 19590–19595. Copyright 2012 National Academy of Sciences, USA.)

### 2.6.3 CHEMICAL PROPERTIES

An important factor in characterizing cryogels is the study of the chemical composition of the cryogels synthesized. This is specifically necessary for composite cryogels synthesized by cross-linking of two polymers/monomer units or where a biomolecule is being added to the cryogel to improve its mechanical/functional properties.

#### 2.6.3.1 Fourier Transform Infrared Spectroscopy (FTIR)

FTIR is a spectroscopic method used to identify the presence of certain functional groups. It is based on bonds or groups of bonds when exposed to an IR beam and can absorb this IR radiation, resulting in excitation to higher vibrational states. In FTIR analysis, the specimen is subjected to a modulated IR beam and the transmittance and reflectance of the beam at different frequencies are plotted. This spectral pattern is then analysed and matched with known signature patterns of materials in the FTIR library. FTIR can be used to study the interaction between functional molecules as well as to study the crystalline behavior of small-sized particles.

Tretinnikov and co-workers made use of ATR-FTIR spectroscopic analysis to study the crystalline phase of PVA cryogels (Tretinnikov et al., 2013). Through this study, a crystallinity band very specific to PVA gels was observed at 1144 $cm^{-1}$. The attenuated total reflection (ATR) method of FTIR is used to minimize any difficulties that would arise due to the absorption of IR by water (Tretinnikov et al., 2013), hence ATR-FTIR is a commonly used method for structural studies in aqueous media. In addition, through this study, it was revealed that there are major significant differences in the IR spectra of PVA in solution and PVA cryogels, showing a rearrangement of the system of hydrogen bonds between water and PVA during gelation. In another similar study, FTIR was used to show the cross-linking of gelatin by glutaraldehyde (Kemençe and Bölgen, 2013). The FTIR 'shoulder-like' peak at 3500 $cm^{-1}$ characteristic of unbound gelatin was not present in the spectrum of the cross-linked gelatin, suggesting that the –OH group responsible for the peak in unbound gelatin may have participated in the cross-linking reaction.

In another study, Paduraru and co-workers used FTIR analysis to characterize PVA-cellulose cryogels, to show the incorporation of cellulose into the PVA cryogel (Păduraru et al., 2012). Apart from the increase in intensity of bands for specific functional groups (–OH group band at 3436 $cm^{-1}$, C–H from alkyl group at 2625 $cm^{-1}$, $CH_2$ groups at 1384 $cm^{-1}$, etc.), the group noted an increase in the energy of H-bond with an increase in cellulose content. In addition, the values for H-bond distances decrease with an increase in cellulose content, clearly indicating the presence of molecular interactions between components of the cryogel.

#### 2.6.3.2 X-Ray Diffraction (XRD)

X-ray crystallography is a technique used to determine the atomic and molecular structure of a material, wherein the crystalline atoms in the material absorb X-ray beams and then diffract them in given specific directions.

XRD is used as an alternative to FTIR analysis, to study the crystallinity of materials. XRD analysis was used to confirm the crystallinity of PVA-cellulose cryogels

(Păduraru et al., 2012) revealed by FTIR. A very strong characteristic peak of crystallinity is observed in the diffractogram for PVA. In addition, a decrease in the main peaks of the PVA gel with a decrease in cellulose content indicates cryogel formation between PVA and cellulose. Also, the crystallinity index (calculated as the area under the crystallinity peak divided by the total area under the diffractogram) for the cryogel had a value that was intermediate to the value for pure PVA and pure cellulose.

In another study, XRD was used to show an increase in crystallinity of chitosan due to cross-linking by glutaraldehyde and freeze–thawing process of cryogelation, rendering the resulting chitosan membrane more stable and making amine groups available on the surface (Orrego and Valencia, 2009).

### 2.6.3.3 XPS and Elemental Analysis

Whereas FTIR and XRD provide structural or conformational information on the active functional groups available in the cryogel, XPS and element analyser are effective tools used for a more detailed study on the specific elements that are present in the cryogel (Filatova and Volkov, 2010; Daoud-Attieh et al., 2013; Emin Çorman et al., 2013; Reichelt et al., 2014).

XPS or X-ray photoelectron spectroscopy is a quantitative spectroscopic technique that is used to measure the elemental composition and empirical formula of the elements that exist in a material. It is a surface chemical analysis technique that provides information on the surface chemistry of a material (Loo et al., 2013).

### 2.6.4 BIOLOGICAL PROPERTIES

For the use of cryogels as tissue engineering scaffolds, thorough biological characterization is needed.

### 2.6.4.1 Protein Adsorption Studies

The first step to characterize cryogels for biological applications is checking for protein adsorption on the surface of the cryogels. It is a well-known fact that implanted materials are immediately coated by proteins from blood and interstitial fluids (Wilson et al., 2005). This adsorbed layer of proteins attracts cells toward the implant and allows for their adhesion onto the implanted material (Tamada and Ikada, 1993). For this reason, the first step in characterizing cryogels as potential biological matrices is to evaluate the amount of protein adsorption occurring on the cryogel. Typically, protein adsorption is examined by incubating the cryogel scaffold in 0.1 M PBS containing 10% FBS. Cryogels are pretreated using ethanol and washed with plain PBS three times. The scaffolds are then incubated in PBS + 10% FBS at 37°C for a known period of time.

$$\textit{Adsorbed protein (mg/ml)} = \textit{Amt of protein before incubation} \\ - \textit{Amt of protein after incubation.}$$

Cryogels of poly(acrylonitrile) (PAN) and poly(acrylamide)-chitosan (PAAC) were found to adsorb a significant amount of protein on incubation with FBS (28.69%) and

(18%), respectively (Jain and Kumar, 2009). The protein adsorption studies also help in understanding hydrophobicity/hydrophilicity of cryogels. In the study described above, PAN cryogel was able to adsorb a higher amount of protein due to better hydrophobicity as compared to polyacrylamide (PAAM) or chitosan. The protein adsorption studies corroborated the surface roughness studies (by SEM) as well as surface area studies (by porosimetry), which indicated PAN to have a rougher surface with increased surface area. PAN is therefore able to attract serum proteins to its surface, due to its roughness and increased surface area, both factors that influence protein adsorption on a surface (Rechendorff et al., 2006; Scopelliti et al., 2010).

### 2.6.4.2 *In Vitro* Biocompatibility

Once protein adsorption is established, the next step in characterizing cryogels for biological properties is to check the morphology, growth and viability of cells seeded within cryogel matrices. A number of techniques are used to study the biocompatibility of cryogel scaffolds.

The morphology of cells seeded on cryogel scaffolds is most often studied microscopically using SEM, confocal laser scanning microscopy. For SEM evaluation, the cryogel sample seeded with cells is fixed using 2.5% glutaraldehyde and then washed using ethanol gradient. The sample is then vacuum dried and sputter coated with gold/palladium.

CLSM is usually used to perform a live/dead cell viability assay. The cell-seeded scaffolds are incubated with fluorescent stains and these fluorescently labeled scaffolds are observed under a confocal microscope.

The viability of cells growing in cell-seeded scaffolds can be assessed using a tetrazolium salt assay (MTT or MTS) (Bhat et al., 2011; Chang et al., 2013). The assay involves converting the tetrazolium salt by viable cells into an insoluble colored product. An inorganic solvent DMSO can dissolve the insoluble formazan crystals, and the optical density quantified spectrophotometrically. An increase in the optical density over a period of time indicates increasing metabolic activity of cells as well as indirectly hinting at the growth/proliferation of cells in the scaffold.

To further confirm the compatibility of the cryogels with cells, cellular functionality may be addressed using assays or stains, for example, oil red O stain for adipocytes (Chang et al., 2013) or ALP assay for osteoblasts (Bhat et al., 2011).

Last, but not the least, *in vitro* biocompatibility of cryogel scaffolds is measured using *in vitro* degradation studies. Tissue engineering aims at regenerating a tissue with the simultaneous degradation of the supporting matrix, that is, the rate of degradation must match the rate of formation of neo-tissue (Dainiak et al., 2010; Bhat et al., 2011). The degree of degradation of the cryogels is calculated as

$$D(\%) = \frac{m_i - m_e}{m_i} * 100,$$

where
  $m_i$ = weight of cryogel before degradation
  $m_e$ = weight of cryogel after degradation

The enzymatic degradation was studied for gelatin-fibrinogen cryogels, to be used as dermal matrices for wound repair (Dainiak et al., 2010). It was observed that for scaffolds containing low amounts of glutaraldehyde (used for cross-linking), degradation occurred almost immediately (within 10–15 minutes), whereas increasing the GA concentration tenfold brought down the degradation rate almost 80-fold (Dainiak et al., 2010), allowing the scaffolds to be used *in vivo*.

In another study, conducting cryogels, synthesized for use as scaffolds for neural tissue engineering, were studied for *in vitro* degradation (Vishnoi and Kumar, 2013). The study revealed that the degradation of the scaffolds was not affected by the incorporation of polypyrrole. In addition, a slow degradation rate was seen as desirable to enable maintenance of the conducting properties of the scaffold over a long period of time.

An important biological property for characterizing cryogels for use in biological applications is the *in vivo* biocompatibility and degradation studies. Generally, tissue implants face a lot of inflammation-induced chronic oxidative stress, which eventually leads to implant rejection (Yu et al., 2011), hence the need to develop reactive oxidative species-responsive biomaterials for biomedical applications. In a study carried out to evaluate the biocompatibility of various polymeric cryogels, it was observed that the synthesized polymeric cryogels, namely poly(vinylcaprolactam), PVA-alginate-bioactive glass composite, PHEMA-gelatin and CAG were not only biocompatible but also degraded in an ROS-independent manner (Shakya et al., 2014).

The degradation kinetics were studied by excising the scaffolds implanted subcutaneously at regular time intervals and using gravimetric analysis to check for changes in the weight of the scaffold before and after implantation. It was further observed that the degradation of the scaffolds did not induce any cytotoxicity, as shown by the negligible levels of TNF-alpha in the sera of mice collected after transplantation. No mast cell infiltration in the surrounding tissue disqualified the presence of any allergic reaction to the polymeric scaffolds. In addition, the host response to cryogels (as observed through microscopic analysis of sections obtained from the implantation site) was similar to that for any other implanted material, suggesting *in vivo* biocompatibility.

## 2.7 CONCLUSION

Macroporous hydrogels—better referred to as *cryogels*—stand apart from their conventional counterparts primarily due to the interconnectivity of their pores. This feature is attributed to the phase separation that occurs during the freezing of the solvent, where the formation of the ice crystals of the solvent results in the formation of a liquid microphase. The ice crystals act as porogens, which when thawed give rise to a complex network of pores. Various methods have been used to synthesize cryogels, ranging from simple physical cross-linking of polymers to the more complex free radical polymerization using different initiators and cross-linkers. Each of these methods significantly affects the pore structure and the overall characteristics of the cryogel formed. Apart from the more obvious factors like monomer concentration and amount of solvent, numerous other factors like freezing rate, freezing temperature, degree of cross-linking, pH and ionic strength of the solvent and so on also affect the structure, morphology and mechanical strength of cryogels. To study these effects and how the features of the cryogels differ from those of conventional hydrogels, numerous techniques have been identified, such as cryoporometry, thermogravimetry, mercury porosimetry and so on. Microscopic techniques, however, remain the most common of these to be used to study the inherent porous structure of these macroporous hydrogels. Based on their physical, chemical and biological features, cryogels have found various applications in the medical, industrial and pharmaceutical fields as reactor units, tissue engineering scaffolds, separation units and immobilization matrices. With an increasing list of areas of applications of these macroporous hydrogels, there is an increasing need to synthesize and characterize as many varied forms of polymeric/composite cryogels as possible.

## ACKNOWLEDGEMENTS

The authors acknowledge the Indian Institute of Technology Kanpur, Department of Biotechnology, and Department of Science and Technology, India, for funding various projects on cryogel application by our team. The authors also acknowledge Archana Sharma, for providing some micro-CT images. We acknowledge Dr. Phil Salmon from Bruker, Belgium for carrying out the micro-CT analysis for the cryogels.

## LIST OF ABBREVIATIONS

| | |
|---|---|
| **2D** | Two-dimensional |
| **3D** | Three-dimensional |
| **AIBN** | 2,2′-azo-*bis*-isobutyronitrile |
| **APS** | Ammonium persulphate |
| **CLSM** | Confocal laser scanning microscopy |
| **CMC** | Carboxymethylcellulose |
| **DMF** | Dimethylformamide |
| **DMSO** | Dimethylsulphoxide |
| **DSC** | Differential scanning calorimetry |
| **ECM** | Extracellular matrix |
| **EDC-NHS** | Ethyl (dimethylaminopropyl) carbodiimide-*N*-hydroxysuccinimide |
| **FBS** | Fetal bovine serum |
| **FTIR** | Fourier transform infra-red |
| **MIP** | Molecularly imprinted polymer |
| **NFLMP** | Non-freezing liquid microphase |
| **NMR** | Nuclear magnetic resonance |
| **pAAm** | Polyacrylamide |
| **PBS** | Phosphate buffer saline |
| **pDMAAm** | Polydimethylacrylamide |
| **pDMAEM** | Polydimethylaminoethyl-methacrylate |
| **PEG** | Polyethylene glycol |
| **pHEMA** | Polyhydroxyethyl-methacrylate |

pNIPAAm     Poly N-isopropylacrylamide
PSD         Pore size distribution
PVA         Polyvinylalcohol
SEM         Scanning electron microscopy
SPE         Solid phase extraction
TEMED       N,N,N',N'-tetramethylethylenediamine
Tg          Glass transition temperature
TGA         Thermogravimetric analysis
TMOS        Tetramethoxysilane
UV          Ultraviolet
XPS         X-ray photoelectron spectroscopy
XRD         X-ray diffraction

## REFERENCES

Andac, M., Plieva, F., Denizli, A., Galaev, I. Y. and Mattiasson, B. (2008). Poly (hydroxyethyl methacrylate)-based macroporous hydrogels with disulfide cross-linker. *Macromolecular Chemistry and Physics* **209**(6): 577–584.

Auriemma, F., De Rosa, C., Ricciardi, R., Lo Celso, F., Triolo, R. and Pipich, V. (2008). Time-resolving analysis of cryotropic gelation of water/poly (vinyl alcohol) solutions via small-angle neutron scattering. *The Journal of Physical Chemistry B* **112**(3): 816–823.

Autissier, A., Visage, C. L., Pouzet, C., Chaubet, F. and Letourneur, D. (2010). Fabrication of porous polysaccharide-based scaffolds using a combined freeze-drying/cross-linking process. *Acta Biomaterialia* **6**(9): 3640–3648.

Bajpai, A. and Saini, R. (2005a). Preparation and characterization of biocompatible spongy cryogels of poly (vinyl alcohol)–gelatin and study of water sorption behaviour. *Polymer International* **54**(9): 1233–1242.

Bajpai, A. and Saini, R. (2005b). Preparation and characterization of spongy cryogels of poly (vinyl alcohol)–casein system: Water sorption and blood compatibility study. *Polymer International* **54**(5): 796–806.

Bandi, S. and Schiraldi, D. A. (2006). Glass transition behavior of clay aerogel/poly(vinyl alcohol) composites. *Macromolecules* **39**(19): 6537–6545.

Belavtseva, E. M., Titova, E. F., Lozinsky, V. I., Vainerman, E. S. and Rogozhin, S. V. (1984). Study of cryostructurization of polymer systems. *Colloid and Polymer Science* **262**(10): 775–779.

Bencherif, S. A., Sands, R. W., Bhatta, D., Arany, P., Verbeke, C. S., Edwards, D. A. and Mooney, D. J. (2012). Injectable preformed scaffolds with shape-memory properties. *Proceedings of the National Academy of Sciences* **109**(48): 19590–19595.

Bernhardt, A., Despang, F., Lode, A., Demmler, A., Hanke, T. and Gelinsky, M. (2009). Proliferation and osteogenic differentiation of human bone marrow stromal cells on alginate–gelatine–hydroxyapatite scaffolds with anisotropic pore structure. *Journal of Tissue Engineering and Regenerative Medicine* **3**(1): 54–62.

Bhat, S., Tripathi, A. and Kumar, A. (2011). Supermacroprous chitosan–agarose–gelatin cryogels: *In vitro* characterization and *in vivo* assessment for cartilage tissue engineering. *Journal of The Royal Society Interface* **8**(57): 540–554.

Bloch, K., Lozinsky, V., Galaev, I. Y., Yavriyanz, K., Vorobeychik, M., Azarov, D., Damshkaln, L., Mattiasson, B. and Vardi, P. (2005). Functional activity of insulinoma cells (INS-1E) and pancreatic islets cultured in agarose cryogel sponges. *Journal of Biomedical Materials Research Part A* **75**(4): 802–809.

Bölgen, N., Plieva, F., Galaev, I. Y., Mattiasson, B. and Pişkin, E. (2007). Cryogelation for preparation of novel biodegradable tissue-engineering scaffolds. *Journal of Biomaterials Science, Polymer Edition* **18**(9): 1165–1179.

Bölgen, N., Vargel, I., Korkusuz, P., Güzel, E., Plieva, F., Galaev, I., Matiasson, B. and Pişkin, E. (2009). Tissue responses to novel tissue engineering biodegradable cryogel scaffolds: An animal model. *Journal of Biomedical Materials Research Part A* **91A**(1): 60–68.

Bott, K., Upton, Z., Schrobback, K., Ehrbar, M., Hubbell, J. A., Lutolf, M. P. and Rizzi, S. C. (2010). The effect of matrix characteristics on fibroblast proliferation in 3D gels *Biomaterials* **31**(32): 8454–8464.

Bovey, F., Schilling, F., Kwei, T. and Frisch, H. (1977). Dynamic carbon-13 NMR measurements on poly (vinylidene fluoride) and poly (methyl methacrylate) and their mixed solutions. *Macromolecules* **10**(3): 559–561.

Bovey, F. A. and Mirau, P. A. (1996). *NMR of Polymers*, Academic Press.

Bower, D. I. (2002). *An Introduction to Polymer Physics*, Cambridge University Press.

Braun, D., Cherdron, H., Ritter, H., Braun, D. and Cherdron, H. (2001). *Polymer Synthesis: Theory and Practice*, Springer.

Bruice, T. C. and Butler, A. R. (1964). Catalysis in water and ice. II. 1 The reaction of thiolactones with morpholine in frozen systems. *Journal of the American Chemical Society* **86**(19): 4104–4108.

Bryant, S. J. and Anseth, K. S. (2002). Hydrogel properties influence ECM production by chondrocytes photoencapsulated in poly (ethylene glycol) hydrogels. *Journal of Biomedical Materials Research* **59**(1): 63–72.

Bryant, S. J. and Anseth, K. S. (2003). Controlling the spatial distribution of ECM components in degradable PEG hydrogels for tissue engineering cartilage. *Journal of Biomedical Materials Research Part A* **64**(1): 70–79.

Bryant, S. J., Nuttelman, C. R. and Anseth, K. S. (2000). Cytocompatibility of UV and visible light photoinitiating systems on cultured NIH/3T3 fibroblasts *in vitro*. *Journal of Biomaterials Science, Polymer Edition* **11**(5): 439–457.

Butler, A. and Bruice, T. (1964). Catalysis in water+ ice. comparison of kinetics of hydrolysis of acetic anhydride beta-propiolactone+ rho-nitrophenyl acetate+ dehydration of 5-hydro-6-hydroxy-deoxyuridine in water and in ice. *Journal of the American Chemical Society* **86**(3): 313.

Cais, R. E. and Kometani, J. M. (1984). Synthesis of pure head-to-tail poly(trifluoroethylenes) and their characterization by 470-MHz fluorine-19 NMR. *Macromolecules* **17**(10): 1932–1939.

Carvalho, B., Da Silva, S., Da Silva, L., Minim, V., Da Silva, M., Carvalho, L. and Minim, L. (2014). Cryogel poly(acrylamide): Synthesis, structure and applications. *Separation & Purification Reviews* **43**(3): 241–262.

Cascone, M., Maltinti, S., Barbani, N. and Laus, M. (1999). Effect of chitosan and dextran on the properties of poly (vinyl alcohol) hydrogels. *Journal of Materials Science: Materials in Medicine* **10**(7): 431–435.

Ceccarelli, J., Cheng, A. and Putnam, A. J. (2012). Mechanical strain controls endothelial patterning during angiogenic sprouting. *Cellular and Molecular Bioengineering* **5**(4): 463–473.

Chang, C. and Zhang, L. (2011). Cellulose-based hydrogels: Present status and application prospects. *Carbohydrate Polymers* **84**(1): 40–53.

Chang, K.-H., Liao, H.-T. and Chen, J.-P. (2013). Preparation and characterization of gelatin/hyaluronic acid cryogels for adipose tissue engineering: *In vitro* and *in vivo* studies. *Acta Biomaterialia* **9**(11): 9012–9026.

Chen, S.-L. and Lee, T.-S. (1998). A study of supercooling phenomenon and freezing probability of water inside horizontal cylinders. *International Journal of Heat and Mass Transfer* **41**(4): 769–783.

Cong, F., Zhang, H. and Zhang, W. (2003). Viscoelasticity of aqueous suspensions of curdlan with different molecular weights. *Acta Polymerica Sinica* (5): 693–698.

Coviello, T., Matricardi, P., Marianecci, C. and Alhaique, F. (2007). Polysaccharide hydrogels for modified release formulations. *Journal of Controlled Release* **119**(1): 5–24.

Dainiak, M. B., Allan, I. U., Savina, I. N., Cornelio, L., James, E. S., James, S. L., Mikhalovsky, S. V., Jungvid, H. and Galaev, I. Y. (2010). Gelatin–fibrinogen cryogel dermal matrices for wound repair: Preparation, optimisation and *in vitro* study. *Biomaterials* **31**(1): 67–76.

Daoud-Attieh, M., Chaib, H., Armutcu, C., Uzun, L., Elkak, A. and Denizli, A. (2013). Immunoglobulin G purification from bovine serum with pseudo-specific supermacroporous cryogels. *Separation and Purification Technology* **118**(0): 816–822.

Decker, C. (1996). Photoinitiated crosslinking polymerisation. *Progress in Polymer Science* **21**(4): 593–650.

Denisov, E. T., Denisova, T. G. and Pokidova, T. S. (2003). *Handbook of Free Radical Initiators*, John Wiley & Sons.

Dinu, M. V., Ozmen, M. M., Dragan, E. S. and Okay, O. (2007). Freezing as a path to build macroporous structures: Superfast responsive polyacrylamide hydrogels. *Polymer* **48**(1): 195–204.

Doycheva, M., Petrova, E., Stamenova, R., Tsvetanov, C. and Riess, G. (2004). UV-Induced Cross-Linking of Poly (ethylene oxide) in Aqueous Solution. *Macromolecular Materials and Engineering* **289**(7): 676–680.

Doyle, J., Giannouli, P., Martin, E., Brooks, M. and Morris, E. (2006). Effect of sugars, galactose content and chainlength on freeze–thaw gelation of galactomannans. *Carbohydrate Polymers* **64**(3): 391–401.

Draye, J.-P., Delaey, B., Van de Voorde, A., Van Den Bulcke, A., Bogdanov, B. and Schacht, E. (1998). *In vitro* release characteristics of bioactive molecules from dextran dialdehyde cross-linked gelatin hydrogel films. *Biomaterials* **19**(1–3): 99–107.

Dubruel, P., Unger, R., Van Vlierberghe, S., Cnudde, V., Jacobs, P. J., Schacht, E. and Kirkpatrick, C. (2007). Porous gelatin hydrogels: 2. *In vitro* cell interaction study. *Biomacromolecules* **8**(2): 338–344.

Eiselt, P., Lee, K. Y. and Mooney, D. J. (1999). Rigidity of two-component hydrogels prepared from alginate and poly (ethylene glycol)-diamines. *Macromolecules* **32**(17): 5561–5566.

El-Hadi, A. A. (2003). Factors affecting the production of prednisolone by immobilization of *Bacillus pumilus* E601 cells in poly (vinyl alcohol) cryogels produced by radiation polymerization. *Process Biochemistry* **38**(12): 1659–1664.

El Fray, M., Pilaszkiewicz, A., Swieszkowski, W. and Kurzydlowski, K. J. (2007). Morphology assessment of chemically modified cryostructured poly (vinyl alcohol) hydrogel. *European Polymer Journal* **43**(5): 2035–2040.

Eliasson, A. C. and Kim, H. R. (1992). Changes in rheological properties of hydroxypropyl potato starch pastes during freeze-thaw treatments i. a rheological approach for evaluation of freeze-thaw stability. *Journal of Texture Studies* **23**(3): 279–295.

Emin Çorman, M., Bereli, N., Özkara, S., Uzun, L. and Denizli, A. (2013). Hydrophobic cryogels for DNA adsorption: Effect of embedding of monosize microbeads into cryogel network on their adsorptive performances. *Biomedical Chromatography* **27**(11): 1524–1531.

Filatova, A. G. and Volkov, I. O. (2010). Investigating the microstructure and surface composition of corn starch cryogel treated with FeCl3 solution. *Bulletin of the Russian Academy of Sciences: Physics* **74**(7): 1037–1038.

Franks, F. (1982). *The Properties of Aqueous Solutions at Subzero Temperatures*, Springer.

Gavrilaş, S., Dumitru, F. and Stănescu, M. D. (2012). Commercial laccase oxidation of phenolic compounds, *UPB Sci. Bull., Series B* **74**: 3–10.

Giannouli, P. and Morris, E. (2003). Cryogelation of xanthan. *Food Hydrocolloids* **17**(4): 495–501.

Goharian, M., Moran, G. R., Wilson, K., Seymour, C., Jegatheesan, A., Hill, M., Thompson, R. T. and Campbell, G. (2007). Modifying the MRI, elastic stiffness and electrical properties of polyvinyl alcohol cryogel using irradiation. *Nuclear Instruments and Methods in Physics Research Section B: Beam Interactions with Materials and Atoms* **263**(1): 239–244.

Guiot, J., Ameduri, B. and Boutevin, B. (2002). Radical homopolymerization of vinylidene fluoride initiated by tert-butyl peroxypivalate. Investigation of the microstructure by 19F and 1H NMR spectroscopies and mechanisms. *Macromolecules* **35**(23): 8694–8707.

Gun'ko, V. M., Mıkhalovska, L. I., Savina, I. N., Shevchenko, R. V., James, S. L., Tomlins, P. E. and Mikhalovsky, S. V. (2010). Characterisation and performance of hydrogel tissue scaffolds. *Soft Matter* **6**(21): 5351–5358.

Gun'ko, V. M., Savina, I. N. and Mikhalovsky, S. V. (2013). Cryogels: Morphological, structural and adsorption characterisation. *Advances in Colloid and Interface Science* **187–188**(0): 1–46.

Gustavsson, P.-E., Tiainen, P. and Larsson, P.-O. (2009). Superporous agarose gels: Production, properties, and applications. In: *Macroporous Polymers: Production Properties and Biotechnological/Biomedical Applications.* Mattiasson, B., Kumar, A., Galaev, I. Yu., Eds. CRC Press, Taylor & Francis Group, US., pp. 155–178.

Hadjipanayi, E., Mudera, V. and Brown, R. A. (2009). Guiding cell migration in 3D: A collagen matrix with graded directional stiffness. *Cell Motility and the Cytoskeleton* **66**(3): 121–128.

Hassan, C. M. and Peppas, N. A. (2000). Structure and morphology of freeze/thawed PVA hydrogels. *Macromolecules* **33**(7): 2472–2479.

Hatakeyama, T., Naoi, S., Iijima, M. and Hatakeyama, H. (2005). Locust bean gum hydrogels formed by freezing and thawing. *Macromolecular Symposia*, Wiley Online Library.

He, X., Yao, K., Shen, S. and Yun, J. (2007). Freezing characteristics of acrylamide-based aqueous solution used for the preparation of supermacroporous cryogels via cryo-copolymerization. *Chemical Engineering Science* **62**(5): 1334–1342.

Henderson, T. M. A., Ladewig, K., Haylock, D. N., McLean, K. M. and O'Connor, A. J. (2013). Cryogels for biomedical applications. *Journal of Materials Chemistry B* **1**(21): 2682–2695.

Hennink, W. and Van Nostrum, C. (2012). Novel crosslinking methods to design hydrogels. *Advanced Drug Delivery Reviews* **64**: 223–236.

Hernández, R., Rusa, M., Rusa, C. C., López, D., Mijangos, C. and Tonelli, A. E. (2004a). Controlling PVA Hydrogels with γ-Cyclodextrin. *Macromolecules* **37**(25): 9620–9625.

Hernández, R., Sarafian, A., López, D. and Mijangos, C. (2004b). Viscoelastic properties of poly(vinyl alcohol) hydrogels and ferrogels obtained through freezing–thawing cycles. *Polymer* **45**(16): 5543–5549.

Hiemstra, C., Zhou, W., Zhong, Z., Wouters, M. and Feijen, J. (2007). Rapidly *in situ* forming biodegradable robust hydrogels by combining stereocomplexation and photopolymerization. *Journal of the American Chemical Society* **129**(32): 9918–9926.

Hong, Y., Song, H., Gong, Y., Mao, Z., Gao, C. and Shen, J. (2007). Covalently crosslinked chitosan hydrogel: Properties of *in vitro* degradation and chondrocyte encapsulation. *Acta Biomaterialia* **3**(1): 23–31.

Hunter, C. J., Mouw, J. K. and Levenston, M. E. (2004). Dynamic compression of chondrocyte-seeded fibrin gels: Effects on matrix accumulation and mechanical stiffness. *Osteoarthritis and Cartilage* **12**(2): 117–130.

Iravani, A., Mueller, J. and Yousefi, A.-M. (2014). Producing homogeneous cryogel phantoms for medical imaging: A finite-element approach. *Journal of Biomaterials Science, Polymer Edition* **25**(2): 181–202.

Ivanov, R. V., Lozinsky, V. I., Noh, S. K., Han, S. S. and Lyoo, W. S. (2007). Preparation and characterization of polyacrylamide cryogels produced from a high-molecular-weight precursor. I. Influence of the reaction temperature and concentration of the crosslinking agent. *Journal of Applied Polymer Science* **106**(3): 1470–1475.

Jain, E. and Kumar, A. (2009). Designing supermacroporous cryogels based on polyacrylonitrile and a polyacrylamide–chitosan semi-interpenetrating network. *Journal of Biomaterials Science, Polymer Edition* **20**(7–8): 877–902.

Jain, E. and Kumar, A. (2013). Disposable polymeric cryogel bioreactor matrix for therapeutic protein production. *Nature Protocols* **8**(5): 821–835.

Jameela, S. and Jayakrishnan, A. (1995). Glutaraldehyde cross-linked chitosan microspheres as a long acting biodegradable drug delivery vehicle: Studies on the *in vitro* release of mitoxantrone and *in vivo* degradation of microspheres in rat muscle. *Biomaterials* **16**(10): 769–775.

Jin, Y., Zhang, H., Yin, Y. and Nishinari, K. (2006). Conformation of curdlan as observed by tapping mode atomic force microscopy. *Colloid and Polymer Science* **284**(12): 1371–1377.

Kaneko, T., Ogomi, D., Mitsugi, R., Serizawa, T. and Akashi, M. (2004). Mechanically drawn hydrogels uniaxially orient hydroxyapatite crystals and cell extension. *Chemistry of Materials* **16**(26): 5596–5601.

Kathuria, N., Tripathi, A., Kar, K. K. and Kumar, A. (2009). Synthesis and characterization of elastic and macroporous chitosan–gelatin cryogels for tissue engineering. *Acta Biomaterialia* **5**(1): 406–418.

Kemençe, N. and Bölgen, N. (2013). Gelatin- and hydroxyapatite-based cryogels for bone tissue engineering: Synthesis, characterization, *in vitro* and *in vivo* biocompatibility. *Journal of Tissue Engineering and Regenerative Medicine* Aug. 28, 2013, doi: 10.1002/term.1813.

Kim, B.-S. and Mooney, D. J. (2000). Scaffolds for engineering smooth muscle under cyclic mechanical strain conditions. *Journal of Biomechanical Engineering* **122**(3): 210–215.

Kim, J. O., Park, J. K., Kim, J. H., Jin, S. G., Yong, C. S., Li, D. X., Choi, J. Y., Woo, J. S., Yoo, B. K. and Lyoo, W. S. (2008). Development of polyvinyl alcohol–sodium alginate gel-matrix-based wound dressing system containing nitrofurazone. *International Journal of Pharmaceutics* **359**(1): 79–86.

Kirsebom, H., Rata, G., Topgaard, D., Mattiasson, B. and Galaev, I. Y. (2009). Mechanism of cryo-polymerization: Diffusion-controlled polymerization in a nonfrozen microphase. An NMR study. *Macromolecules* **42**(14): 5208–5214.

Kisiday, J. D., Jin, M., DiMicco, M. A., Kurz, B. and Grodzinsky, A. J. (2004). Effects of dynamic compressive loading on chondrocyte biosynthesis in self-assembling peptide scaffolds. *Journal of Biomechanics* **37**(5): 595–604.

Kobayashi, M., Ando, I., Ishii, T. and Amiya, S. (1995). Structural study of poly (vinyl alcohol) in the gel state by high-resolution solid-state 13C NMR spectroscopy. *Macromolecules* **28**(19): 6677–6679.

Kuijpers, A., Van Wachem, P., Van Luyn, M., Engbers, G., Krijgsveld, J., Zaat, S., Dankert, J. and Feijen, J. (2000). *In vivo* and *in vitro* release of lysozyme from cross-linked gelatin hydrogels: A model system for the delivery of antibacterial proteins from prosthetic heart valves. *Journal of Controlled Release* **67**(2): 323–336.

Kumar, A. and Srivastava, A. (2010). Cell separation using cryogel-based affinity chromatography. *Nature Protocols* **5**(11): 1737–1747.

Kumar, A., Bansal, V., Nandakumar, K. S., Galaev, I. Y., Roychoudhury, P. K., Holmdahl, R. and Mattiasson, B. (2006). Integrated bioprocess for the production and isolation of urokinase from animal cell culture using supermacroporous cryogel matrices. *Biotechnology and Bioengineering* **93**(4): 636–646.

Kumar, A., Mishra, R., Reinwald, Y. and Bhat, S. (2010). Cryogels: Freezing unveiled by thawing. *Materials Today* **13**(11): 42–44.

Kumar, A., Plieva, F. M., Galaev, I. Y. and Mattiasson, B. (2003). Affinity fractionation of lymphocytes using a monolithic cryogel. *Journal of Immunological Methods* **283**(1–2): 185–194.

Kumar, A., Rodríguez-Caballero, A., Plieva, F. M., Galaev, I. Y., Nandakumar, K. S., Kamihira, M., Holmdahl, R., Orfao, A. and Mattiasson, B. (2005). Affinity binding of cells to cryogel adsorbents with immobilized specific ligands: Effect of ligand coupling and matrix architecture. *Journal of Molecular Recognition* **18**(1): 84–93.

Lazaridou, A. and Biliaderis, C. (2004). Cryogelation of cereal β-glucans: Structure and molecular size effects. *Food Hydrocolloids* **18**(6): 933–947.

Lazaridou, A., Vaikousi, H. and Biliaderis, C. (2008). Effects of polyols on cryostructurization of barley β-glucans. *Food Hydrocolloids* **22**(2): 263–277.

Le Noir, M., Plieva, F., Hey, T., Guieysse, B. and Mattiasson, B. (2007). Macroporous molecularly imprinted polymer/cryogel composite systems for the removal of endocrine disrupting trace contaminants. *Journal of Chromatography A* **1154**(1): 158–164.

Lee, K. Y., Bouhadir, K. H. and Mooney, D. J. (2000). Degradation behavior of covalently cross-linked poly (aldehyde guluronate) hydrogels. *Macromolecules* **33**(1): 97–101.

Li, M., Minoura, N., Dai, L. and Zhang, L. (2001). Preparation of porous poly (vinyl alcohol)-silk fibroin (PVA/SF) blend membranes. *Macromolecular Materials and Engineering* **286**(9): 529–533.

Loo, S.-L., Fane, A. G., Lim, T.-T., Krantz, W. B., Liang, Y.-N., Liu, X. and Hu, X. (2013). Superabsorbent cryogels decorated with silver nanoparticles as a novel water technology for point-of-use disinfection. *Environmental Science & Technology* **47**(16): 9363–9371.

Lozinsky, V., Damshkaln, L., Kurochkin, I. and Kurochkin, I. (2005). Study of cryostructuring of polymer systems: 25. The influence of surfactants on the properties and structure of gas-filled (foamed) poly (vinyl alcohol) cryogels. *Colloid Journal* **67**(5): 589–601.

Lozinsky, V., Damshkaln, L., Kurochkin, I. and Kurochkin, I. (2008). Study of cryostructuring of polymer systems: 28. Physicochemical properties and morphology of poly (vinyl alcohol) cryogels formed by multiple freezing-thawing. *Colloid Journal* **70**(2): 189–198.

Lozinsky, V., Damshkaln, L., Shaskol'skii, B., Babushkina, T., Kurochkin, I. and Kurochkin, I. (2007). Study of cryostructuring of polymer systems: 27. Physicochemical properties of poly (vinyl alcohol) cryogels and specific features of their macroporous morphology. *Colloid Journal* **69**(6): 747–764.

Lozinsky, V., Domotenko, L., Zubov, A. and Simenel, I. (1996). Study of cryostructuration of polymer systems. XII. Poly (vinyl alcohol) cryogels: Influence of low-molecular electrolytes. *Journal of Applied Polymer Science* **61**(11): 1991–1998.

Lozinsky, V. and Plieva, F. (1998). Poly(vinyl alcohol) cryogels employed as matrices for cell immobilization. 3. Overview of recent research and developments. *Enzyme and Microbial Technology* **23**(3): 227–242.

Lozinsky, V. I. (1998). Cryotropic gelation of poly (vinyl alcohol) solutions. *Russian Chemical Reviews* **67**(7): 573–586.

Lozinsky, V. I. (2002). Cryogels on the basis of natural and synthetic polymers: Preparation, properties and application. *Russian Chemical Reviews* **71**(6): 489–511.

Lozinsky, V. I., Bakeeva, I. V., Presnyak, E. P., Damshkaln, L. G. and Zubov, V. P. (2007). Cryostructuring of polymer systems. XXVI. Heterophase organic–inorganic cryogels prepared via freezing–thawing of aqueous solutions of poly (vinyl alcohol) with added tetramethoxysilane. *Journal of Applied Polymer Science* **105**(5): 2689–2702.

Lozinsky, V. I., Damshkaln, L. G., Bloch, K. O., Vardi, P., Grinberg, N. V., Burova, T. V. and Grinberg, V. Y. (2008). Cryostructuring of polymer systems. XXIX. Preparation and characterization of supermacroporous (spongy) agarose-based cryogels used as three-dimensional scaffolds for culturing insulin-producing cell aggregates. *Journal of Applied Polymer Science* **108**(5): 3046–3062.

Lozinsky, V. I., Damshkaln, L. G., Brown, R. and Norton, I. T. (2000a). Study of cryostructuration of polymer systems. XVI. Freeze–thaw-induced effects in the low concentration systems amylopectin–water. *Journal of Applied Polymer Science* **75**(14): 1740–1748.

Lozinsky, V. I., Damshkaln, L. G., Brown, R. and Norton, I. T. (2000b). Study of cryostructuration of polymer systems. XVIII. Freeze–thaw influence on water-solubilized artificial mixtures of amylopectin and amylose. *Journal of Applied Polymer Science* **78**(2): 371–381.

Lozinsky, V. I., Damshkaln, L. G., Brown, R. and Norton, I. T. (2000c). Study of cryostructuring of polymer systems. XIX. On the nature of intermolecular links in the cryogels of locust bean gum. *Polymer International* **49**(11): 1434–1443.

Lozinsky, V. I., Damshkaln, L. G., Brown, R. and Norton, I. T. (2002). Study of cryostructuration of polymer systems. XXI. Cryotropic gel formation of the water–maltodextrin systems. *Journal of Applied Polymer Science* **83**(8): 1658–1667.

Lozinsky, V. I., Damshkaln, L. G., Ezernitskaya, M. G., Glotova, Y. K. and Antonov, Y. A. (2012). Cryostructuring of polymer systems. Wide pore poly (vinyl alcohol) cryogels prepared using a combination of liquid–liquid phase separation and cryotropic gel-formation processes. *Soft Matter* **8**(32): 8493–8504.

Lozinsky, V. I., Damshkaln, L. G., Kurochkin, I. N. and Kurochkin, I. I. (2014). Cryostructuring of polymeric systems. 36. Poly (vinyl alcohol) cryogels prepared from solutions of the polymer in water/low-molecular alcohol mixtures. *European Polymer Journal* **53**: 189–205.

Luan, T., Wu, L., Zhang, H. and Wang, Y. (2012). A study on the nature of intermolecular links in the cryotropic weak gels of hyaluronan. *Carbohydrate Polymers* **87**(3): 2076–2085.

Łukaszczyk, J., Śmiga, M., Jaszcz, K., Adler, H. J. P., Jähne, E. and Kaczmarek, M. (2005). Evaluation of oligo (ethylene glycol) dimethacrylates effects on the properties of new biodegradable bone cement compositions. *Macromolecular Bioscience* **5**(1): 64–69.

Mao, C.-F. and Chen, J.-C. (2006). Interchain association of locust bean gum in sucrose solutions: An interpretation based on thixotropic behavior. *Food Hydrocolloids* **20**(5): 730–739.

Millon, L. and Wan, W. (2006). The polyvinyl alcohol–bacterial cellulose system as a new nanocomposite for biomedical applications. *Journal of Biomedical Materials Research Part B: Applied Biomaterials* **79**(2): 245–253.

Moad, G., Chiefari, J., Mayadunne, R. T., Moad, C. L., Postma, A., Rizzardo, E. and Thang, S. H. (2002). Initiating free radical polymerization. *Macromolecular Symposia*, Wiley Online Library.

Morgan, K. R. and Ofman, D. J. (1998). Glucagel, 1 A gelling β-blucan from barley. *Cereal Chemistry* **75**(6): 879–881.

Morsi, S., Zaki, A. and El-Khyami, M. (1977). The role of charge transfer interactions in the decomposition of organic peroxides—II: Spontaneous and amine-induced decomposition of dibenzoyl peroxide in partially restricted media. *European Polymer Journal* **13**(11): 851–854.

Navarro, A., Martino, M. and Zaritzky, N. (1997). Viscoelastic properties of frozen starch-triglycerides systems. *Journal of Food Engineering* **34**(4): 411–427.

Nishinari, K., Zhang, H. et al. (2009). Curdlan. In: *Handbook of Hydrocolloids*. Phillips, G. O., Williams, P. A., Eds. Woodhead Publishing Ltd., Abington Hall, Abington, 567–591.

O'Driscoll, K., Lyons, P. and Patsiga, R. (1965). Kinetics of polymerization of styrene initiated by substituted benzoyl peroxides. III. Induced decomposition. *Journal of Polymer Science Part A: General Papers* **3**(4): 1567–1571.

Oda, T., Maeshima, T. and Sugiyama, K. (1978). Cage effects in nitrogen radical formation. *Die Makromolekulare Chemie* **179**(9): 2331–2336.

Odian, G. (2004). *Principles of Polymerization,* John Wiley & Sons.

Okamoto, A. and Miyoshi, T. (2002). A biocompatible gel of hyaluronan. In: *Hyaluronan-Chemical, Biochemical and Biological Aspects.* Kennedy, J. F., Phillips, G. O. Williams, P. A. and Hascall, V. C., Eds. Woodhead Publishing Ltd., pp. 285–292.

Omidian, H. and Park, K. (2009). Fast-responsive macroporous hydrogels. In: *Macroporous Polymers: Production Properties and Biotechnological/Biomedical Applications*. Mattiasson, B., Kumar, A., Galaev, I. Yu., Eds. CRC Press, Taylor & Francis Group, US., 179–208.

Orrego, C. and Valencia, J. (2009). Preparation and characterization of chitosan membranes by using a combined freeze gelation and mild crosslinking method. *Bioprocess and Biosystems Engineering* **32**(2): 197–206.

Oster, G. and Yang, N.-L. (1968). Photopolymerization of vinyl monomers. *Chemical Reviews* **68**(2). 125–151.

Ovenall, D. W. and Uschold, R. E. (1991). Detection of branching on poly(vinyl fluoride) by NMR and the effect of synthesis conditions on polymer structure. *Macromolecules* **24**(11): 3235–3237.

Ozmen, M. M. and Okay, O. (2005). Superfast responsive ionic hydrogels with controllable pore size. *Polymer* **46**(19): 8119–8127.

Păduraru, O. M., Ciolacu, D., Darie, R. N. and Vasile, C. (2012). Synthesis and characterization of polyvinyl alcohol/cellulose cryogels and their testing as carriers for a bioactive component. *Materials Science and Engineering: C* **32**(8): 2508–2515.

Papancea, A., Valente, A. J. M. and Patachia, S. (2010). Diffusion and sorption studies of dyes through PVA cryogel membranes. *Journal of Applied Polymer Science* **115**(3): 1445–1453.

Paranhos, C. M., Oliveira, R. N., Soares, B. G. and Pessan, L. A. (2007). Poly(vinyl alcohol)/sulfonated polyester hydrogels produced by freezing and thawing technique: Preparation and characterization. *Materials Research* **10**: 43–46.

Park, H. and Kim, D. (2006). Swelling and mechanical properties of glycol chitosan/poly (vinyl alcohol) IPN-type superporous hydrogels. *Journal of Biomedical Materials Research Part A* **78**(4): 662–667.

Park, Y. D., Tirelli, N. and Hubbell, J. A. (2003). Photopolymerized hyaluronic acid-based hydrogels and interpenetrating networks. *Biomaterials* **24**(6): 893–900.

Patachia, S., Florea, C., Friedrich, C. and Thomann, Y. (2009). Tailoring of poly (vinyl alcohol) cryogels properties by salts addition. *Express Polymer Letters* **3**(5): 320–331.

Perez, P., Plieva, F., Gallardo, A., San Roman, J., Aguilar, M. R., Morfin, I., Ehrburger-Dolle, F., Bley, F., Mikhalovsky, S. and Galaev, I. Y. (2007). Bioresorbable and nonresorbable macroporous thermosensitive hydrogels prepared by cryo-polymerization. Role of the cross-linking agent. *Biomacromolecules* **9**(1): 66–74.

Peters, E. C., Svec, F., Fréchet, J. M., Viklund, C. and Irgum, K. (1999). Control of porous properties and surface chemistry in 'molded' porous polymer monoliths prepared by polymerization in the presence of TEMPO. *Macromolecules* **32**(19): 6377–6379.

Petrov, P., Petrova, E., Stamenova, R., Tsvetanov, C. B. and Riess, G. (2006). Cryogels of cellulose derivatives prepared via UV irradiation of moderately frozen systems. *Polymer* **47**(19): 6481–6484.

Petrov, P., Petrova, E., Tchorbanov, B. and Tsvetanov, C. B. (2007). Synthesis of biodegradable hydroxyethylcellulose cryogels by UV irradiation. *Polymer* **48**(17): 4943–4949.

Pincock, R. E. and Kiovsky, T. E. (1965). Bimolecular reactions in frozen organic solutions1. *Journal of the American Chemical Society* **87**(9): 2072–2073.

Plieva, F., Huiting, X., Galaev, I. Y., Bergenståhl, B. and Mattiasson, B. (2006a). Macroporous elastic polyacrylamide gels prepared at subzero temperatures: Control of porous structure. *Journal of Materials Chemistry* **16**(41): 4065–4073.

Plieva, F., Oknianska, A., Degerman, E., Galaev, I. Y. and Mattiasson, B. (2006b). Novel supermacroporous dextran gels. *Journal of Biomaterials Science, Polymer Edition* **17**(10): 1075–1092.

Plieva, F. M., Andersson, J., Galaev, I. Y. and Mattiasson, B. (2004). Characterization of polyacrylamide based monolithic columns. *Journal of Separation Science* **27**(10–11): 828–836.

Plieva, F. M., Galaev, I. Y. and Mattiasson, B. (2007). Macroporous gels prepared at subzero temperatures as novel materials for chromatography of particulate-containing fluids and cell culture applications. *Journal of Separation Science* **30**(11): 1657–1671.

Plieva, F. M., Karlsson, M., Aguilar, M.-R., Gomez, D., Mikhalovsky, S. and Galaev, I. Y. (2005). Pore structure in supermacroporous polyacrylamide based cryogels. *Soft Matter* **1**(4): 303–309.

Plieva, F. M., Karlsson, M., Aguilar, M. R., Gomez, D., Mikhalovsky, S., Galaev, I. Y. and Mattiasson, B. (2006c). Pore structure of macroporous monolithic cryogels prepared from poly (vinyl alcohol). *Journal of Applied Polymer Science* **100**(2): 1057–1066.

Plieva, F. M., Oknianska, A., Degerman, E. and Mattiasson, B. (2008). Macroporous gel particles as robust macroporous matrices for cell immobilization. *Biotechnology Journal* **3**(3): 410–417.

Plieva, F. M., Savina, I. N., Deraz, S., Andersson, J., Galaev, I. Y. and Mattiasson, B. (2004). Characterization of supermacroporous monolithic polyacrylamide based matrices designed for chromatography of bioparticles. *Journal of Chromatography B* **807**(1): 129–137.

Podorozhko, E., Kurskaya, E., Kulakova, V. and Lozinsky, V. (2000). Cryotropic structuring of aqueous dispersions of fibrous collagen: Influence of the initial pH values. *Food Hydrocolloids* **14**(2): 111–120.

Rafat, M., Li, F., Fagerholm, P., Lagali, N. S., Watsky, M. A., Munger, R., Matsuura, T. and Griffith, M. (2008). PEG-stabilized carbodiimide crosslinked collagen–chitosan hydrogels for corneal tissue engineering. *Biomaterials* **29**(29): 3960–3972.

Rechendorff, K., Hovgaard, M. B., Foss, M., Zhdanov, V. P. and Besenbacher, F. (2006). Enhancement of Protein Adsorption Induced by Surface Roughness. *Langmuir* **22**(26): 10885–10888.

Reichelt, S., Prager, A., Abe, C. and Knolle, W. (2014). Tailoring the structural properties of macroporous electron-beam polymerized cryogels by pore forming agents and the monomer selection. *Radiation Physics and Chemistry* **94**(0): 40–44.

Ricciardi, R., Auriemma, F., Gaillet, C., De Rosa, C. and Lauprêtre, F. (2004). Investigation of the crystallinity of freeze/thaw poly (vinyl alcohol) hydrogels by different techniques. *Macromolecules* **37**(25): 9510–9516.

Rigby, S. P., Barwick, D., Fletcher, R. S. and Riley, S. N. (2003). Interpreting mercury porosimetry data for catalyst supports using semi-empirical alternatives to the Washburn equation. *Applied Catalysis A: General* **238**(2): 303–318.

Ritter, H. L. and Drake, L. C. (1945). Pressure porosimeter and determination of complete macropore-size distributions. *Industrial & Engineering Chemistry Analytical Edition* **17**(12): 782–786.

Rootare, H. M. and Prenzlow, C. F. (1967). Surface areas from mercury porosimeter measurements. *The Journal of Physical Chemistry* **71**(8): 2733–2736.

Rubinstein, M. and Colby, R. H. (2003). *Polymer Physics*, OUP Oxford.

Sarac, A. (1999). Redox polymerization. *Progress in Polymer Science* **24**(8): 1149–1204.

Sato, T., Abe, M. and Otsu, T. (1979a). Application of spin trapping technique to radical polymerization, 16. Photo-decomposition of diphenyl disulfide and initiation mechanism: Evaluation of relative reactivities of vinyl monomers toward phenylthio radical. *Die Makromolekulare Chemie* **180**(5): 1165–1174.

Sato, T., Kita, S. and Otsu, T. (1979b). Application of spin trapping technique to radical polymerization, 18. Initiation mechanism of vinyl polymerization with N, N-dimethylaniline N-oxide/acid anhydride systems. *Die Makromolekulare Chemie* **180**(8): 1911–1916.

Savina, I. N., Cnudde, V., D'hollander, S., Van Hoorebeke, L., Mattiasson, B., Galaev, I. Y. and Du Prez, F. (2007). Cryogels from poly (2-hydroxyethyl methacrylate): Macroporous, interconnected materials with potential as cell scaffolds. *Soft Matter* **3**(9): 1176–1184.

Savina, I. N., Gun'ko, V. M., Turov, V. V., Dainiak, M., Phillips, G. J., Galaev, I. Y. and Mikhalovsky, S. V. (2011). Porous structure and water state in cross-linked polymer and protein cryo-hydrogels. *Soft Matter* **7**(9): 4276–4283.

Savina, I. N., Hanora, A., Plieva, F. M., Galaev, I. Y., Mattiasson, B. and Lozinsky, V. I. (2005). Cryostructuration of polymer systems. XXIV. Poly (vinyl alcohol) cryogels filled with particles of a strong anion exchanger: Properties of the composite materials and potential applications. *Journal of Applied Polymer Science* **95**(3): 529–538.

Scopelliti, P. E., Borgonovo, A., Indrieri, M., Giorgetti, L., Bongiorno, G., Carbone, R., Podestà, A. and Milani, P. (2010). The effect of surface nanometre-scale morphology on protein adsorption. *PLoS ONE* **5**(7): e11862.

Shakya, A. K., Holmdahl, R., Nandakumar, K. S. and Kumar, A. (2014). Polymeric cryogels are biocompatible, and their biodegradation is independent of oxidative radicals. *Journal of Biomedical Materials Research Part A* **102**(10): 3409–3418.

Sharma, A., Bhat, S., Vishnoi, T., Nayak, V. and Kumar, A. (2013). Three-dimensional supermacroporous carrageenan-gelatin cryogel matrix for tissue engineering applications. *BioMed Research International* 2013: 478279.

Sheppard, C. and Mark, H. (1988). *Encyclopedia of Polymer Science and Engineering*. Vol. 11, Wiley-Interscience, New York.

Silva, S. S., Mano, J. F. and Reis, R. L. (2010). Potential applications of natural origin polymer-based systems in soft tissue regeneration. *Critical Reviews in Biotechnology* **30**(3): 200–221.

Simionescu, B. C., Neamtu, A., Balhui, C., Danciu, M., Ivanov, D. and David, G. (2013). Macroporous structures based on biodegradable polymers—Candidates for biomedical application. *Journal of Biomedical Materials Research Part A* **101**(9): 2689–2698.

Spiller, K. L., Laurencin, S. J., Charlton, D., Maher, S. A. and Lowman, A. M. (2008). Superporous hydrogels for cartilage repair: Evaluation of the morphological and mechanical properties. *Acta Biomaterialia* **4**(1): 17–25.

Srivastava, A., Jain, E. and Kumar, A. (2007). The physical characterization of supermacroporous poly(N-isopropylacrylamide) cryogel: Mechanical strength and swelling/deswelling kinetics. *Materials Science and Engineering: A* **464**(1–2): 93–100.

Stauffer, S. R. and Peppast, N. A. (1992). Poly(vinyl alcohol) hydrogels prepared by freezing-thawing cyclic processing. *Polymer* **33**(18): 3932–3936.

Tabata, Y. and Ikada, Y. (1989). Synthesis of gelatin microspheres containing interferon. *Pharmaceutical Research* **6**(5): 422–427.

Takamura, A., Ishii, F. and Hidaka, H. (1992). Drug release from poly(vinyl alcohol) gel prepared by freeze-thaw procedure. *Journal of Controlled Release* **20**(1): 21–27.

Tamada, Y. and Ikada, Y. (1993). Effect of preadsorbed proteins on cell adhesion to polymer surfaces. *Journal of Colloid and Interface Science* **155**(2): 334–339.

Tretinnikov, O. N., Sushko, N. I. and Zagorskaya, S. A. (2013). Detection and quantitative determination of the crystalline phase in poly(vinyl alcohol) cryogels by ATR FTIR spectroscopy. *Polymer Science Series A* **55**(2): 91–97.

Tripathi, A. and Kumar, A. (2011). Multi-featured macroporous agarose–alginate cryogel: Synthesis and characterization for bioengineering applications. *Macromolecular Bioscience* **11**(1): 22–35.

Tripathi, A., Kathuria, N. and Kumar, A. (2009). Elastic and macroporous agarose–gelatin cryogels with isotropic and anisotropic porosity for tissue engineering. *Journal of Biomedical Materials Research Part A* **90**(3): 680–694.

Tsung, L. H., Chang, K.-H. and Chen, J. P. (2011). Osteogenesis of adipose-derived stem cells on three-dimensional, macroporous gelatin–hyaluronic acid cryogels. *Biomedical Engineering: Applications, Basis and Communications* **23**(02): 127–133.

Uygun, M., Şenay, R. H., Avcıbaşı, N. and Akgöl, S. (2014). Poly (HEMA-co-NBMI) Monolithic cyogel columns for IgG adsorption. *Applied Biochemistry and Biotechnology* **172**(3): 1574–1584.

Vaikousi, H. and Biliaderis, C. (2005). Processing and formulation effects on rheological behavior of barley β-glucan aqueous dispersions. *Food Chemistry* **91**(3): 505–516.

Van Vlierberghe, S., Cnudde, V., Dubruel, P., Masschaele, B., Cosijns, A., De Paepe, I., Jacobs, P. J., Van Hoorebeke, L., Remon, J. P. and Schacht, E. (2007). Porous gelatin hydrogels: 1. Cryogenic formation and structure analysis. *Biomacromolecules* **8**(2): 331–337.

Van Vlierberghe, S., Dubruel, P., Lippens, E., Cornelissen, M. and Schacht, E. (2009). Correlation between cryogenic parameters and physico-chemical properties of porous gelatin cryogels. *Journal of Biomaterials Science, Polymer Edition* **20**(10): 1417–1438.

Van Vlierberghe, S., Dubruel, P. and Schacht, E. (2011). Biopolymer-based hydrogels as scaffolds for tissue engineering applications: A review. *Biomacromolecules* **12**(5): 1387–1408.

Varganici, C.-D., Paduraru, O. M., Rosu, L., Rosu, D. and Simionescu, B. C. (2013). Thermal stability of some cryogels based on poly(vinyl alcohol) and cellulose. *Journal of Analytical and Applied Pyrolysis* **104**(0): 77–83.

Velickova, E., Petrov, P., Tsvetanov, C., Kuzmanova, S., Cvetkovska, M. and Winkelhausen, E. (2010). Entrapment of *Saccharomyces cerevisiae* cells in uv crosslinked hydroxyethylcellulose/poly (ethylene oxide) double-layered gels. *Reactive and Functional Polymers* **70**(11): 908–915.

Viklund, C., Svec, F., Fréchet, J. M. and Irgum, K. (1996). Monolithic, molded, porous materials with high flow characteristics for separations, catalysis, or solid-phase chemistry: Control of porous properties during polymerization. *Chemistry of Materials* **8**(3): 744–750.

Vishnoi, T. and Kumar, A. (2013). Conducting cryogel scaffold as a potential biomaterial for cell stimulation and proliferation. *Journal of Materials Science: Materials in Medicine* **24**(2): 447–459.

Vrana, N. E., Cahill, P. A. and McGuinness, G. B. (2010). Endothelialization of PVA/gelatin cryogels for vascular tissue engineering: Effect of disturbed shear stress conditions. *Journal of Biomedical Materials Research Part A* **94**(4): 1080–1090.

Watase, M. and Nishinari, K. (1989). Effect of the degree of saponification on the rheological and thermal properties of poly (vinyl alcohol) gels. *Die Makromolekulare Chemie* **190**(1): 155–163.

Wells, R. G. (2008). The role of matrix stiffness in regulating cell behavior. *Hepatology* **47**(4): 1394–1400.

Welzel, P. B., Grimmer, M., Renneberg, C., Naujox, L., Zschoche, S., Freudenberg, U. and Werner, C. (2012). Macroporous StarPEG-Heparin Cryogels. *Biomacromolecules* **13**(8): 2349–2358.

Westermarck, S., Juppo, A. M., Kervinen, L. and Yliruusi, J. (1998). Pore structure and surface area of mannitol powder, granules and tablets determined with mercury porosimetry and nitrogen adsorption. *European Journal of Pharmaceutics and Biopharmaceutics* **46**(1): 61–68.

Willmott, N., Cummings, J., Stuart, J. and Florence, A. (1985). Adriamycin-loaded albumin microspheres: Preparation, *in vivo* distribution and release in the rat. *Biopharmaceutics & Drug Disposition* **6**(1): 91–104.

Wilson, C. J., Clegg, R. E., Leavesley, D. I. and Pearcy, M. J. (2005). Mediation of biomaterial-cell interactions by adsorbed proteins: A review. *Tissue Engineering* **11**(1-2): 1–18.

Wu, J., Zhang, F. and Zhang, H. (2012). Facile synthesis of carboxymethyl curdlan-capped silver nanoparticles and their application in SERS. *Carbohydrate Polymers* **90**(1): 261–269.

Wu, X. Y., Huang, S. W., Zhang, J. T. and Zhuo, R. X. (2004). Preparation and characterization of novel physically cross-linked hydrogels composed of poly(vinyl alcohol) and amine-terminated polyamidoamine dendrimer. *Macromolecular Bioscience* **4**(2): 71–75.

Xiao, C. and Gao, Y. (2008). Preparation and properties of physically crosslinked sodium carboxymethylcellulose/poly (vinyl alcohol) complex hydrogels. *Journal of Applied Polymer Science* **107**(3): 1568–1572.

Yang, X., Liu, Q., Chen, X., Yu, F. and Zhu, Z. (2008). Investigation of PVA/ws-chitosan hydrogels prepared by combined γ-irradiation and freeze-thawing. *Carbohydrate Polymers* **73**(3): 401–408.

Ye, P., Xu, Z.-K., Che, A.-F., Wu, J. and Seta, P. (2005). Chitosan-tethered poly (acrylonitrile-*co*-maleic acid) hollow fiber membrane for lipase immobilization. *Biomaterials* **26**(32): 6394–6403.

Yokoyama, F., Masada, I., Shimamura, K., Ikawa, T. and Monobe, K. (1986). Morphology and structure of highly elastic poly (vinyl alcohol) hydrogel prepared by repeated freezing-and-melting. *Colloid and Polymer Science* **264**(7): 595–601.

Yu, H., Xu, X., Chen, X., Lu, T., Zhang, P. and Jing, X. (2007). Preparation and antibacterial effects of PVA-PVP hydrogels containing silver nanoparticles. *Journal of Applied Polymer Science* **103**(1): 125–133.

Yu, S. S., Koblin, R. L., Zachman, A. L., Perrien, D. S., Hofmeister, L. H., Giorgio, T. D. and Sung, H.-J. (2011). Physiologically relevant oxidative degradation of oligo(proline) cross-linked polymeric scaffolds. *Biomacromolecules* **12**(12): 4357–4366.

Zhan, X. Y., Lu, D. P., Lin, D. Q. and Yao, S. J. (2013). Preparation and characterization of supermacroporous polyacrylamide cryogel beads for biotechnological application. *Journal of Applied Polymer Science* **130**(5): 3082–3089.

Zhang, H., Nishinari, K., Williams, M. A., Foster, T. J. and Norton, I. T. (2002). A molecular description of the gelation mechanism of curdlan. *International Journal of Biological Macromolecules* **30**(1): 7–16.

Zhang, H., Zhang, F. and Wu, J. (2013). Physically crosslinked hydrogels from polysaccharides prepared by freeze–thaw technique. *Reactive and Functional Polymers* **73**(7): 923–928.

Zhang, R., Tang, M., Bowyer, A., Eisenthal, R. and Hubble, J. (2005). A novel pH-and ionic-strength-sensitive carboxy methyl dextran hydrogel. *Biomaterials* **26**(22): 4677–4683.

Zheng, Y., Gun'ko, V. M., Howell, C. A., Sandeman, S. R., Phillips, G. J., Kozynchenko, O. P., Tennison, S. R., Ivanov, A. E. and Mikhalovsky, S. V. (2012). Composites with Macroporous Poly(vinyl alcohol) Cryogels with Attached Activated Carbon Microparticles with Controlled Accessibility of a Surface. *ACS Applied Materials & Interfaces* **4**(11): 5936–5944.

# 3 Production of Synthetic Cryogels and the Study of Porosity Thereof

*Irina N. Savina\* and Igor Yu. Galaev*

## CONTENTS

3.1  Introduction .................................................................................................92
3.2  Freezing of Aqueous Solutions ....................................................................92
3.3  Mechanism of Cryo-polymerization............................................................94
3.4  The Effect of Solutes on Cryo-polymerization.............................................97
3.5  The Methods for Studying Porosity of Macroporous Polymeric Gels ..........99
3.6  Grafting Polymers on the Pore Surface of Cryogels .................................. 104
3.7  Conclusions............................................................................................... 108
List of Abbreviations............................................................................................ 108
References............................................................................................................ 108

## ABSTRACT

This chapter provides an introduction on the macroporous polymeric gel preparation in semifrozen solutions—cryogelation. The phenomena of freezing of aqueous polymer solution, formation of unfrozen liquid phase, cryo-concentration of solutions and cryo-polymerization are discussed. The formation of nonfrozen liquid phase and its size dependence on the temperature, nature and concentration of solutes was demonstrated using NMR analysis. The applications of various physio-chemical techniques (microscopic techniques, thermogravimetry [TG], differential scanning calorimetry [DSC], low temperature $^1$H NMR spectroscopy and thermally stimulated depolarization [TSD]) for characterization of cryogel porosity are critically discussed. Additionally, graft polymerization inside the porous hydrogel matrix as a useful tool to the cryogel modification is described.

## KEYWORDS

NMR, microscopy, DSC, TSD, graft polymerization

---

\* Corresponding author: E-mail: i.n.savina@brighton.ac.uk; Tel: +441273642034.

## 3.1 INTRODUCTION

This chapter gives a brief introduction on freezing of aqueous solutions and how the phenomenon of cryo-concentration of solutes in an unfrozen liquid phase is used for the production of macroporous polymeric gels, known as cryogels. The methods for studying the porosity of macroporous polymeric gels are critically reviewed and modification of cryogels by grafting secondary polymers at the pore surface is also described.

## 3.2 FREEZING OF AQUEOUS SOLUTIONS

Freezing of water is a well-known phenomenon; however, the process of freezing is far from being either simple or straightforward. For a pure substance such as water, the freezing temperature is well defined and is the same as the melting temperature. It is defined as 0°C. The crystallization of water requires nucleation. This occurs usually at or very near to the cooled surfaces contacting the liquid and at temperatures significantly lower than the bulk freezing point. Nucleation can also be triggered by particles (inherently present as dust particles or deliberately added like AgI crystals to induce water crystallization) or surface defects of the containers. The *nucleation temperature* refers to the temperature at which the first crystal nucleus forms in the liquid and it can be as low as –20°C for water with very low particulate content (Lam and Moore, 2010). Moreover, nucleation is a stochastic process, that is, the probability of nucleus formation and hence freezing increases with the size of the liquid sample. Therefore, water in capillaries could be super cooled to very low temperatures, a phenomenon responsible in the formation of ice lenses in the soil in winter and cavities when the ice melts in warmer periods.

The size of ice crystals formed depends on the rate of heat transfer and setup target temperature of freezing. Fast cooling promotes the formation of numerous small ice crystals, whereas slow cooling results in a smaller number of larger crystals (Figure 3.1).

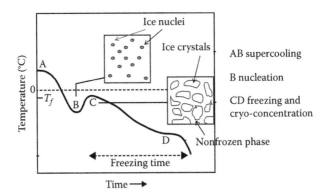

**FIGURE 3.1** Schematic representation of a freezing process shows different stages and nature of frozen material. Cooling starts at A, and the solution supercools to B. Nucleation occurs at B, followed by freezing of the whole mass between C and D, representing the freezing time or duration. Freezing temperature drops during this process as the unfrozen fraction becomes progressively cryo-concentrated. After further removal of sensible heat and cooling to point D (the target temperature), the process is considered complete.

The situation becomes even more complicated when dealing with aqueous solutions. In this case, the *freezing point* reflects the process of *freezing point depression* caused by the solute. Freezing point depression is a colligative property of the solution, and when a value is given, it applies only to the initial solution (Lam and Moore, 2010). The composition of solvent changes during the freezing because the pure solvent freezes first, and the concentration of the solutes increases in the remaining unfrozen solvent.

As an aqueous solution freezes, conversion of water into ice causes progressive freeze-concentration (*cryo-concentration*) of the unfrozen mixture because the growing ice crystals exclude solutes. The increasing concentration of solutes in the unfrozen fraction results in a continuously decreasing freezing point (Figure 3.1, CD section of temperature versus time section). If all the freezable water is converted to ice, then the concentration of solutes can become very high. In the case of, for example, NaCl, reaching a solubility limit during cryo-concentration could cause the salt to crystallize out of solution, forming a eutectic (Figure 3.2). When freezing a 150 mM (~0.9% w/w) NaCl solution, the eutectic at −21.2°C has a concentration of 23.3% w/w, meaning about a 25-fold increase in concentration of NaCl in the unfrozen solution at this temperature. There is a range of temperatures and solute concentrations where ice coexists with concentrated solute solution. Macroscopically it looks like a solid block of ice but it contains significant amounts (up to one quarter) of nonfrozen liquid finely distributed in between the ice crystals.

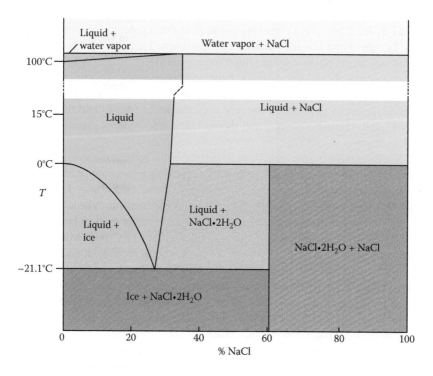

**FIGURE 3.2** Phase diagram of NaCl in water. (Reproduced from https://www.uwgb.edu/dutchs/Petrology/beutect.htm.)

## 3.3 MECHANISM OF CRYO-POLYMERIZATION

With this basic knowledge of freezing of aqueous solutions, one could start discussing the mechanism of cryogel formation by radical polymerization in the frozen state or, more correctly, the semifrozen state, as part of the sample remains nonfrozen due to the reasons presented above. When a 6 wt% mixture of a monomer, dimethylacrylamide (DMMAm) and cross-linker, PEG diacrylate is frozen at –10°C, the system can be either in a supercooled and hence metastable liquid state or in a frozen state. Ice crystallization could be induced by adding a few crystals of AgI, which have a structure similar to an ice crystal and serve as crystallization seeds. AgI is not soluble in the polymerization medium and hence does not affect the reaction; nor does the AgI influence the NMR measurements.

The volume of the nonfrozen phase is determined by the initial concentration of solutes (osmolytes) and the temperature used. For example, the volume of the nonfrozen phase is about one-tenth the total volume of the system with a 6 wt% monomer concentration at –10°C. Solutes are concentrated in the nonfrozen phase until the depression in freezing point due to the increased concentration is equal to the temperature used. Freezing-point depression of an ideal solution can be calculated with the following equation, and can be used as an estimation to describe the system used (Atkins, 2001):

$$\Delta T = K_f b \tag{3.1}$$

where $\Delta T$ is the depression in freezing point, $K_f$ is the cryoscopic constant for water (1.86 K kg mol$^{-1}$) and $b$ is the molality of the solutes. The actual concentrations in the nonfrozen liquid phase were obtained from $^1$H-NMR signals of nonfrozen water and from monomers, and theoretically expected concentrations were calculated as the ratio between the area of the vinyl peak in the NMR spectrum at 6.9 ppm and the area of the water peak assuming that only the water freezes and the monomer remains liquid. The experimental and theoretical concentrations show good agreement (Kirsebom et al., 2009).

The NMR signals from the monomers broadened in the semifrozen state as a result of the increasing rotational correlation time upon decreasing temperature and differences in magnetic susceptibility between the coexisting fluid and solid phases. However, the areas of the peaks appeared to be unchanged upon initial freezing, indicating that most of the monomers were concentrated in the liquid phase rather than embedded in ice. Approximately 8–9% of the water remained unfrozen at –10°C prior to the onset of polymerization. As the polymerization reaction proceeds, the molality of the liquid phase decreases due to the conversion of individual monomer molecules into the cross-linked polymer, and the effect of freezing point depression becomes negligible, according to Equation 3.1. However, approximately 5% of the initial water remained unfrozen even after completion of the polymerization reaction. This amount of nonfrozen water corresponds to approximately 5 water molecules per monomer unit of polymer, and the nonfrozen water is most probably associated with the polymer (Kirsebom et al., 2008).

The cryo-concentration effect when the monomers are concentrated in the liquid phase due to the formation of ice crystals was studied with a 6 wt% monomer concentration at –10°C as it was possible to carry out the polymerization reaction both in (supercooled) liquid and in a semifrozen system. This allows study of the reaction at the same temperature. However, under supercooled conditions, the monomer concentration in the liquid phase is 6 wt% whereas in the semifrozen system, the concentration is approximately 33 wt% in the nonfrozen phase. Thus, the polymerization for these two systems proceeds under substantially different conditions even though the initial condition (before freezing) and the reaction temperature are the same. Not surprisingly, the resulting cryogels were very different. A heterogeneous, spongy, elastic, opaque cryogel was produced in the semifrozen system whereas a homogeneous brittle gel was produced under supercooled conditions (Figure 3.3). The structure and physical properties of the gel produced under supercooled conditions were close to that of a gel produced at ambient temperature, whereas the properties of a cryogel are defined by macropores (produced in place of melted ice crystals) and pore walls composed of concentrated polymer phase (produced as a result of polymerization in cryo-concentrated solution) (Kirsebom et al., 2009).

The reaction times for supercooled and semifrozen systems are similar despite significant differences in the concentration: 6 and 33 wt%, respectively. The 5.5-fold higher monomer concentration in the semifrozen system would make the polymerization considerably faster than that in the supercooled system. However, this is not the case and the reaction times are similar. One possible explanation for this could be related to diffusion in the system. Diffusion of molecules in confined spaces can be considered for either long or short time scales. At short time scales, the short-range diffusion is equal to the bulk diffusion ($D_0$). Diffusion at longer time scales reaches a plateau value ($D_\infty$), which depends on the connectivity of the pore space (Valiullin and Skirda, 2001). Differences in the diffusion in the supercooled and semifrozen system are both for the long-range diffusion (at high $t_{diff}$) and the short-range diffusion (at short $t_{diff}$). The diffusion at both long and short $t_{diff}$ is significantly faster (Figure 3.4) for the supercooled system than for the semifrozen, which could explain the reaction times (see further discussions about the influence of diffusion next).

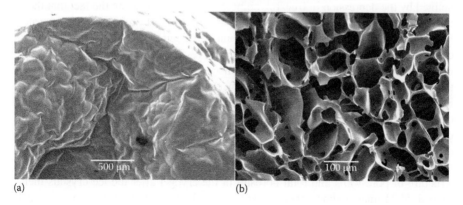

(a)  (b)

**FIGURE 3.3** SEM images of a DMAAm-co-PEG (60:1) gel (a) prepared from a 6 wt% feed at −20°C in a supercooled system, and cryogel prepared from a 6 wt% feed in a semifrozen system (b). (Reprinted with permission from Kirsebom, H., Rata, G., Topgaard, D., Mattiasson, B. and Galaev, I. Yu., *Macromolecules* 42(14), 5208–5214. Copyright 2009 American Chemical Society.)

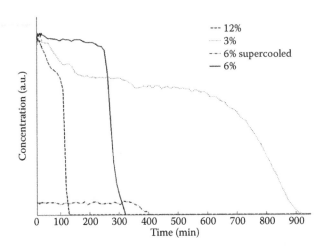

**FIGURE 3.4** Change in monomer concentrations for (cryo)polymerization of DMAAm-co-PEG diacrylate (approximated as the ratio of monomer and water NMR signals) in different (cryo)polymerization systems as a function of time. (Reprinted with permission from Kirsebom, H., Rata, G., Topgaard, D., Mattiasson, B. and Galaev, I. Yu., *Macromolecules* 42(14), 5208–5214. Copyright 2009 American Chemical Society.)

At the studied $t_{diff}$ (50 ms to 1 s) the studied molecules move in the range between 5 and 19 μm depending on the $t_{diff}$ and the sample studied. This corresponds to the observed distances in the nonfrozen microphase, which is visualized as the pore walls in the SEM images (Figure 3.3) and the sizes of the pores (Kirsebom et al., 2009).

Comparing the samples with initial concentrations of 3, 6 and 12 wt%, it is evident that the reactions start at different times and proceed at different rates (Figure 3.4), although the monomer/initiator concentrations were the same in all systems as defined by the depression in freezing point (Equation 3.1). Despite the fact that there were different initial concentrations before freezing, in all three systems the water froze with a concomitant increase in monomer concentration, until a sufficient concentration was reached corresponding to the depression in freezing point. Diffusion in semifrozen samples (with 33 wt% monomer concentration) was found to be slower than that in a similar (33 wt% monomer concentration) supercooled sample at –10°C. Even if the systems studied were at the same temperature and equal, apparent concentration of the monomer, the observed diffusion was different. The difference in diffusion coefficients for three semifrozen systems clearly indicates the existence of different diffusion restrictions in liquid microphase resulting in different pore structures of the cryogels as the volume and shape of the liquid microphase define the structure of pore walls and total porosity in the cryogel after the ice crystals melt. Indeed, SEM images show much larger pores and thinner pore walls in the cryogel sample produced from 3 wt% feed as compared to cryogel produced form 12 wt% feed (Figure 3.5). Smaller pore size and thicker pore walls in cryogels prepared from 12 wt% feed resulted in significantly less flow of liquid through the cryogel, and a significant increase in the mechanical stability (elastic modulus) compared to the cryogels prepared from feeds with lower monomer concentration (Kirsebom et al., 2009).

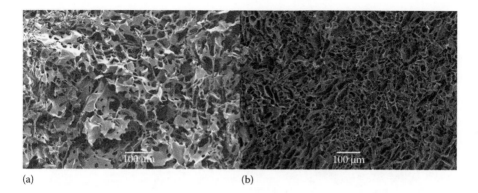

(a)                            (b)

**FIGURE 3.5** SEM images of DMAAm-co-PEG diacrylate cryogels prepared at –10°C from a feed with monomer concentrations of 3 wt% (a) and with 12 wt% (b). (Reprinted with permission from Kirsebom, H., Rata, G., Topgaard, D., Mattiasson, B. and Galaev, I. Yu., *Macromolecules* 42(14), 5208–5214. Copyright 2009 American Chemical Society.)

## 3.4 THE EFFECT OF SOLUTES ON CRYO-POLYMERIZATION

Different inert solutes, when added to a polymerization mixture, affect total solute concentration in the nonfrozen microphase and hence the properties of the gels such as porosity, pore wall thickness and mechanical properties. The concentration of solutes in the nonfrozen phase is determined by the polymerization temperature according to Equation 3.1. Thus, the starting concentration of monomers does not influence the concentration in the nonfrozen phase. It is only the size of the nonfrozen phase that is affected by the starting concentration. NaCl or $CaCl_2$ (with an ion concentration of 0.3 or 0.6 M, assuming complete dissociation) was added to study the effect of increasing the solute concentration. Acetone or methanol was added as water miscible noncharged compounds to study the influence on freezing point depression, and to get insights into the effects of substances that act as poor solvents for the formed polymer (Caykara and Dogmus, 2004). The samples with added solutes clearly have smaller pores and thicker pore walls corresponding to a larger nonfrozen phase during the formation of the cryogels, which in turn was confirmed by the NMR measurements of nonfrozen water (Figure 3.6) (Kirsebom et al., 2010). The formation of the cross-linked polymeric gel occurs in the nonfrozen phase, and the structure of the final gel can thus give an indication of the size of this phase.

In the cryogels produced with the addition of either methanol or acetone, the submicrometer micropores in the walls of macropores were most probably due to the reaction-induced phase separation of synthesized polymer in the presence of organic solvents in nonfrozen microphase. Since both acetone and methanol are poor solvents for polyacrylamide, the growing polymeric chains become insoluble at a certain point, resulting in the formation of a cauliflower-like structure similar to that observed previously for precipitation polymerization (Okay, 2000; Ozmen and Okay, 2008). Gels synthesized at room temperature in the presence or absence of 0.6 M acetone were apparently similar. The pronounced induced phase separation of the cryogels could thus be a result of a

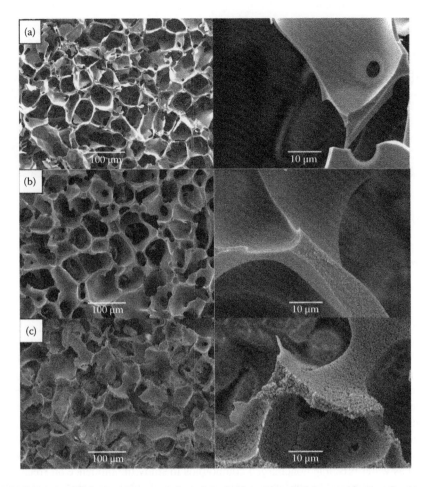

**FIGURE 3.6** SEM of pAAm cryogels at low (left) or high (right) magnification for (a) no added solute, (b) 0.3 M NaCl added, (c) 0.6 M acetone added. (Reprinted with permission from Kirsebom, H., Topgaard, D., Galaev, I. Yu. and Mattiasson, B., *Langmuir* 26(20), 16129–16133. Copyright 2010 American Chemical Society.)

significant increase in the solvent as well as in the monomer concentrations in the nonfrozen microphase.

The addition of NaCl or $CaCl_2$ results in cryogels with a lower pore volume compared to a cryogel with no added solute. These data correlate with the finding that the addition of solutes increases the nonfrozen phase. However, the addition of methanol or acetone at either concentration does not affect the pore volume of the cryogels when measured by cyclohexane uptake. Even if the addition of these solutes results in a cryogel with a larger nonfrozen phase and thicker pore walls, the porosity remains the same as that of a cryogel produced with no added solutes. A possible explanation for this could be that the induced secondary porosity of the pore walls increases the porosity of the cryogels in case this volume inside the pore walls is accessible for the nonsolvent cyclohexane (Kirsebom et al., 2010).

The porosity of cryogels in the completely hydrated state was studied using confocal microscopy after staining cryogels with a fluorescent dye. Figure 3.7 shows the distribution of pore size and pore wall thickness of plain cryogel and cryogels produced with the addition of 0.3 M NaCl or 0.6 M acetone. These results confirm the conclusion obtained from the analysis of SEM images, showing that adding solutes results in thicker pore walls and smaller pores. Interestingly, the cryogel prepared with the addition of 0.6 M acetone shrunk when it was frozen while the plain cryogel and the one produced with the addition of 0.3 M NaCl retained their shape. Since the heat transfer will occur mainly from the side of the tubes, the ice crystallization starts most likely also from the side. When ice crystals start forming in the presence of 0.6 M acetone, the water in the submicrometer pores in the walls does not freeze due to capillary-induced freezing point depression (Enustun et al., 1978). The ice crystals formed at the sides of the sample draw the liquid water from the ice formation in porous materials (Watanabe et al., 2001; Talamucci, 2003). This apparently causes the cryogel to shrink and be compressed by the growing ice crystals. However, the cryogel completely regained its shape after thawing. For the plain cryogels and cryogels prepared with the addition of 0.3 M NaCl, no such effect was observed because the pore walls contain no submicrometer pores. Thus, no capillary-induced freezing point depression occurs and ice crystallization takes place in the macropores (Kirsebom et al., 2010).

## 3.5 THE METHODS FOR STUDYING POROSITY OF MACROPOROUS POLYMERIC GELS

The porosity is the main feature of the cryogels in terms of practical applications. The size of the pores, their interconnectivity and pore surface area together with other structural characteristics such as pore wall thickness and density represent particular interest for the developing cryogel-based materials for biotechnological, medical and environmental applications. Being soft, highly hydrated materials, cryogels are challenging to study. Many techniques routinely used for the porosity assessment such as mercury porosimetry, nitrogen adsorption and scanning electron microscopy require dried samples that are not always suitable for studying hydrogel materials. Cryogels hold considerable amounts of water, which supports their structure and integrity. Removing water is a destructive process and may have considerable impact on the cryogel structure. Applying the high pressure in mercury porosimetry analysis will cause the gel compression and lead to misinterpretation of the data. Thus, techniques capable for the visualization of the hydrated samples are intensively investigated. Confocal laser scanning microscopy (CLSM) and multiphoton microscopy (MPM) were applied for the analysis of the pore size, wall thickness and pore surface area (Savina et al., 2011). Scanning the samples in a 3D regime allows reconstruction of 3D images of the cryogel samples preliminary stained with fluorescent dye and allows for the quantitative analysis of the structural characteristics. The images collected were subjected to the analysis using the ImageJ or Fiji software, which allows obtaining some quantitative characteristics of the cryogel porosity. Such analysis allows getting quantitative characteristics for the pore size distribution (PSD), surface area and wall

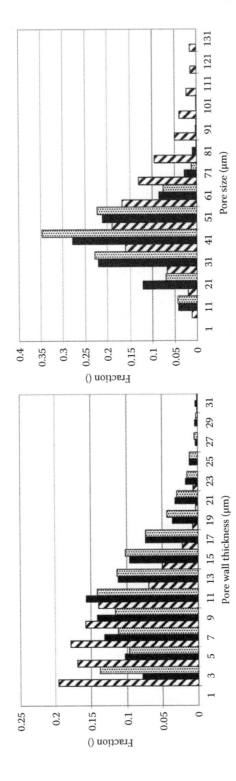

**FIGURE 3.7** Pore wall thickness and pore size distributions for plain cryogel (striped), cryogel produced in the presence of 0.3 M NaCl (black), and cryogel produced in the presence of 0.6 M acetone (dotted). (Reprinted with permission from Kirsebom, H., Topgaard, D., Galaev, I. Yu. and Mattiasson, B., *Langmuir* 26(20), 16129–16133. Copyright 2010 American Chemical Society.)

thickness for hydroxyethyl methacrylate (HEMA) cryogels with different morphology (Figure 3.8). The morphology of the HEMA cryogel varies between the samples depending on the preparation conditions. The higher monomer and initiator concentration in the initial solution results in the formation of the cryogels with smaller pores. The difference was not so obvious and difficult to assess just using images of gel morphology, while the quantitative analysis of the images with ImageJ software allowed assessing quantitatively the difference (Figure 3.8a and b). CLSM, despite its advantage to study fully hydrated samples, is limited to the laser penetration depth of 100–200 μm; thus, only the analysis of small sections of cryogels could be done.

X-ray microcomputed tomography (μ-CT) uses X-rays for scanning the sample and allows collecting a series of 2D images and analysis of the larger sample (Figure 3.9) (Savina et al., 2007). Analysis of the 3D reconstructed cryogel shows that the pores in cryogel are interconnected and form a system of channels rather than isolated voids. This provides a high permeability of the cryogels for a liquid phase that is important for the solutes exchange in biomedical applications or when the cryogels are used as supports for filtration and adsorption.

Yun et al. (2011) suggested the use of a capillary-based model for assessing the pore size distribution, skeleton thickness and pore tortuosity in cryogels. The model is based on the fact that fluid flow and mass transfer within cryogels depend on the complexity of cryogel microstructure. The data calculated using this model were in good agreement with experimental data, SEM images, flow rate performance and nonadsorption breakthrough of proteins; lysozyme, bovine serum albumin and concanavalin A; and allows characterizing both the cryogel structure and performance in chromatography.

The simplest method of analysis of the macropore fraction in cryogel is the gravimetric method. As most of the water is located in large interconnected pores, it could be simply squeezed out from the material and porosity or pore volume could be calculated as

$$V_{macropores} = m_{water\ in\ macropores} / \rho_w \times V_{swollen\ gel} \times 100\%,$$

where $m_{water\ in\ macropores} = m_{swollen\ gel} - m_{swollen\ gel\ walls}$, $\rho_w$ is the density of liquid (water) in the pores, $V_{swollen\ gel}$ is the total volume of the specimen and $V_{macropores}$ is volume of macropores.

This simple experiment shows that most of the water in cryogels is nonbound water, which is located in the big pores. Similar results were obtained using thermoporometric calculations based on studying the water properties at different temperatures. It appears that bulk water and interfacial water will have different temperatures of freezing, melting or evaporation. The water at the interface or in narrow confined spaces, nanopores, will freeze at lower temperatures or evaporate at higher temperatures because of the interaction with the polymer surface. Thermogravimetry (TG), differential scanning calorimetry (DSC), low temperature $^1$H NMR spectroscopy and thermally stimulated depolarization (TSD) have been used for studying the temperature-dependent behaviour of water (Savina et al., 2011; Gun'ko et al., 2013). The amount of nonbound, weakly and strongly bound water was calculated. This provides information about the structure of the cryogels, as the state of the water is dependent on the structure of the material. The water in big macropores behaves as nonbound water and its amount (volume) will give information about macropores.

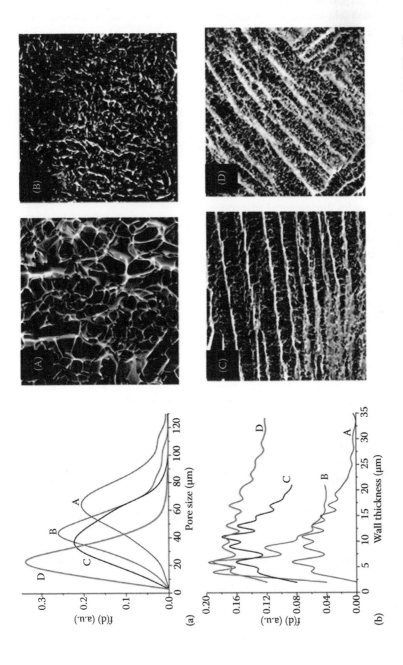

**FIGURE 3.8** Pore size (a) and wall thickness (b) distribution for HEMA cryogels with different morphology (A, B, C and D). The 3D reconstructed CLSM images of the A, B, C and D HEMA cryogels are on the right. Samples A and C were produced from 6% w/v monomer solution; samples B and D –12% w/v; A and B low initiator concentration; C and D high initiator concentration. (Adapted from Savina, I. N., Gun'ko, V. M., Turov, V. V., Dainiak, M., Phillips, G. J., Galaev, I. Yu. and Mikhalovsky, S. V., *Soft Matter* 7, 4276–4283, 2011. Reproduced by permission of The Royal Society of Chemistry.)

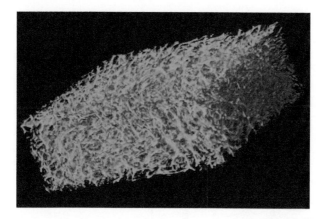

**FIGURE 3.9** 3D reconstruction of the X-ray microcomputed tomography images of HEMA cryogel.

The amount of weak bound and strongly bound water will give information about small pores, nanopores and surface area.

For example, Gibbs-Thomson Equation (Equation 3.2) showing the relationship between the freezing temperature depression of water inside a cylindrical pores and its radius R was used for assessing the pore size based on the DSC measurements:

$$R(nm) = 0.68 - k/(T_m - T_{m0}) \tag{3.2}$$

where $T_m$ and $T_{m0}$ are the melting temperature of confined and bulk water, respectively, and $k = 32.33$ K nm (Gun'ko et al., 2013). The $k$ value depends on the nature of the material.

Low-temperature $^1$H NMR spectroscopy is more sensitive to changes in the water state than DSC and gives a more comprehensive picture of the cryogel morphology, particularly information about nanopores. The gelatin cryogel structure was analysed by recording $^1$H NMR spectra of water in frozen samples. The signal of frozen water (ice crystals) was not contributing to the spectrum because of significant difference in the relaxation time of the nonmobile phase in comparison with the mobile one (liquid water) (Savina et al., 2011). The samples were frozen to 200 K and then $^1$H NMR spectrums were recorded with increase in the temperature (Figure 3.10).

Changes in the Gibbs free energy of bound water and free surface energy, as the modulus of integrated changes of the Gibbs free energy values in the bound water layers, were determined from the temperature dependences of the amounts of unfrozen water at $T = 200-273$ K. An equation similar to Equation 3.2 was used for the calculation of pore size and pore size distribution (Gun'ko et al., 2013). The data obtained were complementary to the CLSM image analysis and provided information about the nanostructure of cryogels (Figure 3.10d).

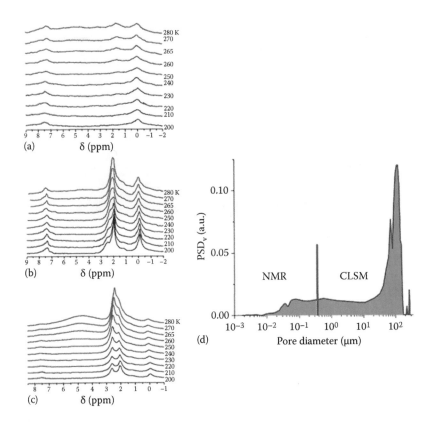

**FIGURE 3.10** $^1$H NMR spectra, recorded at different temperatures, of water adsorbed by gelatine gel: (a) initial freeze-dried (0.3 wt% $H_2O$) in $CDCl_3$; and in a mixture $CDCCl_3$: $CD_3CN$ 3:1 at (b) 0.8 wt%; and (c) 10 wt% of water. (d) Pore size distribution calculated using the $^1$H NMR-cryoporometry and CLSM methods for hydrated collagen gel. (From Gun'ko, V. M., Mikhalovska, L. I., Savina, I. N., Shevchenko, R. V., James, S. L., Tomlins, P. E. and Mikhalovsky, S. V., *Soft Matter* 6, 5351–5358, 2010. Reproduced by permission of The Royal Society of Chemistry.)

## 3.6 GRAFTING POLYMERS ON THE PORE SURFACE OF CRYOGELS

Cryogels with large interconnected pores and high permeability present a particular interest as materials for adsorption and purification. However, there is a direct relation between the pore size and pore surface area available for the adsorption or interaction with solutes during the mass transport of liquid phase through the cryogel material. Increasing pore size results in better permeability but the pore surface area is less in such materials (Figure 3.11). The Cryogels with smaller pores have the largest surface area; however, this has an impact on the flow rate by reducing material permeability.

Grafting polymerization from the surface is a powerful approach for cryogel modification and introduction of functional groups on the cryogel pore surface. Grafting allows introducing of desirable functional groups with control of their

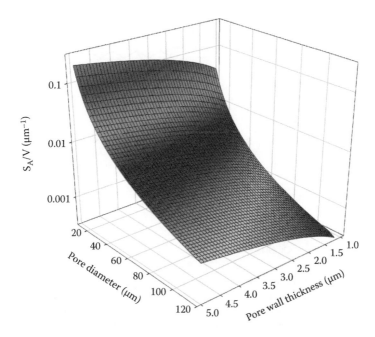

**FIGURE 3.11** Specific surface area of hydrogel as a function of the average pore diameter and wall thickness. (Adapted from Savina, I. N., Gun'ko, V. M., Turov, V. V., Dainiak, M., Phillips, G. J., Galaev, I. Yu. and Mikhalovsky, S. V., *Soft Matter* 7, 4276–4283, 2011. Reproduced by permission of The Royal Society of Chemistry.)

density. Potassium diperiodatocuprate, $K_5[Cu(HIO_6)_2]$, was used for initiating graft polymerization from the pore surface of acrylamide cryogel (AAm gel) (Savina et al., 2006a) (Figure 3.12). Different polymers were successfully grafted with a high degree of grafting (Savina et al., 2005, 2006a,b). Using water-organic solutions (water-DMSO) allows polymerization of water-insoluble monomers such as N-tertbutylacrylamide and glycidyl methacrylate (Savina et al., 2006a). The degree of the polymerization was dependent on many factors such as monomer nature, pH, polymerization time and monomer and initiator concentration. Increase in the monomer concentration results in a higher degree of grafting (Figure 3.13a). The pore structure of cryogel promotes grafting providing ample surface area, good mass transfer inside the matrix and facilitated washing out of the nonreacted reagents and homopolymer formed as a by-product of the grafting reaction.

The structure of the polymer grafted layer and the adsorption of low- and highmolecular weight substances were both found to be dependent on the mode of initiation of polymerization. When the graft polymerization was carried out by adding both initiator and monomer at the same time (one-step graft polymerization), this resulted in the formation of dense polymer brushes. The adsorption capacity for lowmolecular weight substances increased with increase in grafting degree. However, the adsorption of the proteins, large molecular weight compounds, was less affected (Figure 3.14a and b). This variation in the adsorption was explained by different

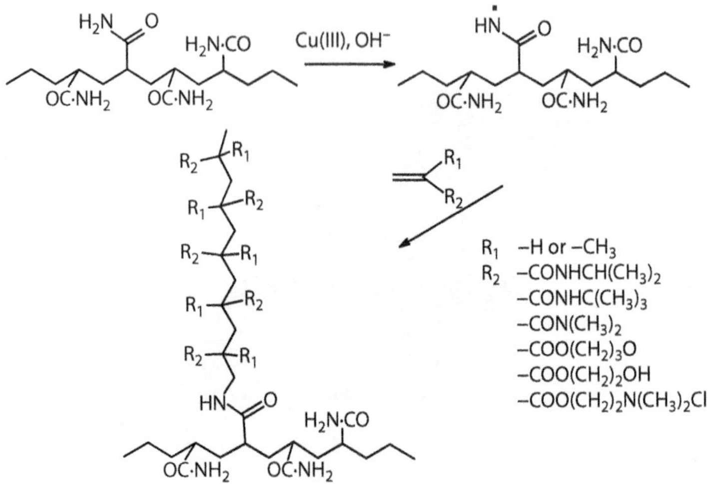

**FIGURE 3.12** Scheme of the graft polymerization of vinyl monomers from an acryl amide cryogel initiated by potassium diperiodatocuprate. (Savina, I. N., Mattiasson, B. and Galaev, I. Yu.: Graft polymerization of vinyl monomers inside macroporous polyacrylamide cryogel, in aqueous and aqueous-organic media initiated by diperiodatocuprate(III) complexes. *Journal of Polymer Science Polymer Chemistry.* 2006. 44. 1952–1963. Copyright Wiley-VCH Verlag GmbH & Co. KGaA. Reproduced with permission.)

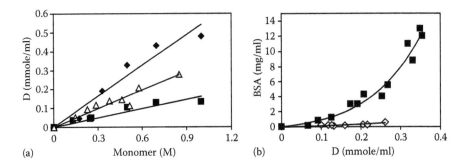

**FIGURE 3.13** (a) The effect of monomer concentration on the grafting density of ion-exchange polymers. Polyacrylate-grafted (closed rhombs), poly(N,N,-dimethylaminoethyl methacrylate) (pDMAEMA-grafted, open triangles) and (2-(methacryloyloxy)ethyl)-trimethyl ammonium chloride (pMETA-grafted, closed squares) pAAm cryogels. Acrylic acid and DMAEMA were grafted using the one-step graft polymerization and META was grafted using the two-step graft polymerization procedure, respectively (for one- and two-step polymerization description, please see text). (b) Binding of the BSA by pDMAEMA-grafted pAAm cryogels prepared by the one-step technique (open rhombs) and two-step technique (closed squares), respectively. The monolith of pDMAEMA-grafted pAAm cryogel were saturated with BSA solution (1 mg/ml in running buffer, 20 mM Tris-HCl buffer, pH 7.0), unbound BSA was washed off with the running buffer, and bound BSA was eluted with 1.5 M NaCl in the running buffer. (From Savina, I. N., Galaev, I. Yu. and Mattiasson, B.: Ion-exchange macroporous hydrophilic gel monolith with grafted polymer brushes. *Journal Molecular Recognition.* 2006. 19. 313–321. Copyright Wiley-VCH Verlag GmbH & Co. KGaA. Reproduced with permission.)

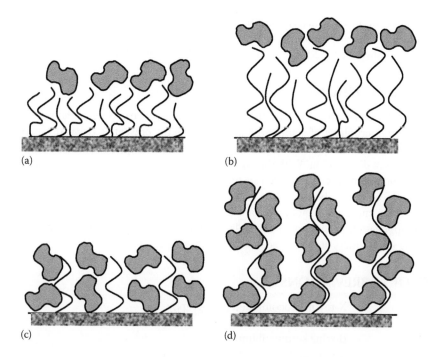

**FIGURE 3.14** Schematic illustration of the BSA adsorption on the grafted polymer surface of different morphology: (a) and (b) polymer was grafted by the one-step procedure and (c) and (d) two-step procedure. Explanation of BSA adsorption difference is in the text.

accessibility of ion-exchange groups. Ion-exchange groups inside the dense grafted layer are less accessible for large molecules, such as proteins, which could not penetrate inside the dense polymer layer (Figure 3.14a and b). The proteins adsorb only on the surface of the grafted polymer brush; thus, their adsorption is limited to the surface area and independent of the number of the ion-exchange groups or grafting degree. A different result was obtained when a two-step polymerization mode was applied. In the two-step approach, the initiator was added first and then partially replaced with a monomer. As a result, fewer radicals formed on the pore surface but polymerization proceeded to a greater extent with the formation of large polymer grafts. As initiator was removed from the system and replaced with monomer solution, the homopolymerization was suppressed and more monomer was available for the growth of grafted polymer. At the same grafting degree (moles of the monomer grafted per gram of cryogel), the structure of grafted layer was different. Apparently when the two-step mode was used, the grafted polymer chains were in less proximity to each other and more accessible for interacting with protein. This results in a so-called 'tentacle' adsorption of protein, when more than one molecule of protein could interact with one molecule of the grafted polymer (Figure 3.14c and d). As a result, the adsorption of the proteins on the grafted cryogels was considerably improved and allows achieveing the capacity 12 times higher compared to a one-step approach (Figure 3.14a and b).

Besides the protein adsorption, the cryogels with grafted polymers have shown the selectivity for adsorption and separation of plasmid DNA from RNA (Hanora et al., 2006; Srivastava et al., 2012) or cell separation (Srivastava et al., 2012).

## 3.7 CONCLUSIONS

This chapter demonstrates clearly that polymerization at subzero temperatures in the nonfrozen microphase of semifrozen aqueous systems presents a powerful method for the production of gels with different porosity, macroporous as well as with bimodal pore distribution. With the knowledge of the mechanisms governing ice formation and polymerization in such systems, the pore morphology of produced cryogels could be precisely controlled and fine-tuned by polymerization temperature, starting concentration of monomers and addition of solutes. The chemical nature and hence application properties of cryogels could be efficiently modified by chemical grafting of polymers at the surface of cryogel pore walls.

## LIST OF ABBREVIATIONS

| | |
|---|---|
| μ-CT | X-ray microcomputed tomography |
| AAm | Acrylamide |
| BSA | Bovine serum albumin |
| CLSM | Confocal laser scanning microscopy |
| DMAAm | Dimethylacrylamide diacrylate |
| DMMAm | Dimethylacrylamide |
| DMSO | Dimethyl sulfoxide |
| DNA | Deoxyribonucleic acid |
| DSC | Differential scanning calorimetry |
| HEMA | Hydroxyethyl methacrylate |
| MPM | Multiphoton microscopy |
| NMR | Nuclear magnetic resonance |
| pAAm | Poly(acrylamide) |
| pDMAEMA | Poly(N,N,-dimethylaminoethyl methacrylate) |
| PEG | Polyethylene glycol |
| pMETA | Poly(2-(methacryloyloxy)ethyl)-trimethyl ammonium chloride |
| PSD | Pore size distribution |
| RNA | Ribonucleic acid |
| TG | Thermogravimetry |
| TSD | Thermally stimulated depolarization |

## REFERENCES

Atkins, P. W., *Physical Chemistry*, 6th ed. Oxford: Oxford University Press, 2001.

Caykara, T. and Dogmus, M. (2004). The effect of solvent composition on swelling and shrinking properties of poly(acrylamide-co-itaconic acid) hydrogels. *European Polymer Journal* **40**(11): 2605–2609.

Enustun, B. V., Senturk, H. S. and Yurdakul, O. (1978). Capillary freezing and melting. *Journal of Colloid Interface Science* **65**(3): 509–516.

Gun'ko, V. M., Savina, I. N. and Mikhalovsky, S. V. (2013). Cryogels: Morphological, structural and adsorption characterisation. *Advances in Colloid and Interface Science* **187–188**: 1–46.

Gun'ko, V. M., Mikhalovska, L. I., Savina, I. N., Shevchenko, R. V., James, S. L., Tomlins, P. E. and Mikhalovsky, S. V. (2010). Characterisation and performance of hydrogel tissue scaffolds. *Soft Matter* **6**: 5351–5358.

Hanora, A., Savina, I., Plieva, F. M., Izumrudov, V. A., Mattiasson, B. and Galaev, I. Yu. (2006). Direct capture of plasmid DNA from non-clarified bacterial lysate using polycation-grafted monoliths. *Journal of Biotechnology* **123**(3): 343–355.

Kirsebom, H., Rata, G., Topgaard, D., Mattiasson, B. and Galaev, I. Yu. (2008). In situ $^1$H-NMR studies of free radical cryo-polymerization. *Polymer* **49**(18): 3855–3858.

Kirsebom, H., Rata, G., Topgaard, D., Mattiasson, B. and Galaev, I. Yu. (2009). Mechanism of cryo-polymerization: Diffusion-controlled polymerization in a non-frozen microphase. An NMR study. *Macromolecules* **42**(14): 5208–5214.

Kirsebom, H., Topgaard, D., Galaev, I. Yu. and Mattiasson, B. (2010). Modulating the porosity of cryogels by influencing the non-frozen liquid phase through addition of inert solutes. *Langmuir* **26**(20): 16129–16133.

Lam, P. and Moore, J. (2010). Freezing Biopharmaceutical Products. In: *Encyclopedia of Industrial Biotechnology: Bioprocess, Bioseparation, and Cell Technology*, pp. 2567–2580, John Wiley & Sons, Online ISBN: 9780470054581.

Okay, O. (2000). Macroporous copolymer networks. *Progress in Polymer Science* **25**(6): 711–779.

Ozmen, M. M. and Okay, O. (2008). Formation of macroporous poly(acrylamide) hydrogels in DMSO/water mixture: Transition from cryogelation to phase separation copolymerization. *Reactive and Functional Polymers* **68**(10): 1467–1475.

Savina, I. N., Mattiasson, B. and Galaev, I. Yu. (2005). Graft polymerization of acrylic acid onto macroporous polyacrylamide gel (cryogel) initiated by potassium diperiodatocuprate. *Polymer* **46**(23): 9596–9603.

Savina, I. N., Mattiasson, B. and Galaev, I. Yu. (2006a). Graft polymerization of vinyl monomers inside macroporous polyacrylamide gel, cryogel, in aqueous and aqueous-organic media initiated by diperiodatocuprate(III) complexes. *Journal of Polymer Science Polymer Chemistry* **44**(6): 1952–1963.

Savina, I. N., Galaev, I. Yu. and Mattiasson, B. (2006b). Ion-exchange macroporous hydrophilic gel monolith with grafted polymer brushes. *Journal of Molecular Recognition* **19**(4): 313–321.

Savina, I. N., Cnudde, V., D'Hollander, S., Van Hoorebeke, L., Mattiasson, B., Galaev, I. Yu. and Du Prez, F. (2007). Cryogels from poly(2-hydroxyethyl methacrylate): Macroporous, interconnected materials with potential as cell scaffolds. *Soft Matter* **3**(9): 1176–1184.

Savina, I. N., Gun'ko, V. M., Turov, V. V., Dainiak, M., Phillips, G. J., Galaev, I. Yu. and Mikhalovsky, S. V. (2011). Porous structure and water state in cross-linked polymer and protein cryo-hydrogels. *Soft Matter* **7**: 4276–4283.

Srivastava, A., Shakya, A. K. and Kumar, A. (2012). Boronate affinity chromatography of cells and biomacromolecules using cryogel matrices. *Enzyme and Microbial Technology* **51**(6–7): 373–381.

Talamucci, F. (2003). Freezing processes in porous media: Formation of ice lenses, swelling of the soil. *Mathematical and Computer Modelling* **37**(5–6): 595–602.

Valiullin, R. and Skirda, V. (2001). Time dependent self-diffusion coefficient of molecules in porous media. *The Journal of Chemical Physics* **114**(2001): 452–458.

Watanabe, K., Muto, Y. and Mizoguchi, M. (2001). Water and solute distributions near an ice lens in a glass-powder medium saturated with sodium chloride solution under unidirectional freezing. *Crystal Growth and Design* **1**(3): 207–211.

Yun, J. X., Jespersen, G. R., Kirsebom, H., Gustavsson, P. E., Mattiasson, B. and Galaev, I. Yu. (2011). An improved capillary model for describing the microstructure characteristics, fluid hydrodynamics and breakthrough performance of proteins in cryogel beds. *Journal of Chromatography* **1218**(32): 5487–5497.

# 4 Fabrication and Characterization of Cryogel Beads and Composite Monoliths

*Junxian Yun,\* Linhong Xu, Dong-Qiang Lin, Kejian Yao and Shan-Jing Yao\**

## CONTENTS

4.1 Introduction ........................................................................................... 112
4.2 Fabrication of Cryogel Beads by Micro-Flow Focusing and Cryo-Polymerization ..................................................................................... 114
    4.2.1 Microchannel Liquid-Flow Focusing and Cryo-Polymerization Method ....................................................................................... 114
    4.2.2 Fabrication of Microchannels for Fluid-Focusing ................... 116
    4.2.3 Generation of Aqueous Drops and Immiscible Slug Flow in Microchannels .................................................................... 118
    4.2.4 Freezing of Aqueous Drops in Water-Immiscible Fluid ......... 119
    4.2.5 Properties of Cryogel Beads .................................................. 120
4.3 Preparation of Composite Monolithic Cryogels via Cryo-Polymerization ... 124
    4.3.1 Preparation of Cryogels with Embedded Macroporous Beads ......... 124
    4.3.2 Grafting Polymerization of Composite Cryogels ............................ 125
    4.3.3 Properties of Composite Cryogels ............................................ 126
4.4 Modelling of Fluid Flow and Mass Transfer for the Characterization of Cryogel Beds ..................................................................................... 129
    4.4.1 Capillary-Based Models for Cryogel-Bead Packed Beds and Monolithic Cryogels ........................................................ 129
        4.4.1.1 Pore Tortuosity and Size Distributions ............................. 130
        4.4.1.2 Bed Porosity and Permeability ........................................ 130
        4.4.1.3 Mass Balance Equation in Mobile Fluid Phase under Nonadsorption Conditions ................................................. 131
        4.4.1.4 Axial Dispersion Coefficient ............................................ 132

---

\* Corresponding authors: E-mail: yunjx@zjut.edu.cn (J.X. Yun); yaosj@zju.edu.cn (S.-J. Yao); Tel: +86-571-88320951.

4.4.1.5 Mass Transfer Coefficient .................................................. 132
4.4.1.6 Adsorption Equilibrium Equation ................................... 132
4.4.2 Numerical Solution of the Model ...................................... 132
4.5 Applications of Cryogel Beads and Composite Cryogels .......... 135
4.6 Conclusions ............................................................................... 135
Acknowledgements ............................................................................ 136
List of Abbreviations .......................................................................... 136
References .......................................................................................... 137

## ABSTRACT

Cryogel beads and monolithic composite cryogels with macroporous particles incorporated in the gel skeleton have been prepared successfully in recent years. These novel particle-form and monolith-form cryogels have analogous chromatographic properties compared with traditional monolithic cryogels and thus are interesting in bioseparation fields. In this chapter, the fabrication of cryogel beads by the microchannel liquid-flow focusing and cryo-polymerization method and the preparation of composite monolithic cryogels by cryo-polymerization of an aqueous mixture suspension containing gel-forming reagents and macroporous beads have been discussed. The capillary-based models for the detailed description of the microstructures, the fluid flow characteristics, the axial dispersion behaviours as well as the mass transfer characteristics of biomolecules like proteins through the monolithic cryogel beds or the cryogel-bead packed beds are introduced. Some examples regarding the applications of these cryogels for the chromatographic adsorption of $\gamma$-globulin and isolation of immunoglobulin G and albumin proteins from serum are also discussed.

## KEYWORDS

Cryogel beads, monolithic composites, bioseparation, microchannel liquid flow focusing

## 4.1 INTRODUCTION

Cryogels are porous cross-linked polymers prepared by cryo-polymerization of an aqueous solution containing gel-forming reagents under freezing conditions. These interesting materials have micron-scale supermacropores and thus high porosity and permeability. Due to their intrinsic properties, cryogels have been suggested as new chromatographic adsorbents, immobilization matrices, cell scaffolds and drug delivery carriers for applications in the biotechnological, biomedical, pharmaceutical and even bioremediation fields (Gun'ko et al., 2013; Kirsebom et al., 2011; Kuyukina et al., 2013; Lozinsky, 2002, 2008; Lozinsky et al., 2003; Mattiasson et al., 2010; Plieva et al., 2007, 2008, 2011).

# Fabrication of Cryogel Beads and Composite Monoliths

Numerous monolithic-form cryogels with functional ligands or with embedded particles have been prepared successfully in laboratory-scale (e.g., Alkan et al., 2009, 2010; Arvidsson et al., 2003; Baydemir et al., 2009; Bereli et al., 2010; Gun'ko et al., 2013; Hajizadeh et al., 2013; Hanora et al., 2005; Jespersen et al., 2013; Jurga et al., 2011; Kirsebom et al., 2011, 2013; Koç et al., 2011; Kumar et al., 2006; Kuyukina et al., 2013; Lozinsky, 2002, 2008; Lozinsky et al., 2003; Mattiasson et al., 2010; Noir et al., 2007; Perçin et al., 2011; Plieva et al., 2006, 2007, 2008, 2011; Savina et al., 2005a,b, 2006; Srivastava et al., 2012; Ünlüer et al., 2013; Üzek et al., 2013; Uzun et al., 2013; Yao et al., 2006a,b, 2007). For industrial bioseparation applications, a monolithic cryogel needs to be fitted with a matched column only with the same diameter as the cryogel itself and thus sometimes the scale-up is not convenient.

Traditional particle-form adsorbents are popular and widely used in downstream processing. These adsorbents can be packed freely in different columns of whatever diameters. In recent years, particle-form cryogels, so called **cryogel beads** as exemplified in Figure 4.1, have been studied (Deng et al., 2013; Orakdogen et al., 2011; Partovinia and Naeimpoor, 2013; Tripathi et al., 2010; Tripathi and Kumar, 2011; Yun et al., 2012, 2013). By integrating cryogel beads or other macroporous polymer particles with monoliths, one can also produce new composite cryogels (Baydemir et al., 2009; Bereli et al., 2010; Hajizadeh et al., 2012, 2013; Koç et al., 2011; Noir et al., 2007; Sun et al., 2012; Ünlüer et al., 2013; Wang and Sun et al., 2013; Wang et al., 2013; Ye et al., 2013). These novel cryogel beads and composite cryogels have analogous chromatographic properties with cryogel monoliths and thus could be potentially applied for a variety of purposes in bioseparation areas. In this chapter, the preparation, properties, applications as well as mathematical modelling regarding these novel cryogels will be discussed.

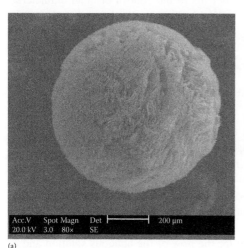

**FIGURE 4.1** (a) Scanning electron microscope (SEM) image and (b) photograph of polyacrylamide (pAAm) cryogel beads.

## 4.2 FABRICATION OF CRYOGEL BEADS BY MICRO-FLOW FOCUSING AND CRYO-POLYMERIZATION

Cryogel beads have particle-form features in morphology shapes, but similar supermacroporous structures within their matrices as those of monolithic cryogels. They can be prepared by freezing–thawing the aqueous droplets containing gel-forming monomers. Different from the pore-generation method using porogens like solvents (Gokmen and Du Prez, 2012; Gustavsson and Larsson, 1996; Tiainen et al., 2007), polymers and oligomers (Gokmen and Du Prez, 2012), solid particles (Du et al., 2010a; Ma et al., 2012; Shi et al., 2005) and ionic liquids (Du et al., 2010b; Tong et al., 2013), the generation of supermacropores within cryogel beads is achieved by the cryo-polymerization approach through solvent (water) crystallization (Deng et al., 2013; Orakdogen et al., 2011; Partovinia and Naeimpoor, 2013; Tripathi et al., 2010; Yun et al., 2012, 2013), similar as that for monolithic cryogels (Gun'ko et al., 2013; Kirsebom et al., 2011; Kuyukina et al., 2013; Lozinsky, 2002, 2008; Lozinsky et al., 2003; Mattiasson et al., 2010; Plieva et al., 2007, 2008, 2011).

The formation of cryogel beads always includes steps to generate aqueous droplets, freeze droplets to form the frozen-state solids, the polymerization of monomers within the solids to form gel matrices and the thawing of the ice crystals within the frozen solids to form supermacropores. Among them, the generation of droplets is crucial to the final diameter distribution of beads. In general, the aqueous droplets can be produced either by a mechanical approach using syringes (Orakdogen et al., 2011; Tripathi and Kumar, 2011) and capillary pipettes (Ariga et al., 1987; Szczęsna-Antczak et al., 2004) or by a liquid jet cutting (Prusse et al., 2000) and splintering (Martins et al., 2003) means. However, some of those techniques suffer from drawbacks such as complex devices or problems in the control of bead diameters. The preparation of cryogel beads with controllable diameters and a narrow diameter distribution is a challenging work.

### 4.2.1 Microchannel Liquid-Flow Focusing and Cryo-Polymerization Method

Recently, the microchannel liquid-flow focusing and cryo-polymerization method has been proposed for the preparation of cryogel beads with a relatively narrow diameter distribution (Deng et al., 2013; Yun et al., 2012, 2013). This novel method is achieved first by generating uniform aqueous droplets containing monomers through micro-flow focusing of the aqueous solution with water-immiscible fluid flow streams in a microchannel, followed by freezing–thawing of the droplets and monomer-polymerization for the formulation of cryogel beads, as schematically shown in Figure 4.2.

A microchannel with a main channel for the flow of the aqueous solution and a cross-junction connecting two branches for the injection of the water-immiscible fluid for flow-focusing is preferred. When the aqueous gel-forming solution is pumped into the main channel, the water-immiscible fluid is injected into the branch channels simultaneously. These immiscible liquid streams meet at the cross and thus

# Fabrication of Cryogel Beads and Composite Monoliths 115

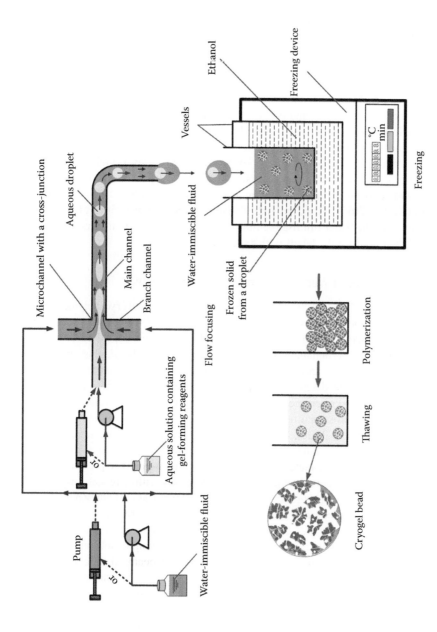

**FIGURE 4.2** A schematic presentation of the microchannel liquid-flow focusing and cryo-polymerization process for the formation of cryogel beads.

the liquid flow focusing occurs to generate uniform-size aqueous droplets. The suspension of the generated droplets flows continuously out the microchannel with the water-immiscible fluid and is frozen *in situ* rapidly in a cold immiscible fluid bulk to form solids. The solids are separated from the immiscible fluid and kept under frozen conditions for cryo-polymerization. During the freezing procedure, ice crystals are formed within the solids, while the unfrozen monomer solution is concentrated around the ice crystals and polymerized to form the gel matrix. After polymerization, the solids are thawed to form the supermacropores and finally the cryogel beads are obtained.

A variety of factors like the microchannel sizes, the flow velocities of fluids, the concentrations of monomers and cross-linkers, the physical properties of fluids, as well as the freezing conditions influence the diameter distribution of cryogel beads, and consequently other physical properties and the chromatographic performance of cryogel-bead beds. However, for certain aqueous and water-immiscible fluids in a given microchannel system under a constant freezing condition, the diameters of cryogel beads depend mainly on the flow hydrodynamics within the microchannel (Yun et al., 2012, 2013). Thus, by simply changing the flow velocities of the aqueous solution and the water-immiscible fluid, one can adjust the diameters of cryogel beads and then produce the beads with expected diameters.

### 4.2.2 Fabrication of Microchannels for Fluid-Focusing

The microchannel is the key device for the micro-flow focusing and cryo-polymerization process, and the fabrication of microchannels with suitable structures is of significant importance. Actually, over the past 20 years there has been considerable interest in the fabrication of microchannels on various materials like silicon, glass, polymers and metals (Ehrfeld et al., 2000). Several effective techniques for micromachining such as electrochemical discharge, high density plasmas, laser, microlithography, injection moulding and micromechanical machining have been exploited and many microchannels or microstructures on different materials have been fabricated successfully (Ehrfeld et al., 2000). Owing to the inert properties of the aqueous solution and water-immiscible fluid as well as their advantages in scale-up for industrial applications, stainless steel materials have received attention as the preferred material for fabrication of microchannels for the micro-flow focusing not only in the production of nano-sized particles (Xu et al., 2012; Yun et al., 2009b; Zhang et al., 2008), but also in the preparation of cryogel beads by the microchannel liquid-flow focusing and cryo-polymerization method (Deng et al., 2013; Yun et al., 2012, 2013; Zhao et al., 2014). However, the precise manufacture of microchannels with repeated sizes and satisfied channel roughness on a stainless steel material by conventional machining techniques is not convenient in some cases and not easy due to limitations like the complex devices used, the excellent skills needed for the operators and the unsatisfied channel sizes and roughness.

The mechanical micro-cutting technique is one effective route for the manufacture of a microchannel on a stainless steel plate or slab. This method utilizes a micro-milling cutter with the diameter of only several hundreds of microns and the fabrication process is achieved in a high-speed computerized control engraving and milling machine (Xu et al., 2012). The quality of the inner channel surfaces and sizes depends on the milling machine and the manufacturing parameters. For an austenitic stainless steel slab, we observed that the cutting speed of 15,000–20,000 rpm, the feed rate of 300–600 mm/min and the cutting depth in each step of 0.005–0.04 mm are preferred by the mechanical micro-cutting technique with micro-milling cutters of diameters about 200–400 μm. Further increasing the spindle speed will give a positive contribution to the improvement of the surface quality of microchannels. An excellent surface quality within the microchannel also can be achieved by the fibre oil-stone polishing process. Figure 4.3 shows examples of microchannels manufactured by the mechanical micro-cutting technique, which can be used in the preparation of cryogel beads.

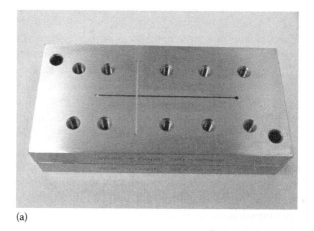

(a)

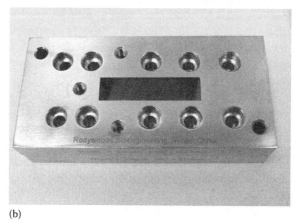

(b)

**FIGURE 4.3** Microchannels for the preparation of cryogel beads by the liquid-flow focusing and cryo-polymerization method. The microchannels (a) and the cover plates (b) were manufactured with the mechanical micro-cutting technique by Wuhan Redywoods Bioengineering Co. Ltd. (Wuhan, China).

### 4.2.3 Generation of Aqueous Drops and Immiscible Slug Flow in Microchannels

Immiscible fluid flow in microchannels or microreactors has been studied by numerous investigators (Günther and Jensen, 2006) and various flow patterns have been observed, such as bubble or droplet flow, segmented or slug flow, liquid ring flow, churn flow, annular flow, rivulet flow and liquid lump flow (Thome et al., 2013; Timung and Mandal, 2013). For the liquid-flow focusing and cryopolymerization process, expected flow pattern is liquid-liquid slug flow, which is characterized by bullet-shaped droplet slugs with surrounding thin immiscible liquid film and segregated by immiscible liquid slugs (Yun et al., 2012, 2013; Zhao et al., 2014).

The formation of uniform-sized droplets is the key step to produce cryogel beads with satisfied diameters. In order to get cryogel beads with an expected narrow diameter distribution, the formation mechanisms of droplets and the hydrodynamics of immiscible liquid-liquid slug flow in microchannels needs to be revealed. Actually, the interfacial force and shear stress play important roles in the formation of droplets for liquid-liquid slug flow within microchannels. In the case that the shear stress predominates, the formation of droplets follows the dripping mechanism due to the different flow rates of immiscible liquids, while in the case that the interfacial force dominates, a local increase of pressure upstream of an emerging droplet leads to the breakup of immiscible threads and the formation of droplets and then follows the squeezing mechanism (Garstecki et al., 2006).

The flow parameters of immiscible liquid-liquid slug flow in microchannels include the aqueous slug or droplet velocity, aqueous slug length, water-immiscible slug length and thickness of the liquid film surrounding the aqueous slugs. Similar to those for the gas-liquid immiscible slug flow, the hydrodynamic parameters of immiscible liquid-liquid slug flow can be correlated to some dimensionless numbers like the Weber number, capillary number, the velocity ratio of immiscible fluids and fluid hold-up fraction (Yun et al., 2010; Zhao et al., 2014). In a rectangle channel with a cross junction for micro-flow focusing, the aqueous slug velocity $U_{TB}$ is a linear function of the total velocity of the aqueous phase $U_w$ and water-immiscible phase $U_{Im}$ in the main channel.

$$U_{TB} = \frac{A}{A_{TB}}(U_w + U_{Im}) \tag{4.1}$$

For the water-immiscible phase of $n$-hexane containing Span 80 and the aqueous phase containing the monomer hydroxyethylmethacrylate and the cross-linker poly(ethylene glycol) diacrylate in the preparation of poly(2-hydroxyethylmethacrylate) (pHEMA) cryogel beads, the cross-area ratio of the aqueous slugs $A_{TB}$ to the main channel $A$ is about 0.92 (Zhao et al., 2014), which is very close to that for the gas-liquid of nitrogen gas and diluted aqueous solution (Yun et al., 2009b, 2010).

The aqueous slug length $L_{TB}$ depends on the velocity ratio of two immiscible phases, the channel geometry (channel depth and width) and capillary number of the continuous phase (Zhao et al., 2014).

$$\frac{L_{TB}}{D_h} = \left(\alpha + \beta \frac{U_w}{U_{Im}}\right) Ca_{Im}^{\gamma} \quad (4.2)$$

where $\alpha$, $\beta$ and $\gamma$ are constants, $D_h$ is the hydraulic diameter of the rectangular microchannel and $Ca_{Im}$ is the capillary number based on the viscosity and surface tension of the water-immiscible phase. The lengths of the aqueous slug nose $L_{TBnose}$ or tail $L_{TBtail}$ to the whole aqueous slug decreased exponentially with the increase of the ratio of the aqueous phase velocity to the overall velocity, similar to those for the gas-miscible liquids system in our previous work (Yun et al., 2010):

$$\frac{L_{TBnose}}{L_{TB}} = a \exp\left(b \frac{U_w}{U_T}\right), \quad \frac{L_{TBtail}}{L_{TB}} = c \exp\left(d \frac{U_w}{U_T}\right) \quad (4.3)$$

where $a$, $b$, $c$ and $d$ are constants for a given immiscible fluid in a certain microchannel.

The water-immiscible slug length $L_{LS}$ is a complicated function of the Reynolds number, Eötvös number, capillary number and the fluid hold-up fraction. In a rectangle microchannel, the ratio of $L_{LS}$ to $L_{TB}$ can be correlated to a function of phase hold-up for the gas-liquid of nitrogen gas and diluted aqueous solution (Yun et al., 2010), or the velocity ratio of two immiscible phases for the water-immiscible phase of n-hexane and the aqueous phase of hydroxyethylmethacrylate and poly(ethylene glycol) diacrylate (Zhao et al., 2014). For a given immiscible liquid-liquid system in a rectangle microchannel, it is possible to estimate the cryogel beads formed from the diameters of the generated droplets.

### 4.2.4 Freezing of Aqueous Drops in Water-Immiscible Fluid

The freezing of aqueous droplets is crucial to the formation of both the gel matrices and the supermacropores within cryogel beads. In actual cases, the immiscible fluids for the micro-flow focusing and cryo-polymerization contain various solutes such as monomers, cross-linkers, catalysts and even a small amount of surfactants. Some of these substances can be dissolved in both aqueous solutions and water-immiscible fluid. The transfer of these solutes from the aqueous phase to the water-immiscible liquids or vice versa could cause only the partial polymerization of the monomers or even the failure of polymerization to form the cryogel beads. Therefore, in order to prevent the mass transfer of solutes between the two liquids, a rapid generation of aqueous droplets in the microchannel and a quick formation of frozen-state solid particles within the immiscible liquid bulk are necessary.

The freezing machines are convenient in most laboratories and they work well in the generation of freezing conditions for the microchannel liquid-flow focusing and cryo-polymerization process. When the freezing temperature for the formation of frozen solid beads is kept at about −20°C to −30°C, polyacrylamide (pAAm) and pHEMA cryogel beads can be prepared successfully (Deng et al., 2013; Yun et al., 2012). However, with the increase of the particle concentration of the frozen beads in the bulk of the water immiscible fluid, some aqueous drops could aggregate together

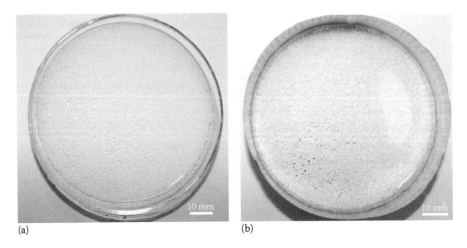
(a) (b)

**FIGURE 4.4** The pAAm cryogel beads by flow-focusing and cryo-polymerization method using a cooling machine at about −23°C (a), and dry ice at about −58°C (b), as the freezing resource, respectively.

or even attach with the frozen solids before being frozen when they flow out the microchannel and drop into the immiscible bulk phase. The reason is that the freezing process has always been insufficiently rapid in this temperature range (Yun et al., 2013). Obviously, this obstacle could induce variations in the diameter distribution of the obtained cryogel beads and, therefore, a large volume ratio of bulk water–immiscible fluid to the particles is always needed. In order to overcome this problem, a rapid freezing procedure using dry ice as the cooling generator instead of the freezing machines has been developed and demonstrated in the preparation of pAAm cryogel beads (Yun et al., 2013). Dry ice can induce a bulk temperature as low as −55°C to −61°C, which is enough for the rapid formation of frozen solids when the gel-forming aqueous drops drip into the water-immiscible bulk. The time to form frozen-state beads could be as short as ~3 seconds. Figure 4.4 shows examples of pAAm cryogel beads prepared by the microchannel liquid-flow focusing and cryo-polymerization method using a cooling machine at about −23°C (Figure 4.4a) and using dry ice as the freezing resource at about −58°C first, followed by keeping at about the same temperature as that of cooling machine. The fluids were pumped by syringes and peristaltic pumps, respectively.

### 4.2.5 Properties of Cryogel Beads

The diameters of cryogel beads depend on the aqueous drop sizes generated in the microchannels and actually, the formation of aqueous drops is a flow rate-controlled process. For a given liquid system in certain microchannels, the bead diameters were mainly influenced by the flow velocities of the aqueous solution and the water-immiscible fluid. In general, increasing the water-immiscible velocity causes the formation of aqueous drops with smaller sizes and thus the decrease of the bead diameters, while increasing the aqueous phase velocity results in an increase of the bead diameters (Yun et al., 2012, 2013). Therefore, by adjusting the flow velocities, the cryogel beads with expected diameters can be prepared. pAAm and pHEMA cryogel beads with diameters in the range of about 500–2000 μm with relatively narrow size distributions have been prepared at aqueous phase velocities from 0.5 to 2.0 cm/min and the total velocities of the water-immiscible phase from 2.0 to 6.5 cm/min (Deng et al., 2013; Yun et al., 2012, 2013). The polydispersity index is in the range of about 0.15–0.22.

Similar to monolithic cryogels, wet cryogel beads have elastic properties and can restore their original shapes. Figure 4.5 shows an example of the restoring of pAAm beads when the dried beads contact with water. When the dried hydrophilic beads are immersed into water (Figure 4.5a), they can restore their original shapes in only several seconds (Figure 4.5b).

Cryogel beads also have similar sponge-like porous structures as those in monolithic cryogels. Figure 4.6 shows an example image of pAAm cryogel beads observed by scanning electron microscope (SEM). In general, the successful fabrication of cryogel beads depends not only on the microchannels, the flow-focusing conditions,

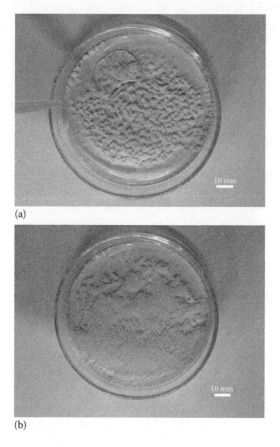

**FIGURE 4.5** The restoring characteristics of pAAm cryogel beads at the beginning (a) and after several seconds of contacting water (b).

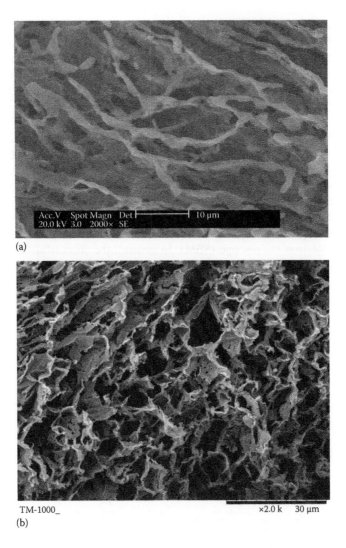

**FIGURE 4.6** Supermacopores within pAAm cryogel beads observed by SEM of XL-30-ESEM (a) or TM-1000 (b).

the freezing strategies and the properties of water-immiscible fluid, but also on the monomer concentration of the aqueous solution. In order to prepare cryogel beads with a satisfied mechanical strength, the monomer concentration for the preparation of cryogel beads is always higher than that for the monolithic cryogels. For example, in the preparation of pAAm monolithic cryogels, the satisfied concentration of monomer and cross-linker is about 7%, while for the cryogel beads, it is recommended to be 9%, which is higher than that for the monoliths (Yun et al., 2012, 2013). In the case that the same monomer concentration is used for both monolithic and bead cryogels, the obtained cryogel beads would sometimes have a low mechanical strength compared with that of monolith because the beads are so small compared with those of monoliths. The freezing temperature employed in the microchannel liquid-flow focusing and cryo-polymerization is always lower than that in the cryo-polymerization for monoliths as mentioned above, which could result in slightly smaller pores within cryogel beads. The pore sizes in monolithic cryogels are in the range from 10 to several hundred microns, while those values are about 3–50 microns in pAAm cryogel beads (Yun et al., 2012) and 10–40 microns in pHEMA cryogel beads (Deng et al., 2013), respectively. These beads have effective porosity of 82–89%, that is, similar high porosity as those of monolithic cryogels.

Cryogel beads can be packed in a column and thus used for bioseparation purposes. The properties of cryogel-bead packed beds such as the fluid permeability, the axial dispersion and the chromatographic adsorption capacity of biomolecules after suitable functionalization are fundamental for an actual separation application (Deng et al., 2013; Yun et al., 2012, 2013). These properties depend not only on the physical properties of cryogel beads, but also on the packing of the bed. Due to the good elastic property, cryogel beads can be packed easily in a given column layer by layer in a manual manner. Figure 4.7 shows examples of cryogel-bead packed beds with pAAm cryogel beads. A tight bed can be obtained by suitable packing though slightly compressing each layer during the packing procedure to decrease those large inter-bead voids as much as possible. In the case that a weak dispersion and high adsorption capacity per bed volume is expected, a tight packing is preferred, while in the situation where a primary adsorption purpose from a crude feedstock containing microbial cells or cell debris, a soft packing is also possible. However, additional

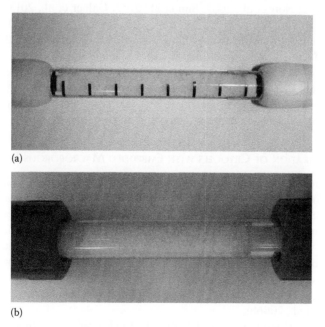

**FIGURE 4.7** The packed beds of pAAm cryogel beads by liquid-flow focusing and cryo-polymerization method using a cooling machine at about −23°C (a), and dry ice at about −58°C (b), as the freezing resource, respectively.

over-tight packing could also distort the intra-bead pores and cause the compression of the bed, which consequently causes a decrease of the fluid permeability.

Cryogel-bead packed beds have a dual-void system, that is, large inter-bead voids with sizes from several hundreds to thousands of microns, and intra-bead supermacropores with sizes from several to 50 microns. Therefore, the cryogel-bead beds always have high bed voidage and high water permeability. For example, packed beds with pAAm cryogel beads have bed voidages of 81–90% and water permeabilities of $10^{-12}$–$10^{-10}$ m$^2$ (Yun et al., 2012, 2013). The axial dispersion coefficient is in the range of $10^{-7}$–$10^{-4}$ m$^2$/s when the velocity is increased from 0.5 to 15 cm/min. These cryogel-bead beds have similar adsorption capacities and chromatographic performance as those observed in monolithic cryogels. Therefore, cryogel beads could be an interesting adsorbent with supermacropores as monolithic cryogels.

## 4.3 PREPARATION OF COMPOSITE MONOLITHIC CRYOGELS VIA CRYO-POLYMERIZATION

Composite cryogels are novel monolithic adsorbents prepared by cryo-polymerization of an aqueous mixture suspension of gel-forming reagents and particles or beads under freezing conditions. The preferred solids can be nano-sized particles (e.g., $Fe_3O_4$ and $SiO_2$), porous polymer beads (e.g., agarose, poly(glycidyl methacrylate-ethylene glycol dimethacrylate) and other molecularly imprinted polymers) and supermacroporous cryogel beads (Baydemir et al., 2009; Bereli et al., 2010; Hajizadeh et al., 2013; Koç et al., 2011; Noir et al., 2007; Sun et al., 2012; Ünlüer et al., 2013; Wang and Sun et al., 2013; Wang et al., 2013; Yao et al., 2006a,b; Ye et al., 2013). The purpose of incorporating solid particles or polymer beads into the matrices of the cryogels includes enhancing the adsorption capacity of the biomolecules, improving mechanical strength as well as introducing certain functions for removing substances and isolating targets by affinity or a molecularly imprinted mechanism from complex resources. These composite cryogels have different properties in gel skeleton, pore system and active sites, and thus are interesting in separation areas.

### 4.3.1 Preparation of Cryogels with Embedded Macroporous Beads

The embedding of inorganic particles without pores or those beads with nanoscale pores in cryogels is always achieved by directly freezing the mixture suspension containing reactive gel-forming monomers and solids, as a similar procedure as that for usual monolith cryogels. The formation of those composite monoliths is a complex process combining water crystallization, particles embedding and monomer polymerization. During this process, the heat of the mixture suspension is removed through the column wall by the freezing media. For different suspensions, the freezing strategy could be varied and the optimization of the preparation conditions is needed (Yao et al., 2006b).

Incorporating macroporous beads with microscale pores into a monolith cryogel is a new route to produce composite cryogels. The macroporous beads can be fabricated by either traditional methods like suspension polymerization or the microflow focusing approach. Recently, macroporous cellulose beads have been prepared

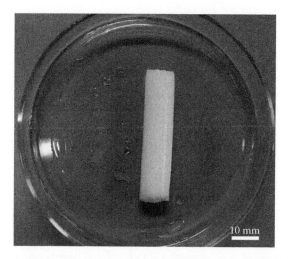

**FIGURE 4.8** A pHEMA monolithic cryogel embedded with cellulose beads.

successfully using the micro-flow focusing method. In this procedure, microcrystalline cellulose can be dissolved into an ionic liquid as one phase, which can be segmented to drops by flow-focusing with an immiscible oil-phase in microchannels (Tong et al., 2013; Ye et al., 2013). These drops were *in situ* suspended into the oil-phase bulk at subzero temperature. The frozen beads were filtered and washed with suitable solvents for thawing and removing solvents to get macroporous cellulose beads. The preparation of composite cryogels with embedded macroporous beads can be conducted in a similar way as those for embedding particles without pores. However, for hydrophilic macroporous beads like cellulose, the aqueous monomer solution can penetrate into the pores of beads when they are mixed with a gel-forming solution and thus the complicated cryo-polymerization could occur. Therefore, the beads always need to be saturated completely with the aqueous solution containing gel-forming monomers before the suspension is frozen and the cryo-polymerization is induced by the initiators. Following this procedure, a composite monolithic pHEMA cryogel embedding with macroporous cellulose beads can be prepared successfully (Ye et al., 2013). Figure 4.8 shows a composite pHEMA cryogel embedded with cellulose beads.

### 4.3.2 Grafting Polymerization of Composite Cryogels

As for other monolithic cryogels, functionalization of a composite cryogel can be achieved either by coupling the functional ligands directly on the beads before the cryo-polymerization (Baydemir et al., 2009; Bereli et al., 2010; Hajizadeh et al., 2013; Koç et al., 2011; Noir et al., 2007; Ünlüer et al., 2013; Wang and Sun et al., 2013; Wang et al., 2013) or grafting suitable ligands after the cryo-polymerization (Ye et al., 2013). In the former case, the active functional groups are expected only in the beads, while for the latter case, the groups are distributed in both the cryogel skeleton and the bead matrix. For the pHEMA or pAAm composite cryogels embedded with

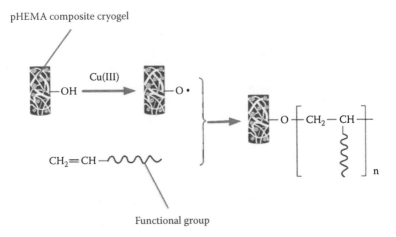

**FIGURE 4.9** Grafting of vinyl monomers with functional ligands onto the pHEMA composite cryogel embedded with cellulose beads.

cellulose beads in the latter case, graft polymerization can be achieved using vinyl monomers containing the desired functional groups like anion or cation exchange, hydrophobic and even mix-mode groups. The graft can be initiated by potassium diperiodatocuprate, which can induce the formation of active or radical sites on the matrices with $-C(=O)NH_2$ or $-OH$ groups and thus the effective coupling of the functional ligands onto the gel matrices (Ye et al., 2013; Deng et al., 2013). Figure 4.9 shows schematically the mechanism of grafting vinyl monomers with functional ligands onto a pHEMA composite cryogel initiated by potassium diperiodatocuprate and alkaline (Ye et al., 2013). A similar mechanism could be expected for the grafting of monomers onto pAAm-based composite cryogels.

### 4.3.3 Properties of Composite Cryogels

Composite cryogels have multiscale pores with sizes from nanoscale to about 50–100 μm, depending on the preparation procedure, the gel-formation system and the embedded beads. For composite monolithic cryogels embedded with cryogel beads, the formation of a pore system is actually complex because the gel-forming solution within the beads forms a new gel matrix within pore voids. During the cryogel fabrication procedure, cryo-copolymerization occurs not only in the bulk monomer solution outside the beads, but also within the original macropores of the beads. The solvent (water) crystallization of the gel-forming solution within the pores of the beads could generate much smaller pores with sizes from sub-microns to about several microns, while the water crystallization of the reactive solution outside the beads could induce the formation of supermacropores with sizes from several to 50–100 μm, as exemplified in Figure 4.10. Those small pores can provide big functional inner surfaces for binding the biomolecules, while these supermacropores outside the beads can permit the easy pass-flowing of crude fluids and free passage of microbial cells or cell debris. The gel skeleton covering the beads has numerous

# Fabrication of Cryogel Beads and Composite Monoliths

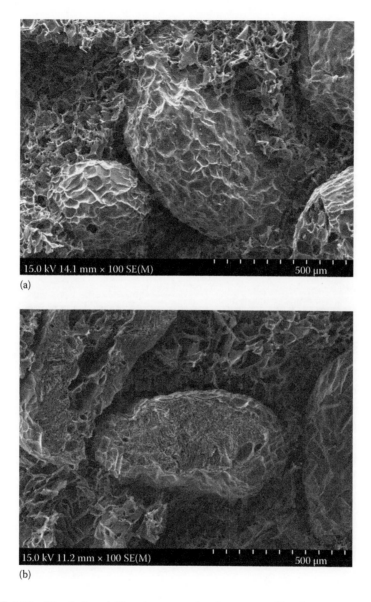

**FIGURE 4.10** Morphology (a,b) and cross-sectional structure of beads. (*Continued*)

interconnected pores with size of about 50–100 nm, which distribute on the external surface of the beads and can serve as an obstacle for the migration of cells or cell debris into the beads. Therefore, this cryogel is favourable as a new separation media in the isolation of biomolecules from fermentation broths or other crude feedstocks (Ye et al., 2013).

The composite cryogels possess high permeabilities close to those of monolithic cryogels as well as high porosities. For example, the composite anion-exchange

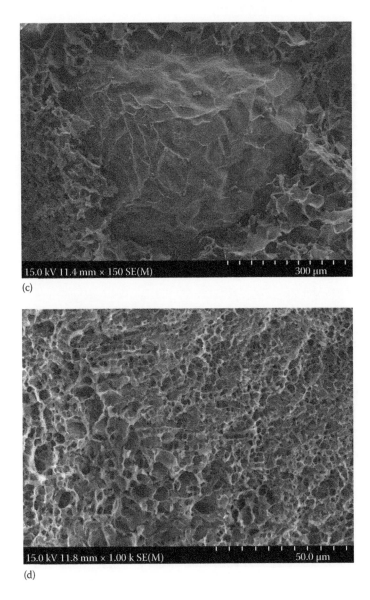

**FIGURE 4.10 (CONTINUED)** Morphology (c) and cross-sectional structure of beads (d) in a composite pHEMA cryogel embedded with cellulose beads by SEM.

pHEMA cryogel embedded cellulose beads has a water permeability of $8.9 \times 10^{-13}$ m², porosity of 80.1–90.4% and adsorption capacity of the cryogel for bovine serum albumin of 1.1 mg/mL cryogel bed even at the high flow velocity of 4.3 cm/min. The axial dispersion coefficients are in the range of $10^{-7}$–$10^{-5}$ m²/s for liquid flow velocity increases from 0.8 to 6.6 cm/min, indicating the liquid dispersion in the cryogel bed is not strong (Ye et al., 2013).

## 4.4 MODELLING OF FLUID FLOW AND MASS TRANSFER FOR THE CHARACTERIZATION OF CRYOGEL BEDS

Different from those traditional chromatographic beds, the pore sizes in cryogels are in the range from several to hundreds of microns, which are much larger than are those within the ordinary packed beds or monolithic beds and, thus, convection flow and mass transfer always occur. The microstructures, the fluid flow characteristics, the axial dispersion behaviours as well as the mass transfer characteristics of biomolecules like proteins through supermacroporous cryogel beds of monoliths or cryogel-bead packed beds can be described by so-called capillary-based models (Persson et al., 2004; Yun et al., 2009a, 2011, 2013). These models are used because the cryogels are porous with interconnected pores surrounded by density polymer wall.

### 4.4.1 Capillary-Based Models for Cryogel-Bead Packed Beds and Monolithic Cryogels

Cryogel-bead packed beds and monolithic cryogels have similar supermacroporous structures and thus similar fluid flow and mass transfer characteristics. Therefore, it is always assumed that a monolithic cryogel bed or a cryogel-bead packed bed can be represented by a series of tortuous capillaries with gel skeleton. The interconnectivity among capillaries was ignored. In order to simplify the model, the polymer skeleton outside each capillary is assumed to have a constant thickness. To simplify purposes, the capillaries can be assumed to have constant thickness, but with different diameters, as schematically shown in Figure 4.11 (Yun et al., 2009a, 2011, 2013).

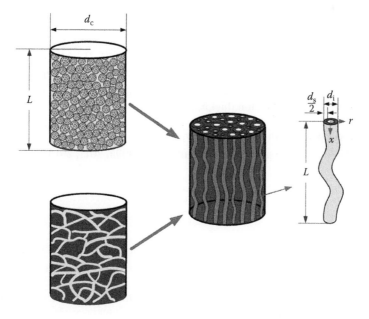

**FIGURE 4.11** Capillary pore structure in cryogel beads.

### 4.4.1.1 Pore Tortuosity and Size Distributions

The diameters of these capillaries can be described by a normal distribution with the following probability density (Yun et al., 2005, 2009a, 2011, 2013):

$$f(d_i) = \frac{\frac{1}{\sqrt{2\pi}\sigma}\exp\left[-\frac{(d_i-d_m)^2}{2\sigma^2}\right]}{1-\int_{-\infty}^{d_{min}}\frac{1}{\sqrt{2\pi}\sigma}\exp\left[-\frac{(d_i-d_m)^2}{2\sigma^2}\right]\delta d_i - \int_{d_{max}}^{+\infty}\frac{1}{\sqrt{2\pi}\sigma}\exp\left[-\frac{(d_i-d_m)^2}{2\sigma^2}\right]\delta d_i} \quad (4.4)$$

where $d_i$ is the diameter of capillary i, $\sigma$ is the standard deviation, $d_{max}$ is the maximum capillary diameter, $d_{min}$ is the minimum capillary diameter and $d_m$ is the mean diameter of capillaries in the cryogel bed.

For a given capillary i, the tortuous $\tau_i$ is defined as

$$\tau_i = \frac{L_i}{L} \quad (4.5)$$

where $L_i$ and $L$ are the capillary length and the cryogel height, respectively. The tortuosity can be assumed as a linear function of the capillary diameter (Yun et al., 2009a, 2011):

$$\tau_i = \tau_{dmin} + \frac{(d_i - d_{min})}{(d_{max} - d_{min})}\left(\sqrt{\frac{t_{dmax}U_L}{32kL}}d_{max} - \tau_{dmin}\right) \quad (4.6)$$

where $U_L = Q/A$ is the liquid flow velocity in the cryogel bed with the flow rate $Q$, and $\tau_{dmin}$ ($\geq 1$) the tortuosity of the capillary with diameter $d_{min}$, which can be determined $\tau_{dmin}$ from the residence time test. In a given group, capillaries have the same diameter, tortuosity and skeleton thickness.

### 4.4.1.2 Bed Porosity and Permeability

The cryogel bed porosity $\varphi$ is given by (Yun et al., 2009a, 2011, 2013)

$$\varphi = \frac{\pi}{4A}\sum_{i=1}^{N_g}n_i d_i^2 \tau_i \quad (4.7)$$

where $A$ ($= \pi d_c^2/4$, $d_c$ is the cryogel diameter) is the cross-area of the cryogel bed, $N_g$ is the total number of capillary groups with the same diameter and $n_i$ is the number of capillaries in group i.

The bed volume is determined by the total volume of capillaries and skeleton walls and therefore (Yun et al., 2009a, 2011, 2013)

$$\frac{\pi}{4A}\sum_{i=1}^{N_g} n_i(d_i+d_s)^2 \tau_i = 1 \qquad (4.8)$$

The flow rate in the cryogel bed can be given by combining the Hagen-Poiseuille equation in each capillary and Darcy's equation across the whole cryogel bed and thus the following equation is obtained (Yun et al., 2009a, 2011, 2013):

$$\frac{\pi}{128kA}\sum_{i=1}^{N_g}\frac{n_i d_i^4}{\tau_i} = 1 \qquad (4.9)$$

where $k$ is the fluid permeability of the cryogel bed.

### 4.4.1.3 Mass Balance Equation in Mobile Fluid Phase under Nonadsorption Conditions

For the breakthrough process of biomolecules in a cryogel bed under nonadsorption conditions, the differential mass balance equation of biomolecules in the mobile fluid phase of capillary i can be written as (Persson et al., 2004; Yun et al., 2009a, 2011, 2013)

$$\frac{\partial C_{Di}(x_D,t_{Di})}{\partial t_{Di}} = \frac{1}{Pe_i}\frac{\partial^2 C_{Di}(x_D,t_{Di})}{\partial x_D^2} - \frac{\partial C_{Di}(x_D,t_{Di})}{\partial x_D} \qquad (4.10)$$

with the following initial and boundary conditions

$$C_{Di}(x_D,t_{Di})\Big|_{t_{Di}=0, x_D>0} = 0 \qquad (4.11)$$

$$C_{Di}(x_D,t_{Di})\Big|_{x_D=0} = 1 \qquad (4.12)$$

$$\frac{\partial C_{Di}(x_D,t_{Di})}{\partial x_D}\Big|_{x_D=1} = 0 \qquad (4.13)$$

where $C_{Di}(x_D, t_{Di})$ is the dimensionless bulk-phase concentration of biomolecules in the capillary i, $x_D$ is the dimensionless distance from the inlet along the capillary length, $t_{Di}$ is the dimensionless time and $Pe_i$ is the axial Peclet number. The dimensionless variables are defined by $C_{Di} = C_i(x, t)/C_0$, $x_D = x/\tau_i L$, $t_{Di} = tU_i/\tau_i L$ and $Pe_i = \tau_i L U_i/D_{axi}$, respectively. In these expressions, $C_i(x, t)$ is the bulk-phase concentration, $C_0$ is the inlet concentration, $x$ is the distance, $t$ is the time, $U_i = U_L d_i^2/32k\tau_i$ is the velocity and $D_{axi}$ is the axial liquid dispersion coefficient in capillary i.

#### 4.4.1.4 Axial Dispersion Coefficient

The axial dispersion coefficient $D_{axi}$ in capillary i can be estimated using the correlation in fixed beds (Gutsche and Bunke, 2008) based on the Taylor-Aris correlation and written as (Yun et al., 2011, 2013)

$$D_{axi} = \frac{D_{AB}}{\tau_i} + \frac{1}{\psi}\frac{U_i^2 d_i^2}{192 D_{AB}} \tag{4.14}$$

where $D_{AB}$ is the molecular diffusion coefficient of biomolecules and the parameter $\psi = 0.018 Pe_{ABi}^{0.775}$ with the molecule Peclet number in capillary i $Pe_{ABi} = d_i U_i / D_{AB}$.

#### 4.4.1.5 Mass Transfer Coefficient

For the breakthrough process of biomolecules in a cryogel bed under adsorption conditions, the differential mass balance equation can be extended by modifying Equation 4.10 and complex differential equations need to be developed. In that case, the mass transfer coefficient in capillary i can also be calculated by (Seguin et al., 1996; Yun et al., 2009a)

$$k_{fi} = \gamma \frac{D_{AB}}{d_i}\left(Re_i Sc_L \frac{d_i}{\tau_i L}\right)^{\frac{1}{3}} \tag{4.15}$$

where $\gamma$ is a constant depending on the capillary system within the cryogel and can be determined by fitting the model with experimental breakthrough data under adsorption conditions, $Re_i = \rho_L U_i d_i / \mu_L$, the Reynolds number in capillary i with the liquid density $\rho_L$ and viscosity $\mu_L$, $Sc_L = \rho_L \mu_L / D_{AB}$ the Schmidt number of liquid in capillary i.

#### 4.4.1.6 Adsorption Equilibrium Equation

Langmuir isotherm equilibrium adsorption is assumed for the binding of biomolecules within capillary i (Persson et al., 2004; Yun et al., 2009a):

$$q_i = \frac{q_{max} C_{fi}(x,t)}{K_d + C_{fi}(x,t)} \tag{4.16}$$

where $q_{max}$ is the adsorption capacity and $K_d$ is the dissociation constant.

### 4.4.2 NUMERICAL SOLUTION OF THE MODEL

The mass balance equation in the model can be solved by the finite difference method in a similar procedure as those reported in references (Özdural et al., 2004; Yun et al., 2005). The terms of $\partial C_{Di}/\partial x_D$ and $\partial^2 C_{Di}/\partial x_D^2$ can be discretized by the central difference approximation

$$\frac{\partial C_{Di}}{\partial x_D} = \frac{C_{Di,j+1}^{n+1} - C_{Di,j-1}^{n+1}}{2\Delta x_D} \qquad (4.17)$$

$$\frac{\partial^2 C_{Di}}{\partial x_D^2} = \frac{C_{Di,j+1}^{n+1} - 2C_{Di,j}^{n+1} + C_{Di,j-1}^{n+1}}{\Delta x_D^2} \qquad (4.18)$$

and the term $\partial C_{Di}/\partial t_{Di}$ can be discretized by the implicit backward difference approximation

$$\frac{\partial C_{Di}}{\partial t_{Di}} = \frac{C_{Di,j}^{n+1} - C_{Di,j}^{n}}{\Delta t_{Di}} \qquad (4.19)$$

Then, the mass balance equation can be discretized as

$$\frac{C_{Di,j}^{n+1} - C_{Di,j}^{n}}{\Delta t_{Di}} = \frac{1}{Pe_i} \frac{C_{Di,j+1}^{n+1} - 2C_{Di,j}^{n+1} + C_{Di,j-1}^{n+1}}{\Delta x_D^2} - \frac{C_{Di,j+1}^{n+1} - C_{Di,j-1}^{n+1}}{2\Delta x_D} \qquad (4.20)$$

Let $\alpha_{Di} = \dfrac{\Delta t_{Di}}{2\Delta x_D}$ and $\lambda_{Di} = \dfrac{\Delta t_{Di}}{Pe_i \Delta x_D^2}$. Then the mass balance equation can be written as

$$-(\alpha_{Di} + \lambda_{Di})C_{Di,j-1}^{n+1} + (1 + 2\lambda_{Di})C_{Di,j}^{n+1} + (\alpha_{Di} - \lambda_{Di})C_{Di,j+1}^{n+1} = C_{Di,j}^{n} \qquad (4.21)$$

where $j = 1, 2, 3, \ldots N_x$, $n = 1, 2, 3, \ldots N_t$. $N_x$ and $N_t$ are the discretization numbers along the axial length and time domains, respectively. The concentration at the point $j = N_x + 1$ is obtained by extrapolating $j = N_x - 1$ and $j = N_x$ (Özdural et al., 2004; Yun et al., 2005):

$$C_{Di,N_x+1}^{n+1} - 2C_{Di,N_x}^{n+1} + C_{Di,N_x-1}^{n+1} = 0 \qquad (4.22)$$

Considering the initial and boundary conditions, the concentration at each discretization point can be determined by solving the following equations:

$$\begin{pmatrix} 1 & 0 & 0 & \cdots & 0 & 0 \\ -(\alpha_{Di}+\lambda_{Di}) & 1+2\lambda_{Di} & \alpha_{Di}-\lambda_{Di} & 0 & \cdots & 0 \\ 0 & -(\alpha_{Di}+\lambda_{Di}) & 1+2\lambda_{Di} & \alpha_{Di}-\lambda_{Di} & 0 & \cdots \\ \vdots & \vdots & \vdots & \vdots & \vdots & \vdots \\ 0 & 0 & 0 & 1 & -2 & 1 \end{pmatrix} \begin{pmatrix} C_{Di,1}^{n+1} \\ C_{Di,2}^{n+1} \\ C_{Di,3}^{n+1} \\ \vdots \\ C_{Di,j}^{N_x+1} \end{pmatrix} = \begin{pmatrix} 1 \\ C_{Di,2}^{n} \\ C_{Di,3}^{n} \\ \vdots \\ 0 \end{pmatrix}$$

$$(4.23)$$

Then, the concentration at the cryogel bed outlet can be calculated by

$$C = \frac{\sum_{i}^{N_g} C_i n_i Q_i}{Q} \quad (4.24)$$

The deviations of the model predictions from the experimental data of the porosity, permeability, bed volume and protein breakthrough can be estimated by comparing the predicted (subscript symbol is 'cal') and experimental data (subscript symbol is 'exp'). The relative difference of the porosity can be estimated by (Yun et al., 2011)

$$\delta_\varphi = \frac{|\varphi_{cal} - \varphi_{exp}|}{\varphi_{exp}} \quad (4.25)$$

The relative difference of the permeability can be calculated by

$$\delta_k = \frac{|k_{cal} - k_{exp}|}{k_{exp}} \quad (4.26)$$

The relative difference of the bed volume can be determined by

$$\delta_V = \frac{|V_{cal} - V_{exp}|}{V_{exp}} \quad (4.27)$$

and the mean dimensionless difference of the breakthrough data can be calculated by

$$\delta_C = \frac{\sum_{j=1}^{N_j} |C_{cal,j} - C_{exp,j}|}{N_j C_0} \quad (4.28)$$

where $V_{cal}$ and $V_{exp}$ are the predicted and experimental volumes of the cryogel and $N_j$ the total data number of the experimental breakthrough.

Capillary-based models could be applied to characterize the microstructures focusing on the pore size distribution and the gel skeleton and it is possible to be used in the prediction of the breakthrough behaviours under different flow conditions. Several examples have been reported in references (Persson et al., 2004; Yun et al., 2009a, 2011, 2013) and the detailed contents and results can be found in these references. However, for actual cryogels that are more complex some aspects in these models need to be modified further. For examples, the pore shape was only assumed regular, the gel skeleton thickness was assumed a constant while the thickness distribution was not considered, the chromatographic adsorption for multi-component biomolecules was not described and some necessary improvements on the mass transfer and axial dispersion coefficients in each pore are needed. These are the future work.

## 4.5 APPLICATIONS OF CRYOGEL BEADS AND COMPOSITE CRYOGELS

Cryogel-beads and composite cryogels have potential applications in biological areas. Some examples for the adsorption and separation of immunoglobulin and albumin proteins have been demonstrated (Ye et al., 2013; Yun et al., 2013).

Ion-exchange cryogel beads are useful as packing adsorbents in chromatographic adsorption of γ-globulin (Yun et al., 2013). For the packed bed of pAAm-based anion-exchange cryogel beads, the binding capacity of γ-globulin in the bed depends not only on the physical properties of the cryogel-beads, but also on the chromatographic conditions like the velocity and loading volume of protein solution (Yun et al., 2013). Due to the wide pore size distribution of the large interbead pores and intrabead supermacropores in the cryogel-bead packed bed, the axial dispersion within the bed voids is slightly strong, which could impact the dynamic adsorption capacity of γ-globulin during the chromatographic adsorption process. The convective mass transfer in the interbead pores and those large-size intrabead pores is predominated and thus the mass transfer within those voids is rapid and a short time is needed for γ-globulin to be transferred from the bulk to the binding sites. However, the in-pore diffusion of proteins from the bulk liquid within some small intrabead pores to the pore walls is slow and thus sometimes could influence the whole mass transfer process of protein molecules and consequently the breakthrough behaviours.

By suitable graft modification, the composite cryogels embedded with macroporous beads can be used in the isolation of immunoglobulin G and albumin proteins from serum samples. By employing a pHEMA-based composite cryogel embedded with cellulose beads, immunoglobulin G with a purity of 83% and albumin with a purity of 98% can be obtained with the salt step-elution strategy (Ye et al., 2013). Due to the cryo-polymerization of the gel-forming agents within the bead pores, the composite cryogel has multiscale pores, which is much different from those widely used commercial ion-exchange adsorbents like Sepharose adsorbents. Those pores with large sizes permit the free passage of the viscous serum, while those small-size pores within the embedded beads provide remarkable binding sites and permit the adsorption of immunoglobulin G and albumin proteins. This multiscale pore system benefits positively to the efficient chromatographic separation of immunoglobulin G and albumin proteins and thus the high purity of the target biomolecules.

## 4.6 CONCLUSIONS

Cryogel beads have sponge-like porous structures like those in monolithic cryogels and thus interesting properties such as good elasticity, high porosity and high permeability. These particle-form cryogels can be fabricated by the microchannel liquid-flow focusing and cryo-polymerization methods. The precise manufacture of useful and easy-scale-up microchannels for this method can be achieved effectively on stainless steel slabs by a mechanical micro-cutting technique utilizing micromilling cutters in high-speed engraving and milling machines followed by a polishing process. The formation of cryogel beads with this technique is influenced by the microchannels, the micro-flow focusing conditions, the freezing strategies and the properties of water-immiscible liquid as well as the concentration of gel-forming monomers in an aqueous solution during cryo-polymerization. By investigating the hydrodynamics of immiscible liquid-liquid slug flow in microchannels, it is possible to get insights on the formation mechanisms of droplets with expected sizes and consequently the cryogel beads with satisfied diameters. Cryogel beads can be packed in different columns conveniently and the obtained packed beds have similar chromatographic properties as those monolithic cryogels. Therefore, these adsorbents have potential in a variety of applications regarding biological and biomedical areas.

Composite cryogels embedded with macroporous beads are novel adsorbents prepared by cryo-polymerization. They have different properties in gel skeleton, pore system and active sites, and thus are interesting in separation areas. Composite cryogels always have multiscale pores with sizes from nanoscale within the embedded beads to about 50–100 μm in the cryogel matrix outside these beads, which could be helpful in isolating target biomolecules from complex feedstock.

Monolithic cryogels and cryogel-bead packed beds can be characterized by capillary-based models. These models are based on the assumption that a cryogel bed is made up of tortuous capillaries with a diameter distribution, surrounded by gel skeleton with a constant thin thickness. Convection flow and mass transfer are predominated within supermacropores. These models can be solved numerically. Under given conditions of bed sizes, porosity, water permeability and other experimental chromatographic conditions, these models can give a detailed description regarding the microstructures, the fluid flow characteristics, the axial dispersion behaviours as well as the mass transfer characteristics of biomolecules through the cryogel bed, which is helpful and useful in the characterization of cryogels and predictions of the chromatographic performance in actual applications.

## ACKNOWLEDGEMENTS

The authors gratefully acknowledge the financial support partially by the National Natural Science Foundation of China (Nos. 21036005, 20876145, 20606031, 21576240), the International Science & Technology Cooperation Program between China and Europe Country's Governments from the Ministry of Science and Technology of China (No. 1017 with S2010GR0616) and the Zhejiang Provincial Natural Science Foundation of China (LZ14B060001). The authors also acknowledge Dr. Songhong Zhang, Dr. Shaochuan Shen, Changming Tu, Wei Zhao, Jialei Ye, Maomao Pan, and Xiuhong Chen for their contributions in experiments.

## LIST OF ABBREVIATIONS

**pAAm**     Polyacrylamide
**pHEMA**   Poly(2-hydroxyethylmethacrylate)
**SEM**      Scanning electron microscope

## REFERENCES

Alkan, H., Bereli, N., Baysal, Z., and Denizli, A. (2009). Antibody purification with protein A attached supermacroporous poly(hydroxyethyl methacrylate) cryogel. *Biochem. Eng. J.* **45**(3): 201–208.

Alkan, H., Bereli, N., Baysal, Z., and Denizli, A. (2010). Selective removal of the autoantibodies from rheumatoid arthritis patient plasma using protein A carrying affinity cryogels. *Biochem. Eng. J.* **51**: 153–159.

Ariga, O., Takagi, H., Nishizawa, H., and Sano, Y. (1987). Immobilization of microorganisms with PVA hardened by iterative freezing and thawing. *J. Ferment. Technol.* **65**(6): 651–658.

Arvidsson, P., Plieva, F.M., Lozinsky, V.I., Galaev, I.Y., and Mattiasson, B. (2003). Direct chromatographic capture of enzyme from crude homogenate using immobilized metal affinity chromatography on a continuous supermacroporous adsorbent. *J. Chromatogr. A* **986**(2): 275–290.

Baydemir, G., Bereli, N., Andaç, M., Say, R., Galaev, I.Y., and Denizli, A. (2009). Supermacroporous poly(hydroxyethyl methacrylate) based cryogel with embedded bilirubin imprinted particles. *React. Funct. Polym.* **69**(1): 36–42.

Bereli, N., Şener, G., Altıntaş, E.B., Yavuz, H., and Denizli, A. (2010). Poly(glycidyl methacrylate) beads embedded cryogels for pseudo-specific affinity depletion of albumin and immunoglobulin G. *Mat Sci Eng C* **30**: 323–329.

Deng, W.P., Zhang, G.H., Xu, L.H., Shen, S.C., Yun, J.X., and Yao, K.J. (2013). Preparation of poly(2-hydroxyethyl methacrylate) cryogel microbeads and the isolation of adenosine triphosphate from *Saccharomyces cerevisiae* broths thereafter. *J. Chem. Eng. Chinese Univ.* **27**(1): 119–124. (In Chinese)

Du, K.-F., Bai, S., Dong, X.-Y., and Sun, Y. (2010a). Fabrication of superporous agarose beads for protein adsorption: Effect of $CaCO_3$ granules content. *J. Chromatogr. A* **1217**(37): 5508–5516.

Du, K.-F., Yan, M., Wang, Q.-Y., and Song, H. (2010b). Preparation and characterization of novel macroporous cellulose beads regenerated from ionic liquid for fast chromatography. *J. Chromatogr. A* **1217**(8): 1298–1304.

Ehrfeld, W., Hessel, V., and Löwe, H. (2000). *Microreactors: New Technology for Modern Chemistry*. Weinheim: Wiley-VCH Verlag GmbH.

Garstecki, P., Fuerstman, M.J., Stone, H.A., and Whitesides, G.M. (2006). Formation of droplets and bubbles in a microfluidic T-junction—Scaling and mechanism of break-up. *Lab Chip* **6**: 437–446.

Gokmen, M.T., and Du Prez, F.E. (2012). Porous polymer particles–A comprehensive guide to synthesis, characterization, functionalization and applications. *Prog. Polym. Sci.* **37**(3): 365–405.

Gun'ko, V.M., Savina, I.N., and Mikhalovsky, S.V. (2013). Cryogels: Morphological, structural and adsorption characterization. *Adv. Colloid. Interfac.* **187–188**: 1–46.

Gustavsson, P.-E., and Larsson, P.-O. (1996). Superporous agarose, a new material for chromatography. *J. Chromatogr. A* **734**(2): 231–240.

Gutsche, R., and Bunke, G. (2008). Modelling the liquid-phase adsorption in packed beds at low Reynolds numbers: An improved hydrodynamic model. *Chem. Eng. Sci.* **63**(16): 4203–4217.

Günther, A., and Jensen, K.F. (2006). Multiphase microfluidics: From flow characteristics to chemical and materials synthesis. *Lab Chip* **6**(12): 1487–1503.

Hajizadeh, S., Kirsebom, H., Leistner, A., and Mattiasson, B. (2012). Composite cryogel with immobilized concanavalin A for affinity chromatography of glycoproteins. *J. Sep. Sci.* **35**(21): 2978–2985.

Hajizadeh, S., Xu, C.G., Kirsebom, H., Ye, L., and Mattiasson, B. (2013). Cryogelation of molecularly imprinted nanoparticles: A macroporous structure as affinity chromatography column for removal of β-blockers from complex samples. *J. Chromatogr. A* **1274**: 6–12.

Hanora, A., Plieva, F.M., Hedström, M., Galaev, I.Y., and Mattiasson, B. (2005). Capture of bacterial endotoxins using a supermacroporous monolithic matrix with immobilized polyethyleneimine, lysozyme or polymyxin B. *J. Biotechnol.* **118**(4): 421–433.

Jespersen, G.R., Nielsen, A.L., Matthiesen, F., Andersen, H.S., and Kirsebom, H. (2013). Dual application of cryogel as solid support in peptide synthesis and subsequent protein-capture. *J. Appl. Polym. Sci.* **130**(6): 4383–4391.

Jurga, M., Dainiak, M.B., Sarnowska, A. et al. (2011). The performance of laminin-containing cryogel scaffolds in neural tissue regeneration. *Biomaterials* **32**(13): 3423–3434.

Kirsebom, H., Elowsson, L., Berillo, D. et al. (2013). Enzyme-catalyzed crosslinking in a partly frozen state: A new way to produce supermacroporous protein structures. *Macromol. Biosci.* **13**(1): 67–76.

Kirsebom, H., and Mattiasson, B. (2011). Cryostructuration as a tool for preparing highly porous polymer materials. *Polym. Chem.* **2**(5): 1059–1062.

Koç, İ., Baydemir, G., Bayram, E., Yavuz, H., and Denizli, A. (2011). Selective removal of 17β-estradiol with molecularly imprinted particle-embedded cryogel systems. *J. Hazard. Mater.* **192**(3): 1819–1826.

Kumar, A., Bansal, V., Nandakumar, K.S. et al. (2006). Integrated bioprocess for the production and isolation of urokinase from animal cell culture using supermacroporous cryogel matrices. *Biotechnol. Bioeng.* **93**(4): 636–646.

Kuyukina, M.S., Ivshina, I.B., Kamenskikh, T.N., Bulicheva, M.V., and Stukova, G.I. (2013). Survival of cryogel-immobilized *Rhodococcus* strains in crude oil-contaminated soil and their impact on biodegradation efficiency. *Int. Biodeter. Biodegr.* **84**: 118–125.

Lozinsky, V.I. (2008). Polymeric cryogels as a new family of macroporous and supermacroporous materials for biotechnological purposes. *Russ. Chem. Bull.* **57**(5): 1015–1032.

Lozinsky, V.I. (2002). Cryogels on the basis of natural and synthetic polymers: Preparation, properties and applications. *Russ. Chem. Rev.* **71**(6): 489–511.

Lozinsky, V.I., Galaev, I.Y., Plieva, F.M., Savina, I.N., Jungvid, H., and Mattiasson, B. (2003). Polymeric cryogels as promising materials of biotechnological interest. *Trends Biotechnol.* **21**(10): 445–451.

Ma, C.Y., Lin, D.-Q., and Yao, S.-J. (2012). Preparation of macroporous β-cyclodextrin matrix and its adsorption behaviors for rutin. *CIESC J.* **63**: 133–138. (In Chinese)

Martins, R.F.R.F., Plieva, F.M., Santos, A., and Hatti-Kaul, R. (2003). Integrated immobilized cell reactor-adsorption system for β-cyclodextrin production: A model study using PVA-cryogel entrapped bacillus *Agaradhaerens* cells. *Biotechnol. Lett.* **25**(18): 1537–1543.

Mattiasson, B., Kumar, A., and Galaev, I.Y. (2010). *Macroporous Polymers: Production and Biotechnological/Biomedical Applications*. Boca Raton, FL: CRC Press, Taylor & Francis Group.

Noir, M.L., Plieva, F., Hey, T., Guieysse, B., and Mattiasson, B. (2007). Macroporous molecularly imprinted polymer/cryogel composite systems for the removal of endocrine disrupting trace contaminants. *J. Chromatogr. A* **1154**(1–2): 158–164.

Orakdogen, N., Karacan, P., and Okay, O. (2011). Macroporous, responsive DNA cryogel beads. *React. Funct. Polym.* **71**(8): 782–790.

Özdural, A.R., Alkan, A., and Kerkhof, P.J.A.M. (2004). Modeling chromatographic columns: Non-equilibrium packed-bed adsorption with non-linear adsorption isotherms. *J. Chromatogr. A* **1041**(1–2): 77–85.

Partovinia, A., and Naeimpoor, F. (2013). Phenanthrene biodegradation by immobilized microbial consortium in polyvinyl alcohol cryogel beads. *Int. Biodeter. Biodegr.* **85**: 337–344.

Perçin, I., Sağlar, E., Yavuz, H., Aksöz, E., and Denizli, A. (2011). Poly(hydroxyethyl methacrylate) based affinity cryogel for plasmid DNA purification. *Int. J. Biol. Macromol.* **48**(4): 577–582.

Persson, P., Baybak, O., Plieva, F.M. et al. (2004). Characterization of a continuous supermacroporous monolithic matrix for chromatographic separation of large bioparticles. *Biotechnol. Bioeng.* **88**(2): 224–236.

Plieva, F.M., Bober, B., Dainiak, M., Galaev, I.Y., and Mattiasson, B. (2006). Macroporous polyacrylamide monolithic gels with immobilized metal affinity ligands: The effect of porous structure and ligand coupling chemistry on protein binding. *J. Mol. Recogn.* **19**(4): 305–312.

Plieva, F.M., Galaev, I.Y., and Mattiasson, B. (2007). Macroporous gels prepared at subzero temperatures as novel materials for chromatography of particulate-containing fluids and cell culture applications. *J. Sep. Sci.* **30**(11): 1657–1671.

Plieva, F.M., Galaev, I.Y., Noppe, W., and Mattiasson, B. (2008). Cryogel Applications in Microbiology. *Trends. Microbiol.* **16**(11): 543–551.

Plieva, F.M., Kirsebom, H., and Mattiasson, B. (2011). Preparation of macroporous cryostructurated gel monoliths, their characterization and main applications. *J. Sep. Sci.* **34**(16–17): 2164–2172.

Prusse, U., Dalluhn, J., Breford, J., and Vorlop, K.D. (2000). Production of spherical beads by JetCutting. *Chem. Eng. Technol.* **23**(12): 1105–1110.

Savina, I.N., Galaev, I.Y., and Mattiasson, B. (2005a). Anion-exchange supermacroporous monolithic matrices with grafted polymer brushes of N,N-dimethylaminoethyl-methacrylate. *J. Chromatogr. A* **1092**(2):199–205.

Savina, I.N., Galaev, I.Y., and Mattiasson, B. (2006). Ion-exchange macroporous hydrophilic gel monolith with grafted polymer brushes. *J. Mol. Recogn.* **19**(4): 313–321.

Savina, I.N., Mattiasson, B., and Galaev, I.Y. (2005b). Graft polymerization of acrylic acid onto macroporous polyacrylamide gel (cryogel) initiated by potassium diperiodatocuprate. *Polymer.* **46**(23): 9596–9603.

Seguin, D., Montillet, A., Bmnjail, D., and Corniti, J. (1996). Liquid–solid mass transfer in packed beds of variously shaped particles at low reynolds numbers: Experiments and model. *Chem. Eng. J. Biochem. Eng. J.* **63**(1): 1–9.

Shi, Q.-H., Zhou, X., Sun, Y. (2005). A novel superporous agarose medium for high-speed protein chromatography. *Biotechn. Bioeng.* **92**(5): 643–651.

Sun, S.J., Tang, Y.H., Fu, Q. et al. (2012). Monolithic cryogels made of agarose–chitosan composite and loaded with agarose beads for purification of immunoglobulin G. *Int. J. Biol. Macromol.* **50**(4): 1002–1007.

Srivastava, A., Shakya, A.K., and Kumar, A. (2012). Boronate affinity chromatography of cells and biomacromolecules using cryogel matrices. *Enzyme Microb. Tech.* **51**(6–7): 373–381.

Szczęsna-Antczak, M., Galas, E., and Bielecki, S. (2004). Stability of extracellular proteinase productivity by bacillus subtilis cells immobilized in PVA-cryogel. *J. Mol. Catal. B: Enzyme* **11**(2): 168–176.

Thome, J.R., Bar-Cohen, A., Revellin, R., and Zun, I. (2013). Unified mechanistic multiscale mapping of two-phase flow patterns in microchannels. *Exp. Therm. Fluid Sci.* **44**: 1–22.

Tiainen, P., Gustavsson, P.-E., Ljunglöf, A., and Larsson, P.-O. (2007). Superporous agarose anion exchangers for plasmid isolation. *J. Chromatogr. A.* **1138**(1–2): 84–94.

Timung, S., and Mandal, T.K. (2013). Prediction of flow pattern of gas–liquid flow through circular microchannel using probabilistic neural network. *Appl. Soft Comput.* **13**(4): 1674–1685.

Tong, F.L., Lin, D.-Q., Liu, C., Yun, J.X., and Yao, S.J. (2013). Preparation of cellulose-based chromatography matrix with uniform size by cross-flow microchannel chip. *CIESC J.* **64**(2):676–682. (In Chinese)

Tripathi, A., and Kumar, A. (2011). Multi-featured macroporous agarose-alginate cryogel: Synthesis and characterization for bioengineering applications. *Macromol. Biosci.* **11**(1): 22–35.

Tripathi, A., Sami, H., Jain, S.R., Viloria-Cols, M., Zhuravleva, N., Nilsson, G., Jungvid, H., Kumar, A. (2010). Improved bio-catalytic conversion by novel immobilization process using cryogel beads to increase solvent production. *Enzyme Microb. Tech.* **47**(1–2): 44–51.

Ünlüer, Ö.B., Ersöz, A., Denizli, A., Demirel, R., and Say, R. (2013). Separation and purification of hyaluronic acid by embedded glucuronic acid imprinted polymers into cryogel. *J. Chromatogr. B* **934**: 46–52.

Üzek, R., Uzun, L., Şenel, S., and Denizli, A. (2013). Nanospines incorporation into the structure of the hydrophobic cryogels via novel cryogelation method: An alternative sorbent for plasmid DNA purification. *Colloids Surf. B. Biointerfaces* **102**: 243–250.

Uzun, L., Armutcu, C., Biçen, Ö., Ersöz, A., Say, R., and Denizli, A. (2013). Simultaneous depletion of immunoglobulin g and albumin from human plasma using novel monolithic cryogel columns. *Colloids Surf. B. Biointerfaces* **112**: 1–8.

Wang, C., Dong, X.-Y., Jiang, Z.Y., and Sun, Y. (2013). Enhanced adsorption capacity of cryogel bed by incorporating polymeric resin particles. *J. Chromatogr. A* **1272**: 20–25.

Wang, C., and Sun, Y. (2013). Double sequential modifications of composite cryogel beds for enhanced ion-exchange capacity of protein. *J. Chromatogr. A* **1307**: 73–79.

Xu, L.H., Tan, X., Yun, J.X. et al. (2012). Formulation of poorly water-soluble compound loaded solid lipid nanoparticles in a microchannel system fabricated by mechanical microcutting method: Puerarin as a model drug. *Ind. Eng. Chem. Res.* **51**(35): 11373–11380.

Yao, K.J., Shen, S.C., Yun, J.X., Wang, L.H., He, X.J., and Yu, X.M. (2006a). Preparation of polyacrylamide-based supermacroporous monolithic cryogel beds under freezing-temperature variation conditions. *Chem. Eng. Sci.* **61**(20): 6701–6708.

Yao, K.J., Yun, J.X., Shen, S.C., and Chen, F. (2007). *In-situ* graft-polymerization preparation of cation-exchange supermacroporous cryogel with sulfo groups in glass columns. *J. Chromatogr. A* **1157**(1–2): 246–251.

Yao, K.J., Yun, J.X., Shen, S.C., Wang, L.H., He, X.J., and Yu, X.M. (2006b). Characterization of a novel continuous supermacroporous monolithic cryogel embedded with nanoparticles for protein chromatography. *J. Chromatogr. A* **1109**(1):103–110.

Ye, J.L., Yun, J.X., Lin, D.Q. et al. (2013). Poly(hydroxyethyl methacrylate)-based composite cryogel with embedded macroporous cellulose beads for separation of human serum immunoglobulin and albumin. *J. Sep. Sci.* **36**(24): 3813–3820.

Yun, J.X., Dafoe, J.T., Peterson, E., Xu, L.H., Yao, S.J., and Daugulis, A.J. (2013). Rapid freezing cryo-polymerization and microchannel liquid-flow focusing for cryogel beads: Adsorbent preparation and characterization of supermacroporous bead-packed bed. *J. Chromatogr. A* **1284**: 148–154.

Yun, J.X., Jespersen, G.R., Kirsebom, H., Gustavsson, P.E., Mattiasson, B., and Galaev, I.Yu. (2011). An improved capillary model for describing the microstructure characteristics, fluid hydrodynamics and breakthrough performance of proteins in cryogel beds. *J. Chromatogr. A* **1218**(32): 5487–5497.

Yun, J.X., Kirsebom, H., Galaev, I.Y., and Mattiasson, B. (2009a). Modeling of protein breakthrough performance in cryogel columns by taking into account the overall axial dispersion. *J. Sep. Sci.* **32**(15–16): 2601–2607.

Yun, J.X., Lei, Q., Zhang, S.H., Shen, S.C., and Yao, K.J. (2010). Slug flow characteristics of gas-miscible liquids in a rectangular microchannel with cross and t-shaped junctions. *Chem. Eng. Sci.* **65**(18): 5256–5263.

Yun, J.X., Lin, D.-Q., and Yao, S.J. (2005). Predictive modeling of protein adsorption along the bed height by taking into account the axial nonuniform liquid dispersion and particle classification in expanded beds. *J. Chromatogr. A* **1095**(1–2): 16–26.

Yun, J.X., Tu, C.M., Lin, D.Q. et al. (2012). Microchannel liquid-flow focusing and cryopolymerization preparation of supermacroporous cryogel beads for bioseparation. *J. Chromatogr. A* **1247**: 81–88.

Yun, J.X., Zhang, S.H., Shen, S.C., Chen, Z., Yao, K.J., and Chen, J.Z. (2009b). Continuous production of solid lipid nanoparticles by liquid flow-focusing aand gas displacing method in microchannels. *Chem. Eng. Sci.* **64**(19): 4115–4122.

Zhang, S.H., Yun, J.X., Shen, S.C. et al. (2008). Formation of solid lipid nanoparticles in a microchannel system with a cross-shaped junction. *Chem. Eng. Sci.* **63**(23): 5600–5605.

Zhao, W., Zhang, S.H., Lu, M.Z. et al. (2014). Immiscible liquid-liquid slug flow characteristics in the generation of aqueous drops within a rectangular microchannel for preparation of poly(2-hydroxyethylmethacrylate) cryogel beads. *Chem. Eng. Res. Des.* **92**(11): 2182–2190.

# Section II

*Application of Supermacroporous Cryogels in Biomedical Engineering*

# 5 Cryogel Tissue Phantoms with Uniform Elasticity for Medical Imaging

*Azizeh-Mitra Yousefi,\* Corina S. Drapaca and Colin J. Kazina*

## CONTENTS

| | | |
|---|---|---|
| 5.1 | Introduction | 146 |
| 5.2 | Elastography Techniques for Medical Imaging | 147 |
| | 5.2.1 Ultrasound Elastography | 147 |
| | 5.2.2 Optical Coherence Elastography | 148 |
| | 5.2.3 Magnetic Resonance Elastography | 149 |
| 5.3 | Need for Elastography Phantoms | 150 |
| | 5.3.1 Liver Phantoms | 151 |
| | 5.3.2 Breast Phantoms | 152 |
| | 5.3.3 Brain Phantoms | 154 |
| 5.4 | Cryogel Phantoms | 155 |
| | 5.4.1 Polyvinyl Alcohol Homopolymer | 156 |
| | 5.4.2 Other Polymers | 158 |
| 5.5 | Cryogel Phantom Characterization and Homogeneity | 160 |
| | 5.5.1 Mechanical Properties | 160 |
| | 5.5.2 Wave Propagation and MRI Properties | 166 |
| 5.6 | Conclusions | 166 |
| Acknowledgements | | 167 |
| List of Abbreviations | | 167 |
| References | | 167 |

## ABSTRACT

Elastography-based medical imaging techniques such as ultrasound elastography, optical coherence elastography and magnetic resonance elastography are valuable tools to quantitatively assess the mechanical properties of biological tissues *in vivo*. Estimating the local tissue stiffness or elasticity can be an effective noninvasive way of detecting tumors, particularly for parts of the body that are not accessible for physical examination by palpation. However, there are significant differences in the absolute values of the reported modulus for biological

---

\* Corresponding author: E-mail: yousefiam@MiamiOH.edu; Tel: 001-513-529-0766.

tissues in clinical elastography studies. This is usually attributed to differences in experimental methodology and reconstruction algorithms, as well as the age and gender of volunteers. Tissue-mimicking phantoms with well-defined properties can help in identifying the potential weaknesses in elastography systems. This is primarily because a calibrated phantom can be characterized by independent measurements to directly estimate its mechanical properties. This chapter gives an overview of the current strategies in developing tissue-mimicking materials (TMMs) as elastography phantoms, with a focus on polymeric cryogels. Among these materials, polyvinyl alcohol cryogel (PVA-C) has been extensively used as a phantom material by medical imaging researchers. Since large cryogel phantoms suffer from variations in properties due to the inhomogeneous thawing rates during freeze–thaw cycles, this chapter also briefly covers some of the recent efforts to improve the homogeneity of cryogel phantoms.

**KEYWORDS**

Cryogel phantoms, elastography, medical imaging, PVA cryogel, tissue phantoms

## 5.1 INTRODUCTION

Palpation is a simple clinical method for diagnostic assessment of certain diseases. This is mainly because many disease processes are associated with significant changes in tissue mechanical properties. To improve upon this well-established clinical method, elastography-based imaging techniques have been developed for diagnostic and prognostic purposes, which are partly inspired from a desire to 'palpate by imaging' (Glaser et al., 2012). Ultrasound elastography (USE), optical coherence elastography (OCE), as well as magnetic resonance elastography (MRE) are among the elastography techniques that have shown promise for clinical applications (e.g. tumor detection, traumatic brain injury, Alzheimer's disease) (Murphy et al., 2012; Simon et al., 2013; Zhang et al., 2011).

A healthy biological tissue may exhibit a range of properties among different individuals, which could pose a challenge for diagnostic assessments by elastography techniques. For example, a cutoff of 2.93 kPa for shear stiffness has been reported as a threshold for differentiating a fibrotic liver from a healthy one (Mariappan et al., 2010; Yin et al., 2007). In cases of brain pathologies, the consistency of different tumor types may fall within the range for healthy brain tissue (Xu et al., 2007). Thus, the magnitude of error in elastography may seriously impact its diagnostic capabilities for certain biological tissues.

A limited number of studies have objectively assessed the full capabilities of the different elastography approaches, partly due to the scarcity of suitable tissue-mimicking materials to validate the accuracy of the measurements (Cournane et al., 2012a). In a recent review, different elastography techniques have been compared in terms of their respective spectrum of contrast mechanisms (Mariappan et al., 2010). The shear modulus in MRE has variations over five orders of magnitude among various physiological states of normal and pathologic tissues, which is the largest when compared with the attenuation coefficient in computed tomography (CT) and bulk modulus in

USE. This emphasizes the need for well-characterized phantoms with homogeneous properties. Any improvement in phantom homogeneity can greatly contribute to the optimization of elastography systems for clinical applications.

Hydrogels are water-swollen polymeric materials that can maintain a distinct three-dimensional structure, and were the first biomaterials used in the human body (Kopeček, 2008). Hydrogels prepared by the freeze–thaw technique (cryogels) are potential candidates for mimicking the mechanical properties of soft tissues, and have been extensively used in elastography techniques. However, the size of cryogel phantoms produced by the freeze–thaw technique is limited due to the need for a uniform thawing rate throughout to ensure a uniform stiffness (Cournane et al., 2010). On the other hand, the use of sufficiently large phantoms is important to enable a more accurate assessment of certain elastography systems. Assuming that a propagated transient wave is reflected by a boundary, the volume of the phantom should be sufficient such that the path travelled by the reflected wave is long enough to minimize any interference during acquisition (Cournane et al., 2010). Any variations in mechanical properties across the phantom can limit its suitability for elastography applications.

Obtaining adequate statistical significance to assess the clinical utility of new diagnostic imaging systems requires clinical trials involving many patients and considerable time and cost (Madsen et al., 2006). For example, MRE of the human brain can greatly benefit from brain-sized phantoms replicating the properties of white and gray matter, thus enabling a more realistic evaluation of the MRE technique for brain pathologies. However, developing large and homogeneous tissue-mimicking phantoms poses a challenge to medical imaging researchers. This chapter gives an overview of the current strategies in developing phantoms for elastography techniques, as well as the recent efforts for improving the homogeneity and shelf-life of cryogel phantoms.

## 5.2 ELASTOGRAPHY TECHNIQUES FOR MEDICAL IMAGING

### 5.2.1 Ultrasound Elastography

Ultrasound (US) was the first imaging technique used to perform elastography. Ultrasound elastography (USE) requires a mechanical driver that locally applies (quasi-) static or dynamic loading on the tissue under investigation, a Doppler ultrasound system and an algorithm that combines Doppler detection algorithms, mechanical models of tissue deformation, finite element solvers and fitting methods.

Dickinson and Hill (1982) and Wilson and Robinson (1982) published the first studies on static USE. A combination of US imaging and natural internal mechanical stimuli such as cardiac contractions and pulsatile blood flow was used to track echo-movements near blood vessels. Since the measured displacements and deformations have small amplitudes and vary slowly in time, this approach describes a quasi-static response. Almost a decade later, Ophir et al. (1991) used externally applied compression on tissues to create strain images under static conditions and called the approach (strain imaging) elastography. A cross-correlation analysis was used to compare the echoes before and after compression, and estimate displacement. This method has the advantage that the US scanning transducer can be used to produce the localized compression near the region of interest (Parker et al., 2011). The implementation of compression-based static USE on B-scan imagers as real-time signal processing or as postprocessing of the before and after compression signals facilitated the use of the method in clinical applications (Varghese and Ophir, 1997). Recently, Dietrich et al. (2014) proposed performing real-time static USE during endoscopic US examinations to provide a better characterization of pathology under investigation. The scanning echoendoscopes can be used simultaneously for imaging and local compression.

The first studies of dynamic USE were reported by Lerner and Parker (1987, 1988). The approach uses externally applied low-frequency vibrations on the tissue under investigation to generate internal waves propagating through the tissue. These waves are detected by the Doppler signal, and a real-time vibration image can be made. This original USE method was called vibration amplitude sonoelastography (sonoelastography or sonoelasticity imaging). By using color Doppler scanners, regions of higher stiffness (that correlate to pathology) inside a soft tissue can be found from the distribution of low-velocity color. This implies that both pathological regions and regions without signal are represented by holes in the sonoelastogram, and thus special care must be taken in analysing the results (Greenleaf et al., 2003). Yamakoshi et al. (1990) developed a vibration phase gradient approach that uses simultaneously transmitted ultrasonic waves. In this way, both the amplitude and the phase of the low-frequency waves propagating through tissue can be measured from which the shear wave propagation velocity and dispersion properties are derived (the imaging system was slow enough so that the fast propagating longitudinal waves were not observed). By modeling the tissue as a linear visco-elastic Voigt material, the dynamic stiffness and attenuation can be further calculated. Promising results on transient USE using one or rapidly fired multiple impulses of acoustic radiation force generated shear waves have also been reported (e.g. McAleavey et al., 2009; Sarvazyan et al., 1998; Tanter et al., 2008).

### 5.2.2 Optical Coherence Elastography

Optical coherence elastography (OCE) is another noninvasive imaging technique that can be used to image tissues during mechanical stimulation and create elastograms. Correlation and Doppler techniques used in US imaging can also be applied for OCE (Parker et al., 2011). OCE measures the intensity of back-reflected infrared light when very short light pulses of low coherent light are applied on the tissue. Distances are measured using the echo delay time (time for the light to be reflected back), and then the intensity of back-reflected light is plotted as a function of depth (Rogowska et al., 2004). By scanning the entire tissue, an OCE image made of one-dimensional light versus depth plots can be obtained. Because the imaging technique is based on light, OCE images have very high spatial resolution (of several microns) and provide information at a tissue's depth of several millimeters. Thus, OCE can estimate biomechanical properties of tissues at microscopic scales but only in a shallow depth over which the measurements can be obtained (Liang et al., 2008; Schmitt, 1998).

Schmitt (1998) was the first to report static OCE results using compression tests in gelatin phantoms, pork meat and intact skin. The proposed approach uses two-dimensional cross-correlation speckle tracking to measure displacements in tissues under applied compression. In OCE images of highly scattering tissues, speckle is a sparkling effect caused by the cross-interference of random phase fields, size and temporal coherence of the light source and the aperture of the detector (Stella and Trivedi, 2011). The random granular pattern of the high intensity speckle noise alters the quality of the image, making it harder to differentiate between highly scattering structures. Since speckle noise contains information about tissue microstructure, tracking speckle locations before and after compression can therefore be used to measure displacements and calculate strains. However, tracking speckle is affected by the fact that speckles do not necessarily have the same signal amplitude or shape before and after tissue deformation and thus static OCE is not able to provide quantitative stiffness elastograms. To address this issue, Liang et al. (2008) proposed a dynamic OCE approach that uses phase-resolved (Doppler-like) imaging, a mechanical wave driver and the solution to the wave equation describing wave propagation through homogeneous, isotropic, linear elastic materials. The measured displacement is fitted to the solution of the wave equation to calculate stiffness values. In this way, a dynamic stiffness elastogram at the microscopic scale can be obtained.

Other studies that use Doppler and correlation methods to estimate the response of tissues during static or dynamic mechanical excitation have been reported by Wang et al. (2006), Kirkpatrick et al. (2006), Kennedy et al. (2009, 2014). Recently, Sampson et al. (2013) developed an annular piezoelectric loading transducer that can simultaneously vibrate the tissue and image it, which could potentially be used to perform *in vivo* dynamic OCE on human subjects.

### 5.2.3 MAGNETIC RESONANCE ELASTOGRAPHY

Almost two decades ago, Fowlkes et al. (1995) and Muthupillai et al. (1995) proposed independently to use magnetic resonance imaging (MRI) technology to create maps of mechanical properties (elastograms) of tissues *in vivo*. Magnetic resonance elastography (MRE) requires the following three components (Parker et al., 2011): (1) an MR-compatible mechanical driver, (2) an MRI pulse sequence and (3) an inversion algorithm. Mechanical drivers consist of electromechanical coils, piezoelectric elements or pneumatically powered actuators (Glaser et al., 2012). They are located on the surface of the tissue under consideration and apply local (quasi-) static or dynamic (harmonic or transient) mechanical excitation to the tissue. The common MRI pulse sequences used in MRE are spin-echo, gradient-echo or echo-planar. The inversion algorithm combines classic inversion methods used to create MR images, image processing techniques (such as denoising, deblurring and registration) and equations from continuum mechanics that describe the response to an applied load of almost incompressible, locally homogeneous and isotropic, linear (visco-) elastic soft materials such as biological tissues and phantoms. Recently, Drapaca (2010) proposed an inversion method that uses a nonlocal mechanical model for materials with microstructures. The method has the advantage that it does not require the pre- or postprocessing of the MR images and stiffness elastograms.

The quasi-static approach to MRE proposed by Fowlkes et al. (1995) uses the displacement of a spatial grid caused by spin saturation to estimate tissue strain. In order to improve the spatial resolution of the MR images, the phase-contrast MRI technique has been utilized by other authors (Plewes et al., 1995; Hardy et al., 2005) to provide more accurate estimations of qausi-static strains. Static stiffness elastograms have also been created by combining MR image registration techniques and finite element solvers used in classic mechanical analysis (Miga, 2003; Van Houten et al., 2003; Weaver et al., 2001; Weis et al., 2013).

The dynamic MRE approach by Muthupillai et al. (1995, 1996) combines the phase-contrast MRI technique and applied time-harmonic, shear excitation at one or more acoustic frequencies to create stiffness elastograms. In this case, more than one cycle of the mechanical motion and the MR pulse sequence are synchronized by including in the pulse sequence additional motion-encoding gradients. The mechanical motion is therefore encoded in the phase image. Transient MRE proposed by McCracken et al. (2005) is a variant of dynamic MRE in which only one cycle of a time-windowed sinusoidal excitation motion is applied and sensitized by only one rectangular bipolar gradient. This approach could be used to investigate the response of brain tissues to traumatic injuries.

## 5.3 NEED FOR ELASTOGRAPHY PHANTOMS

Phantoms that have similar physical characteristics to *in vivo* biological tissues are essential for the proper characterization and calibration of imaging elastography modalities, the development of guided surgical procedures and for biopsy and surgical training. However, most of the tissue-mimicking materials (TMMs) used in imaging elastography are merely for the validation of proposed elastography methods, and therefore their mechanical, US and magnetic resonance (MR) properties may differ from those of human soft tissues. In addition, only a few commercially available elastography phantoms reflect the lesion-to-background elasticity ratio represented in the clinical realm. Therefore, these phantoms may not fully challenge the capabilities of the different elastography techniques, proving limited for quality control or training purposes (Cournane et al., 2012a,b). Table 5.1 provides a list of proposed tissue-mimicking properties for specific applications, based on reported mechanical properties in a number of clinical studies. The background properties usually represent the approximate stiffness values of healthy tissue, while the range of the target stiffness provides the typical values of pathological lesions in the respective clinical application. Table 5.1 also includes the minimum target dimensions detected, as well as the elastic contrast ratio of the target stiffness to that of background material.

This section provides an overview on liver, breast and brain TMMs (phantoms) for elastography, whereas the following section gives an overview of cryogel phantoms and their mechanical and imaging properties.

## TABLE 5.1
Proposed Mechanical Properties (Shear Modulus) for Tissue-Mimicking Materials and the Typical Target Dimensions for Each of the Respective Clinical Applications

| Clinical Applications | Mechanical Properties | | | Typical Range of Target Dimensions (mm) |
|---|---|---|---|---|
| | Background (kPa) | Target (kPa) | Elastic Contrast | |
| Liver | 3 | 3–16 | 1–5.3 | 10–80 |
| Breast | 25 | 30–200 | 1.2–8 | 1–20 |
| Brain[a] | 2–14 | – | – | 31–75 |
| Prostate | 15 | 10–40 | 0.7–2.7 | 5–40 |
| Thyroid | 10 | 15–180 | 1.5–18 | 10–40 |
| Liver | 3 | 3–16 | 1–5.3 | 10–80 |

*Source:* Modified from Cournane, S., Fagan, A.J., and Browne, J.E. (2012). *Ultrasound* 20: 16–23 Copyright (2012): Sage Publications Ltd. With permission.

[a] Xu et al., 2007; Zhang et al., 2011.

### 5.3.1 Liver Phantoms

Some commonly used materials for elastography phantoms are (Cournane et al., 2010) oil-in-agar, oil-in-gelatin, agar dispersions (Madsen et al., 2003), copolymer in oil (Oudry et al., 2009), polyacrylamide gels (Luo and Konofagou 2009) and polyvinyl alcohol cryogel (PVA-C) (Surry et al., 2004). Liver phantoms have been made using several of the above-mentioned TMMs developed for general soft tissue phantoms. The TMM phantoms must have similar physical characteristics to those of the human liver tissue reported in clinical studies. The acoustic properties of liver that are recommended for liver phantoms are US speed of 1595 m/s, density of 1060 kg/m$^3$, US attenuation coefficient of 0.5 dB/cm MHz and acoustic impedance of 1.69 MRayl (Culjat et al., 2010). The T1 and T2 relaxation times for liver in an MR scanner with a main magnetic field of 1.5 Tesla are approximately 490 ms and 40 ms, respectively. The mechanical properties of liver proposed to be used for liver phantoms were obtained from MRE and transient USE studies and are listed in Table 5.1 (Cournane et al., 2012a,b). There are no commercially available liver phantoms for elastography studies, and the few liver TMM phantoms that are reported in the literature have been made by the research groups developing imaging elastography tools. Figure 5.1 shows an elastogram of a PVA-C liver phantom developed by Cournane et al. (2010).

In Madsen et al. (2003), a TMM phantom for USE was made of gelatin with a 50% suspension of microscopic safflower oil droplets and 4 g/L of type 3000E Potters glass bead scatterers. A 2-cm diameter cylindrical inclusion made of gelatin and 20-g/L glass beads were implanted in the phantom. The safflower oil, bead scatterers and

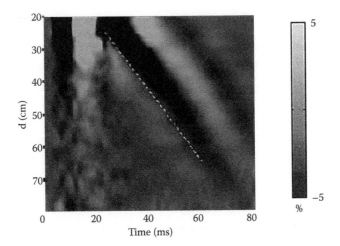

**FIGURE 5.1** Elastogram of the liver phantom developed by Cournane et al. (2010). The elastogram is made by the Fibroscan® system. The image shows the amplitude of the strains induced in the phantom as a function of depth d and time. Fibroscan® estimates the shear wave speed from a linear fit to the shear wave represented by the white line. (Reprinted from Cournane, S, Cannon, L., Browne, J.E, and Fagan, A.J., *Physics in Medicine and Biology* 55: 5965–83. Copyright (2010). SAGE Publications Ltd. With permission.)

hot gelatin were mixed together with cross-linking formaldehyde. The mixture was baked in a 50°C oven for 0, 5, 10, 15, 20 and 25 days and then tested at room temperature. While the US speed, US attenuation and T1 relaxation time for this phantom were within the limits of healthy liver tissues, the T2 relaxation time and stiffness values were not. The T1 relaxation time for the inclusion was 1610 ms, which is in the range of 784–1080 ms for T1 of liver hepatomas. The stiffness of the inclusion was approximately 30 kPa, which is the upper limit for stiffness values reported for liver malignant lesions. However, the T2 relaxation time of the inclusion was almost 5 times the T2 value for liver hepatomas of 84 ± 26 ms.

### 5.3.2 Breast Phantoms

The USE phantoms are made to match the acoustic properties of soft tissues and integrate some of the mechanical properties of the tissues as well. These physical parameters of tissues are based on measurements reported in clinical studies. The mechanical properties proposed for breast TMM are listed in Table 5.1 (Cournane et al., 2012a,b). Commercially available breast phantoms are produced by Computerized Imaging Reference Systems (CIRS) Inc. and Blue Phantom™. CIRS model 059 phantom is made of Zerdine™ gel, and has the anatomical, ultrasound and mechanical properties of breast tissue. Therefore, the phantom has a background stiffness of about 20 kPa, an acoustic velocity of 1540 ± 10 m/s and an attenuation coefficient of 0.5 dB/cm/MHz. In addition, the phantom contains many randomly distributed lesion-mimicking masses of varying sizes (2–10 mm in diameter) and of isoechoic nature. They are three times stiffer than the background material (Cournane et al., 2012a). A recent study of the internal structure of CIRS model 059 phantom has shown comparable (size and location) heterogeneities inside it on images obtained using US transmission tomography, computed tomography and MRI (Opielinski et al., 2013). This finding suggests the possibility of using US transmission tomography imaging for early detection of malignant breast lesions. The breast phantom made by Blue Phantom™ contains lesion-mimicking masses of varying echogenicity and sizes (6–11 mm in diameter). However, the mechanical and acoustical properties of this phantom are not specified, making it difficult to assess its ability to mimic the physical properties of breast tissue (Cournane et al., 2012a).

Many breast phantoms reported in the literature have been developed for elastography research. For instance, Madsen et al. (2003) proposed a breast phantom for USE made of gelatin with a 50% suspension of microscopic safflower oil droplets and 4 gm/l of type 3000E Potters glass bead scatterers. The phantom has a 2-cm diameter cylindrical inclusion made of gelatin. The safflower oil and glass bead scatterers are mixed in the hot gelatin, and formaldehyde is added for cross-linking. The mixture is then cooled and put into containers for congealing. The stiffness values reported for this phantom are within the stiffness limits reported by Krouskop et al. (1998) for healthy breast glandular tissue (28 ± 14 kPa) at 5% compression and invasive ductal carcinoma (106 ± 32 kPa). Madsen et al. (2006) proposed using the same recipe for making breast phantoms except that the glass bead scatterers were replaced by propylene glycol to increase US propagation speeds (Figure 5.2). The mechanical properties and T1 and T2 relaxation times of these phantoms are in good agreement with the values of breast tissues reported in the literature. The elastic properties of the phantoms have good stability and the phantoms can be used for

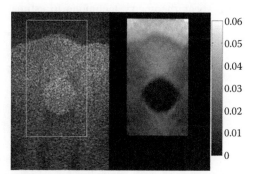

**FIGURE 5.2** US image (left) and elastogram (right) of the breast phantom with a 14-mm diameter stiffer inclusion reported in Madsen et al. (2006). The storage (elastic) modulus values of the healthy surrounding breast TMM and of the inclusion mimicking fibroadenoma are 39.9 ± 1.3 kPa and 97.8 ± 2.9 kPa, respectively. (Reprinted from *Ultrasound in Medicine and Biology* 32, Madsen, E.L., Hobson, M.A., Frank, G.R. et al., Anthropomorphic Breast Phantoms for Testing Elastography Systems, 857–74. Copyright 2006, with permission from Elsevier.)

USE as well as MRE. Lastly, it is worth mentioning that although agarose gel-based TMMs are the most widely used, they appear not to be able to match the mechanical properties of breast tissue. A breast phantom developed by Madsen et al. (1982) made of agar, gelatin and oil has the attenuation, speed of sound and density of breast tissue, but cannot be used in elastography studies for strains less than 10%. This is because for strains greater than 2%, the stiffness of agar increases much faster with strain than the stiffness of breast tissue (Madsen et al., 2003).

### 5.3.3 Brain Phantoms

Brain TMM phantoms for elastography must match the acoustic, mechanical and MR characteristics of the human brain tissue reported in clinical studies. Culjat et al. (2010) proposed the following standard acoustic properties for brain phantoms: US speed of 1560 m/s, density of 1040 kg/m$^3$, US attenuation coefficient of 0.6 db/cm MHz and acoustic impedance of 1.62 MRayl. The reported T1 and T2 relaxation times of brain tissue in an MR scanner with the main magnetic field of 1.5 Tesla are as follows: T1 = 780 ms and T2 = 90 ms for white matter, and T1 = 920 ms and T2 = 100 ms for gray matter. Dynamic MRE studies (Sack et al., 2008; Zhang et al., 2011) provide information on viscoelastic parameters of *in vivo* human brain tissue, which may be standardized for the preparation of brain phantoms. For instance, the storage modulus values of gray matter and white matter are 2.34 ± 0.22 kPa and 2.41 ± 23 kPa at 80 Hz, respectively, and the loss modulus values of gray matter and white matter are 1.11 ± 0.03 kPa and 1.21 ± 0.21 kPa at 80 Hz, respectively (Zhang et al., 2011).

There are very few brain TMM phantoms available for elastography research. They are made from either agarose gel or PVA-C (Figure 5.3) (Surry et al., 2004). Deepthi et al. (2010) made a brain phantom from agar gel material of different concentrations (0.5%–0.8%) and used a shear rheometry technique to estimate the viscoelastic

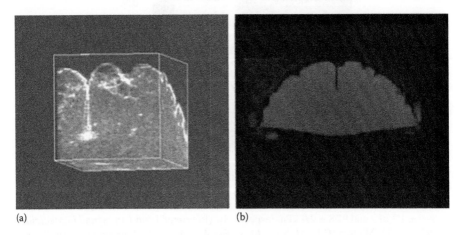

(a)  (b)

**FIGURE 5.3** (a) 3D US image and (b) MRI showing a coronal slice of a PVA-C brain phantom. (Reprinted from Surry, K.J.M., Austin, H.J.B., Fenster, A., and Peters, T.M., *Physics in Medicine and Biology* 49: 5529–46. Copyright (2004), IOP Publishing Ltd. With permission.)

parameters of the phantom. The complex modulus values of the phantoms of gel concentrations 0.5% and 0.6% were comparable to the corresponding values of human brain tissue found using dynamic MRE at 25 Hz and 50 Hz (Sack et al., 2008). However, no acoustic T1 and T2 relaxation parameters were measured for these phantoms and therefore their usefulness in imaging elastography studies still needs to be established.

## 5.4 CRYOGEL PHANTOMS

**Cryogel phantoms** are perhaps one of the most promising and widely used TMMs for imaging, due to the fact that their stiffness and scattering coefficient can be tailored for specific applications by controlling the parameters of the freeze–thaw cycles (FTCs) (Pogue and Patterson, 2006). Each FTC consists of freezing an initially homogeneous polymer solution at low temperatures, storing it in the frozen state for a certain amount of time and then thawing the sample by bringing it back to room temperature. During an FTC, crystal nuclei are generated in the freezing stage, and upon thawing, these nuclei grow into crystallites (Cournane et al., 2010). As the polymer concentration increases due to the conversion of water to ice, the alignment of macromolecular chains provides a mechanism for the formation of side-by-side cross-linking sites (Zhang et al., 2013). Schematics of the cryogel formation process are shown in Figure 5.4.

A system of large interconnected pores is the main characteristic feature of a cryogel. The size of macropores within cryogels varies from a few micrometers to tens or even hundreds of micrometers (Lozinsky et al., 2003). The following subsections cover some of the most widely used cryogel formulations as tissue phantoms for medical imaging, as well as their mechanical properties and imaging characteristics.

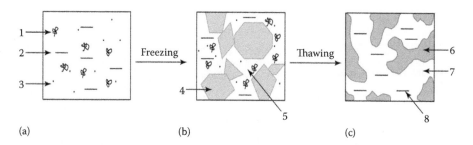

**FIGURE 5.4** Cryogel formation process; 1, macromolecules in a solution; 2, solvent; 3, low-molecular solutes; 4, polycrystals of frozen solvent; 5, unfrozen liquid microphase; 6, polymeric framework of a cryogel; 7, macropores; 8, solvent. (a) Initial system, (b) frozen system and (c) thawed cryogel. (Reprinted from *Trends in Biotechnology* 21, Lozinsky, V.I, Galaev, I.Y., Plieva, F.M., Savina, I.N., Jungvid, H., and Mattiasson, B., Polymeric Cryogels as Promising Materials of Biotechnological Interest, 445–51. Copyright 2003, with permission from Elsevier.)

### 5.4.1 Polyvinyl Alcohol Homopolymer

Polyvinyl alcohol (PVA) is of particularly great interest for various pharmaceutical and biomedical applications because of its many desirable characteristics. The physical characteristics of PVA depend on the degree of polymerization and the degree of hydrolysis. It has been reported that PVA grades with high degrees of hydrolysis have low solubility in water (Hassan and Peppas, 2000). The preparation of PVA hydrogels by the freeze-thaw technique (cryogels) has many advantages. This method not only addresses toxicity issues associated with chemical cross-linking, it also leads to gels with higher mechanical strength than gels cross-linked by chemical or irradiative techniques. **Polyvinyl alcohol cryogel (PVA-C)** is one of the most common materials used by medical imaging researchers to make tissue phantoms. PVA-C is a good candidate for such studies because it can be tailored to mimic many soft tissues in terms of its water content and mechanical properties (e.g. compressibility and elasticity) (Chen et al., 2012; Duboeuf et al., 2009).

During an FTC, the increase in solute concentration in unfrozen regions of the specimen promotes the interactions of PVA chains and favors intermolecular hydrogen bonding (Domotenko et al., 1988). In order to improve the strength of cryogels, the FTC is repeated a number of times. By increasing the number of FTCs, the degree of polymer phase separation, crystallite formation and hydrogen bonding are increased (Lozinsky et al., 2012; Ricciardi et al., 2004). Figure 5.5 shows micrographs of 20 wt% PVA-C in the hydrated state produced using different numbers of FTCs. Differences in microstructure can be observed even after four and five cycles, demonstrating the macroscopic remodeling of the polymer-rich phase with the number of FTCs. The porosity as well as pore size appears to increase with FTCs

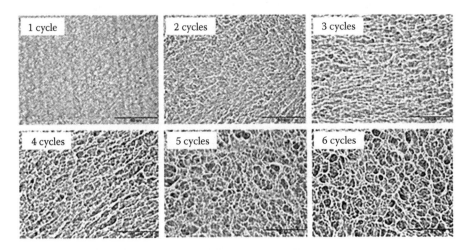

**FIGURE 5.5** Micrographs of 20 wt% PVA cryogels produced with different numbers of FTCs. Scale bars: 80 μm. (Reprinted from *Acta Biomaterialia* 6, Holloway, J.L., Lowman, A.M., and Palmese, G.R., Mechanical Evaluation of Poly(vinyl Alcohol)-Based Fibrous Composites as Biomaterials for Meniscal Tissue Replacement, 4716–24. Copyright 2010, with permission from Elsevier.)

(Holloway et al., 2010). When cryogels are used as elastography phantoms, a higher number of FTCs can be used to produce stiffer embedded inclusions for quantifying the localizability of lesions, both in terms of size and stiffness extent (De Craene et al., 2013; Heyde et al., 2012).

The freezing temperature, the rate of freezing–thawing and the permanence time of the solution at low temperatures are the other key parameters that affect the mechanical properties and the stability of PVA-C (Hassan and Peppas 2000; Lozinsky et al., 2012). Cryogels with high mechanical properties can be obtained by imposing a single FTC either by the prolonged storage of the samples at freezing state or with the use of low thawing rates. This is because the formation of cryogel takes place at the thawing stage (Lozinsky, 2002), where ice melting increases the mobility of macromolecules, and therefore enhances the efficiency of the intermolecular interactions. The lower the thawing rate, the longer the specimen remains at the temperatures optimal for gel network formation (Pazos et al., 2009). This explains the variations in mechanical properties reported for large phantoms, as the different regions of the phantom experience different freezing times and thawing rates.

A modified FTC featuring multiple isotherms has been proposed recently to enhance the homogeneity of large cryogel phantoms (Minton et al., 2012). Figure 5.6a and b shows the temperature profiles for a conventional and a modified FTC, respectively,

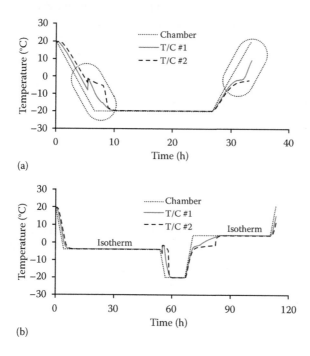

**FIGURE 5.6** Temperature evolution during an FTC at two locations within a 10-cm diameter PVA-C; (a) conventional FTC; (b) modified FTC. The dotted ellipses identify the zones with significant temperature gradient. (Modified from Minton, J., Iravani, A., and Yousefi, A.M., *Medical Physics* 39: 6796–807. Copyright (2012). American Association of Physicists in Medicine. With permission.)

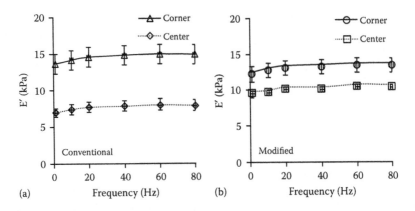

**FIGURE 5.7** Storage modulus of the samples extracted from the corner and center of a 10-cm diameter PVA-C; (a) conventional process, (b) modified process. (Modified from Minton, J., Iravani, A., and Yousefi, A.M., *Medical Physics* 39: 6796–807. Copyright (2012), American Association of Physicists in Medicine. With permission.)

measured at two locations inside a 10-cm diameter PVA-C phantom. Two phenomena can be seen in Figure 5.6a: (1) a sudden temperature increase from about −7.5°C to −2°C during the freezing stage, and (2) a small slowing down of temperature increase around 0°C during the thawing stage. The first observation corresponds to crystallization, whereas the second one indicates ice melting. In case of the modified process (Figure 5.6b), the isotherm at −4°C ensures that the entire solution is completely frozen by holding the temperature below the freezing point (0°C), while the isotherm at 4°C enables a complete thawing above the phase change temperature.

Minton et al. (2012) reported that the storage modulus of the PVA-C produced by the conventional FTC (Figure 5.6a) revealed a variation of 47% across the phantom, ranging from 7.9 ± 0.6 kPa at the center to 14.9 ± 1.3 kPa at the corner (Figure 5.7a). The measured modulus for the PVA-C produced by the modified FTC (Figure 5.6b) was 10.4 ± 0.3 kPa at the center and 13.4 ± 1.0 kPa at the corner, indicating an improvement in homogeneity by over 50% (Figure 5.7b). The observed variations in modulus closely correlated with the variations in the enthalpy of fusion measured by differential scanning calorimetry (Minton et al., 2012). In general, phantoms ranging between 6 and 10 cm in height/diameter are commonly used for elastography techniques (Cournane et al., 2010; Iravani et al., 2014; Latta et al., 2011). The freeze–thaw technique may not be suitable to produce larger phantoms (e.g. 13 cm in diameter) due to incomplete gelation observed at the center of cryogels (Cournane et al., 2010).

### 5.4.2 Other Polymers

Network stability, mechanical properties and water uptake of PVA-C can be altered by interpolymer complexation with other polymers. Blending PVA with natural polymers has been used by a number of groups for a wide range of biomedical applications. These studies include blending PVA with gelatin (Bajpai and Saini 2005; Liu et al., 2009, 2010; Moscato et al., 2008), collagen (Liu et al., 2009; Pon-On et al., 2014), starch (Liu et al., 2009; Zhang et al., 2013), chitosan (Cascone et al., 1999; Liu et al., 2009; Ranjha and Khan, 2013; Silva, 2007), dextran (Cascone et al., 1999; Fathi et al., 2011), albumin (Bajpai and Saini, 2006) and alginate (Chhatri et al., 2011). Since cryogels are usually stored in water to avoid dehydration, the water uptake over time (swelling) is a parameter that may influence the reproducibility of elastography data obtained for cryogel phantoms. Therefore, a phantom with long-term stability is desired to serve as a test bed for elasticity imaging method development and testing (Pavan et al., 2012).

Minton et al. (2012) used blends of PVA with polyvinyl pyrrolidone (PVP), agarose (AGS) and polyacrylic acid (PAA) as potential brain-mimicking phantom materials (Table 5.2). Using an overall polymer concentration of 10%, the effect of cryogel composition on its swelling was investigated for up to 10 days after immersion in water. All the gels demonstrated an initial increase in swelling (Figure 5.8a), which is usually associated with the leach-out of polymer chains that are not incorporated in the crystalline structure (Hassan et al., 2002). The high swelling for PVA-C was attributed to the hydrophilic nature of the PVA molecule (Hassan and Peppas, 2000; Mallapragada and Peppas, 1996). The lower amount of swelling for the gel containing PVP was an indication of a more stable network structure. This was attributed to hydrogen bonding between hydroxyl groups on PVA chains and carbonyl groups on PVP chains (Thomas et al., 2003). Increasing the PVP content resulted in a higher swelling for the gels (not shown). In general, the blends prepared with higher ratios of PVP/PVA have shown significant polymer dissolution due to the interruption of PVA crystallite formation with bulky pyrrolidone rings (Thomas et al., 2003). Figure 5.8b shows that the effect of PAA on swelling was not statistically significant (day 10), whereas both PVP and AGS reduced the water uptake at the same concentration ($p < 0.05$) (Table 5.2).

Chhatri et al. (2011) investigated the effect of alginate (ALG) on the equilibrium swelling of PVA/ALG blend cryogels (PVA/ALG ratios of 40/60 to 80/20). When

**TABLE 5.2**
**Blends of PVA with Other Polymers Proposed as Phantom Formulations**

| Formulation | Polymer-2 | Polymer-2/PVA Ratio (w/w) | Total Concentration (% w/w) | Swelling (Day 10) (%) |
|---|---|---|---|---|
| Pure PVA | – | 0 | 10 | 74.7 ± 7.3 |
| AGS/PVA | Agarose | 1/9 | 10 | 50.6 ± 3.3 |
| PAA/PVA | Polyacrylic acid | 1/9 | 10 | 71.4 ± 14.6 |
| PVP/PVA | Polyvinyl pyrrolidone | 1/9 | 10 | 23.0 ± 1.9 |
| PVP/PVA-2 | Polyvinyl pyrrolidone | 2/8 | 10 | – |

*Source:* Modified from Minton, J., Iravani, A., and Yousefi, A.M. (2012). *Medical Physics* 39: 6796–807. Copyright (2012). American Association of Physicists in Medicine. With permission.

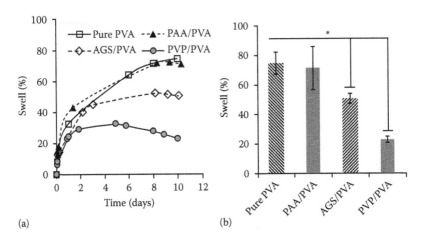

**FIGURE 5.8** (a) Swelling versus time for the different cryogel formulations, and (b) summary of the swelling results at day 10. (Modified from Minton, J., Iravani, A., and Yousefi, A.M., *Medical Physics* 39: 6796–807. Copyright (2012), American Association of Physicists in Medicine. With permission.)

the PVA/ALG ratio increased from 40/60 to 66/34, the overall hydrophilicity of the cryogel increased due to increasing the PVA content and, therefore, the equilibrium swelling also increased. In addition, it was reported that at the experimental pH (7.4), high fractions of PVA led to a fully expanded conformation of nonionic PVA and resulted in a porous cryogel with large pores. On the other hand, at low PVA/ALG ratios, ALG chains were present as fully extended conformation due to the existing repulsion forces between the anionic—COO–groups present along the ALG chains.

## 5.5 CRYOGEL PHANTOM CHARACTERIZATION AND HOMOGENEITY

### 5.5.1 Mechanical Properties

Selecting an appropriate PVA concentration and number of FTCs is essential for developing tissue-mimicking phantoms with desired mechanical properties. Figure 5.9 shows the effect of PVA concentration and the number of FTCs on both compressive and tensile moduli of PVA-C (Holloway et al., 2011). Linear relationships can be seen between the number of FTCs and both compressive and tensile moduli through the first six cycles, whereas the properties remain almost constant at higher FTCs.

PVA solution at a concentration of 10% has been widely used for producing PVA-C phantoms (Fromageau et al., 2007; Herbert et al., 2009; Minton et al., 2012; Surry et al., 2004). The PVA solution at this concentration has a low viscosity at room temperature, making it versatile for filling irregularly shaped molds, while allowing the replication of anatomical features of biological tissues. In addition, the phantoms produced with 10% PVA have a low shrinkage after FTCs and therefore maintain their shape after fabrication (Surry et al., 2004). As for the mechanical

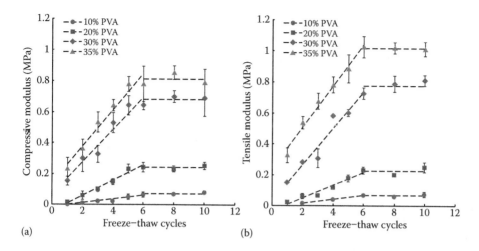

**FIGURE 5.9** (a) The compressive modulus, and (b) the tensile modulus of 10, 20, 30 and 35 wt% PVA-C with increasing FTCs. (Reprinted from *Acta Biomaterialia* 7, Holloway, J.L., Spiller, K.L., Lowman, A.M., and Palmese, G.R., Analysis of the *in vitro* Swelling Behavior of Poly(vinyl Alcohol) Hydrogels in Osmotic Pressure Solution for Soft Tissue Replacement, 2477–82. Copyright 2011, with permission from Elsevier.)

properties, a wide range of modulus has been reported at this concentration due to the differences in cryogel fabrication parameters.

Fromageau et al. (2007) investigated the effect of FTCs (1–10 cycles) on the modulus of PVA-C phantoms prepared using 10 wt% PVA solution. The moduli measured by both tensile testing and three elastography techniques revealed an increasing trend as a function of FTCs (Figure 5.10). This wide range of modulus (20–600 kPa) satisfies the observed Young's moduli for human tissues, ranging from 28 kPa for a normal glandular tissue of breast (Krouskop et al., 1998) to 630 kPa for a healthy carotid artery (Fromageau et al., 2007; Selzer et al., 2001).

Cournane et al. (2010) used cryogel phantoms made of PVA solutions at 5–6 wt% concentration to assess the accuracy of a USE system. The mechanical properties of the phantoms were investigated using a classic compression testing as well as the Fibroscan® transient USE system. The study reported compressive Young's moduli between 1.6 and 16.1 kPa, which mimic those observed in liver tissue. However, the reported stiffness values by the transient elastography system overestimated Young's modulus values representative of the progressive stages of liver fibrosis by up to 32%. In this study, the Young's modulus measured for the phantoms of varying sizes revealed an increasing inhomogeneity as the diameter increased, ranging from 12.5 ± 0.7 kPa, 10.3 ± 2.1 kPa and 9.9 ± 3.1 kPa for the 6.7, 9.2 and 11 cm diameter phantoms, respectively (Cournane et al., 2010). The decrease in the homogeneity was attributed to inhomogeneous thawing rates throughout the phantoms.

Chen et al. (2012) used the assessment of a neurosurgeon as a means of fine-tuning the mechanical properties of PVA-C to those of live human brain. The surgeon was asked to palpate an array of PVA-C samples prepared with 1, 2 or 3 FTCs and 4%–8% PVA solutions (Figure 5.11). Following this initial assessment, the

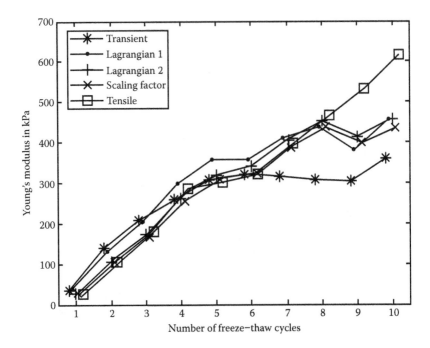

**FIGURE 5.10** Mean Young's moduli of PVA-C estimated by four different elastography methods as a function of FTCs. (From Fromageau, J., Gennisson, J.-L., Schmitt, C., Maurice, R.L., Mongrain, R., and Cloutier, G., *IEEE Transactions on Ultrasonics, Ferroelectrics, and Frequency Control* 54: 498–509, © 2007 IEEE, © 2007 Fromageau et al.)

**FIGURE 5.11** Samples of PVA-C from 4%–8% (left to right) and 1–3 FTCs (top to bottom). (From Chen S.J., Hellier, P., Marchal, M., Gauvrit, J.Y., Carpentier, R., Morandi, X., Collins, D.L., *Medical Physics* 39: 554–61. Copyright (2012), American Association of Physicists in Medicine. With permission.)

cryogel made of 6% PVA with 1 FTC was selected and subjected to direct rheological measurements (Chen et al., 2012). The estimated Young's modulus of 4.6 kPa ± 0.5% was found to be in the range reported for human brain (Metz et al., 1970). However, it should be noted that the reported data on the modulus of brain tissue are inconsistent, mainly because most estimates have been obtained *ex vivo*, from specimens without blood pressure and metabolic activity (Figure 5.12). Therefore, valid quantitative measurements of elastic properties may be useful in the characterization of focal brain lesions, as well as in the development of neurosurgery simulation techniques (Kruse et al., 2008). On the other hand, there are differences in the absolute values of the reported shear modulus for brain tissues in elastography studies. This is usually attributed to differences in experimental methodology and reconstruction algorithms, as well as the age and gender of the volunteers (Sack et al., 2009; Zhang et al., 2011). In light of this, elastography techniques should be verified using a calibrated phantom in order to obtain the range of modulus with independent measurements (Green et al., 2008).

Typically, soft biological tissues are viscoelastic; therefore, they exhibit both solid-like and fluid-like characteristics (Qin et al., 2013). The effect of blending PVA with other water-soluble polymers (e.g. PVP, PAA and AGS) on viscoelastic characteristics of produced hydrogels has been investigated by a number of research groups (Liu and Ovaert 2011; Minton et al., 2012; Zainuddin et al., 2002). Minton et al. (2012) created an improved brain phantom by changing both the composition of the

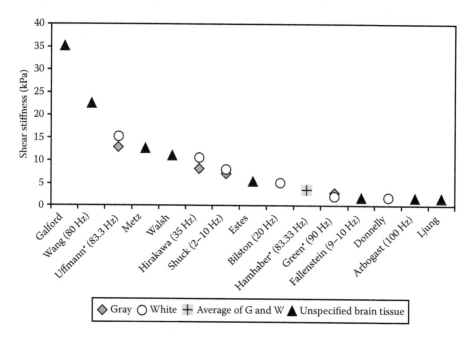

**FIGURE 5.12** The reported data in the literature for shear stiffness measurements of mammalian brain tissue. (Reprinted from *NeuroImage* 39, Kruse, S.A., Rose, G.H., Glaser, K.J., Manduca, A., Felmlee, J.P., Jack, C.R., and Ehman, R.L., Magnetic Resonance Elastography of the Brain, 231–37. Copyright 2008, with permission from Elsevier.)

gel and the freeze-thaw process. Dynamic unconfined compression tests at ambient temperature were performed in displacement control with a sinusoidal waveform between 1 and 80 Hz and a resulted strain of 5%. The dynamic storage modulus of the PVP/PVA phantom was found to be 9.78 ± 1.05 kPa at 80 Hz (Figure 5.13a), which was comparable to the estimated value of 7.09 ± 0.68 kPa at 80 Hz for brain white matter (Minton et al., 2012). These findings suggest that this brain TMM phantom has desirable features for a successful use in dynamic MRE studies of brain tissue. In addition, stress relaxation tests were performed for 30 min following a 5% compressive strain, and the results were compared with that of bovine white matter (Cheng and Bilston, 2007) (Figure 5.13b). The relaxation profile for PVP/PVA cryogel appeared to closely follow that of bovine brain tissue. It should be noted that the ratios of the modulus at the peak to the modulus at equilibrium ($E_{peak}/E_{equil}$) measured for the different formulations were not statistically significant. This was attributed to flow-induced viscoelastic characteristics for these formulations. As a further improvement, the team proposed to tailor the T1 and T2 relaxation times of the phantoms to those of brain tissues and typical brain tumors.

In order to understand the role played by fatty deposits on the accuracy of the Fibroscan® liver transient USE system, Cournane et al. (2012a,b) made fatty liver phantoms and fat layers overlaying healthy liver phantoms by adding an olive oil component to PVA mixtures (90 wt% olive oil, 9 wt% degassed water and 1 wt% Synperonic A7 surfactant). The overlaying fat layers had different thicknesses ranging from 15 mm to 55 mm. The compression tested shear modulus values of the no-fat liver phantoms and the fat-layered liver phantoms with layer thickness up to 35 mm that were between 4 kPa and 8.1 kPa, which agree not only with the shear modulus values found using the Fibroscan system but also with the values found in the early stages of liver fibrosis by the Fibroscan system. However, the Fibroscan

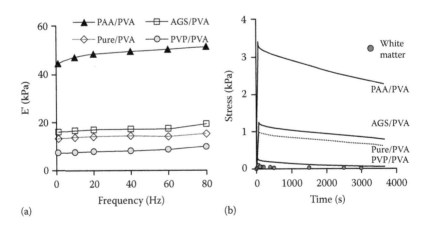

**FIGURE 5.13** (a) Storage modulus versus frequency for different cryogel formulations, and (b) stress relaxation of the cryogels compared with brain white matter. (Modified from Minton, J., Iravani, A., and Yousefi, A.M., *Medical Physics* 39: 6796–807. Copyright (2012), American Association of Physicists in Medicine. With permission.)

measurements of shear modulus of fatty liver phantoms showed a consistent overestimation of about 54% when compared to the values found by compression testing. These findings suggest that special care must be exercised when interpreting the results from the Fibroscan liver transient USE system in clinical practice. T1 and T2 relaxation times for these liver phantoms have not been reported yet (Cournane et al., 2012a,b).

Most elastography studies assume isotropic mechanical properties for reconstructing viscoelastic parameters. However, some biological tissues exhibit anisotropic mechanical properties that vary in different spatial directions. In general, this characteristic is related to the local tissue microstructure such as fiber alignment (Qin et al., 2013). A combination of MRE and diffusion tensor imaging (DTI) has been proposed to assess anisotropic elasticity. The anisotropic phantoms used in the study were produced by embedding elastic Spandex fibers in PVA-C. The MRE/DTI results showed that the ratio of storage moduli parallel and perpendicular to the local fiber axis in anisotropic phantoms and bovine skeletal muscle samples closely matched the rheometry results. Moreover, the MRE/DTI technique was able to distinguish the degrees of mechanical anisotropy between different phantoms and bovine muscle samples (Qin et al., 2013).

Further advancements in designing tissue-mimicking cryogel phantoms for elastography require a better insight into the effect of FTC parameters on phantom properties. Finite-element modeling of heat transfer during cryogel fabrication process may contribute to developing homogeneous phantoms with minimal variations in properties across volume. In a recent paper, Iravani et al. (2014) used finite-element simulations in ABAQUS and inverse modeling to estimate the thermodynamic properties of their brain TMM phantoms at different stages of the freeze–thaw process. Figure 5.14 compares the predicted temperature distribution at the center of a 6-cm diameter cryogel with experimental measurements (Iravani et al., 2014). This computational tool can be used as a virtual controller to find the optimal chamber temperature profile for different phantom sizes. The predicted chamber profile allows

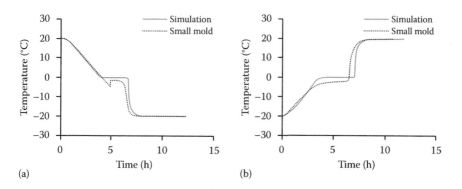

**FIGURE 5.14** Comparison between the simulated and measured temperature at the center of a 6-cm PVA-C during (a) freezing and (b) thawing stage. (Modified from Iravani, A., Mueller, J., and Yousefi, A.M., *Journal of Biomaterials Science. Polymer Edition* 25: 181–202. Copyright (2014). VSP. With permission.)

the center of the phantom reach the freezing/thawing temperature during an isotherm, before initiating the subsequent temperature ramp. It was reported that the virtual controller was capable of providing accurate estimates for the duration of the isotherms required for two different mold sizes, and could potentially be used for PVA concentrations ranging between 5 and 15% (Iravani et al., 2014).

### 5.5.2 Wave Propagation and MRI Properties

Studies on PVA-C have shown the many desirable physical characteristics of this cryogel. In addition to tunable mechanical properties reported for PVA-C, its versatile acoustic and MR properties make it easy to use in multimodality elastography studies. The physical characteristics of cryogel phantoms made of PVA-C have been rigorously investigated by several research groups. Surry et al. (2004) have shown that PVA-C can be used to make breast phantoms for US imaging and MRI. They combined 40 g of PVA powder and 360 g of deionized water to make 10 wt% PVA liquid. A brain mold was created using a stereo-lithography apparatus and a laser deposition of plastic. The PVA liquid was then poured into the mold and subjected to a single FTC. The US speed and T1 relaxation values of this phantom were found to be 1520–1540 m/s and 718–1034 ms, respectively, which are similar to values reported for gray and white matter. However, the T2 relaxation and acoustic attenuation values did not match those of the brain tissue.

Fromageau et al. (2007) found a logarithmic increasing relationship for the acoustic speeds when the number of FTC increased (PVA-C 10%). The reported values were between 1525 and 1560 m/s, which are similar to what can be observed in biological tissues. Two recent studies (Cournane et al., 2010, 2012b) have used PVA-C to make liver TMM phantoms in order to test the accuracy of the Fibroscan® liver transient USE system used in clinical practice. PVA solutions at 5–6% concentration, 0.05% by weight concentration benzalkonium chloride, aluminium oxide acoustic scatterers and 0–16% by weight concentration glycerol were mixed together with deionized water, heated to 90°C, homogenized, cooled at room temperature and then subjected to 1–6 FTCs (Cournane et al., 2010). A scanning acoustic microscope was used to find the US speed and attenuation coefficient of the phantom. The study reported acoustic speeds of 1540–1570 m/s, which mimic those observed in liver tissue. Cournane et al. (2012a,b) measured the values of the acoustic speed for their fatty liver phantoms. The same acoustic and mechanical tests run by Cournane et al. (2010) were performed on these new liver phantoms. The reported acoustic speeds were 1563 ± 4 m/s for no-fat liver phantoms, 1500 ± 4 m/s for fatty liver phantoms and 1491 ± 2 m/s for the fat layers.

## 5.6 CONCLUSIONS

The assessment of elastography-based imaging systems and their diagnostic capabilities requires clinical studies involving many patients and considerable time and cost to obtain adequate statistical significance. Objective assessment of elastography approaches will pave the way to their more effective use in clinical applications. Therefore, phantoms with well-defined properties can help to eliminate the potential weaknesses in elastography systems and would contribute to early tumor detection in patients. The more closely a phantom mimics a healthy or pathologic tissue, the more effective the phantom is for validating imaging techniques. For example, the MRI characteristics of the phantoms can be tailored to meet the T1 and T2 relaxation requirements of healthy biological tissues and typical tumors. Thus, the various components of a phantom should ideally mimic the target tissue not only with respect to mechanical properties, but also with respect to imaging characteristics.

Developing homogeneous phantoms with rigorously controlled mechanical properties and imaging characteristics can fill the gap in the scientific literature with regard to tissue-mimicking phantom materials. In addition, this can serve as a solid foundation for designing sophisticated phantom systems with anatomical features (e.g. white and gray matter in human brain as well as tumor boundaries). Reflecting the lesion-to-background elasticity ratio observed in clinical scanning is also essential to realistically replicate the disease state. Future studies should also look into formulating cryogel phantoms with highly controlled nonlinear stress-strain characteristics, as well as viscoelastic and anisotropic properties mimicking biological tissues.

## ACKNOWLEDGEMENTS

This work was partially supported by The Ohio Board of Regents and the Ohio Third Frontier Program Grant Entitled 'Ohio Research Scholars in Layered Sensing'. The authors thank Joshua Minton, Amin Iravani and Dr. Jens Mueller, as well as the Research Computing Support and the Instrumentation Lab at Miami University for their technical assistance.

## LIST OF ABBREVIATIONS

| | |
|---|---|
| **AGS** | Agarose |
| **CT** | Computed tomography |
| **FTC** | Freeze–thaw cycle |
| **MRE** | Magnetic resonance elastography |
| **OCE** | Optical coherence elastography |
| **PAA** | Polyacrylic acid |
| **PVA** | Polyvinyl alcohol |
| **PVA-C** | Polyvinyl alcohol cryogel |
| **PVP** | Polyvinyl pyrrolidone |
| **TMM** | Tissue-mimicking material |
| **USE** | Ultrasound elastography |

## REFERENCES

Bajpai, A.K., and Saini, R. (2005). Preparation and characterization of biocompatible spongy cryogels of poly(vinyl alcohol)-gelatin and study of water sorption behaviour. *Polymer International* 54: 1233–42.

Bajpai, A.K., and Saini, R. (2006). Preparation and characterization of novel biocompatible cryogels of poly (vinyl alcohol) and egg-albumin and their water sorption study. *Journal of Materials Science. Materials in Medicine* 17: 49–61.

Cascone, M.G., Maltinti, S., Barbani, N., and Laus, M. (1999). Effect of chitosan and dextran on the properties of poly(vinyl alcohol) hydrogels. *Journal of Materials Science. Materials in Medicine* 10: 431–35.

Chen S.J., Hellier, P., Marchal, M., Gauvrit, J.Y., Carpentier, R., Morandi, X., and Collins, D.L. (2012). An anthropomorphic polyvinyl alcohol brain phantom based on colin27 for use in multimodal imaging. *Medical Physics* 39: 554–61.

Cheng, S., and Bilston, L.E. (2007). Unconfined compression of white matter. *Journal of Biomechanics* 40: 117–24.

Chhatri, A., Bajpai, J., Bajpai, A.K., Sandhu, S.S., Jain, N., and Biswas, J. (2011). Cryogenic fabrication of savlon loaded macroporous blends of alginate and polyvinyl alcohol (PVA). Swelling, deswelling and antibacterial behaviors. *Carbohydrate Polymers* 83: 876–82.

Cournane, S., Browne, J.E, and Fagan, A.J. (2012a). The effects of fatty deposits on the accuracy of the Fibroscan® liver transient elastography ultrasound system. *Physics in Medicine and Biology* 57: 3901–14.

Cournane, S, Cannon, L., Browne, J.E, and Fagan, A.J. (2010). Assessment of the accuracy of an ultrasound elastography liver scanning system using a PVA-cryogel phantom with optimal acoustic and mechanical properties. *Physics in Medicine and Biology* 55: 5965–83.

Cournane, S., Fagan, A.J., and Browne, J.E. (2012b). Review of ultrasound elastography quality control and training test phantoms. *Ultrasound* 20: 16–23.

Culjat, M.O., Goldenberg, D., Tewari P., and Singh, R.S. (2010). A review of tissue substitutes for ultrasound imaging. *Ultrasound in Medicine & Biology* 36: 861–73.

De Craene, M., Marchesseau, S., Heyde, B. et al. (2013). 3D strain assessment in ultrasound (straus): A synthetic comparison of five tracking methodologies. *IEEE Transactions on Medical Imaging* 32: 1632–46.

Deepthi, R., Bhargavi, R., Jagadeesh, K., and Vijaya, M.S. (2010). Rheometric studies on agarose gel–a brain mimic material. *SAS Tech Journal* 9: 27–30.

Dickinson. R.J., and Hill, C.R. (1982). Measurement of soft tissue motion using correlation between A-scans. *Ultrasound in Medicine & Biology* 8: 263–71.

Dietrich, C.F., Saftoiu, A., and Jenssen, C. (2014). Real time elasotgraphy endoscopic ultrasound (RTE-EUS), a comprehensive review. *European Journal of Radiology* 83: 405–14.

Domotenko, L.V., Lozinskii, V.I., Vainerman, Ye., S., and Rogozhin, S.V. (1988). Effect of freezing conditions of dilute solutions of polyvinyl alcohol and conditions of defreezing samples on properties of cryogels obtained. *Polymer Science U.S.S.R.* 30: 1758–64.

Drapaca, C.S. (2010). A novel mechanical model for magnetic resonance elastography. *Revue Roumaine des Sciences Techniques-Série Mécanique Apliquée* 55: 3–18.

Duboeuf, F., Basarab, A., Liebgott, H., Brusseau, E., Delachartre, P., and Vray, D. (2009). Investigation of PVA cryogel young's modulus stability with time, controlled by a simple reliable technique. *Medical Physics* 36: 656–61.

Fathi, E., Atyabi, N., Imani, M., and Alinejad, Z. (2011). Physically crosslinked polyvinyl alcohol–dextran blend xerogels: Morphology and thermal behavior. *Carbohydrate Polymers* 84: 145–52.

Fowlkes, J.B., Emelianov, S.Y., Pipe, J.G., Skovoroda, A.R., Carson, P.L., Adler, R.S., and Sarvazyan, A.P. (1995). Magnetic-resonance imaging techniques for detection of elasticity variation. *Medical Physics* 22: 1771–78.

Fromageau, J., Gennisson, J.-L., Schmitt, C., Maurice, R.L., Mongrain, R., and Cloutier, G. (2007). Estimation of polyvinyl alcohol cryogel mechanical properties with four ultrasound elastography methods and comparison with gold standard testings. *IEEE Transactions on Ultrasonics, Ferroelectrics, and Frequency Control* 54: 498–509.

Glaser, K.J., Manduca, A., and Ehman, R.L. (2012). Review of MR elastography applications and recent developments. *Journal of Magnetic Resonance Imaging* 36: 757–74.

Green, M.A., Bilston, L.E., and Sinkus, R. (2008). In vivo brain viscoelastic properties measured by magnetic resonance elastography. *NMR in Biomedicine* 2008: 755–64.

Greenleaf, J.F., Fatemi, M., and Insana, M. (2003). Selected methods for imaging elastic properties of biological tissues. *Annual Review of Biomedical Engineering* 5: 57–78.

Hardy, P.A., Ridler, A.C., Chiarot, C.B., Plewes, D.B., and Henkelman, R.M. (2005). Imaging articular cartilage under compression-cartilage elasograpy. *Magnetic Resonance Imaging* 53: 1065–1073.

Hassan, C.M., and Peppas, N.A. (2000). Structure and applications of poly (vinyl alcohol) hydrogels produced by conventional crosslinking or by freezing/thawing methods. *Advances in Polymer Science* 153: 37–65.

Hassan, C.M., Trakampan, P., and Peppas, N.A. (2002). Water solubility characteristics of poly(vinyl alcohol) and gels prepared by freezing/thawing processes. In: *Water Soluble Polymers*, Amjad, Z., Ed. Springer US, pp. 31–40.

Herbert, E., Pernot, M., Montaldo, G., Fink, M., and Tanter, M. (2009). Energy-based adaptive focusing of waves: Application to noninvasive aberration correction of ultrasonic wavefields. *IEEE Transactions on Ultrasonics, Ferroelectrics, and Frequency Control* 56: 2388–99.

Heyde, B., Cygan, S., Choi, H.F. et al. (2012). Regional cardiac motion and strain estimation in three-dimensional echocardiography: A validation study in thick-walled univentricular phantoms. *IEEE Transactions on Ultrasonics, Ferroelectrics, and Frequency Control* 59: 668–82.

Holloway, J.L., Lowman, A.M., and Palmese, G.R. (2010). Mechanical evaluation of poly(vinyl alcohol)-based fibrous composites as biomaterials for meniscal tissue replacement. *Acta Biomaterialia* 6: 4716–24.

Holloway, J.L., Spiller, K.L., Lowman, A.M., and Palmese, G.R. (2011). Analysis of the *in vitro* swelling behavior of poly(vinyl alcohol) hydrogels in osmotic pressure solution for soft tissue replacement. *Acta Biomaterialia* 7: 2477–82.

Iravani, A., Mueller, J., and Yousefi, A.M. (2014). Producing homogeneous cryogel phantoms for medical imaging: A finite-element approach. *Journal of Biomaterials Science. Polymer Edition* 25: 181–202.

Kennedy, B.F., Hillman, T.R., McLaughlin, R.A., Quirk, B.C., and Sampson, D.D. (2009). *In vivo* dynamic optical coherence elastography using a ring actuator. *Optics Express* 17: 21762–72.

Kennedy, B.F., Kennedy, K.M., and Sampson, D.D. (2014). A review of optical coherence elastography: Fundamentals, techniques and prospects. *IEEE Journal on Selected Topics in Quantum Electronics* 20.

Kirkpatrick, S.J., Wang, R.K., and Duncan, D.D. (2006). OCT-based elastography for large and small deformations. *Optics Express* 14: 11585–97.

Kopeček, J. (2008). Hydrogel biomaterials: A smart future? *Biomaterials* 28: 5185–92.

Krouskop, T.A., Wheeler, T.M., Kallel, F., Garra, B.S., and Hall, T. (1998). Elastic moduli of breast and prostate tissues under compression. *Ultrasonic Imaging* 20: 260–74.

Kruse, S.A., Rose, G.H., Glaser, K.J., Manduca, A., Felmlee, J.P., Jack, C.R., and Ehman, R.L. (2008). Magnetic resonance elastography of the brain. *NeuroImage* 39: 231–37.

Latta, P., Gruwel, M.L.H., Debergue, P., Matwiy, B., Sboto-Frankenstein, U.N., and Tomanek, B. (2011). Convertible pneumatic actuator for magnetic resonance elastography of the brain. *Magnetic Resonance Imaging* 29: 147–52.

Lerner, R.M., and Parker, K.J. (1987). Sonoelasticity images derived from ultrasound signals in mechanically vibrated targets. *Proceedings of the 7th European Communities Workshop,* Oct. 1987, Nijmegen, the Netherlands.

Lerner, R.M., Parker, K.J., Holen, J., Gramiak, R., and Waag, R.C. (1988). Sonoelasticity: Medical elasticity images derived from ultrasound signals in mechanically vibrated targets. *Acoustic Imaging* 16: 317–327.

Liang, X., Oldenburg, A.L., Crecea, V., Chaney, E.J., and Boppart, S.A. (2008). Optical micro-scale mapping of dynamic biomechanical tissue properties. *Optics Express* 16: 11052–65.

Liu, K., and Ovaert, T.C. (2011). Poro-viscoelastic constitutive modeling of unconfined creep of hydrogels using finite element analysis with integrated optimization method. *Journal of the Mechanical Behavior of Biomedical Materials* 4: 440–50.

Liu, Y., Geever, L.M., Kennedy, J.E., Higginbotham, C.L, Cahill, P.A., and McGuinness, G.B. (2010). Thermal behavior and mechanical properties of physically crosslinked PVA/gelatin hydrogels. *Journal of the Mechanical Behavior of Biomedical Materials* 3: 203–9.

Liu, Y., Vrana, N.E., Cahill, P.A., and McGuinness, G.B. (2009). Physically crosslinked composite hydrogels of PVA with natural macromolecules: Structure, mechanical properties, and endothelial cell compatibility. *Journal of Biomedical Materials Research. Part B, Applied Biomaterials* 90: 492–502.

Lozinsky, V.I., Damshkaln, L.G., Kurochkin, I.N., and Kurochkin, I.I. (2012). Study of cryo-structuring of polymer systems. 33. Effect of rate of chilling aqueous poly(vinyl alcohol) solutions during their freezing on physicochemical properties and porous structure of resulting cryogels. *Colloid Journal* 74: 319–27.

Lozinsky, V.I. (2002). Cryogels on the basis of natural and synthetic polymers: Preparation, properties and application. *Russian Chemical Review* 71: 489–511.

Lozinsky, V.I., Galaev, I.Y., Plieva, F.M., Savina, I.N., Jungvid, H., and Mattiasson, B. (2003). Polymeric cryogels as promising materials of biotechnological interest. *Trends in Biotechnology* 21: 445–51.

Luo, J., and Konofagou, E. (2009). Effects of various parameters on lateral displacement estimation in ultrasound elastography. *Ultrasound in Medicine and Biology* 35: 1352–66.

Madsen, E.L., Zagzebski, J.A., and Frank, G.R. (1982). An anthropomorphic ultrasound breast phantom containing intermediate-sized scatterers. *Ultrasound in Medicine and Biology* 8: 381–92.

Madsen, E.L., Frank, G.R., Krouskop, T.A., Varghese, T., Kallel, F., and Ophir, J. (2003). Tissue-mimicking oil-in-gelatin dispersions for use in heterogeneous elastography phantoms. *Ultrasonic Imaging* 25: 17–38.

Madsen, E.L., Hobson, M.A., Frank, G.R. et al. (2006). Anthropomorphic breast phantoms for testing elastography systems. *Ultrasound in Medicine and Biology* 32: 857–74.

Mallapragada, S.K., and Peppas, N.A. (1996). Dissolution mechanism of semicrystalline poly (vinyl alcohol) in water. *Journal of Polymer Science Part B: Polymer Physics* 34: 1339–46.

Mariappan, Y.K., Glaser, K.J., and Ehman, R.L. (2010). Magnetic resonance elastography: A review. *Clinical Anatomy* 23: 497–511.

McAleavey, S., Collins, E., Kelly, J., Elegbe, E., and Menon, M. (2009). Validation of SMURF estimation of shear modulus in hydrogels. *Ultrasonic Imaging* 31: 131–50.

McCracken, P.J., Manduca, A., Felmlee, J., and Ehman, R.L. (2005). Mechanical transient-based magnetic resonance elastography. *Magnetic Resonance in. Medicine* 53: 628–39.

Metz, H., McElhaney, J., and Ommaya, A.K. (1970). A comparison of the elasticity of live, dead, and fixed brain tissue. *Journal of Biomechanics* 3: 453–58.

Miga, M.I. (2003). A new approach to elastography using mutual information and finite elements. *Physics in Medicine and Biology* 48: 467–80.

Minton, J., Iravani, A., and Yousefi, A.M. (2012). Improving the homogeneity of tissue-mimicking cryogel phantoms for medical imaging. *Medical Physics* 39: 6796–807.

Moscato, S., Mattii, L., D'Alessandro, D. et al. (2008). Interaction of human gingival fibroblasts with PVA/gelatine sponges. *Micron* 39: 569–79.

Murphy, M.C., Curran, G.L., Glaser, K.J. et al. (2012). Magnetic resonance elastography of the brain in a mouse model of Alzheimer's disease: Initial results. *Magnetic Resonance Imaging* 30: 535–39.

Muthupillai, R., Lomas, D.J., Rossman, P.J., Greenleaf, J.F., Manduca, A., and Ehman, R.L. (1995). Magnetic resonance elastography by direct visualization of propagating acoustic waves, *Science* 269: 1854–857.

Muthupillai, R., Rossman, P.J., Lomas, D.J., Greenleaf, J.F., Riederer, S.J., and Ehman, R.L. (1996). Magnetic resonance imaging of transverse acoustic strain waves. *Magnetic Resonance in Medicine* 36: 266–74.

Ophir, J., Cespedes, I., Ponnekanti, H., Yazdi, Y., and Li, X. (1991). Elastography: A Quantitative method for imaging the elasticity of biological tissues. *Ultrasonic Imaging* 13: 111–34.

Opielinski, K.J., Pruchnicki, P., Gudra, T. et al. (2013). Ultrasound transmission tomography imaging of structure of breast elastography phantom compared to US, CT and MRI. *Archives of Acoustics* 38: 321–34.

Oudry, J., Bastard, C., Miette, V., Willinger, R., and Sandrin, L. (2009). Copolymer-in-oil phantom materials for elastography. *Ultrasound in Medicine and Biology* 35: 1185–197.

Parker, K.J., Doyley, M.M., and Rubens, D.J. (2011). Imaging the elastic properties of tissue: The 20 year perspective. *Physics in Medicine and Biology* 56: R1–R29.

Pavan, T.Z., Madsen, E.L., Frank, G.R. et al. (2012). A nonlinear elasticity phantom containing spherical inclusions. *Physics in Medicine and Biology* 57: 4787–804.

Pazos, V., Mongrain, R., and Tardif, J.C. (2009). Polyvinyl alcohol cryogel: Optimizing the parameters of cryogenic treatment using hyperelastic models. *Journal of the Mechanical. Behavior of Biomedical Materials* 2: 542–49.

Plewes, D.B., Betty, I., Urchuk, S.N., and Soutar I. (1995). Visualizing tissue compliance with MR imaging. *Magnetic Resonance Imaging* 5: 733–38.

Pon-On, W., Charoenphandhu, N., Teerapornpuntakit, J. et al. (2014). Mechanical properties, biological activity and protein controlled release by poly(vinyl alcohol)-bioglass/chitosan-collagen composite scaffolds: A bone tissue engineering applications. *Materials Science and Engineering: C* 38: 63–72.

Pogue, B.W., and Patterson, M.S. (2006). Review of tissue simulating phantoms for optical spectroscopy, imaging and dosimetry. *Journal of Biomedical Optics* 11: 041102.

Qin, E.C., Sinkus, R., Geng, G. et al. (2013). Combining MR elastography and diffusion tensor imaging for the assessment of anisotropic mechanical properties: A phantom study. *Journal of Magnetic Resonance Imaging* 37: 217–26.

Ranjha, N.M., and Khan, S. (2013). Chitosan/poly (vinyl alcohol) based hydrogels for biomedical applications: A review. *Journal of Pharmacy and Alternative Medicine* 2: 30–42.

Ricciardi, R., Auriemma, F., De Rosa, C., and Lauprêtre, F. (2004). X-ray diffraction analysis of poly (vinyl alcohol) hydrogels, obtained by freezing and thawing techniques. *Macromolecules* 37: 1921–927.

Rogowska, J., Patel, N.A., Fujimoto, J.G., and Brezinski, M.E. (2004). Optical coherence tomographic elastography technique for measuring deformation and strain of atherosclerotic tissues. *Heart* 90: 556–62.

Sack, I., Beierbach, B., Hamhaber, U., Klatt, D., and Braun, J. (2008). Non-invasive measurement of brain viscoelasticity using magnetic resonance elastography. *NMR in Biomedicine* 21: 265–71.

Sack, I., Beierbach, B., Wuerfel, J. et al. (2009). The impact of aging and gender on brain viscoelasticity. *NeuroImage* 46: 652–57.

Sampson, D., Kennedy, K., McLaughlin, R., and Kennedy, B. (2013). Optical elastography probes mechanical properties of tissue at high resolution. *Society of Photographic Instrumentation Engineers Newsroom* January 11, 2013. doi: 10.1117/2.1201212.004605.

Sarvazyan A.P., Rudenko, O.V., Swanson, S.D., Fowlkes, J.B., and Emelianov, S.Y. (1998). Shear wave elasticity imaging: A new ultrasonic technology of medical diagnostics. *Ultrasound in Medicine and Biology* 24: 1419–435.

Schmitt, J.M. (1998). OCT elastography: Imaging microscopic deformation and strain of tissue. *Optics Express* 3: 199–211.

Selzer, R.H., Mack, W.J., Lee, P.L., Kwong-Fu, H., and Hodis, H.N. (2001). Improved common carotid elasticity and intima-media thickness measurements from computer analysis of sequential ultrasound frames. *Atherosclerosis* 154: 185–93.

Silva, M.J. (2007). Biomechanics of osteoporotic fractures. *Injury* 38 Suppl 3: S69–76.

Simon, M., Guo, J., Papazoglou, S. et al. (2013). Non-invasive characterization of intracranial tumors by magnetic resonance elastography. *New Journal of Physics* 15: 085024.

Stella, A., and Trivedi, B. (2011). A review on the use of optical coherence tomography in medical imaging. *International Journal of Advanced Research in Computer Science* 2: 542–44.

Surry, K.J.M., Austin, H.J.B., Fenster, A., and Peters, T.M. (2004). Poly(vinyl alcohol) cryogel phantoms for use in ultrasound and MR imaging. *Physics in Medicine and Biology* 49: 5529–46.

Tanter, M., Bercoff, J., Athanasiou, A. et al. (2008). Quantitative assessment of breast lesion viscoelasticity: Initial clinical results using supersonic shear imaging. *Ultrasound in Medicine Biololgy* 34: 1373–386.

Thomas, J., Lowman, A., and Marcolongo, M. (2003). Novel associated hydrogels for nucleus pulposus replacement. *Journal of Biomedical Materials Research Part A* 67A: 1329–337.

Van Houten, E.E., Doyley, M.M., Kennedy, F.E., Weaver, J.B., and Paulsen, K.D. (2003). Initial *in vivo* experience with steady-state subzone-based MR elastography of the human breast. *Journal of Magnetic Resonance Imaging* 17: 72–85.

Varghese, T., and Ophir, J. (1997). A theoretical framework for performance characterization of elastography: The strain filter. *IEEE Transactions on Ultrasonics, Ferroelectrics, and Frequency Control* 44: 164–72.

Wang, R.K, Ma, A., and Kirkpatrick, S.J. (2006). Tissue Doppler optical coherence elastography for real time strain rate and strain mapping of soft tissue. *Applied Physics Letters* 89: 144103.

Weaver, J.B, Van Houten, E.E., Miga, M.I., Kennedy, F.E., and Paulsen, K.D. (2001). Magnetic resonance elasography using 3D gradient echo measurements of steady-state motion. *Medical Physics* 28: 1620–628.

Weis J.A., Yankeelov, T.E., Munoz, S.A. et al. (2013). A consistent pre-clinical/clinical elastography approach for assessing tumor mechanical properties in therapeutic systems. *Medical Imaging 2013: Biomedical Applications in Molecular, Structural, and Functional Imaging, Proc. of the SPIE*, 8672, 86721F: 1–7.

Wilson, L.S., and Robinson, D.E. (1982). Ultrasonic measurement of small displacements and deformations of tissue. *Ultrasonic Imaging* 4: 71–82.

Xu, L., Lin, Y., Ha1, J.C., Xi, Z.N., Shen, H., and Gao, P.Y. (2007). Magnetic resonance elastography of brain tumors: Preliminary results. *Acta Radiologica* 48: 327–30.

Yamakoshi, Y., Sato, J., and Sato, T. (1990). Ultrasonic-imaging of internal vibration of soft-tissue under forced vibration. *IEEE Transactions on Ultrasonics, Ferroelectrics, and Frequency Control* 37: 45–53.

Yin, M., Talwalkar, J.A., Glaser, K.J. et al. (2007). Assessment of hepatic fibrosis with magnetic resonance elastography. *Clinical Gastroenterology and Hepatology* 5: 1207–1213.e2.

Zainuddin, Cooper-White, J.J., and Hill, D.J. (2012). Viscoelasticity of radiation-formed PVA/PVP hydrogel. *Journal of Biomaterials Science. Polymer Edition* 13: 1007–20.

Zhang, H., Zhang, F., and Wu, J. (2013). Physically crosslinked hydrogels from polysaccharides prepared by freeze–thaw technique. *Reactive and Functional Polymers* 73: 923–28.

Zhang, J., Green, M.A., Sinkus, R., and Bilston, L.E. (2011). Viscoelastic properties of human cerebellum using magnetic resonance elastography. *Journal of Biomechanics* 44: 1909–13.

# 6 Cryogels in Regenerative Medicine

*Irina N. Savina,\* Rostislav V. Shevchenko, Iain U. Allan, Matthew Illsley and Sergey V. Mikhalovsky*

## CONTENTS

| | | |
|---|---|---|
| 6.1 | Introduction | 176 |
| 6.2 | Cryogel Properties | 176 |
| 6.3 | Cryogels for Dermal Tissue Regeneration | 177 |
| 6.4 | Cryogels as a Component of Topical Wound Dressings | 184 |
| 6.5 | Cryogels for Cartilage Repair and Regeneration | 186 |
| 6.6 | Cryogels for the Regeneration of Bone | 188 |
| 6.7 | Cryogels for the Regeneration of Neural Tissue | 189 |
| 6.8 | Cryogels for the Regeneration of Myocardial Tissue | 189 |
| 6.9 | Cryogels for Regeneration of the Liver and Pancreas | 189 |
| 6.10 | Microcryogels for Cell Delivery | 190 |
| 6.11 | Conclusions | 190 |
| Acknowledgements | | 191 |
| List of Abbreviations | | 191 |
| References | | 191 |

## ABSTRACT

**Cryogels** are hydrogel materials produced in semifrozen conditions and have a unique structure of large interconnected pores and many features of biomaterials. In this chapter, the cryogel properties with particular focus on applications in regenerative medicine are discussed. Recent studies in dermal tissue regeneration, topical wound dressing and emerging targets such as cartilage, bone, neural, cardiac, liver and pancreas tissue are reviewed. The use of microcryogels for cell delivery is also discussed.

## KEYWORDS

Regenerative medicine, cryogel, tissue engineering, microcryogel, cell delivery

---

\* Corresponding author: E-mail: i.n.savina@brighton.ac.uk; Tel: +441273642034.

## 6.1 INTRODUCTION

The method of cryogelation is often used by bioengineers to produce highly porous materials with large interconnected pores and thick polymer walls to provide the elasticity, as well as the osmotic and mechanical stability of a scaffold (Bolgen et al., 2007; Dainiak et al., 2010; Jurga et al., 2011). The cryogelation technique does not require the use of any organic solvents and it allows the porosity of the engineered scaffold to be controlled and therefore adjusted for a particular application (Lozinsky et al., 2003). Cryogels manufactured in aqueous solutions have large pores up to 100–150 µm, favourable for the infiltration, migration and proliferation of living cells (Savina et al., 2009a,b, 2011; Dainiak et al., 2010). Cryogels have also been shown to possess shape memory—those made from biological components can be squeezed to expel water from their pores and then recover their original shape when immersed in water or aqueous fluids (Savina et al., 2007). Cryogel water is mainly nonbound or weakly bound, and it does not strongly interfere with the bioprocesses within the cryogel, making it an attractive material for biomedical applications (Savina et al., 2011; Gun'ko et al., 2013). This chapter focuses on cryogel applications in regenerative medicine, with particular reference to the skin and emerging targets such as cartilage, bone and neural and cardiac tissue.

## 6.2 CRYOGEL PROPERTIES

**Cryogels** are hydrogels produced from semifrozen solutions when a large part of the solvent is crystallized. Like hydrogels, cryogels are formed by the physical or chemical cross-linking of polymers or by polymerization. The solvent crystals in the reagent mixture aggregate, filling up cavities in the cryogel matrix, thus playing the role of pore-forming substances. As water is commonly used as the solvent, the pores are filled with water after defrosting, which can be easily expelled by squeezing. A wide range of synthetic and natural polymers has been used for cryogel synthesis. For regenerative medicine, biocompatible (nontoxic) polymers are required. Naturally occurring polymers such as collagen, gelatin, chitosan, dextran and agarose have been explored for the production of macroporous matrices. An excellent review of biopolymers used for hydrogel production and their application in tissue engineering has been published by Van Vlierberghe and co-workers (Van Vlierberghe et al., 2007, 2011). Synthetic polymers are often used for optimization of the cryogel matrix for certain cell culturing conditions. Polyacrylamide, poly-(2-hydroxyethyl methacrylate) and polyethylene glycol cryogels have been developed as scaffolds for cell growth (Savina et al., 2007). In general, synthetic polymers have superior properties in terms of chemical and mechanical stability; however, they are poorly biointegrated. Glutaraldehyde is commonly used for cross-linking both synthetic and natural polymers. Glutaraldehyde is known to be toxic and hence unreacted glutaraldehyde should be completely removed from the matrix before biomaterial application. Increasing concentrations of glutaraldehyde in gelatin cryogels were associated with a decrease in the biocompatibility of the cryogel material and decrease in cell proliferation in gelatin cryogels (Dainiak et al., 2010).

An alternative cross-linker is 1-ethyl-3-3-dimethylaminopropylcarbodiimide hydrochloride (EDC) (Sharma et al., 2013). Collagen, gelatin, dextran and hyaluronic acid were cross-linked using EDC. Some polymers can form a gel through physical cross-linking. Polymers like poly(vinyl alcohol) (PVA) and agarose form gels as a result of the intermolecular interactions between their respective polymer chains. These cryogels are thermoreversible, that is, they melt at a higher temperature and are stable at the biological temperature of 37°C. The use of physically cross-linked cryogels is very attractive for the applications where the addition of chemicals is not desirable. To avoid adding the cross-linker or initiators for polymerization, polysaccharide cryogels were made using an electron beam that initiates cross-linking (Reichelt et al., 2014). The advantages of this method are short reaction time (10 min), low cost and ability to produce materials of high purity. As no initiator or cross-linker is used, there is no contamination and no need for washing. The method has no limitation to the sample size.

The cryogel properties are dependent on the cryogelation conditions, nature of the polymer, its concentration and the presence of additives. Composite cryogels containing solid particles were reported (Sandeman et al., 2011; Gun'ko et al., 2013). The composites combine the properties of both the hydrogel base material and the filler. Activated carbon particles are added to improve the adsorption properties of the cryogels, which could be used as a wound dressing. Hydroxyapatite could be added to the cryogels to facilitate bone regeneration. The incorporation of hard particles in the cryogel modifies its mechanical properties and can be used to produce mechanically stronger materials. Highly porous materials with large interconnected pores are produced for cell cultures. Skin may be repaired using cryogels made from components of, or similar to, the dermal extracellular matrix. Improved mechanical properties are required for cryogels to be used in bone replacement or cartilage repair with some promising developments (Kemençe and Bölgen, 2013; Gonzalez and Alvarez, 2014; Mishra et al., 2014). The cryogelation process provides for flexibility of material design and optimization of cryogel characteristics—tailored for the particular application.

## 6.3 CRYOGELS FOR DERMAL TISSUE REGENERATION

Burns and scalds, acute trauma, genetic skin disorders and chronic wounds can result in skin loss. Extensive, deep wounds with substantial areas of skin damage have a limited ability for skin regeneration, and therefore need to be treated surgically with the underlying bulk of dermal tissue requiring to be replaced. The wound must also be closed using surgical skin grafting procedures using a sample of the patient's own uninjured epithelium (upper layer of skin) (Papini, 2004), or increasingly, a preparation of autologous keratinocyte cells (covered with a protective dressing until such time as a functional epithelium has developed). To replace bulk dermis, which underlies the thinner outer epithelial layer of skin, bioengineered skin substitutes can be used (MacNeil, 2007; Shevchenko et al., 2008, 2010). Significant progress has been made recently in the development, marketing and clinical use of skin substitution products (Horch et al., 2005; Clark et al., 2007; MacNeil, 2007; Shevchenko et al., 2010; Zhong et al., 2010; van der Veen et al., 2011; Almeida et al., 2013).

The consensus is that the clinical 'gold standard' material for dermal substitution in full thickness wounds is Integra® (Heimbach et al., 2003; Yannas et al., 2011; Muller et al., 2013). It is collagen-based sponge sheet, laminated with a detachable pseudo-epidermal silicone layer, designed to biointegrate and promote controlled neodermis formation to achieve a safe and effective burn treatment. Unfortunately, its high cost and long biointegration time could be preventive factors for many patients. Collagen, the main component of Integra, is the natural and most abundant component of the extracellular matrix (ECM) of skin (Badylak, 2007). Gelatin, on the other hand, is a product of collagen hydrolysis and chemically, in many ways, is similar to collagen (Ward and Courts, 1977). Being inexpensive, biocompatible, biodegradable and nonimmunogenic, it is widely used in clinics as a component of, for example, wound dressings, vascular stent modifying materials, implantable antibiotic carriers and in dental practice, as well as for neurosurgical applications (Kharitonova et al., 1990; Choi et al., 2001; Nakayama et al., 2003; Yang et al., 2010; Pal et al., 2013). Gelatin cryogels were successfully made and analysed as materials supporting the growth of skin cells, indicating their potential for use in dermal tissue regeneration (Dubruel et al., 2007; Shevchenko et al., 2008, 2014; Dainiak et al., 2010; Savina et al., 2011).

Many of the physical, chemical and biological features associated with cryogels make them ideal for use as dermal regeneration scaffolds (Savina et al., 2009b; Dainiak et al., 2010). Cryogel scaffolds, synthesized from natural polymers such as chitosan, agarose or gelatin and seeded with fibroblasts were reported to support cellular proliferation within the matrix (Bhat and Kumar, 2012; Shevchenko et al., 2014). It was noted that such scaffolds provided surfaces favourable for cell proliferation and cellular synthesis of extracellular matrix including the deposition of collagen on the scaffold surface.

A variety of approaches to apply skin substitutes include scaffolds relying on recruiting host cells *in vivo*, scaffolds preseeded with allogenic nonviable cells to release cytokines to aid wound healing and scaffolds preseeded with viable autologous cells to produce an immediate skin cover. Cryogels are no exception to these approaches. Another step toward ready-to-use 3D scaffolds for regenerative medicine was taken when adhesion-based cryopreservation of human mesenchymal stem cells (hMSCs) within alginate-gelatin cryogel scaffolds was reported (Katsen-Globa et al., 2013). The authors analysed different cultivation times prior to freezing and the effect on cell viability and functionality upon thawing. The results of the work suggested that adherent hMSCs on alginate-gelatin scaffolds can be cryopreserved successfully without any further preparation steps, and cryopreservation success depends on the cultivation time (with minimal cell spreading) before cryopreservation in order to recover adherent cells with maintained functionality. This informs research toward the development of tissue regeneration matrices preloaded with cells to be stored frozen, ready for almost immediate use.

Macroporous gelatin cryogel scaffolds cross-linked with glutaraldehyde (GL-GA) in a sheet form were synthesized and a silicone pseudoepidermal layer (GL-GA-S) applied to mimic a bilayered skin structure (Shevchenko et al., 2014). The resulting prototype construct was assessed for wound repair potential by studying the porosity and structure, biomechanical, cytotoxic, bioconductive, biosynthetic, proinflammatory and biotransformation properties of the scaffolds when seeded with human skin fibroblasts or keratinocytes. Further assessment included an *in vivo* preclinical evaluation in a porcine wound healing model. The tissue-engineered cryogel material was compared with the clinically proven Integra dermal regeneration template, which is widely used in clinics for extensive burn management (Heimbach et al., 2003; Yannas et al., 2011; Muller et al., 2013).

These cryogel scaffolds revealed a supermacroporous anisotropic structure with smaller pores at the upper surface and larger pores at the opposite bottom surface (Figure 6.1) unlike Integra, in which the mean pore size does not have a defined gradient across the thickness of the material, as confirmed by confocal microscopy in the hydrated state.

This anisotropic arrangement, with smaller pores at the upper surface, could be of potential benefit for clinical use, allowing subsequent keratinocyte application to form a continuous epithelial layer on the upper surface of a gel already infiltrated with dermal cells in the underlying bulk phase. The keratinocytes would be restricted to this layer by an inability to migrate through the small restrictive pores underneath. This would form a natural barrier similar to that found in real skin where epithelial cells form a distinct interface with the underlying dermal cells. *In vitro* results support this statement, as a continuous epithelial layer was formed over a cryogel surface, when seeded with human keratinocytes suspensions. This was not the case with Integra scaffold (without a silicone layer) (Figure 6.2).

An additional benefit of larger pores at the bottom of the material is its increased flexibility to adjust to the roughness and curvature of the wound surface. The failure of skin substitute biomaterials to adjust to the wound surface often results in suboptimal wound healing (Burke et al., 1981).

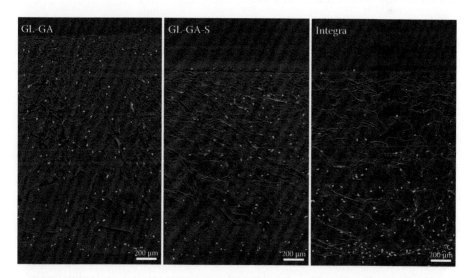

**FIGURE 6.1** Supermacroporous anisotropic structure of cryogel and no pore size gradient within Integra, as assessed by confocal microscopy in a natural hydrated state of a scaffold. The biomaterials were populated with human dermal fibroblasts (green), seeded on the bottom surface, and allowed to proliferate for 14 days. Bar: 200 μm.

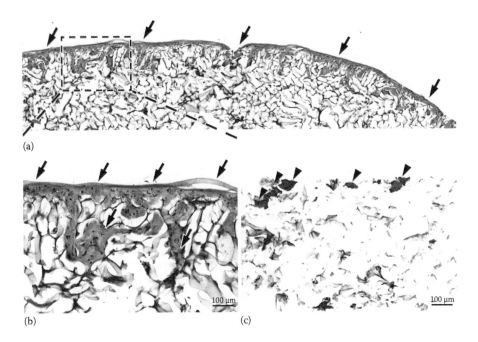

**FIGURE 6.2** Epithelialization of GL-GA cryogel and Integra scaffolds. After 17 days *in vitro*, a continuous epithelial layer formed over GL-GA cryogel (a, arrows) which seemed to be more structured and mature when compared with epithelial islands (arrowheads) formed over Integra (b). Keratinocyte migration (c) into the pores of the topmost layer of GL-GA scaffold was also evident. H&E, Bar: 100μm.

Gelatin is a product of collagen hydrolysis and it consists of peptides and proteins, where molecular bonds between individual collagen strands and hydrogen bonds, which stabilize the collagen helix, are broken down (Ward and Courts, 1977). Gelatin cryogels do not exhibit significant cytotoxic effects, as confirmed by MTS assay, when cultured with human keratinocytes for 42 h (Figure 6.3).

These findings are also supported by histological observations, where there are no signs of toxic effects, that is, keratinocyte necrosis, cell shrinkage, vacuolization, membrane rupture, loss of junction attachments between cells or epithelial layer detachment. Gelatin cryogel is also a suitable substrate for dermal fibroblast growth. Cells readily populate 3D matrix and proliferate *in vitro*, as assessed at days 0, 1, 7 and 28 (Figure 6.4), confirming the nontoxic nature of the cryogel scaffold.

Biochemical viability assessments, such as the MTT assay, allow the estimation of the total number of viable cells within a scaffold such as gelatin cryogel; however, they do not allow measurement of the cell distribution, crucial for characterization of the performance of potential skin substitute biomaterials. Penetration of tissue engineered scaffold materials by host cells is a basic requirement for the important initial step of scaffold colonization and subsequent infiltration, leading to successful constructive ECM remodelling, neodermis formation and wound healing (Badylak, 2007). Direct visualization of cells within scaffolds using confocal laser scanning microscopy (e.g. Allan et al., 2009) is a useful supplement to biochemical

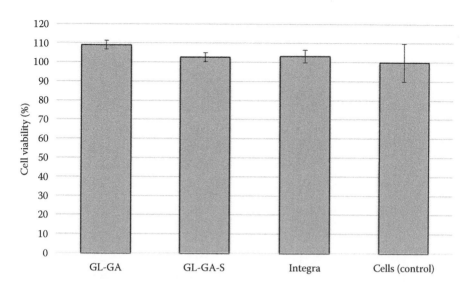

**FIGURE 6.3**  MTS assay with culture medium, where scaffolds were cultured with multi-layered human epidermal model for 42 h. Values are mean n = 3, error bars represent S.D.

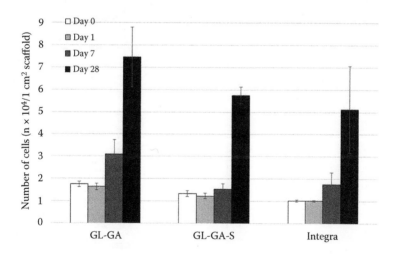

**FIGURE 6.4**  Proliferation of normal primary human dermal fibroblasts within GL-GA, GL-GA-S and Integra scaffolds over 28 days *in vitro*. MTT assay, values are the number of cells ×10$^4$ per 1 cm$^2$ of scaffold, mean n = 3, error bars represent S.D.

viability tests, providing spatial information concerning cell infiltration versus time and indeed direct viability measurements (in conjunction with appropriate fluorescent viability stains). In the Shevchenko at al. (2014) study, confocal microscopy supported the MTT assay findings of active cellular proliferation within matrices (Figure 6.1). Moreover, it showed crucial differences between cryogel scaffolds and Integra under the conditions of the experiment: cells were evenly distributed across

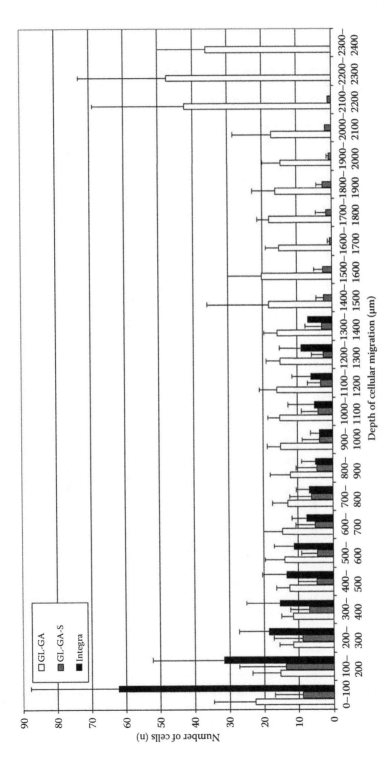

**FIGURE 6.5** Normal primary human dermal fibroblasts distribution within GL-GA, GL-GA-S and Integra scaffolds after 28 days *in vitro*. Mean n = 9, error bars represent S.D.

the entire thickness of the cryogel scaffolds, whereas in Integra, the cells were predominantly located at the lower portion of the scaffold (Figures 6.1 and 6.5).

Gelatin cryogel scaffolds also do not seem to elicit a pro-inflammatory response. The GL-GA, GL-GA-S and Integra matrices were exposed to a multilayered human epidermal model for 42 h to assess IL1-$\alpha$ production, as a pro-inflammatory cytokine, which plays a crucial role in wound healing (Gillitzer and Goebeler, 2001; Jones et al., 2003; Hu et al., 2010). When IL1-$\alpha$ was measured in the supernatant, the concentrations were the following: The GL-GA 1.18 ± 14.55 pg/ml; GL-GA-S 13.64 ± 29.21 pg/ml; Integra 9.04 ± 11.41 pg/ml (Figure 6.6). The GL-GA-S cryogel scaffold induced the highest IL1-$\alpha$ production in the test, probably due to the silicone rubber; however, the results did not differ from other scaffolds statistically, and were similar to controls. This may indicate that the assessed biomaterial scaffolds are not likely to cause an inflammatory response and can be used for *in vivo* application without inducing an excessive inflammatory reaction, although it would be prudent to monitor the cellular production of different cytokines, also associated with inflammation, in response to exposure to the cryogels.

Cells infiltrating and proliferating within cryogels (and other 3D matrices) can be probed for the expression of a variety of proteins, as a means to assess their physiological state and, at the point of sampling, their influence on the remodelling of the material to become more like real skin (Allan et al., 2009).

Synthesis of a chitosan derivative cryogel was also performed and these were assessed in preclinical rodent models (Takei et al., 2012). Synthesized cryogels, when implanted in mice, promoted the accumulation of inflammatory cells such as polymorphonuclear leukocytes, which have the potential to release chemical mediators effective for wound healing. When the same chitosan-based cryogels were used topically to treat full-thickness skin wounds in rats, they were reported to accelerate the healing of the wounds. Further work from the same group of authors assessed chitosan-gluconic acid conjugate/poly(vinyl alcohol) cryogels, which possessed improved mechanical strength, enhanced water retention, and resistance to degradation of the gels by lysozyme. The incorporation of basic fibroblast growth factor into the cryogels accelerated wound healing on partial-thickness wounds in diabetic rats (Takei et al., 2013).

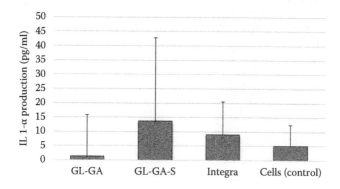

**FIGURE 6.6** IL1-$\alpha$ production by keratinocytes as response to exposure of GL-GA, GL-GA-S and Integra matrices for 42 h, ELISA assay. Mean n = 3, error bars represent S.D.

The majority of work toward developing cryogels for tissue regeneration has focused on dermal regeneration. However, further research is now being directed to include other body sites.

## 6.4 CRYOGELS AS A COMPONENT OF TOPICAL WOUND DRESSINGS

Wound dressings are used to accelerate healing of wounds, and there are multiple rationale behind the mechanisms used to achieve this goal including hydration, protection, prevention of symptoms, for example, itching and infection. Since infection is the major mechanism by which healing is delayed, most dressings include an antimicrobial. However, in certain cases another dressing type maybe indicated. In our laboratory, we have addressed the issue of chemical burn injuries where the injury-causing agent continues to cause damage for a significant time.

A cryogel comprised of polyacrylamide was produced that was impregnated with activated carbon (Figure 6.7a). SEM imaging of this dressing showed that the activated carbon was loose within the pores of the cryogel, free to absorb any wound contaminant (Figure 6.7b). This dressing was applied to sulphur mustard wounds on a large white pig model 1 h after injury occurred. The effect of absorbing the chemical agent on the speed of wound healing was examined.

The treatment of slow healing chemical burns by an activated carbon-cryogel composite dressing was demonstrated to be better than control wounds (Figure 6.8); in some instances, there was complete healing in the carbon-cryogel composite dressings (Figure 6.8c). Because of the major features of cryogels, hydration, mass swelling, shape memory, robustness after multiple drying/swelling events, and, when made from synthetic polymers, the nonintegration into the wound, these polymers have the potential to be used as a primary wound healing dressing. Further cryogels are adaptable enough for special formulations to be used for treatment of unusual or persistent wounds, such as the sulphur mustard wounds presented here.

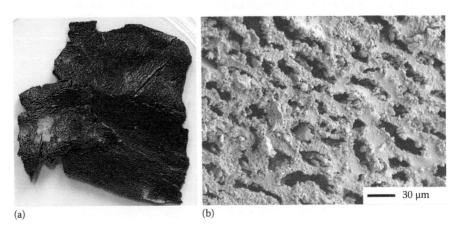

**FIGURE 6.7** Activated carbon impregnated polyacrylamide cryogels were created (a) and observed by SEM (b) to have freely available carbon within the pores of the cryogel.

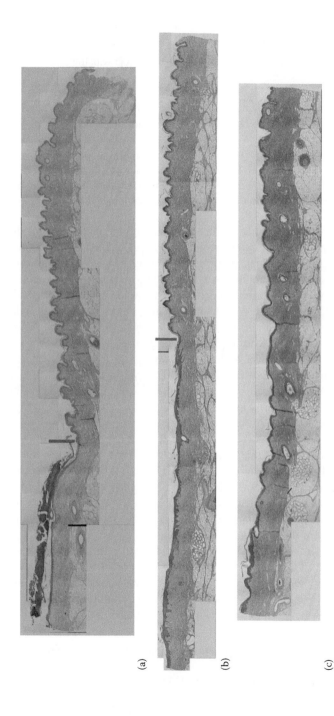

**FIGURE 6.8** Composite photomicrographs (×40) of H&E stained histological sections of large white pig sulphur mustard wounds 19 days postwounding after treatment with (a) control gauze treatment; (b) 'gold standard' control—Acticoat 7; and (c) University of Brighton activated carbon-cryogel dressing. The red line bisecting the sections indicates where the wound margin begins; to the right, the wound is unchallenged with SM, to the right of the treated challenged wound, this margin is not apparent in (c).

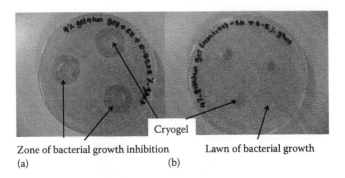

Zone of bacterial growth inhibition    Lawn of bacterial growth
(a)                                     (b)

**FIGURE 6.9**  (a) A 4% gelatin cryogel + 0.2% CX + 0.0625% glutaraldehyde ('low') with large zones of inhibition against *S. aureus*; (b) a 4% gelatin gel without CX but cross-linked with 0.5% v/v glutaraldehyde with no zones of inhibition against *S. aureus*.

Preliminary tests infusing cryogels with the antimicrobial agent chlorhexidine (CX) were undertaken in this laboratory with a view to evaluating the potential of cryogels for delivering antimicrobials in a wound dressing. This agent is commonly found in antibacterial scrubs and mouthwashes. It is anticipated that controlled release of CX from gelatin cryogels is a promising goal in the development of an antibacterial wound dressing. Cryogels made from bovine gelatin (4% w/v) were cross-linked with glutaraldehyde at concentrations in the range of 0.0625–0.5% v/v and infused with or without 0.2% w/v chlorhexidine. Tests investigating any bacteriostatic (i.e., inhibition of bacterial growth) effect were undertaken. Figure 6.9 shows agar plates inoculated with *Staphylococcus aureus* MRSA NCTC15, with the cryogels (4% gelatin w/v) neatly housed in wells excised from the agar. Following overnight incubation at 37°C, heavy golden bacterial growth is visible on the agar. Near the cryogels infused with 0.2% CX, large clear zones of bacterial growth inhibition are visible. This was also apparent when the cryogels were tested against *Pseudomonas aeruginosa* (not shown). The cryogels that contained glutaraldehyde only did not exert a bacteriostatic effect. The presence of glutaraldehyde neither impeded the release of CX out of the gel, nor inactivated it (as evidenced by the bacteriostatic activity of CX where present). It was also notable that when using the relatively low final concentration of 0.0625% (v/v) glutaraldehyde in the 4% gelatin cryogels, they remained intact at 37°C in the presence of the bacteria.

Other groups have synthesized cryogels for wound healing purposes. A polysaccharide-based savlon-loaded antibacterial cryogel was reported to be effective for wound healing applications (Chhatri et al., 2011), as it was shown to possess antibacterial properties being effective against *E. coli in vitro*. This could find its use in wound dressings; however, no preclinical data are yet available.

## 6.5 CRYOGELS FOR CARTILAGE REPAIR AND REGENERATION

The degradation of articulating cartilage (such as that of the knee and hip) is a growing problem as the population ages. This soft tissue provides a lubricated interface smoothing the movement of the underlying bone plates, and absorbs the impacts of walking and running. Like skin, an artificial cartilage must be biocompatible; however, it must also be able to withstand strong mechanical forces. Cartilage is composed primarily of water and the ECM components collagen and glycosaminoglycans. The principal cells present are chondrocytes. These cells have low motility and this, together with a lack of vascularisation to enable blood flow, means cartilage does not readily repair or regenerate naturally (Baker et al., 2012).

Polyvinyl alcohol (PVA) has been studied quite extensively as a candidate cartilage-replacement material. It is nontoxic, can be prepared with a similar tensile strength and compressive modulus to native cartilage, and with similar water content. Hydrogels of PVA will present a smooth surface, which, although it offers some lubrication, cannot be readily infiltrated with cells. One major problem associated with long-term implantation of soft or hard biomaterials is erosion and degradation on their surface—the resulting micron-sized molecules stimulate a macrophage response, which can escalate to extended inflammation and failure of the material. Thus, for long-term implant success, any cartilage replacement material should ultimately be replaced with regenerated native cartilage. This could occur *in situ* by recruitment of chondrocytes from the surrounding cartilage to an implanted cryogel or by preseeding cryogels with these cells in the laboratory. In both cases, the cryogel should be gradually remodelled by cellular processes to resemble native cartilage.

In cryogel form, polymers such as PVA present a highly porous network with dimensions suitable for cellular infiltration. This would be a prerequisite for subsequent cell-mediated remodelling of the material and native cartilage regeneration. However, PVA needs to be augmented with suitable biomimetic molecules for any cellular attachment to occur. Candidates that have been assessed to create composite cryogels with PVA include hydroxyapatite, gelatin, collagen and other ECM-derived components. Gonzalez and Alvarez (2014) investigated the mechanical properties of PVA and hydroxyapatite (HA) composite cryogels as a preliminary assessment of their suitability as cartilage replacement materials. HA is the mineral component of bone and is therefore biocompatible and bioactive. The addition of HA at defined concentrations was found to improve the tensile properties of the PVA cryogels provided that HA particles did not agglomerate and that the pore size within the cryogels was directly proportional to the concentration of HA. This latter finding has some potential to create cryogels with optimal geometry for cellular infiltration. No studies were undertaken to assess cellular attachment to these materials and subsequent proliferation and infiltration.

Bhat et al. (2013) describe an *in vitro* assessment of cellular compatibility with a cryogel composite composed of the natural biopolymers gelatin, chitosan and agarose. Goat chondrocytes were found to proliferate on this composite and produce a 'neo-cartilage' that was easily removed from the surface of the cryogels. Biochemically, the neo-cartilage was broadly similar to the native state with similar concentrations of GAG, DNA and collagen present. However, collagen alignment was found to be less organized in the neo-collagen as indicated by the intensity of direct red staining. The mechanical properties of the neo- and native cartilage also differed, with the stored elasticity (storage modulus) greater with the latter. However, it is suggested that if such cryogels seeded with chondrocytes are implanted *in vivo*, the local mechanical forces may stimulate these cells to produce ECM with a similar density, orientation and mechanical properties to native cartilage. Cells are well known to greatly alter their production of ECM in response to mechanical stimuli.

## 6.6 CRYOGELS FOR THE REGENERATION OF BONE

Bone replacement and repair is another area considered for intervention using cryogels. Like cartilage, a load-bearing bone is subjected to strong mechanical forces. Therefore, any materials used to replace such bone must be capable of withstanding these forces. Some bone types are more responsive to non-load bearing treatments with cryogels, for example, craniofacial bones do not have to withstand such high force loads as the rest of the skeleton.

Trabecular bone is highly porous with high blood flow, comprised of interlocking flat planar lamellae, each lamella being never more than 180 µm thick to allow osteoclasts to penetrate the whole structure. Therefore, any regenerative osteogenic material would require cryogel-like properties: high flow rate, planar pore walls and high strength-to-weight ratio.

Bone regenerative materials are generally comprised of calcium hydroxyapatite (HA) since the major constituents of bone are mineralised collagen with HA. Most materials in this field do not match the tensile strength of cortical or trabecular bone and are therefore not generally very useful for healing of large bone defects in the load-bearing skeleton.

Cryogels have properties similar to those of a non-load bearing trabecular bone and may provide an interesting class of bone regenerative materials for craniofacial reconstruction and orthodontic drug delivery.

For bone repair, again the goal should be an ability of the implanted material to regenerate native tissue with eventual replacement of the implant by noninflammatory resorption. Cryogel composites that allow the growth and differentiation of osteoblasts to ultimately produce mineralized bone are the focus of some research. Various cryogel composite formulations have been examined. Ak et al. (2013) produced a cryogel from silk fibroin that was aligned to a beta sheet structure by adding ethylene glycol diglycidyl ether during the cryogelation process. They showed that the resultant material had a very large compressive modulus (50 MPa) so that it could be deformed without cracking, expel the water from its pores and return quickly to its original shape when the force was removed. The cryogel material was therefore considered a promising bone implant material.

Polyvinyl alcohol-tetraethylorthosilicate-alginate-calcium oxide (PTAC) composite cryogels were found to support the growth and differentiation of osteoblasts (bone-producing cells) in cranial bone defects in a rat model *in vivo* (Mishra et al., 2014). Microcomputed tomography, hematoxylin and eosin staining, scanning electron microscopy and X-ray spectroscopy all revealed new bone formation while real-time PCR showed mRNA expression of markers of osteoblastic activity.

Kemence and Bolgen (2013) produced gelatin and HA cryogels cross-linked with glutaraldehyde, examined the morphological and mechanical properties of these and found that these did not elicit a cytotoxic response *in vitro* or *in vivo*. Other researchers have loaded vascular endothelial growth factor (VEGF) into similar cryogel composites of gelatin and HA in an attempt to promote blood vessel formation at the sites of critical-sized bone defects in the tibiae of rabbits (Ozturk et al., 2013). Radiological and histological results indicated that the cryogels with and without VEGF improved the healing process compared with no intervention. The presence of VEGF improved healing over the first 6 weeks but no difference was found between the cryogels with and without VEGF after 12 weeks.

## 6.7 CRYOGELS FOR THE REGENERATION OF NEURAL TISSUE

One interesting line of research was to use cryogels as a substratum for neural tissue growth to repair brain injuries (Jurga et al., 2011). Cryogels composed of gelatin or dextran linked with laminin, which is the primary ECM component of the brain, were found to support the growth and differentiation of human cord blood-derived stem cells *in vitro*. Non-neurally committed stem cells attached optimally to cryogels of pore size 80–100 μm forming colonies that ultimately differentiated into mature 3D networks of neurons (MAP$^{2+}$) and glia (S100beta+) cells. These laminin-rich cryogels were implanted into a rat brain model and did not induce inflammation or glial scarring. They were found to promote the infiltration of neuroblasts, indicating good neuro-regenerative potential.

## 6.8 CRYOGELS FOR THE REGENERATION OF MYOCARDIAL TISSUE

A myocardial infarction (commonly known as a heart attack) damages the muscle tissue of the heart and increases the likelihood of another episode. Biomaterials offer an opportunity to replace and then regenerate this tissue. Cryogels of polyhydroxyethyl methacrylate (pHEMA)-gelatin have been constructed that support the growth of myoskeletal cell lines (Singh et al., 2010). Chitosan-agarose-gelatin cryogel matrices were shown to support the growth and proliferation of a mouse myoblast cell line, C2C12 and a mouse cardiomyocyte cell line, HL1 (Bhat and Kumar, 2012). The cell number reached confluence on the eighth day and was considered sufficient for the successful utilization of the scaffold. Further mechanical stimulation was suggested by authors to mimic the *in vivo* condition and improve tissue development.

## 6.9 CRYOGELS FOR REGENERATION OF THE LIVER AND PANCREAS

Agarose-based cryogels were shown to have potential for growth and transplantation of insulin producing cells (Bloch et al., 2005; Lozinsky et al., 2008). RINm cells seeded on the agarose scaffold did not adhere to the matrix but formed spherical aggregates 50–100 μm in size after overnight incubation. The RINm pseudoislets did not increase significantly in size even after 10 days of incubation. Histological examination shows the signs of necrosis caused by limitation in oxygen and nutrients during the prolonged incubation. Insufficient oxygenation inside of the scaffold is a very common problem, which is difficult to overcome without proper vascularisation of the matrix. The response of two types of cells with different oxygen demands was studied by Bloch et al., (2005). Rat tumour INS-1E cells, which have low oxygen consumption rates and mouse pancreatic islets, with a significantly higher respiration activity, were selected. To increase the cells adhesion to matrix agarose, cryogels were modified by grafting gelatin. Insulinoma cells attached to the scaffold and grew as a monolayer in the gelatin-grafted agarose cryogels. A normal insulin secretion and content was observed for the cells cultured for 2 weeks. However, cultivation of adult pancreatic islets results in impaired insulin response to glucose and decreased intracellular insulin content probably due to limited oxygen supply. The authors believed that neovascularisation could improve the viability and biological activity of pancreatic insulin-producing cells.

Silk fibroin cryogels were synthesized at different temperatures: −20°C and −80°C (Kundu and Kundu, 2013). The cryogel matrices supported growth of hepatocarcinoma cells *in vitro*. Cryogels made at −20°C had higher cell growth rate compared with cryogels made at −80°C. The formation of cell monolayers occurred initially and was followed by formation of cell aggregates over time. The enhanced cell growth was attributed to larger pores in the gel synthesized at −20°C. It was suggested that the proteinaceous nature of the silk fibroin and the mechanical stability of the scaffold are advantageous for long-term cell cultivation and supporting liver tissue development.

## 6.10 MICROCRYOGELS FOR CELL DELIVERY

An interesting approach has been suggested to use microcryogel particles as cell carriers to be delivered *in vivo* by injection (Koshy et al., 2014; Liu et al., 2014). Microparticles (2–3 mm) of the polyethylene glycol diacrylate cryogel (Liu et al., 2014) were seeded with NIH/3T3 cells and injected subcutaneously into mice. Similar work was reported using gelatin-based cryogels (Koshy et al., 2014). The cryogel provided a supporting platform for cells and protected them from the shear-induced damage that occurred by direct injection of the cells, which was found to cause necrosis. The cryogel-carrier approach increased cell retention, survival and reintegration in tissue. Initial results showed a great potential of microcryogels for the local delivery of cells that could tackle some challenges of cell therapies of the future.

## 6.11 CONCLUSIONS

**Cryogels** are a versatile supermacroporous biomaterial containing many relevant features of regenerative biomaterials. They have a high flow rate indicating high connectivity within the pores; the ability to tailor the chemical composition for intended usage combined with a bespoke pore size range excellent for encouragement of each cell type infiltration and remodelling are relevant to many of the challenging uses of biomaterials. These features of cryogels are being exploited for use in soft tissue reconstruction and, when combined with HA, some nonload bearing hard tissues.

As well as their potential application in the reconstruction and regeneration of damaged tissues, cryogels have excellent shape memory and are capable of multiple drying/swelling cycles, properties suitable for their use as dressings. When made from nonadherent polymers, they do not integrate into the healing wound and can be easily loaded with antimicrobials, odour control agents, or produced specifically for wound type and even made the correct shape for treatment of joints.

## ACKNOWLEDGEMENTS

Some of this work was funded by FP6 MTKI-CT-2006-042768-MATISS and FP7 PERG08-GA-2010-276954-Bio-Smart projects; M. Helias and Dr. J. Salvage are acknowledged for helping with SEM. S.V. Mikhalovsky, I.U. Allan and M. Illsley acknowledge the financial support of the Longevity (Ageing) Programme of the Ministry of Education and Science of the Republic of Kazakhstan, project 'Carbon-Polymer Dressings for the Treatment of Chronic Wounds'.

## LIST OF ABBREVIATIONS

| | |
|---|---|
| 3D | Three dimensional |
| CX | Chlorhexidine |
| DNA | Deoxyribonucleic acid |
| ECM | Extracellular matrix |
| EDC | 1-ethyl-3-(3-dimethylaminopropyl)carbodiimide hydrochloride |
| GAG | Glycosaminoglycan |
| GL-GA | Gelatin cryogel scaffolds cross-linked with glutaraldehyde |
| GL-GA-S | Gelatin cryogel scaffolds cross-linked with glutaraldehyde with silicone layer |
| HA | Hydroxyapatite |
| hMSCs | Human Mesenchymal Stem Cells |
| IL1-$\alpha$ | Interleukin-1 alpha |
| mRNA | Messenger ribonucleic acid |
| PCR | Polymerase chain reaction |
| pHEMA | Poly-hydroxyethyl methacrylate |
| PTAC | Polyvinyl alcohol-tetraethylorthosilicate-alginate-calcium oxide |
| PVA | Poly(vinyl alcohol) |
| SEM | Scanning electron microscopy |
| VEGF | Vascular endothelial growth factor |

## REFERENCES

Ak, F., Oztoprak, Z., Karakutuk, I., and Okay, O. (2013). Macroporous silk fibroin cryogels. *Biomacromolecules* 14: 719–27.

Allan, I.U., Shevchenko, R., Rowshanravan, B., Kara, B., Jahoda, C.A., and James, S.E. (2009). The use of confocal laser scanning microscopy to assess the potential suitability of 3-D scaffolds for tissue regeneration, by monitoring extra-cellular matrix deposition and by quantifying cellular infiltration and proliferation. *Soft Material* 7: 319–41.

Almeida, L.R., Martins, A.R., Fernandes, E.M. et al. (2013). New biotextiles for tissue engineering: Development, characterization and *in vitro* cellular viability. *Acta Biomaterialia*. 9: 8167–81.

Badylak, S.F. (2007). The extracellular matrix as a biologic scaffold material. *Biomaterials* 28: 3587–93.

Baker, M.I., Walsh, S.P., Schwartz, Z., and Boyan, B.D. (2012). A review of polyvinyl alcohol and its uses in cartilage and orthopedic applications. *Journal of Biomedical Material Research Part B* 100: 1451–57.

Bhat, S., Lidgren, L., and Kumar, A. (2013). In vitro neo-cartilage formation on a three-dimensional composite polymeric cryogel matrix. *Macromolecular Biosciences* 13: 827–37.

Bhat, S., and Kumar, A. (2012). Cell proliferation on three-dimensional chitosan-agarose-gelatin cryogel scaffolds for tissue engineering applications. *Journal of Bioscience Bioengineering* 114: 663–70.

Bloch, K., Lozinsky, V.I., Galaev, I.Y. et al. (2005). Functional activity of insulinoma cells (INS-1E) and pancreatic islets cultured in agarose cryogel sponges. *Journal of Biomedical Material Research, Part A* 75A: 802–09.

Bolgen, N., Plieva, F., Galaev, I.Y., Mattiasson, B., and Piskin, E. (2007). Cryogelation for preparation of novel biodegradable tissue-engineering scaffolds. *Journal of Biomaterial Science Polymer Edition* 18: 1165–79.

Burke, J.F., Yannas, I.V., Quinby, W.C., Bondoc, Jr., C.C., and Jung, W.K. (1981). Successful use of a physiologically acceptable artificial skin in the treatment of extensive burn injury. *Annals of Surgery* 194: 413–28.

Chhatri, A., Bajpai, J., and Bajpai, A.K. (2011). Designing polysaccharide-based antibacterial biomaterials for wound healing applications. *Biomatter* 1: 189–97.

Choi, Y.S., Lee, S.B., Hong, S.R. et al. (2001). Studies on gelatin-based sponges. Part III: A comparative study of cross-linked gelatin/alginate, gelatin/hyaluronate and chitosan/hyaluronate sponges and their application as a wound dressing in full-thickness skin defect of rat. *Journal of Material Science. Material in Medicine.* 12: 67–73.

Clark, R.A., Ghosh, K., and Tonnesen, M.G. (2007). Tissue engineering for cutaneous wounds. *The Journal of Investigative Dermatology.* 127: 1018–29.

Dainiak, M.B., Allan, I.U., Savina, I.N. et al. (2010). Gelatin-fibrinogen cryogel dermal matrices for wound repair: Preparation, optimisation and in vitro study. *Biomaterials.* 31: 67–76.

Dubruel, P., Unger, R., Van Vlierberghe, S. et al. (2007). Porous gelatin hydrogels: 2: *In vitro* cell interaction study. *Biomacromolecules* 8: 338–44.

Gillitzer, R., and Goebeler, M. (2001). Chemokines in cutaneous wound healing. *Journal of Leukocyte Biology* 69: 513–21.

Gonzalez, J.S., and Alvarez, V.A. (2014). Mechanical properties of polyvinylalcohol/hydroxyapatite cryogel as potential artificial cartilage. *Journal of the Mechanical Behaviour of Biomedical Materials* 34:47–56.

Gun'ko, V.M., Savina, I.N., and Mikhalovsky, S.V. (2013). Cryogels: Morphological, structural and adsorption characterisation. *Advanced Colloid Interface Sciences* 187: 1–46.

Heimbach, D.M., Warden, G.D., Luterman, A. et al. (2003). Multicenter post approval clinical trial of Integra dermal regeneration template for burn treatment. *The Journal of Burn Care and Rehabilitation.* 24: 42–48.

Horch, R.E., Kopp, J., Kneser, U., Beier, J., and Bach, A.D. (2005). Tissue engineering of cultured skin substitutes. *Journal of Cellular and Molecular Medicine* 9(2005): 592–08.

Hu, Y., Liang, D., Li, X. et al. (2010). The role of interleukin-1 in wound biology. Part II: *In vivo* and human translational studies. *Anesthesia and Analgesia.* 111: 1534–42.

Jones, I., James, S.E., Rubin, P., and Martin, R. (2003). Upward migration of cultured autologous keratinocytes in Integra artificial skin: A preliminary report. *Wound Repair and Regeneration* 11: 132–38.

Jurga, M., Dainiak, M.B., Sarnowska, A. et al. (2011). The performance of laminin-containing cryogel scaffolds in neural tissue regeneration. *Biomaterials* 32: 3423–34.

Katsen-Globa, A., Meiser, I., Petrenko, Y.A. et al. (2013). Towards ready-to-use 3-D scaffolds for regenerative medicine: Adhesion-based cryopreservation of human mesenchymal stem cells attached and spread within alginate-gelatin cryogel scaffolds. *Journal of Material Sciences. Materials in Medicine* 25: 857–71.

Kemençe, N., and Bölgen, N. (2013). Gelatin- and hydroxyapatite-based cryogels for bone tissue engineering: Synthesis, characterization, *in vitro* and *in vivo* biocompatibility. *Journal of Tissue Engineering and Regenerative Medicine* doi: 10.1002/term.1813.

Kharitonova, K.I., Rodiukova, E.N., Simonovich, A.E., and Stupak, V.V. (1990). The experimental and clinical use of gelatin sponges with kanamycin and gentamycin for the prevention and treatment of suppurative complications in neurosurgery. *Zhurnal Voprosy Neirokhirurgii Imeni N N Burdenko* 3: 9–11.

Koshy, S.T., Ferrante, T.C., Lewin, S.A., and Mooney, D.J. (2014). Injectable, porous, and cell-responsive gelatin cryogels. *Biomaterials* 35: 2477–87.

Kundu, B., and Kundu, S.C. (2013). Bio-inspired fabrication of fibroin cryogels from the muga silkworm *Antheraea assamensis* for liver tissue engineering. *Biomedical Materials* 8: doi:10.1088/1748-6041/8/5/055003.

Liu, W., Li, Y., Zeng, Y., Zhang, X. et al. (2014). Microcryogels as injectable 3-D cellular microniches for site-directed and augmented cell delivery. *Acta Biomaterialia* 10: 1864–75.

Lozinsky, V.I., Damshkaln, L.G. Bloch, K.O. et al. (2008). Cryostructuring of polymer systems. XXIX. Preparation and characterization of supermacroporous (spongy) agarose-based cryogels used as three-dimensional scaffolds for culturing insulin-producing cell aggregates. *Journal of Applied Polymer Science* 108: 3046–62.

Lozinsky, V.I., Galaev, I.Y., Plieva, F.M. et al. (2003). Polymeric cryogels as promising materials of biotechnological interest. *Trends in Biotechnology*. 21: 445–51.

MacNeil, S. (2007). Progress and opportunities for tissue-engineered skin. *Nature*. 445: 874–80.

Mishra, R., Goel, S.K., Gupta, K.C., and Kumar, A. (2014). Biocomposite cryogels as tissue-engineered biomaterials for regeneration of critical-sized cranial bone defects. *Tissue Engineering. Part A*. 20: 751–62.

Muller, C.S., Schiekofer, C., Korner, R., Pfohler, C., and Vogt, T. (2013). Improved patient-centered care with effective use of Integra(R) in dermatologic reconstructive surgery. *Journal of the German Society of Dermatology*. 11: 537–48.

Nakayama, Y., Nishi, S., Ueda-Ishibashi, H., and Matsuda, T. (2003). Fabrication of micropored elastomeric film-covered stents and acute-phase performances. *Journal of Biomedical Materials Research*. 64: 52–61.

Ozturk, B.Y., Inci, I., Egri, S. et al. (2013). The treatment of segmental bone defects in rabbit tibiae with vascular endothelial growth factor (VEGF)-loaded gelatin/hydroxyapatite 'cryogel' scaffold. *European Journal of Orthopaedic Surgery and Traumatology* 23: 767–74.

Pal, U.S., Singh, B.P., and Verma, V. (2013). Comparative evaluation of zinc oxide eugenol versus gelatin sponge soaked in plasma rich in growth factor in the treatment of dry socket: An initial study. *Contemporary Clinical Dentistry* 4: 37–41.

Papini, R. (2004). Management of burn injuries of various depths. *BMJ* 329: 158–60.

Reichelt, S., Becher, J. Weisser, J. et al. (2014). Biocompatible polysaccharide-based cryogels. *Material Science and Engineering C* 35: 164–70.

Sandeman, S.R., Gun'ko, V.M., Bakalinska, O.M. et al. (2011). Adsorption of anionic and cationic dyes by activated carbons, PVA hydrogels, and PVA/AC composite. *Journal of Colloid Interface Science* 358: 582–92.

Savina, I.N., Dainiak, M., Jungvid, H. et al. (2009a). Biomimetic macroporous hydrogels: Protein ligand distribution and cell response to the ligand architecture in the scaffold. *Journal of Biomaterial Science. Polymer Edition* 20: 1781–95.

Savina, I.N., Cnudde, V., D'Hollander, S. et al. (2007). Cryogels from poly(2-hydroxyethyl methacrylate): Macroporous, interconnected materials with potential as cell scaffolds. *Soft Matter* 3: 1176–84.

Savina, I.N., Gun'ko, V.M., Turov, V.V. et al. (2011). Porous structure and water state in cross-linked polymer and protein cryo-hydrogels. *Soft Matter* 7: 4276–83.

Savina, I.N., Tomlins, P.E. Mikhalovsky, S.V. et al. (2009b). Characterization of macroporous gels. In: *Macroporous Polymers: Production, Properties and Biotechnological/Biomedical Applications*. Mattiasson, B., Kumar, A. and Galaev, I.Y., Eds. Boca Raton, FL: CRC Press, pp. 211–235.

Sharma, A., Bhat, S. Vishnoi, T. et al. (2013). Three-dimensional supermacroporous carrageenan-gelatin cryogel matrix for tissue engineering applications. *BioMed Research International* ID 478279, http://dx.doi.org/10.1155/2013/478279.

Shevchenko, R.V., Sibbons, P.D., Sharpe, J.R. et al. (2008). Use of a novel porcine collagen paste as a dermal substitute in full-thickness wounds. *Wound Repair and Regeneration* 16: 198–07.

Shevchenko, R.V., James, S.L., and James, S.E. (2010). A review of tissue-engineered skin bioconstructs available for skin reconstruction. *Journal of Royal Society, Interface* 7: 229–58.

Shevchenko, R.V., Eeman, M., Rowshanravan, B. et al. (2014). The *in vitro* characterization of a gelatin scaffold, prepared by cryogelation and assessed *in vivo* as a dermal replacement in wound repair. *Acta Biomaterialia* 10: 3156–66.

Singh, D., Nayak, V., and Kumar, A. (2010). Proliferation of myoblast skeletal cells on three-dimensional supermacroporous cryogels. *International Journal of Biological Sciences* 6: 371–81.

Takei, T., Nakahara, H., Ijima, H. et al. (2012). Synthesis of a chitosan derivative soluble at neutral pH and gellable by freeze-thawing, and its application in wound care. *Acta Biomaterialia* 8: 686–93.

Takei, T., Nakahara, H., Tanaka, S. et al. (2013). Effect of chitosan-gluconic acid conjugate/poly(vinyl alcohol) cryogels as wound dressing on partial-thickness wounds in diabetic rats. *Journal of Material Science. Materials in Medicine* 24: 2479–87.

van der Veen, V.C., Boekema, B.K., Ulrich, M.M. et al. (2011). New dermal substitutes. *Wound Repair and Regeneration* 19: doi: 10.1111/j.1524-475X.2011.00713.x.

Van Vlierberghe, S., Dubruel, P., and Schacht, E. (2011). Biopolymer-based hydrogels as scaffolds for tissue engineering applications: A review. *Biomacromolecules* 12: 1387–08.

Van Vlierberghe, S., Cnudde, V. Dubruel, P. et al. (2007). Porous gelatin hydrogels: 1. Cryogenic formation and structure analysis. *Biomacromolecules* 8: 331–37.

Ward, A.G., and Courts, A. (1977). *The science and technology of gelatin*. London: Academic Press.

Yang, J., Woo, S.L., Yang, G. et al. (2010). Construction and clinical application of a human tissue-engineered epidermal membrane. *Plastic and Reconstructive Surgery* 125: 901–09.

Yannas, I.V., Orgill, D.P., and Burke, J.F. (2011). Template for skin regeneration. *Plastic and Reconstructive Surgery*. 127, Suppl 1: 60S–70S. doi: 10.1097/PRS.0b013e318200a44d.

Zhong, S.P., Zhang, Y.Z., and Lim, C.T. (2010). Tissue scaffolds for skin wound healing and dermal reconstruction. *Wiley Interdisciplinary Reviews. Nanomedicine and Nanobiotechnology* 2: 510–25.

# 7 Biocompatibility of Macroporous Cryogel Materials

*Akhilesh Kumar Shakya and Ashok Kumar\**

## CONTENTS

7.1 Introduction .................................................................................. 196
7.2 Factors Affecting Material Biocompatibility ............................... 197
7.3 *In Vitro* Biocompatibility ............................................................. 200
7.4 *In Vivo* Biocompatibility ............................................................. 205
7.5 Conclusion ................................................................................... 209
List of Abbreviations ........................................................................... 210
References ........................................................................................... 210

### ABSTRACT

Biocompatibility is one of the imperative and mandatory evaluations of materials for biomedical applications and can be assessed through either *in vitro* or *in vivo* ways. *In vitro* biocompatibility evaluates cytotoxicity, protein adsorption and blood clotting ability of the material, while *in vivo* assessment tells about localized and systemic toxicities caused by the material. Biocompatibility of a material depends on a number of factors such as the physicochemical nature, surface roughness, nature and density of attached polymers on the surface and balance of hydrophobocity and hydrophilicity. Macroporous cryogels are promising cell scaffold biomaterials for different tissue engineering applications. Cryogels are highly porous, are mechanically strong and provide large surface areas for adhesion and growth of the cells. Cryogels are biocompatible under *in vitro* and *in vivo* cellular environments. They support the growth of mammalian cells under *in vitro* culture and completely integrate within the host tissues, when implanted inside the body. This chapter focuses on different aspects of biocompatibility and types of cryogel biocompatibility.

### KEYWORDS

Biocompatibility, cytotoxicity, cryogel, host tissue response

---

\* Corresponding author: E-mail: ashokkum@iitk.ac.in; Tel: +91-512-2594051.

## 7.1 INTRODUCTION

Generally, regeneration of damaged organs or tissues involves a transplantation procedure—either the implantation of organ/tissue from one individual to another, or surgically transferring tissue from one site to another site in the same individual. Though these procedures are common in practice, they are limited in use due to organ donor shortage, transplant rejection and severe side effects. Under this outlook, in the last 2–3 decades, tissue engineering has emerged as a new branch of biomedical science showing promising potential to restore damaged tissue/organ by minimally affecting the host. Tissue engineering involves *in vitro* growth of mammalian cells over a cell scaffold, and implanting them back to the donor site (Langer and Vacanti, 1993; Vacanti, 2003). Growth of cells over a scaffold is one of the important parameters in tissue engineering. For efficient growth of the cells, the scaffold should be biodegradable, biocompatible and porous for proper exchange of nutrients for the cells in the scaffold. An ideal scaffold should avoid long and permanent chronic inflammation and other surgical complications. Therefore, biomaterial biocompatibility is one of the important aspects while designing scaffolds for tissue engineering purposes. **Biocompatibility** is the potential of a material to perform appropriately inside the body and this can be assessed by *in vitro* and *in vivo* evaluations. *In vitro* material biocompatibility can be evaluated through protein adsorption, blood clot formation, hemolysis and cytotoxicity assays (Hussain and Rajab, 2009). After *in vitro* evaluation, the host tissue response of materials can be checked through implanting in different animal models. *In vivo* biocompatibility gives some initial knowledge about material performance inside the body before developing them in to the final products. Various biomaterials-based three-dimensional constructs have been developed as cell scaffolds for proper cell growth. Cell scaffolds can be synthesized through different methods such as solvent casting, particulate leaching, gas foaming, freeze-drying, melt molding and so on. Although these methods are efficient, they are associated with limitations of uncontrolled pore size, their heterogeneous distribution and the absence of interconnectivity between the pores (Shakya et al., 2013).

Recently, a new technique has evolved for synthesizing cell scaffolds known as cryogelation, which can synthesize polymeric scaffolds at subzero temperature through physical or chemical gelation of polymers or polymerization of their respective monomers (Lozinsky et al., 2001, 2003). **Cryogels** are a kind of hydrogel which have high porosity and high water retention capacity. During synthesis of cryogel at subzero temperature, the solvent freezes and forms ice crystals. Polymerization or gelation starts around these crystals. Ice crystals act as porogens, which on thawing give rise to interconnected pores (Kumar and Srivastava, 2010).

For biocompatibility, degradation of material is a necessary event inside the body and it should occur in a controlled manner. The degradation rate of materials depends on various factors such as physico-chemical nature, synthesis condition, mechanical strength, porosity and site of implantation. For long-term applications, scaffolds should degrade slowly without causing any outward disease symptoms, while for short-term applications scaffold should degrade quickly into nontoxic and nonallergic by-products. Despite the difference in applications, the initial body response against the implanted material is similar for all biomaterials. On implantation, the host reacts against the material in a cascade of events, which ends up with the formation of giant foreign body cells (Brodbeck and Anderson, 2009). Immediately after implantation, the implant is covered by plasma proteins, which enable interactions of immune cells with the material. Initially, an acute inflammatory response takes place due to injury at the time of surgery, which mediates by recruitment of effector cells such as neutrophils. Neutrophils secrete different cytokines and activate the macrophages. Activated macrophages try to clear off the material from the implantation site either by phagocytosis or degradation of material into small fragments. Moreover, neutrophils also secrete different proteases, lysozymes and reactive free radicals, which help in material degradation. If the implant is degraded quickly, then inflammation ends earlier and the body returns to its original state, whereas in the case of slow material degradation, chronic inflammation occurs, which is characterized by a continuous release of proteases and activation of macrophages (Cobelli et al., 2011). Phagocytic cells such as monocytes and macrophages also adhere on the implant surface with the help of integrin molecules. Macrophages are able to phagocytose small micron-sized particles easily; in the presence of large implants, macrophages fuse with each other and form foreign body giant cells that degrade the material into small parts first and then phagocytose them (Figure 7.1). Sometimes in chronic inflammatory responses, the surrounding tissue near the implant is severely damaged and implant rejection takes place.

The type of immune response against the material is determined by the interaction of immune cells with the material, and material properties such as surface roughness, hydrophilicity and hydrophobocity balance and charge on the surface. Biocompatible materials behave friendly inside the body and integrate well with the surrounding tissue, whereas toxic materials activate immune cells to produce cytokines such as TNF-α, which cause speedy material degradation or implant rejection (Gardner et al., 2013).

## 7.2 FACTORS AFFECTING MATERIAL BIOCOMPATIBILITY

Factors which govern material biocompatibility are not yet well understood. However, the material surface dominantly determines material biocompatibility. In material surface, the topography, interfacial free energy, balance of hydrophilicity and hydrophobicity, density of attached functional groups and roughness affect the biocompatibility (Wang et al., 2004). On exposure of a material inside the body, an interface forms between the cells and the material surface. This interface is a kind of thermodynamic force for adhesion of cells on the surface and depends on water availability. A high amount of water provides low interfacial energy for protein adsorption and cell interaction (Ratner et al., 1979). Surfaces with hydrophilic polymers or polar polymers attached on nonpolar surfaces are considered biocompatible that allows low binding of the cells. For example, ungrafted polyethylene (PE), polytetrafluroethylene (PTFE) and polyvinyl chloride (PVC) surfaces showed high platelet adhesion in comparison to grafted PE with *N, N*-dimethylacrylamide (DMAA), PTFE with DMAA and PVC with acrylic acid (AA) (de Queiroz et al., 1997). Grafting of hydrophilic monomers like DMAA and AA reduces the interface energy; consequently, less platelets bind on the surface. In some applications, high interface energy is required for increasing the binding of cells. For example, Puleo

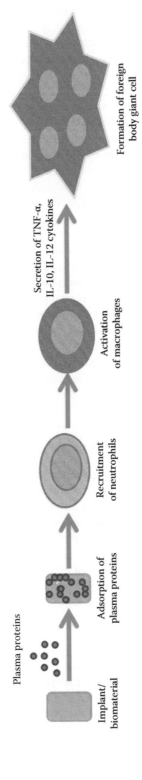

**FIGURE 7.1** Cascade of events during exposure of biomaterial inside the body.

and co-workers studied the modulation of interface free energy of material surface for bone tissue engineering (Puleo and Nanci, 1999).

The type and configuration of adsorbed proteins on the material's surface determines the interface energy (Ruckenstein and Gourisankar, 1986). For example, on exposure of a material to blood, rapid adsorption of plasma proteins occurs, which is the foremost step that enables platelet adhesion and their activation. This activation of platelets initiates the coagulation process and the formation of blood clots. Studies show that highly hydrophobic surfaces adsorb large amounts of plasma proteins compared to highly hydrophilic surfaces (Elbert and Hubbell, 1996) which can ultimately affect the biocompatibility. Hence, to improve biocompatibility, hydrophilic polymers such as polyethylene glycol (PEG), AA and polyvinylpyrrolidone (PVP) can be chemically attached on the surface of biomaterials through grafting or by radiation. In another example, polysulphone (PSF) hollow fibres grafted with PVP showed lower plasma protein adsorption than the polysulphone membrane (Higuchi et al., 2002). Long chains of PVP provide high hydrophilicty on the PSF surface. Alternatively, the surface can also chemically modify through interpenetration of a hydrophilic polymer in the material. In this direction, several IPNs were developed by Roman and colleagues by use of polyurethrane (PU), biospan (BS) and vinylpyrrolidone dimethylacrylamide copolymers (Abraham et al., 2001). A high ratio of vinylpyrrolidone in these IPNs reduces adsorption of fibrinogen and globulin proteins, while adhesion of albumin proteins is increased. After adhesion, sometimes protein conformation changes, which may affect the material biocompatibility. These conformational changes are progressive and irreversible (Kim et al., 2002). Atomic force microscopy (AFM) and circular dichroism (CD) techniques are well known methods to study protein conformational changes occur due to material interaction. Tanaka and colleagues have used the CD technique for study of change in α-helix and random fraction or β-helix after protein adsorption on a polymer surface (Tanaka et al., 2000). From studies, it has been shown that different polymers induce different degrees of conformational change in proteins and this change can affect cells binding on the material surface; for example, binding of platelets inhibited by prior deposition of albumin proteins on the surface and enhanced by prior adsorption of IgG or fibrinogen proteins (Meumer et al., 1995). In another study, Lee and Lee (1998) studied the adsorption of proteins and platelet adhesion on polyethylene surfaces with gradient of wettability properties. Platelet binding increases as wettability increases, in the absence of plasma proteins, while in the presence of plasma proteins, the binding of cells decreases as wettability decreases. In order to achieve a biocompatible surface, a balance of hydrophilicity and hydrophobicity is required. For example, a block polymer of PEG-*b*-PPG-, a combination of hydrophilic and hydrophobic components has increased surface biocompatibility because it reduces adsorptions of proteins (Lee et al., 1989). In this block polymer, hydrophilic PEG blocks orient toward plasma proteins, while hydrophobic PPG binds to the material surface.

Surface morphology is another factor for deciding the response of cells against the material and normally it directs the cell growth. Surfaces with grooves direct better cell movement than the smoother surfaces do. Surfaces with groove sizes less than 0.5 microns are effective for growth and alignment of cells such as fibroblasts and epithelial cells. In addition, a surface with closely spaced grooves is better for cell growth than a widely spaced grooved surface (Brunnette, 1998). Surface patterns/ architectures generate different cellular responses in biological system. To achieve an optimum level of cell response, the surface topography of the biomaterial should match to *in vivo*. Soft lithography is a powerful tool used to achieve optimum surface topography. Whitesides et al. have done significant work in this direction and they have developed different methods for well-defined patterning of 3D nanostructures (Kane et al., 1999). Microcontact printing is another powerful technique for controlling surface chemistry at the molecular level. Moreover, patterning of various proteins or cells can be possible now by using of microfluidic channels. Whitesides and co-workers patterned cell adhesion peptide in combination with hexaethylene glycol thiolate for specific cell adhesion and movement (Chapman et al., 2000). These types of microstructure surfaces do not allow nonspecific binding of other plasma proteins and contribute to achieving the optimum biocompatibility. These versatile surface engineering techniques can design the biocompatible surfaces for different biomedical applications.

Apart from surface morphology, porosity of material also affects biocompatibility. For example, biocompatibility of hydrogels/cryogels materials is better than low porous materials. High porosity allows good nutrient flow across the cryogel, which is good for growth of the cells (Shakya et al., 2013). In another example, porous chitosan scaffolds elicited lower lymphocyte reactions than the nonporous scaffold (VandeVord et al., 2002).

## 7.3 *IN VITRO* BIOCOMPATIBILITY

Generally, before the *in vivo* assessment of biocompatibility of materials, they are subjected to *in vitro* toxicity characterization to get an insight on the compatibility of the material outside the body. *In vitro* characterization includes haemolytic activity, protein adsorption and cytotoxicity assays. According to the American Society for Testing and Materials (ASTM F756-00, 2000), a material is classified as nonhaemolytic if it shows less than 2% haemolysis; slightly haemolytic if the haemolysis index is between 2 and 5%; and haemolytic if more than 5% haemolysis is observed. Macroporous materials such as cryogels are hemocompatible and do not show any significant toxicity when exposed directly to blood. For example, polyvinyl alcohol (PVA)-haemoglobin cryogels are haemocompatible and do not cause significant RBC lysis (Bajpai and Saini, 2009). The haemolytic activity of PVA-haemoglobin decreased as PVA and casein amount increased in the feed mixture and this is due to the hydrophilic nature of PVA and casein. At high concentrations of PVA and casein, the cryogel surface becomes smooth and hence reduces the rate of haemolysis. By increasing freeze and thaw cycles during cryogel synthesis, protein adsorption and haemolysis activity of the cryogel increases due to a reduction in the cryogel's hydrophilicity (Bajpai and Saini, 2009). In another example, hybrid cryogels consisting of synthetic polymer poly-ε-caprolactone and natural polymers like collagen or hylauronic acid showed less than 5% blood clotting (Simionescu et al., 2013).

Direct contact assay is a test for material cytotoxicity, which is based on change in a cell's morphology after coming in contact with material. Generally, this test is performed using human or mouse fibroblast cell lines. Mouse odontoblast, mouse macrophages, rat submandibular salivary gland acinar cells and other primary cell types such as goat chondrocytes and human osteoblasts are the different cell types which are used for checking material cytotoxicity. Materials can show their toxic effects on adjacent cells. In the assay, live cells can be fixed and stained by cytochemical stains, while dead cells are detached from the surface due to material toxicity. Cell damage is visualized in the form of swelling and formation of vacuoles. Moreover, this test does not need a long time to perform or other extra preparations. Apart from direct contact test, colony-forming assay and agar diffusion test are other methods for assessment of *in vitro* biocompatibility (Hussain and Rajab, 2009).

Cell proliferation assay is most widely accepted and quite sensitive to estimate material cytotoxicity using 3-(4,5-dimethylthiazol-2-yl)-2,5-diphenyltetrazolium bromide (MTT) dye. This assay is based on the principal of oxidizing capability of live cells. Mitochondrial succinic dehydrogenase enzyme of the live cells oxidizes yellow colour MTT dye to a violet colour. In this assay, cells grown over a material surface for a few days and their proliferation can be assessed by adding MTT dye to cell culture medium. Live cells convert MTT dye into violet colour formazon crystals, which cannot cross through the cell membrane and hence an organic solvent such as DMSO is used to dissolve them. The extent of crystal formation depends on the number and the metabolic activity of the cells. Cytotoxicity evaluation through MTT is an adequately fast and cost-effective method. The cytotoxicity of different cryogels through MTT assay has been assessed using various cell lines (Table 7.1). For example, cytotoxicity of gelatin-hydroxyapatite-based cryogels was assessed on mouse fibroblasts L929 cell line and the cryogel well supported the growth of mouse fibroblasts (Kemence and Bolgen, 2013). In another study, biocompatibility of gelatin-based injectable cryogels was studied using NIH3T3 cell line. Cells were distributed uniformly throughout the cryogel. After 1 day of culture, cells take a spindle-shape morphology and form a monolayer. Cells secreted a lot of extracellular matrix and covered the whole cryogel surface over the period of the culture time as studied by fluorescence microscopy. Slight contraction was observed in the cryogel on the third day of culture and this is due to traction forces applied by the growing cells. Cryogel scaffold retained more than 90% viable cells over the whole culture period. Cells numbers and their metabolic rates were also increased over the period as studied by DNA synthesis. Moreover, good proliferation of cells was seen on cryogels through histological studies (Koshy et al., 2014) (Figure 7.2).

In another study, the cytotoxicity of a gelatin/hyaluronic acid (GH10) cryogel was demonstrated by the growth of adipose derived stem cells (ADSCs) in normal and differentiation media. At different time points, cell numbers were observed to increase showing continuous growth in control medium unlike in differentiation medium where cell growth was arrested at 28th day of culture because of differentiation of ADSCs into adipocyte lineage. Initially, ADSCs have a spindle-shaped morphology which slowly changes into spherical as they differentiate. ADSCs take up a fibroblast-like morphology in control medium. Cryogels provide a large surface for adhesion and spreading of these cells. In the early stages of culture, a change in the cytoskeletal proteins, such as actin and tubulin, which aid stem cells to bind on cryogel surface, was observed (Chang et al., 2013). In this progression, cytotoxicity of gelatin-laminin (GL) based cryogels was studied by culture of the stem cells.

## TABLE 7.1
### *In Vitro* Biocompatibility of Different Cryogels

| Cryogel | Model Cell Line | Biocompatibility | Compatibility |
|---|---|---|---|
| Poly-*N*-vinylcaprolactam (Srivastava and Kumar, 2009) | COS-7 | Hemocompatible, cytocompatible | Tissue engineering |
| pHEMA-gelatin (Singh et al., 2010) | Primary chondrocytes | Cytocompatible | Cartilage tissue engineering |
| PTAC (Mishra and Kumar, 2010) | L929, MG-63, human osteoblasts | Cytocompatible | Bone tissue engineering |
| CAG (Bhat et al., 2010) | Primary chondrocytes | Cytocompatible | Cartilage tissue engineering |
| Polysaccharide based cryogels (Reichelt et al., 2014) | 3T3 | Cytocompatible | Tissue engineering |
| Gelatin-hyaluronic acid (Chang et al., 2013) | Porcine adipose derived stem cells (ADSCs) | Cytocompatible | Adipose tissue engineering |
| Carrageenan-gelatin (Sharma et al., 2013) | COS-7 | Cytocompatible | Tissue engineering |
| Conducting cryogel (Vishnoi and Kumar, 2012) | Neuro 2a, cardiac muscle C2C12 | Cytocompatible | Neural tissue engineering |
| Gelatin-hydroxyapatite (Kemence and Bolgen, 2013) | L929 mouse fibroblast cell line | Cytocompatible | Bone tissue engineering |
| Polyacrylonitrile and Polyacrylamide-chitosan semi-interpenetrating (Jain and Kumar, 2009) | 6A4D7, B7B10, H9E10 cell lines | Cytocompatible | Bioreactor |
| Chitosan gelatin (Kathuria et al., 2009) | COS-7 | Cytocompatible | Cartilage tissue engineering |
| Agarose gelatin (Tripathi et al., 2009) | COS-7 | Cytocompatible | Cartilage tissue engineering |
| Cryogel-microparticle (C-MP) composite (Sami and Kumar, 2013) | Human cervical cancer cells (HeLa) | Cytocompatible | Tissue engineering |
| Gelatin/oxide-dextran (Odabaş et al., 2012) | Primary chondrocytes | Cytocompatible | Cell/tissue engineering |
| pHEMA-lys (Spina et al., 2014) | Human serum samples | Cytocompatible | Extracorporeal heparin removal |
| Poly-*N*-isopropylacrylamide loaded with chitosan/bemiparin nanoparticles (Peniche et al., 2013) | BaF32 cell line | Cytocompatible | Tissue engineering |
| Polyethylene glycol (Hwang et al., 2010) | Bovine chondrocytes, human mesenchymal stem cells (hMSCs) | Cytocompatible | Cell/tissue engineering |

*(Continued)*

## TABLE 7.1 (CONTINUED)
### *In Vitro* Biocompatibility of Different Cryogels

| Cryogel | Model Cell Line | Biocompatibility | Compatibility |
| --- | --- | --- | --- |
| Poly(hydroxyethyl methacrylate) (Percin et al., 2013) | Human plasma | Hemocompatible | Purification of fibronectin |
| Cell responsive gelatin cryogel (Koshy et al., 2014) | NIH 3T3 | Cytocompatible | Tissue engineering |
| Polyvinylalcohol-casein (Bajpai and Saini, 2005) | Blood plasma | Hemocompatible | Biomedical applications |

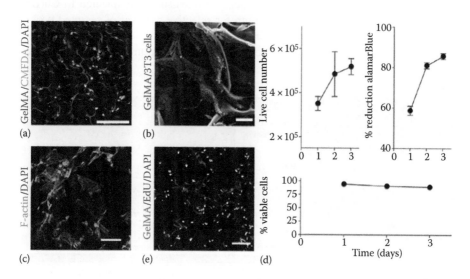

**FIGURE 7.2** Cell cytotoxicity of injectable the cell responsive cryogels. Growth and proliferation of NIH3T3 cell lines on gelatin cryogels *in vitro* (a–c). Analysis of the number of live cells (d), staining of cells for new DNA synthesis within the scaffold using EdU incorporation (e). (Reprinted from *Biomaterials* 35(8), Koshy, S. T., Ferrante, T. C., Lewin, S. A. and Mooney, D. J., Injectable, porous, and cell-responsive gelatin cryogels. 2477–2487. Copyright 2014, with permission from Elsevier.)

Migration and differentiation of the cells were observed in GL cryogels cross-linked by low (0.1%, v/v) and medium (0.3%, v/v) glutaraldehyde concentration. GL cryogels cross-linked with 0.1% cross-linker concentration showed better cell growth and proliferation than the cryogels cross-linked with medium cross-linker concentration. Although 0.1%, v/v cross-linked cryogels are fragile, they can be easily transplanted in animals without affecting the seeded cells morphology. The 0.3% cross-linked cryogels are mechanically strong and provide good protective environment for growth of stem cells (Jurga et al., 2011).

Cryogel's biocompatibility has also been proved in the bioreactor format. For example, polyacrylamide (PAAm) cryogel well supported the growth of mouse hybridoma cells (M2139) for monoclonal antibody production. These cryogels have continuous interconnected macropores, which allow convective nutrient exchange to the M2139 cells which secrete mouse-specific IgG2b antibodies. These antibodies induced arthritis when injected in DBA/1 mice, which reflects the functionality of the antibodies secreted from M2139 cell line (Nilsang et al., 2007) (Figure 7.3).

In another study, gelatin-coupled PAAm cryogel was used for growth of human fibrosarcoma (HT1080) and human colon cancer (HCT116) cells for urokinase (a therapeutic enzyme) production. Cells were attached on a cryogel matrix after 4–6 h of incubation. Continuous production of urokinase into circulating medium is a direct evidence of cryogel biocompatibility and viability of cells inside the cryogel bioreactor. Gelatin-coupled PAAm cryogel was downstream connected to Cu(II)-iminodiacetic acid-PAAm cryogel column for selective removal of secreted urokinase enzyme from the first bioreactor column. After 32 days of culture, 152,600 plough units of urokinase were recovered from 500 ml culture medium (Kumar et

**FIGURE 7.3** Growth of M2139 hybridoma cells on PAAm cryogel matrix. SEM pictures of cryogel matrix after day 1 (a), 7 (b), 36 (c) and 57 (d) of culture. (From Nilsang, S., Nandakumar, K. S., Galaev, I. Y., Rakshit, S. K., Holmdahl, R., Mattiasson, B. and Kumar, A.: Monoclonal antibody production using a new supermacroporous cryogel bioreactor. *Biotechnol Prog.* 2007. 23(4). 932–939. Copyright Wiley-VCH Verlag GmbH & Co. KGaA. Reproduced with permission.)

al., 2006). In this direction, recently different cryogel matrices polyacrylamide-chitosan (PAAC), poly(N-isopropylacrylamide)-chitosan, polyacrylonitrile (PAN) and poly(N-isopropylacrylamide) show excellent biocompatibility for growth of different cell lines such as 6A4D7, B7B10 and H9E10 for efficient production of monoclonal antibodies. Out of these cryogels, PAAC proved relatively better than others in terms of cell growth and antibody production. Monolith PAAC bioreactor support 6A4D7 cell line growth efficiently and produced 57.5 mg antibody in 500 ml culture media after 30 days of culture (Jain et al., 2010; Jain and Kumar, 2013).

## 7.4 IN VIVO BIOCOMPATIBILITY

Biomaterials can be harmful inside the body either by causing toxicity at the site of implantation (also known as localized toxicity) or they can leach out into smaller fragments and cause systemic toxicity. Through blood circulation, they may affect the other organs. Localized toxicity is characterized by wound formation, severe inflammation and high execute formation (a drained fluid at wound site) at site of implantation, whereas systemic toxicity is characterized by nausea, fever, acute or chronic allergy and other illness symptoms. Systemic toxicities can be further categorized into acute response (within 24 h), subchronic (14–28 days) and chronic (long term) on the basis of appearance of illness symptoms. In general, localized effects can be studied through histological analysis of the surrounding tissues, while systemic toxicities studied by change in expression of inflammatory markers such as TNF-$\alpha$ after material implantation. Change in the body weight is another parameter to study systemic toxicity. Localized and systemic toxicities of different cryogel materials have been studied in different animal models (Table 7.2).

For example, *in vivo* responses of new hybrid cryogels consisting of synthetic polymer poly-$\varepsilon$-caprolactone and natural polymers atelocollagen or hylauronic acid derivative were evaluated after implanting them subcutaneously in a Wister rat model. Inflammatory response, neoangiogenesis, cryogel degradation and host tissue formation events were studied at different time points 4 days, 2 and 4 weeks after implantation. After 4 days of implantation, the cryogel was well integrated in the surrounding tissues with minimal inflammatory response. A thin layer of connective tissues surrounded the cryogels with neovascularisation. New vessels formation were proved by the presence of distinct endothelium around the implant. In the interstitial space, different lymphocytes and phagocytes were infiltrated. No trace of cryogel was observed after 2 weeks of implantation (Simionescu et al., 2013). Overall, cryogel implantation did not damage surrounding tissues significantly. In another study, host tissue response and inflammatory response of dextran modified oligolactide cryogels HEMA-LLA-D/HEMA-LLA were studied through implantation in rats at different places, namely, dorsal subcutaneously, iliac submandibular and auricular and calvarial sites. Initially, implanted cryogels showed a mild inflammatory response characterized by infiltration of phagocytic cells. Foreign body giant cells were also observed in soft and hard tissues near the implantation site. Host immune response was significantly low in auricular and calvarial places in comparison to subcutaneous and submuscular places. At cranial defects, significant growth of connective tissues and neovascularisation were also observed (Bolgen et al., 2009). Very recently, biocompatibility of four different cryogels—polyvinylcaprolactam (pVCl),

## TABLE 7.2
### In Vivo Biocompatibility of Different Cryogels

| Cryogel | Animal Model | Toxicity |
| --- | --- | --- |
| Poly-N-vinylcaprolactam (Shakya et al., 2013) | B10Q, B10Q.Ncf1 mice | Systemic, localized toxicity |
| Polyvinyl alcohol-alginate-bioactive glass composite (Mishra et al., 2013; Shakya et al., 2013) | B10Q, B10Q.Ncf1 mice, Wistar rats | Systemic, localized toxicity, in vivo bone regeneration |
| Polyhydroxyethylmethacrylate-gelatin (Shakya et al., 2013) | B10Q, B10Q.Ncf1 mice | Systemic, localized toxicity |
| Chitosan-agarose-gelatin (Shakya et al., 2013) | B10Q, B10Q.Ncf1 mice | Systemic, localized toxicity |
| Gelatin-laminin (Jurga et al., 2011) | Wistar rats | In vivo biocompatibility in brain region |
| Cell responsive gelatin cryogel (Koshy et al., 2014) | C57BL/6J, C57BL/6J-Tyr$^{c-2J}$ mice | Localized toxicity and integration of cryogels |
| Agarose-gelatin (Bloch et al., 2009) | ICR mice | Effect of diabetes on cryogel integration and neovascularisation |
| Gelatin/hyaluronic acid (Chang et al., 2013) | BALB/c nude mice, porcine model | Localized toxicity |
| Hydroxyethylmethacrylate -l-lactide-dextran (HEMA-LLA-D/HEMA-LLA) (Bolgen et al., 2009) | Swiss-Albino rats | Localized toxicity |
| Gelatin-hydroxyapatite (Kemence and Bolgen, 2013) | Sprague Dawley rats | Localized toxicity |

polyvinylalcohol-alginate-bioactive glass composite (pTAC), polyhydroxyethylmethacrylate-gelatin (pHEMA–gelatin) and chitosan–agarose–gelatine (CAG)—was evaluated in B10Q and B10Q.*ncf1* mice at different time points by implanting them subcutaneously. Cryogels integrated well in surrounding tissues with formation of new blood vessels. Later cryogels were covered by thin fibrous tissue with a lot of extracellular matrix. A number of macrophages were observed infiltrating inside the pores of cryogels as well as at the interface of the cryogel and the tissue (Figure 7.4).

Further, *in vivo* biocompatibility of composite gelatin-hydroxyapatite-based cryogels was evaluated in a rat model by implanting them at bone and muscular tissue sites. After 4 and 12 weeks of implantation, neither inflammation nor degeneration of surrounding tissues was observed. At 4 weeks postimplantation, there was no significant change in the cryogel size, while after 12 weeks cryogel degradation was noticed significantly. Implanted cryogels were evaluated according to the standard of ISO10993-6. Implanted cryogels were nonallergic and nontoxic (Kemence and Bolgen, 2013). In another study, *in vivo* cell recruitment in injectable gelatin cryogel was studied. For cell recruitment, granulocyte macrophage colony-stimulating factor (GM-CSF) was mixed into a polymer solution during cryogel synthesis. GM-CSF is a chemotactic protein that directs cellular infiltration in the cryogel. GM-CSF

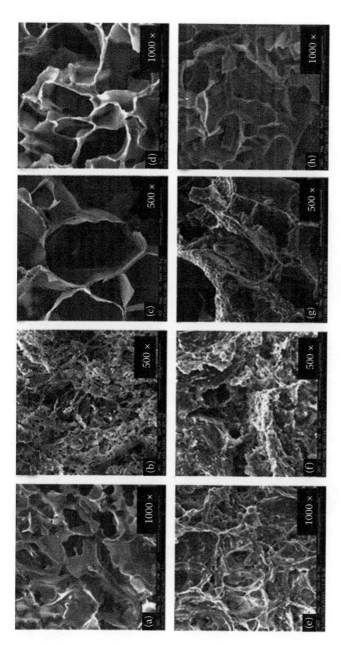

**FIGURE 7.4** Surface morphology of cryogels before and after implantation. SEM images of polyvinylcaprolactam (pVCl) (a), polyvinyl alcohol–alginate–bioactive glass composite (PTAC) (b), chitosan agarose gelatin (CAG) (c) and polyhydroxyethylmethacrylate–gelatin (pHEMA–gelatin) (d) before implantation. pVCl (e), pTAC (f), CAG (g) and pHEMA–gelatin (h) after sixth week of implantation. (From Shakya. A. K., Holmdahl, R., Nandakumar, K. S. and Kumar, A.: Polymeric cryogels are biocompatible and their biodegradation is independent of oxidative radicals. *J Biomed Mater Res A*. 2013. 102(10). 3409–3418. Copyright Wiley-VCH Verlag GmbH & Co. KGaA. Reproduced with permission.)

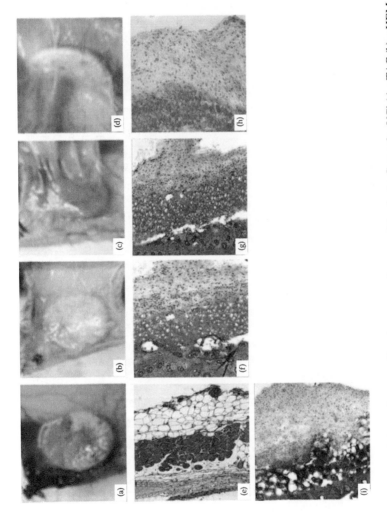

**FIGURE 7.5** Changes in the morphology of surrounding tissues near the cryogel implants. Cryogels pVCl (a), pTAC (b), pHEMA–gelatin (c) and CAG (d) cryogels were well integrated in surrounding tissues with new blood vessel formation. Hematoxylin and eosin staining of surrounding tissues in control mice (e) and mice implanted by pVCl (f), pTAC (g), CAG (h) and pHEMA–gelatin (i) cryogels. (Shakya, A. K., Holmdahl, R., Nandakumar, K. S. and Kumar, A.: Polymeric cryogels are biocompatible and their biodegradation is independent of oxidative radicals. *J Biomed Mater Res A*. 2013. 102(10). 3409–3418. Copyright Wiley-VCH Verlag GmbH & Co. KGaA. Reproduced with permission.)

encapsulated cryogel was injected into C57/Bl6J mice subcutaneously and the effect of the chemotactic protein was studied at different time points. Two weeks postinjection, a thick fibrous capsule with infiltration of primary granulocyte was observed around the cryogel as compared to cryogels without GM-CSF. This study demonstrated the ability of cryogels to deliver growth factor *in vivo* to get better cell response (Koshy et al., 2014). Furthermore, *in vivo* biocompatibility of gelatin cryogels (GH10) was studied in nude mouse and porcine models. Both animal models were healthy before and after implantation. There were no signs of inflammation, infection or skin necrosis observed near the implanting site. Gelatin/hyaluronic acid (GH) cryogels are able to differentiate the adipose tissue derived stem cells (ADSCs) into adipocyte lineage as studied by gene expression analysis through quantitative reverse transcriptase polymerase chain reaction (qRT-PCR). Adipose-like tissue formation was observed in the pores of implanted cryogels seeded with ADSCs, while only fibrous tissue formation or empty pores were observed in implanted cryogels seeded without any cells. Moreover, stem cell seeded cryogels showed more angiogenesis than the acellular cryogels (Chang et al., 2013). Further, *in vivo* biocompatibility of gelatin-laminin-based cryogels was evaluated by transplantation of cryogels in the rat cortex region. Fast degradation of low cross-linked cryogels (0.1%, v/v glutaraldehyde) was observed in comparison to medium cross-linked glutaraldehyde (0.3%, v/v). Neuroblast cells were migrated into the scaffolds and the scaffolds were found to be well integrated in the host tissue (Jurga et al., 2011). Recently, biocompatibility and integration of chitosan-agarose-gelatin (CAG) cryogel in subchondral defect of rabbit was studied. The cryogel implanted site showed well integration of the cryogel with the surrounding tissues. Neither localized nor systemic toxicities were observed after implantation of cryogel (Gupta et al., 2014). Overall, cryogels were well integrated with host tissues without any significant toxicity.

At the cellular level, host tissue response after cryogel implantation follows major inflammation events. The first phase (1–2 weeks) was characterized by the presence of little tissue exudates and infiltration of phagocytic cells followed by the subacute phase (3–4 weeks), characterized by infiltration of macrophages and proliferation of fibroblasts. Four to six weeks postimplantation is the late chronic phase, which is characterized by infiltration of macrophages and dendritic cells. Macroscopically, cryogels were covered by thin connective tissue with the presence of new blood vessels (Figure 7.5).

Due to high porosity, cryogels provide large surface area for cell binding and allow better exchange of nutrients unlike in nonporous material. Cryogels degraded significantly after 6 weeks of implantation. Generally, the degradation of cryogels initiates by infiltration of active macrophages. These active macrophages produce an acidic environment, which helps in degradation. Despite degradation by cells, other mechanisms such as oxidation and hydrolysis help in cryogel degradation (Shakya et al., 2013).

## 7.5 CONCLUSION

Biocompatibility is one of the important assessments while designing of biomaterials. Macroporous cryogels are promising materials in tissue engineering and they well support growth of mammalian cells. High porosity in cryogels provides good exchange of nutrients across the cryogel with minimal transfer resistance. Cryogels are biocompatible and biodegradable when implanted in an animal body. They integrate well in surrounding tissues and are covered by thin vascular fibrous tissue. In the future, they can be used for other biomedical applications.

## LIST OF ABBREVIATIONS

| | |
|---|---|
| AA | Acrylic acid |
| ADSCs | Adipose derived stem cells |
| ASTM | American Society for Testing and Materials |
| BS | Biospan |
| CAG | Chitosan–agarose–gelatin |
| DMAA | $N,N$-dimethylacrylamide |
| GH | Gelatin hyaluronic acid |
| GL | Gelatin-laminin |
| GM-CSF | Granulocyte macrophage colony stimulating factor |
| HEMA-LLA-D | Hydroxyethylmethacrylate-$l$-lactide-dextran |
| MTT | 3-(4,5-dimethylthiazol-2-yl)-2,5-diphenyltetrazolium bromide |
| PE | Polyethylene |
| PEG | Polyethylene glycol |
| pHEMA-gelatin | Polyhydroxyethylmethacrylate–gelatin |
| PPG | Polypropylene glycol |
| PSF | Polysulphone |
| pTAC | Polyvinylalcohol-alginate-bioactive glass composite |
| PTFE | Polytetrafluroethylene |
| PU | Polyurethrane |
| PVC | Polyvinyl chloride |
| pVCl | Polyvinylcaprolactam |
| PVP | Polyvinylpyrrolidone |
| qRT-PCR | Quantitative reverse transcriptase polymerase chain reaction |

## REFERENCES

Abraham, G. A., de Queiroz, A. A. and Roman, J. S. (2001). Hydrophilic hybrid IPNs of segmented polyurethanes and copolymers of vinylpyrrolidone for applications in medicine. *Biomaterials* **22**(14): 1971–1985.

Bajpai, A. and Saini, R. (2005). Preparation and characterization of spongy cryogels of poly(vinyl alcohol)–casein system: Water sorption and blood compatibility study. *Polym Int* **54**: 796–806.

Bajpai, A. K. and Saini, R. (2009). Designing of macroporous biocompatible cryogels of PVA-haemoglobin and their water sorption study. *J Mater Sci Mater Med* **20**(10): 2063–2074.

Bhat, S., Tripathi, A. and Kumar, A. (2010). Supermacroprous chitosan-agarose-gelatin cryogels: *In vitro* characterization and *in vivo* assessment for cartilage tissue engineering. *J R Soc Interface* **8**(57): 540–554.

Bloch, K., Vanichkin, A., Damshkaln, L. G., Lozinsky, V. I. and Vardi, P. (2009). Vascularization of wide pore agarose-gelatin cryogel scaffolds implanted subcutaneously in diabetic and non-diabetic mice. *Acta Biomater* **6**(3): 1200–1205.

Bolgen, N., Vargel, I., Korkusuz, P., Guzel, E., Plieva, F., Galaev, I., Matiasson, B. and Piskin, E. (2009). Tissue responses to novel tissue engineering biodegradable cryogel scaffolds: An animal model. *J Biomed Mater Res A* **91**(1): 60–68.

Brodbeck, W. G. and Anderson, J. M. (2009). Giant cell formation and function. *Curr Opin Hematol* **16**(1): 53–57.

Brunnette, D. M. (1998). The effect of surface topography on cell migration and adhesion. In: *Surface Characterization of Biomaterials*. R. D. Ratner, Ed. Amsterdam, Elsevier, pp. 203–217.

Chang, K. H., Liao, H. T. and Chen, J. P. (2013). Preparation and characterization of gelatin/hyaluronic acid cryogels for adipose tissue engineering: *In vitro* and *in vivo* studies. *Acta Biomater* **9**(11): 9012–9026.

Chapman, R. C., Ostuni, E., Yan, L. and Whitesides, G. M. (2000). Preparation of mixed self-assembled monolayers (SAMs) that resist adsorption of proteins using the reaction of amines with a SAM that present interchain carboxylic anhydride groups. *Langmuir* **16**: 6927–6936.

Cobelli, N., Scharf, B., Crisi, G. M., Hardin, J. and Santambrogio, L. (2011). Mediators of the inflammatory response to joint replacement devices. *Nat Rev Rheumatol* **7**(10): 600–608.

de Queiroz, A. A. A., Barrak, E. R. and de Castro, S. C. (1997). Thermodynamic analysis of the surface of biomaterials. *J Mol Struct* **394**: 271–279.

Elbert, D. L. and Hubbell, J. A. (1996). Surface treatments of polymers for biocompatibility. *Annu Rev Mater Sci* **26**: 365–394.

Gardner, A. B., Lee, S. K., Woods, E. C. and Acharya, A. P. (2013). Biomaterials-based modulation of the immune system. *Biomed Res Int* **2013**: 732182.

Gupta, A., Bhat, S., Jagdale, P. R., Chaudhari, B. P., Lidgren, L., Gupta, K. C. and Kumar, A. (2014). Evaluation of three-dimensional chitosan-agarose-gelatin cryogel scaffold for the repair of subchondral cartilage defects: An *in vivo* study in a rabbit model. *Tissue Eng Part A* **20**(23–24): 3101–3111.

Higuchi, A., Shirano, K., Harashima, M., Yoon, B. O., Hara, M., Hattori, M. and Imamura, K. (2002). Chemically modified polysulfone hollow fibers with vinylpyrrolidone having improved blood compatibility. *Biomaterials* **23**(13): 2659–2666.

Hussain, S. and Rajab, N. F. (2009). *In vitro* testing of biomaterials toxicity and biocompatibility. In: *Cellular Response to Biomaterials*. L. D. Silvio, Ed. Boca Raton, FL, CRC Press, pp. 508–537.

Hwang, Y., Zhang, C. and Varghese, S. (2010). Poly(ethylene glycol) cryogels as potential cell scaffolds: Effect of polymerization conditions on cryogel microstructure and properties. *J Mater Chem* **20**: 345–351.

Jain, E., Karande, A. A. and Kumar, A. (2010). Supermacroporous polymer-based cryogel bioreactor for monoclonal antibody production in continuous culture using hybridoma cells. *Biotechnol Prog* **27**(1): 170–180.

Jain, E. and Kumar, A. (2009). Designing supermacroporous cryogels based on polyacrylonitrile and a polyacrylamide-chitosan semi-interpenetrating network. *J Biomater Sci Polym Ed* **20**(7–8): 877–902.

Jain, E. and Kumar, A. (2013). Disposable polymeric cryogel bioreactor matrix for therapeutic protein production. *Nat Protoc* **8**(5): 821–835.

Jurga, M., Dainiak, M. B., Sarnowska, A., Jablonska, A., Tripathi, A., Plieva, F. M., Savina, I. N., Strojek, L., Jungvid, H., Kumar, A., Lukomska, B., Domanska-Janik, K., Forraz, N. and McGuckin, C. P. (2011). The performance of laminin-containing cryogel scaffolds in neural tissue regeneration. *Biomaterials* **32**(13): 3423–3434.

Kane, R. S., Takayama, S., Ostuni, E., Ingber, D. E. and Whitesides, G. M. (1999). Patterning proteins and cells using soft lithography. *Biomaterials* **20**(23–24): 2363–2376.

Kathuria, N., Tripathi, A., Kar, K. K. and Kumar, A. (2009). Synthesis and characterization of elastic and macroporous chitosan-gelatin cryogels for tissue engineering. *Acta Biomater* **5**(1): 406–418.

Kemence, N. and Bolgen, N. (2013). Gelatin- and hydroxyapatite-based cryogels for bone tissue engineering: Synthesis, characterization, *in vitro* and *in vivo* biocompatibility. *J Tissue Eng Regen Med* Aug 28. doi: 10.1002/term.1813. [Epub ahead of print].

Kim, D. T., Blanch, H. W. and Radke, C. J. (2002). Direct imaging of lysozyme adsorption onto mica by atomic force microscope. *Langmuir* **18**: 5841–5850.

Koshy, S. T., Ferrante, T. C., Lewin, S. A. and Mooney, D. J. (2014). Injectable, porous, and cell-responsive gelatin cryogels. *Biomaterials* **35**(8): 2477–2487.

Kumar, A., Bansal, V., Nandakumar, K. S., Galaev, I. Y., Roychoudhury, P. K., Holmdahl, R. and Mattiasson, B. (2006). Integrated bioprocess for the production and isolation of urokinase from animal cell culture using supermacroporous cryogel matrices. *Biotechnol Bioeng* **93**(4): 636–646.

Kumar, A. and Srivastava, A. (2010). Cell separation using cryogel-based affinity chromatography. *Nat Protoc* **5**(11): 1737–1747.

Langer, R. and Vacanti, J. P. (1993). Tissue engineering. *Science* **260**(5110): 920–926.

Lee, H. J., Kopecek, J. and Andrade, J. D. (1989). Protein resistant surface prepared by PEO containing block copolymer surfactant. *J Biomed Mater Res* **23**: 351–368.

Lee, J. H. and Lee, H. B. (1998). Platelet adhesion onto wettability gradient surfaces in the absence and presence of plasma proteins. *J Biomed Mater Res* **41**(2): 304–311.

Lozinsky, V. I., Galaev, I. Y., Plieva, F. M., Savina, I. N., Jungvid, H. and Mattiasson, B. (2003). Polymeric cryogels as promising materials of biotechnological interest. *Trends Biotechnol* **21**(10): 445–451.

Lozinsky, V. I., Plieva, F. M., Galaev, I. Y. and Mattiasson, B. (2001). The potential of polymeric cryogels in bioseparation. *Bioseparation* **10**(4–5): 163–188.

Meumer, S., Heijnen, H. F. G., Ijsseldijk, M. J. W., Orlando, E., de Groot, P. G. and Sixma, J. J. (1995). Pletelet adhesion to fibronectin in flow: The importance of von Willebrand factor and glycoprotein Ib. *Blood* **86**: 3452–3460.

Mishra, R., Goel, S. K., Gupta, K. C. and Kumar, A. (2013). Biocomposite cryogels as tissue-engineered biomaterials for regeneration of critical-sized cranial bone defects. *Tissue Eng Part A* **20**(3–4): 751–762.

Mishra, R. and Kumar, A. (2010). Inorganic/organic biocomposite cryogels for regeneration of bony tissues. *J Biomater Sci Polym Ed* **22**(16): 2107–2126.

Nilsang, S., Nandakumar, K. S., Galaev, I. Y., Rakshit, S. K., Holmdahl, R., Mattiasson, B. and Kumar, A. (2007). Monoclonal antibody production using a new supermacroporous cryogel bioreactor. *Biotechnol Prog* **23**(4): 932–939.

Odabaş, S., İnci, I. and Pişkin, E. (2012). Gelatin/oxide-dextran cryogels: In-vitro biocompatibility evaluations. *Hacettepe J Biol Chem* **40**: 409–417.

Peniche, H., Reyes-Ortega, F., Aguilar, M. R., Rodriguez, G., Abradelo, C., Garcia-Fernandez, L., Peniche, C. and San Roman, J. (2013). Thermosensitive macroporous cryogels functionalized with bioactive chitosan/bemiparin nanoparticles. *Macromol Biosci* **13**(11): 1556–1567.

Percin, I., Aksoz, E. and Denizli, A. (2013). Gelatin-immobilised poly(hydroxyethyl methacrylate) cryogel for affinity purification of fibronectin. *Appl Biochem Biotechnol* **171**(2): 352–365.

Puleo, D. A. and Nanci, A. (1999). Understanding and controlling the bone-implant interface. *Biomaterials* **20**(23–24): 2311–2321.

Ratner, B. D., Hoffman, A. S., Hanson, S. R., Harker, L. A. and Whiffen, J. D. (1979). Blood-compatibility-water-content relationships for radiation-grafted hydrogels. *J Polym Sci Polym Symp* **66**: 363–372.

Reichelt, S., Becher, J., Weisser, J., Prager, A., Decker, U., Moller, S., Berg, A. and Schnabelrauch, M. (2014). Biocompatible polysaccharide-based cryogels. *Mater Sci Eng C Mater Biol Appl* **35**: 164–170.

Ruckenstein, E. and Gourisankar, S. V. (1986). Preparation and characterization of thin film surface coatings for biological environments. *Biomaterials* **7**(6): 403–422.

Sami, H. and Kumar, A. (2013). Tunable hybrid cryogels functionalized with microparticles as supermacroporous multifunctional biomaterial scaffolds. *J Biomater Sci Polym Ed* **24**(10): 1165–1184.

Shakya, A. K., Holmdahl, R., Nandakumar, K. S. and Kumar, A. (2013). Polymeric cryogels are biocompatible and their biodegradation is independent of oxidative radicals. *J Biomed Mater Res A* **102**(10): 3409–3418.

Sharma, A., Bhat, S., Vishnoi, T., Nayak, V. and Kumar, A. (2013). Three-dimensional supermacroporous carrageenan-gelatin cryogel matrix for tissue engineering applications. *Biomed Res Int* **2013**: 478279.

Simionescu, B. C., Neamtu, A., Balhui, C., Danciu, M., Ivanov, D. and David, G. (2013). Macroporous structures based on biodegradable polymers—Candidates for biomedical application. *J Biomed Mater Res A* **101**(9): 2689–2698.

Singh, D., Nayak, V. and Kumar, A. (2010). Proliferation of myoblast skeletal cells on three-dimensional supermacroporous cryogels. *Int J Biol Sci* **6**(4): 371–381.

Spina, R. L., Tripisciano, C., Mecca, T., Cunsolo, F., Weber, V. and Mattiasson, B. (2014). Chemically modified poly(2-hydroxyethyl methacrylate) cryogel for the adsorption of heparin. *J Biomed Mater Res B Appl Biomater* **102**(6): 1207–1216.

Srivastava, A. and Kumar, A. (2009). Synthesis and characterization of a temperature-responsive biocompatible poly(N-vinylcaprolactam) cryogel: A step towards designing a novel cell scaffold. *J Biomater Sci Polym Ed* **20**(10): 1393–1415.

Tanaka, M., Motomura, T., Kawada, M., Anzai, T., Kasori, Y., Shiroya, T., Shimura, K., Onishi, M. and Mochizuki, A. (2000). Blood compatible aspects of poly(2-methoxyethylacrylate) (PMEA)—Relationship between protein adsorption and platelet adhesion on PMEA surface. *Biomaterials* **21**(14): 1471–1481.

Tripathi, A., Kathuria, N. and Kumar, A. (2009). Elastic and macroporous agarose-gelatin cryogels with isotropic and anisotropic porosity for tissue engineering. *J Biomed Mater Res A* **90**(3): 680–694.

Vacanti, J. P. (2003). Tissue and organ engineering: Can we build intestine and vital organs? *J Gastrointest Surg* **7**(7): 831–835.

VandeVord, P. J., Matthew, H. W., DeSilva, S. P., Mayton, L., Wu, B. and Wooley, P. H. (2002). Evaluation of the biocompatibility of a chitosan scaffold in mice. *J Biomed Mater Res* **59**(3): 585–590.

Vishnoi, T. and Kumar, A. (2012). Conducting cryogel scaffold as a potential biomaterial for cell stimulation and proliferation. *J Mater Sci Mater Med* **24**(2): 447–459.

Wang, Y. X., Robertson, J. L., Spillman, W. B., Jr. and Claus, R. O. (2004). Effects of the chemical structure and the surface properties of polymeric biomaterials on their biocompatibility. *Pharm Res* **21**(8): 1362–1373.

# 8 Cryogel Biomaterials for Musculoskeletal Tissue Engineering

*Ruchi Mishra, Sumrita Bhat and Ashok Kumar**

## CONTENTS

| | | |
|---|---|---|
| 8.1 | Introduction | 216 |
| 8.2 | Prevalence of Musculoskeletal Conditions | 218 |
| 8.3 | Biomaterials for Musculoskeletal Tissue Engineering | 218 |
| | 8.3.1 Cartilage Tissue Engineering | 219 |
| | 8.3.2 Bone Tissue Engineering | 225 |
| | 8.3.3 Muscle Tissue Engineering | 229 |
| | 8.3.4 Tendon Tissue Engineering | 229 |
| | 8.3.5 Ligament Tissue Engineering | 229 |
| 8.4 | Significance of Cryogel Technology in Musculoskeletal Tissue Engineering | 230 |
| 8.5 | Cryogels for Bone Tissue Engineering | 230 |
| | 8.5.1 Polymeric Cryogels for Bone Tissue Engineering | 231 |
| |     8.5.1.1 Gelatin–Based Cryogels | 231 |
| |     8.5.1.2 Polyethylene Glycol–Based Cryogels | 231 |
| |     8.5.1.3 Silk Fibroin–Based Cryogels | 232 |
| |     8.5.1.4 Hydroxyethyl Methacrylate (HEMA)–Based Cryogels | 232 |
| | 8.5.2 Polymeric-Ceramic/Bioactive Glass Composite Cryogels for Bone Tissue Engineering | 232 |
| 8.6 | Cryogels in Cartilage Tissue Engineering | 236 |
| | 8.6.1 Cryogels Fabricated Using Natural Polymers for Cartilage Tissue Engineering | 237 |
| | 8.6.2 Cryogels Fabricated Using Synthetic Polymers for Cartilage Tissue Engineering | 243 |
| 8.7 | Conclusions | 243 |
| Acknowledgements | | 243 |
| List of Abbreviations | | 244 |
| References | | 244 |

---

* Corresponding author: E-mail: ashokkum@iitk.ac.in; Tel: +91-512-2594051.

## ABSTRACT

Musculoskeletal defects can occur due to the loss/defect of musculoskeletal tissue arising from tumour, congenital defects, trauma, fracture or injuries arising from sports, accidents and so on. The current treatment strategies for these defects are autografts and allografts. Due to the limitations associated with these conventional methods such as donor site morbidity, pain during autografting, immune rejection and infection during allografting, tissue-engineered biomaterials appear to be a more feasible option for musculoskeletal repair and regeneration. The fabrication methods currently used for musculoskeletal tissue-engineered biomaterials development are thermally induced phase separation, melt moulding, membrane lamination, supercritical-fluid technology, solid freeform fabrication and so on. To these conventional methods, cryogelation technology has recently been introduced as a method having certain advantages such as aqueous medium based synthesis at subzero temperatures avoiding the production of any toxic intermediates, formation of supermacroporous cryogels with faster polymerization occurring in the unfrozen liquid microphases as well as ease of fabrication. In this chapter, various cryogel biomaterials that have been used for musculoskeletal tissue engineering have been explored.

## KEYWORDS

Biomaterials, bone, cartilage, chondrocytes, composites, cryogel, tissue engineering

## 8.1 INTRODUCTION

The musculoskeletal system comprises muscles, bones, cartilage, tendons, ligaments and other connective tissues. All these tissues differ from each other in the degree of their inherent repair capacity (Henson and Getgood, 2011). For example, cartilage has a limited self-repair capacity due to its avascular and aneural nature, whereas bone has high regeneration capability with vascular and neural supply throughout the tissue. Due to its limited healing, even partial or superficial cartilage defects do not heal, and in the case of deep or subchondral defects, the healing process is facilitated by the recruitment of bone marrow derived stem cells from the underlying bone. The repaired tissue formed in this case, however, is a disorganized fibrocartilage instead of a regular one, that is, hyaline cartilage (O'Driscoll, 1998; Buckwalter, 2002). Another limitation of the self-repair strategy is the limitation with the defect size. Defects of only specific sizes can be healed on their own while others remain unhealed leading to restricted joint movement and frequent pain. Other tissues like tendons, ligaments and menisci also have a limited healing capacity or they heal with the formation of a disorganized tissue. These limitations can be dealt with using three-dimensional tissue engineered scaffolds, which can be an excellent strategy for the repair of both superficial and deep cartilage defects (Molloy et al., 2003; Chu et al., 2010). Among all the listed tissues, bone and muscles have a superior healing capacity. However, self-repair fails when the size of the defect is larger than the critical size, that is, the size at or beyond which the tissues cannot heal on their own during the lifespan of any individual (Schmitz and Hollinger, 1986), for example, treatment of bone or muscle neoplasia (Pelled et al., 2010). In this case, too, scaffolds can play a pivotal role in bridging the gap and, thus, facilitating the process of healing.

A scaffolding material to be used for musculoskeletal tissue engineering should possess properties similar to the native tissue. Most apposite seems to be the autologous tissue recovered for the patient or the use of 'off-the-shelf' allogenic or xenogenic decellularized materials. Although autografts suffer from limitations like donor site morbidity and pain, allografts and xenografts face the problem of immune rejection and infection of pathogens. Considering these limitations, use of polymeric materials/biomaterials is the most practical alternative. Scaffolds intended for musculoskeletal tissue engineering applications should meet certain criteria such as: they should be biodegradable, biocompatible and bioresorbable; they should not elicit any adverse host immune reaction; they should possess a congenial surface and architecture for cell adhesion and proliferation; they should be mechanically stable to withstand the load; and they should possess good handling properties, that is, they should give a possibility to be customized in accordance to the defect size (Getgood et al., 2009). Earlier, scaffolds fabricated from simple synthetic materials like polyethylene meshes, carbon implants or carbon-polylactic acid polymers were used for skeletal tissue regeneration (Parrish et al., 1978; Aragona et al., 1981; Robinson et al., 1993). The focus is now shifting toward the use of natural polymers like proteins and carbohydrates for the fabrication of scaffolds. Collagen type I/III is being commercially utilized for matrix associated chondrocyte implantation by Genzyme Inc., Oxford, UK. Scaffolds fabricated from collagen are cultured with chondrocytes *ex vivo* followed by their implantation at the articular cartilage defect area (Bartlett et al., 2005). Collagen is also being modified by the addition of chondroitin sulphate and is marketed as Novocart 3D (Tetec, Germany). For the fabrication of osteochondral scaffolds, ceramics such as calcium phosphate have been used. Examples include chondromimetic (Tigenix, Cambridge, UK) and Maioregen (Finceramica, Italy) (Henson and Getgood, 2011). Carbohydrate-based scaffolds have also shown utility for musculoskeletal tissue engineering and a few of them are commercially available. For example, BST-cargel system (Bio-Syntec Inc., Canada) uses scaffolds fabricated from chitosan infused with coagulated whole blood before the implantation into the lesion (Shive et al., 2006). Another example is fibrin gel, marketed as autologous chondrocyte implantation device or chondron by Sewon Cellontech, Korea. In addition to these natural products, scaffolds fabricated from synthetic materials are now commercially available. For example, Bioceed-C contains chondrocytes in a biodegradable polyglycolic/polylactic acid/polydioxanane fabricated matrix. Advantages of both synthetic and biological materials are harvested by fabricating a composite scaffold, for example, bioglass-collagen-phosphatidylserine scaffolds, collagen-PLA composites and bioactive glass/polycaprolactone/biphasic calcium phosphate scaffolds. Biphasic hybrid scaffolds composed of type I collagen and $\beta$-tri calcium phosphate ceramic suspended in a polylactic acid lattice are commercially available as Osseofit for use as bone void filler (Getgood et al., 2009). Of late, bioactive scaffolds have found applicability in the area of musculoskeletal tissue engineering. In this concept, selection of scaffolding material is an important parameter and is designed in such a manner that can enhance the tissue formation by controlling the local and surrounding milieu (Place et al., 2009).

Additionally the scaffold surface can be modified by incorporating cell adhesion peptides or growth factor proteins. Initial work done by the immobilization of recombinant or synthetic molecules onto poly (α-hydroxy esters) has shown potential for enhancing bioactivity, promoting adherence and proliferation of cells as well as synthesis of extracellular matrix. Scaffolds fortified with stimulating peptides, for example, stromal derived factors 1-α (SDF 1-α), have been shown to recellularize the implanted scaffolds (Shen et al., 2010; Thevenot et al., 2010). Other types of growth factors like TGF-β loaded scaffolds have been employed to recruit progenitor cells into an unseeded scaffold. Lee et al. (2010) fabricated scaffolds pervaded with TGF-β3 and implanted them acellulary. These vitalized scaffolds showed uniformly distributed chondrocytes in a matrix majorly composed of collagen type II and aggrecan, indicating that bioactive scaffolds can recruit and maintain cellular functionality. Furthermore, altered physiochemical properties of modified materials can regulate local microenvironment under *in vivo* conditions. For example, calcium phosphate biomaterials having modifications in chemical stricture and crystallinity exhibit better protein adsorption, can release ionic species and have the ability to promote ectopic bone growth (Yuan et al., 2010). Lately, decellularized matrices have found immense applicability for *in vivo* tissue regeneration. Scaffolds fabricated from decellularized matrices are bioactive and thus can trigger the release of peptides and stimulators (Badylak et al., 2009). An emerging class of materials for musculoskeletal tissue regeneration is the self-assembling peptide amphiphile (PA) system. Shah et al. demonstrated the effectiveness of PA nanofiber gels incorporated with TGF-binding sequence in the repair of full thickness articular cartilage defects in rabbits (Shah et al., 2010).

## 8.2 PREVALENCE OF MUSCULOSKELETAL CONDITIONS

Disorders of the musculoskeletal system affect millions of people worldwide and are the main cause of long-term pain and physical disability. They include large numbers of conditions like osteoarthritis, rheumatoid arthritis, osteoporosis, low trauma fractures, tumours, congenital defects and so on. These conditions affect approximately 20% of adults worldwide. Different surveys have revealed that musculoskeletal conditions caused 40% of all chronic conditions, 54% of all long-term disability and 24% of all restricted activity days (Badley et al., 1994). Musculoskeletal conditions are the major cause of absence from work in developed countries, which puts its burden on the world economy (Tellnes et al., 1989). This burden is supposed to increase due to changes in lifestyle with more urbanization. There is an instantaneous demand for a novel treatment regime for the management of musculoskeletal conditions; therefore, tissue engineering using scaffolds has emerged as a viable alternative.

## 8.3 BIOMATERIALS FOR MUSCULOSKELETAL TISSUE ENGINEERING

Various types of biomaterials used for the repair/regeneration of defects in different types of musculoskeletal tissues such as cartilage, bone, tendon, ligament and muscle are discussed briefly in the following subsections.

## 8.3.1 Cartilage Tissue Engineering

Scaffolds play a very important role in cartilage tissue engineering, as they act as three-dimensional supports for the adhesion and growth of chondrocytes. They also provide an initial platform for the production of cartilaginous tissue. Numerous scaffold materials have been used for cell delivery in cartilage regeneration. However, the primary focus has been on polymeric materials in forms of hydrogels, sponges, fibrous meshes, nanofibers and so on. Among the other prerequisites, degradation is a very important property of the scaffold, which can occur hydrolytically or enzymatically and by controlling the degradation, the scaffold can enhance and direct new tissue growth. It has been shown that scaffolds with both degradable and nondegradable units show improved ECM distribution compared to completely nondegradable scaffolds (Bryant and Anseth, 2003). However, there should always be a balance as slow degradation may impede new cartilagenous ECM production, whereas fast degradation may compromise structural support and shape retention. Other very important factors that should be considered while designing a scaffold are the cell seeding density and seeding methods. It is very important to seed an appropriate cell number to ensure adequate cell-to-cell interactions. It has been seen that high initial seeding densities tend to facilitate greater ECM synthesis and deposition, which may be due to better cell–cell interactions (Freed et al., 1994; Iwasa et al., 2003; Huang et al., 2009). The method of seeding (statically or dynamically) also plays an important role in dictating the cell distribution and infiltration of cells into the scaffold. It has been seen that in sponge and mesh scaffolds, dynamic seeding can improve the cellular distribution, whereas hydrogels typically support uniform cell distributions if cells are sufficiently suspended during the gelation process.

To date, a wide range of natural and synthetic materials has been investigated as scaffolding materials for cartilage repair. Natural polymers that have not been explored as bioactive scaffolds for cartilage tissue engineering include alginate, agarose, gelatin, fibrin and so on. On the other hand, synthetic polymers that are currently being explored for cartilage tissue engineering include poly($\alpha$-hydroxy esters), PEG, poly(NiPAAm), poly(propylene fumarates) and polyurethanes. Natural polymers can often interact with the cells via cell surface receptors, which regulate or direct the cell functions. However, due to this interaction, these polymers may also elicit an immune response. Therefore, antigenicity and disease transfer are of concern when using these biomaterials at the site, which is guarded by the immune system. Other concerns with natural polymers may be inferior mechanical stability and fast degradation. On the other hand, synthetic polymers are more controllable and predictable; properties of these polymers can be tuned to alter the mechanical and degradation characteristics. However, unless specifically incorporated, synthetic polymers do not benefit from direct cell–scaffold interactions, which can play an important role in adhesion, cell signalling and matrix remodelling. In addition, the degradation products of synthetic polymers can be toxic or can even elicit an inflammatory response in the host. Finally, the scaffold architecture plays an important role in cell growth and proliferation. Table 8.1 gives a comprehensive list of the different types of polymers and methods of fabrication used in the area of cartilage tissue engineering.

## TABLE 8.1
### List of Polymers Used for Cartilage Tissue Engineering

| Fabrication | Method | Cell Source | Outcome |
|---|---|---|---|
| *1. Natural Polymers* | | | |
| **a. Alginate** | | | |
| Chondrocytes in suspension with 2% sodium alginate | *In vivo*; 500 µl of suspension injected subcutaneously into dorsa of nude mice. Calcium chloride then injected into this area to stimulate cross-linking of the scaffold. Cartilage harvested from 14 to 38 weeks. | Human nasal septal chondrocytes | Gross analysis showed that 14/15 constructs resembled native human cartilage. Six of the explants had histologically homogenous resemblance to native cartilage. The neo-constructs stained positively for Col II. |
| 3D alginate scaffold prepared by freeze drying | *In vitro*; cells were cultured in the alginate for 1–4 weeks in a bioreactor. | Porcine articular chondrocytes | RT-PCR analysis showed the cells maintained their differentiated phenotype for up to 4 weeks. |
| 3D alginate gels | *In vitro*; cell/gel constructs were cultured for 0, 6, 12, 18 and 24 days. | Human MSCs | Results of qRT-PCR analysis provided a temporal analysis for marker expression during chondrogenesis. |
| Alginate gel layer | *In vitro*; to evaluate the effect of low intensity ultrasound (LIUS) on cell viability during chondrogenic differentiation. | Human MSCs | When the cell/alginate construct was cultured with TGF-1, cell viability decreased. However, addition of LIUS enhanced viability and inhibited apoptosis under the same conditions. |
| Hydrogel | *In vitro*; chondrocytes were seeded onto alginate after 1, 2 and 3 passages in a monolayer. | Human nasal septal chondrocytes | Alginate stimulated GAG and Col I deposition supporting the chondrocytic phenotype. |
| **b. Chitosan** | | | |
| Fibrous scaffold vs. sponge | *In vitro*; constructs analysed 3, 10 and 21 days after cell seeding. | Mouse BMSC Line | At 10 and 21 days, the cells were embedded but did not aggregate, with fibrous scaffolds containing more ECM. |

*(Continued)*

## TABLE 8.1 (CONTINUED)
### List of Polymers Used for Cartilage Tissue Engineering

| Fabrication | Method | Cell Source | Outcome |
|---|---|---|---|
| Chitosan scaffold and chitosan microspheres | *In vitro*; scaffold and microspheres used as TGF-β1 carrier to see the effect of this growth factor on chondrogenic potential. | Rabbit articular chondrocytes | Encapsulation efficiency of TGF-1 was 90.1%. TGF-1 was released from chitosan in a multiphase fashion. TGF-β1 loaded microspheres significantly improved cell proliferation rate and Col II production, compared with controls with no microspheres or controlled TGF-β1 release. |
| Chitosan scaffold synthesized via freeze-drying | *In vitro*; cells seeded onto chitosan of varying porosity; <10 μm, 10–50 μm and 70–120 μm. Cultured for 28 days in a rotating bioreactor. | Porcine articular chondrocytes | Chitosan scaffolds remained intact compared with the positive control PGA. However, cartilage-specific DNA levels and GAG were lower in the chitosan groups compared with PGA. Chitosan also had the largest pores, with more chondrocytes. |
| **c. *Collagen*** | | | |
| PLGA mesh and collagen sponge | *In vitro*; hybrid disks of PLGA/collagen scaffold with different structures. *In vivo*; week-old cultured constructs implanted into dorsa of athymic nude mice and harvested after 2, 4 and 8 weeks. | Bovine articular chondrocytes | Homogenous cell distribution with natural chondrocytes morphology. Abundant ECM production. Levels of GAG and collagen II DNA, and aggrecan mRNA increased on the scaffolds with more collagen. |
| Hydrogel | *In vivo*: comparison of collagen hydrogel and collagen-alginate hydrogel. Gel injected subcutaneously into rabbit backs. | BM-MSC | Homogenous distribution of cells with chondrocyte characteristics demonstrated the chondrogenic differentiation of BM-MSCs. Both collagen hydrogel and collagen alginate hydrogel may induce chondrogenesis. |

(*Continued*)

## TABLE 8.1 (CONTINUED)
### List of Polymers Used for Cartilage Tissue Engineering

| Fabrication | Method | Cell Source | Outcome |
|---|---|---|---|
| *d. Fibrin* | | | |
| Fibrin gel | *In vivo*: ACL on 30 patients using minimally invasive injection techniques. Mix of fibrin gel and chondrocytes. | Autologous adult chondrocytes | Patients evaluated 24 months postoperatively using the Cincinnati knee ligament rating scores, for which 10 patients had excellent results, 17 with good results, 2 fair and 1 poor result. |
| PLGA/fibrin hybrid scaffold | *In vitro*; PLGA scaffold soaked in chondrocyte-fibrin suspension (polymerized by thrombin CaCl2 solution), constructs were cultured for a maximum of 21 days. | Rabbit articular chondrocytes | Cell proliferation increased steadily until day 14, but declined by day 21. Cartilage formation evident at day 14, confirmed by the presence of cartilaginous cells embedded in basophilic ECM filled lacunae. |
| *e. Hyaluronic Acid (HA)* | | | |
| Hyaff®-11, biodegradable polymer, nonwoven mesh | *In vitro*; chondrocytes were harvested from OA patients and seeded onto Hyaff®. Constructs remained in culture for 28 days, analysed on day 0, 7, 14, 21 and 28. | Human autologous chondrocytes | Viability and proliferation of OA chondrocytes similar to cells from normal subjects. |
| HA immobilized on surface of PLGA scaffold | *In vitro*; biodegradable macroporous PLGA scaffolds chemically conjugated to the surface-exposed amine groups of the PLGA. | Bovine articular chondrocytes | Enhanced cellular attachment was observed compared with PLGA controls. GAG and total Col synthesis was significantly increased for HA/PLGA compared to the control. |

(*Continued*)

## TABLE 8.1 (CONTINUED)
### List of Polymers Used for Cartilage Tissue Engineering

| Fabrication | Method | Cell Source | Outcome |
|---|---|---|---|
| **2. Synthetic** | | | |
| *a. PLGA-Poly(lactic-co-glycolic) Acid* | | | |
| PLGA scaffolds | *In vivo*: PLGA scaffolds were seeded with AD-MSC, cultured in TGF1 containing medium for 3 weeks, prior to implantation in the subcutaneous pockets of nude mice for 8 weeks. | Human ADMSC | RT-PCR demonstrated the increased expression profiles of chondrospecific marker mRNA, compared with control samples after 3 weeks *in vitro* and 8 weeks *in vivo*. |
| PLGA microspheres | *In vivo*; PLGA microsphere seeded with rabbit chondrocytes injected subcutaneously into dorsa of athymic female mice. | Autologous rabbit chondrocytes | The PLGA microsphere permitted cell adhesion. Four and 9 weeks postimplantation there was macroscopic and histological evidence of cartilage formation on the seeded PLGA microsphere compared with nothing on the PLGA and chondrocyte controls. |
| *PCL-Poly(carprolactone)* | | | |
| Electrospun 3D nanofibrous scaffold | *In vitro*; MSC seeded onto prefabricated nanofibrous scaffold for 21 days. | Human BMMSC | The cartilage specific gene profile (Aggrecan, Col I,I and Col X) was low, but improved significantly in chondrogenic medium with TGFβ1. |

(*Continued*)

## TABLE 8.1 (CONTINUED)
### List of Polymers Used for Cartilage Tissue Engineering

| Fabrication | Method | Cell Source | Outcome |
|---|---|---|---|
| *PGA-Polyglycolic Acid* | | | |
| PGA-HA composite scaffold | *In vivo*; MSC were seeded onto the PGA-HA and co-cultured for 72 h. These were then implanted into full thickness cartilage defects in the intercondylar fossa of rabbit femurs. Constructs were then harvested after 16 or 32 weeks of surgery. | Rabbit MSC | Grossly, the constructs demonstrated hyaline cartilage formation and at 16 weeks, there appeared to be integration with surrounding normal cartilage and subchondral bone. |
| *PEG-Poly(ethylene glycol)* | | | |
| Hydrogels | *In vitro*; cells were encapsulated in the PEG hydrogel and allowed to free swell for 24 h. | Bovine temporomandibular chondrocytes | Condylar chondrocyte viability was maintained within the constructs during cell culture. RT-PCR analysis showed the expression of cartilage specific markers, namely Col II, aggrecan and Col I were maintained. |

*Source:* Reprinted from Oseni, A.O. et al., in *Tissue Engineering for Tissue and Organ Regeneration*, 2011. InTech, Croatia—European Union, doi: 10.5772/22453. With permission.

Currently, the major focus is on the development of cost-effective, time-efficient scaffolds that could be used with or without cells. There is a lot of research work focused on the development of these scaffolds as cellular or acellular matrices, but focus is on the development of the best selection for both the surgeon and the patient. Few of the scaffolds are in the final phase of clinical trials. Table 8.2 lists the products that are currently being monitored for clinical success by biomedical corporations.

## 8.3.2 BONE TISSUE ENGINEERING

The treatment methods for bone defects can either be replacement based, for example, autografts, allografts, xenografts and polymeric/ceramic/metal prosthetics (Blom, 2007) or regeneration based, that is, bone tissue engineered biomaterials. Bone tissue engineering aims at the treatment of bone defects formed during the loss of bone tissue due to various reasons such as trauma, fracture, genetic deformities and so on (Salgado et al., 2004; Kneser et al., 2006; Drosse et al., 2008). Although autografts are considered the gold standard among all these methods, the donor site morbidity and the risk of infection are some of the drawbacks that limit their use and indicate for the search of another alternative (Younger and Chapman, 1989; Banwart et al., 1995; Ringe et al., 2002). Bone tissue engineered biomaterials based on polymeric, ceramic/bioactive glasses or polymer-ceramic composite components can form that alternative provided they assist in forming a bone tissue that remodels to provide the properties similar to that of the surrounding native bone tissue. The three main categories of biomaterials used for bone tissue engineering are mentioned next along with examples:

1. Polymeric biomaterials: These contain organic components that can be of two different origins, that is, natural or synthetic. Natural polymers used for bone regeneration are collagen, hyaluronic acid, alginate, chitosan, agarose and so on (Alsberg et al., 2001; Lin and Yeh, 2004; Li et al., 2005; Kim et al., 2005); synthetic polymers can be polylactic acid (PLA), polylactide-co-glycolide (PLGA), polyglycolic acid (PGA), polycaprolactone (PCL), polypropylene fumarate (PPF), polyvinyl alcohol (PVA) and so on (Hutmacher et al., 2001; Behravesh and Mikos, 2003; Dean et al., 2003).
2. Ceramic/bioactive glass biomaterials: In the field of bone tissue engineering, the inorganic materials used are mainly calcium salt based as many of them have bioactive and osteoinductive potential. Some of them are beta-tricalcum phosphate, hydroxyapatite, 70S30C bioactive glass, 45S5 bioactive glass and so on (Hollinger and Battistone, 1986; Yang et al., 1996; Friedman et al., 1998; De Diego et al., 2000; Knabe et al., 2000; Xynos et al., 2000a,b; Vogel et al., 2001; Amaral et al., 2002; Dorozhkin and Epple, 2002; Bosetti et al., 2003; Gough et al., 2004; Lai et al., 2005; Yoshikawa and Myoui, 2005; Nair et al., 2006).
3. Polymer-ceramic/bioactive glass composites: These composites may have the optimum combination of favourable properties coming from both the inorganic and the organic component. Especially, their significance as

## TABLE 8.2
### Clinical Status of Selected Scaffolds with Their Characteristics

| Product Name | Scaffold Type | Trials | Cell-Free? |
|---|---|---|---|
| NeoCart (Histiogenics Corporation, Waltham, MA) | Protein-based (bovine type I collagen) | • Phase II studies complete<br>• Ongoing phase III study | No—Harvested chondrocytes grown into collagen and secured via bioadhesive |
| VeriCart (Histiogenics Corporation, Waltham, MA) | Protein-based (double structured collagen scaffold | • Seeking regulatory clearance as candidate product in European Union | Yes—One step, off the shelf product populated with Mfx cells or BM aspirate |
| CaReS-1S (Arthro-Kinetics, Esslingen, Germany) | Protein-based (rat tail type I collagen) | • Animal Trials<br>• Case study of 15 humans<br>• No comparative human trials to date | Yes—Shown in animal studies to have equivalent repair tissue compared with cellular technique |
| Hyalograft C (Fidia Advanced Biopolymers, Abano Terme, Italy) | Carbohydrate-based (HYAFF 11-esterified derivative of hyaluronate) | • Improved patient outcomes in clinical studies<br>• Multiple comparative studies with MFx showing improved results at 5 years | No—Combined with autologous articular chondrocytes |
| Cartipatch (Tissue Bank of France, Lyon, France) | Carbohydrate-based (agarose-alginate) | • Phase II multicentre study showed significant improvements<br>• Two phase III trials comparing mosaicplasty and MFx ongoing in Europe, Asia and Middle East | No—Autologous chondrocytes implanted into the 3D scaffold |
| TrueFit Plug (Smith & Nephew, Andover, MA) | Synthetic-based (calcium sulfate and bilayered poly-D,L-lactide-co-glycolide fibers [PLGA]) | • Human trials with conflicting results<br>• Not currently recommended for OCD in active patients<br>• Approved in Europe for acute OCD and in the United States only for back filling of OC autograft sites | Yes—Inserted in osteochondral bone and promotes cancellous replacement in sub-chondral bone with fibrocartilage on top |

*(Continued)*

## TABLE 8.2 (CONTINUED)
### Clinical Status of Selected Scaffolds with Their Characteristics

| Product Name | Scaffold Type | Trials | Cell-Free? |
|---|---|---|---|
| Bioseed C (Bio Tissue Technologies GmbH, Freiburg, Germany) | Combined scaffold (fibrin glue combined with a copolymer of PGA, PLA and PDS) | • Phase III trials comparing with ACI alone showed improved outcomes sores and comparable outcome levels<br>• Caution in that ACI only group had larger defect and longer follow-up | No—Harvested autologous chondrocytes are implanted onto the 3D scaffold |
| Chondux (Biomet, Warsaw, IN) | Combined scaffold (hydrogel combining synthetic polyethylene glycol to bioadhesive chondroitin sulfate) | • Clinical trial recently suspended in Europe<br>• Not currently available in the United States | Yes—Applied as a liquid that solidifies on polymerization when exposed to UV light in area prepared with bone marrow stimulation techniques |
| Chondro-Gide (Geistlich Biomaterials, Wolhusen, Switzerland) | Combined scaffold (bilayered collagen membrane) | • Outcome score improvement with adjunct to CCI<br>• Positive clinical improvement and MRI improvement when used as adjunct to MFx<br>• No current studies to compare long-term results of ACI or vs. MFx alone | Both acellular and cellular techniques |

*Source:* Reprinted from Irion, V.H., Flanigan, D.C., *Oper. Tech. Sports Med.* 21: 125–37, 2013. With permission.

bone tissue engineered biomaterials is more than the other types because the bone extracellular matrix itself is a combination of inorganic components as hydroxyapatite and organic components as collagen. Some of the composites that have been studied are HA-collagen, polyvinyl alcohol-tetraethylorthosilicate-alginate-calcium oxide (PTAC), polycaprolactone/nanohydroxyapatite/collagen (PCL/nHA/Col) (Venugopal et al., 2007), polyvinyl alcohol-tetraethylorthosilicate-agarose-calcium chloride (PTAgC) (Mishra et al., 2009; Mishra and Kumar, 2011), hydroxyapatite-polyvinyl alcohol (Wiria et al., 2008), silica-chitosan (Reis et al., 2007), bioactive glass/polyhydroxybutyrate (Paiva et al., 2006), chitosan–alginate (Li et al., 2005), gelatin–siloxane (Ren et al., 2002) and TEOS-PVA (polyvinyl alcohol) (Costa et al., 2007).

### TABLE 8.3
### Commercially Available Biomaterial Based Products in the Field of Bone Tissue Engineering

| Biomaterials | Product | Company |
|---|---|---|
| Collagen | 1. Infuse | 1. Medtronic Inc. |
| | 2. OP-1 | 2. Olympus Biotech |
| Demineralized bone matrix | 1. SYMPHONY® I/C Graft Chambers | 1. DePuy Orthopaedics |
| | 2. Magnifuse Bone Graft | 2. Medtronic Inc. |
| | 3. OsteoSelect Demineralized Bone Matrix Putty | 3. Bacterin International, Inc. |
| Silicate substituted calcium phosphate | Actifuse | Baxter International Inc. |
| Calcium sulphate | CallosPromodelBone Void Filler | Skeletal Kinetics, LLC |
| Tricalcium phosphate | 1. CONDUIT® TCP Granules | 1. DePuy Orthopaedics |
| | 2. Polyfuse—Neurobone | 2. Atpac Medical LLC |
| Bioactive glass | 1. NovaBoneMacroPor-Si$^+$— Bioactive Bone Graft | 1. NovaBone Products, LLC |
| | 2. Vitoss | 2. Orthovita Inc. |
| | 3. Biosphere Bioactive Bone Graft Putty | 3. Synergy Biomedical, LLC |
| Polycaprolactone (PCL) | TRS Cranial Bone Void Filler (TRS C-BVF) | Tissue Regeneration Systems, Inc. |
| Hydroxyapatite | 1. ReproBone | 1. Ceramisys Ltd. |
| | 2. BiceraResorbable Bone Substitute | 2. Wiltrom Corporation Ltd. |
| | 3. Apaceram Bone Graft Substitute | 3. HOYA Corporation |
| Polylactic acid | Kensey Nash Bone Void Filler XC | Kensey Nash Corporation |
| Sodium hyaluronate | MTF new bone void filler | Musculoskeletal transplant foundation |

*Source:* http://www.fda.gov.

The commercially available biomaterials in the field of bone tissue engineering are either allograftic material such as decellularized and demineralized bone matrix (Cheng et al., 2014) or natural or synthetic biomaterials such as collagen, bioactive glass, tricalcium phosphate and so on. Some of these materials approved by the Food and Drug Administration (available at the website http://www.fda.gov) are enlisted along with the examples of the products and their company in Table 8.3.

### 8.3.3 Muscle Tissue Engineering

Skeletal muscle tissue is the most abundant tissue of the human body comprising 45% of the total body weight. The muscle/myogenic cells are aligned to form myofibres, which provide contractile function with the help of surrounding tissue including extracellular matrix, blood vessels and nerves. Under normal conditions, muscles have a high capacity for regeneration, although, in cases such as severe congenital defects or irreparable injuries, this regeneration capability is altered or lost, leading to muscle defects. The available treatments in such cases are tissue transfer from autologous or allograftic sources, although, these suffer from similar limitations as discussed previously in other musculoskeletal tissues. Therefore, tissue engineering seems to be a better option for repair. The natural and synthetic biomaterials used for skeletal muscle tissue engineering are collagen (Kin et al., 2007), polycaprolactone (PCL) (Choi et al., 2008), PLLA (Engelhardt et al., 2011), gelatin (Kim et al., 2010) and so on.

### 8.3.4 Tendon Tissue Engineering

A tendon is a fibrous connective tissue that connects muscle to bone (Awad, 2012). It is mainly composed of tenocyte cells and fascicles of collagen type I fibres. The most common types of tendons suffering injuries are Achilles tendon, rotator cuff tendon and the hand's flexor and extensor tendons (Shearn et al., 2011). The gold standard for tendon treatment is considered to be primary surgical repair, but it is often weak and suffers from reinjury (Butler et al., 2008). Therefore, biomaterial substitutes that can provide mechanical stability as well as regeneration of the injury are required. Some examples of scaffolds/biomaterial grafts used for tendon repair are restore graft (Depuy, Warsaw, IN), polyglycolic acid (PGA), polylactic acid (PLA) and so on (Longo et al., 2012).

### 8.3.5 Ligament Tissue Engineering

These are fibrous connective tissue (similar to tendon) connecting bone to bone. The anterior cruciate ligament (ACL) is the most commonly ruptured ligament tissue with over 200,000 patients diagnosed per year (Beynnon and Fleming, 1998; Albright et al., 1999; Pennisi, 2002; Laurencin and Freeman, 2005). ACL tissue is mainly composed of collagen types I, III and V, and ligament fibroblasts (Laurencin and Freeman, 2005; Kuo et al., 2010). The current strategies for ligament repair are based on autografts or allografts, which suffer from donor site morbidity and pain or infection and graft rejection, respectively (Kuo et al., 2010). Mostly the autogenous tissue harvested for the repair of ACL is patellar or hamstring tendon grafts (Butler et al., 2008). Examples of biomaterials used in ligament tissue engineering are poly(ε-caprolactone) (PCL), PCL/poly(DL-lactide)(PLA) (50:50), poly(DL-lactide-co-glycolide) (PLGA) (Ouyang et al., 2002), PGA, PLGA and poly(L-lactic acid) (PLLA) (Lu et al., 2005) and PLGA-collagen type I hybrids (Chen et al., 2000).

## 8.4 SIGNIFICANCE OF CRYOGEL TECHNOLOGY IN MUSCULOSKELETAL TISSUE ENGINEERING

Until now, many fabrication methods have been applied toward the development of musculoskeletal tissue engineered biomaterials; these include thermally induced phase separation, melt moulding, membrane lamination, supercritical-fluid technology, solid freeform fabrication and so on. All these methods have certain associated advantages and limitations such as some of them require the use of organic solvents that might interfere with the biocompatibility of the materials, some require high synthesis temperatures that might lead to the production of toxic intermediates and so forth.

Cryogelation technology presents an interesting alternative to many of these drawbacks due to its inherent advantages such as the use of aqueous solvents, low temperature freezing, which avoids production of toxic intermediates, and the ability to produce supermacroporous structures, which is essentially required for tissue engineering. Another advantage is that during cryogelation, the monomer solute becomes restricted to the unfrozen liquid microphases causing greater probability of interaction leading to better and faster polymerization. Therefore, cryogels have better mechanical stability than their hydrogel counterparts do. Our lab at the Indian Institute of Technology Kanpur is extensively working toward the use of indigenously fabricated cryogels for the repair of musculoskeletal injuries. Few of the scaffolds fabricated are already being scrutinized for preclinical trials. Although currently the number of tissue-engineered biomaterials prepared by cryogelation method is not very large, due to these associated advantages, cryogelation is emerging as a prospective scaffold fabrication technology, which has an immense potential in the area of regenerative medicine.

## 8.5 CRYOGELS FOR BONE TISSUE ENGINEERING

One of the important prerequisites for bone tissue engineering is a supermacroporous interconnected structure to allow proliferation and maturation of osteoblasts as well as neovascularisation for the vascular supply of oxygen and nutrients. Therefore, due to their capability of forming macroporous structures, cryogels can form an attractive option for fabrication of bone tissue engineered biomaterials. As an exception of the other types of fabrication methods for bone tissue engineered biomaterials which have used all the three types of biomaterial combination, that is, ceramic, polymeric and polymer-ceramic composite, cryogels for bone tissue engineering have only been fabricated until now by either polymeric (organic) or polymeric-ceramic/bioactive glass (inorganic-organic) composite materials. Therefore, the two different categories of cryogels used for bone tissue engineering are discussed as follows.

## 8.5.1 Polymeric Cryogels for Bone Tissue Engineering

Numerous types of polymers have been used in bone tissue engineering like gelatin, alginate, polypropylene fumarate, chitosan, polyglycolic acid, polylactic acid and so on. Many of these polymers have also been applied for the fabrication of polymeric cryogels for bone tissue engineering. Some of these polymers are discussed as follows.

### 8.5.1.1 Gelatin–Based Cryogels

Gelatin is a protein-based polymer that has the ability to mimic the extracellular matrix because it is derived via hydrolysis from one of the components present in the extracellular matrix, that is, collagen. Therefore, gelatin cryogels are attractive biomaterials for tissue engineering because of their biocompatibility (Fassina et al., 2010a,b). The differentiation of bone marrow stromal cells (BMSCs) toward bone lineage on gelatin cryogel matrices has been studied by Fassina and co-workers and they found that different markers related to bone differentiation, that is, decorin, osteocalcin, osteopontin, type-I collagen and type-III collagen, were expressed in higher amounts and coated the gelatin cryogel matrix in the presence of differentiation medium as compared to the control medium (Fassina et al., 2010a).

In some studies, the effect of electromagnetic stimulation on the differentiation of bone marrow derived stromal cells on gelatin cryogels was analysed (Saino et al., 2011). An electromagnetic stimulation comprising a magnetic field of 2 mT and frequency of 75 Hz was applied and it was found that the electromagnetic stimulation had a stimulatory effect on BMSC differentiation, wherein higher cell proliferation and differentiation along with higher coating of bone extracellular matrix proteins was observed in the case of stimulated cryogels. Another study observed the positive effect of electromagnetic and ultrasound stimulus on the cellular activities of Saos-2 human osteoblasts (Fassina et al., 2010b). In a recent study (Katsen-Globa et al., 2014), cryogels prepared via cross-linking of gelatin with alginate have also been proposed for bone tissue engineering applications. These cryogels have been developed in the form of ready-to-use transplantation units for cryo-preservation of mesenchymal stem cells for tissue repair.

### 8.5.1.2 Polyethylene Glycol–Based Cryogels

Polyethylene glycol is a hydrophilic polyether polymer of ethylene oxide. Polyethylene glycol diacrylate (PEGDA) is an attractive candidate for tissue engineering due to photopolymerizability, nontoxicity, injectability and nonimmunogenicity. However, in its native form it forms a hydrated layer around its surface, which leads to low adsorption of cell adhesion protein causing lower cell adhesion. Therefore, this hydrogel is chemically modified with other polymers in order to improve cell adhesion (Tan et al., 2012). In a recent study (Phadke et al., 2013), cryogels of polyethylene glycol diacrylate-co-N-acryloyl 6-aminocaproic acid have been studied for bone tissue engineering. These cryogels were prepared in different forms, that is, spongy and columnar, and the effect of these cryogel microarchitectures on the differentiation of mesenchymal stem cells has been studied. The spongy cryogels had disoriented/randomly oriented porous structure, while the columnar cryogels had oriented pores called 'lamellar columns'. The authors found that the osteogenic differentiation of mesenchymal stem cells was higher on the spongy cryogels than the columnar cryogels on 21 days of culture as denoted by higher alkaline phosphatase production and expression of osteoblastic genes. In addition, the mineralized spongy and columnar cryogels showed ectopic bone formation with and without exogenous mesenchymal stem cells on subcutaneous implantation in nude rats.

### 8.5.1.3 Silk Fibroin–Based Cryogels

Silk fibroin polymer derived from natural silk is a polymer with high mechanical strength; therefore, it is an important biomaterial for bone tissue engineering. Macroporous silk fibroin based cryogels have been prepared recently by Ak and colleagues (Ak et al., 2013) using ethylene glycol diglycidyl ether (EGDE) as a cross-linker, which caused the transition of fibroin structure from random coil to beta-sheets. These cryogels at 12.6% fibroin have very high mechanical properties, that is, a compressive modulus of 50 MPa. Therefore, these cryogels can be suggested as good candidates for bone regeneration.

### 8.5.1.4 Hydroxyethyl Methacrylate (HEMA)–Based Cryogels

Hydroxyethyl methacrylate has been used approximately 2 decades ago in a study where it was used in a constrained space between bone and implant for exerting force on swelling in order to fix the implant/prosthesis (Netti et al., 1993). It is a hydrophilic and biocompatible polymer that does not promote cell adhesion in its native form due to the formation of a hydrated layer on its surface on coming in contact with aqueous solutions. Therefore, it is used for bone tissue engineering in modified forms (Cetin et al., 2011). Lately, hydroxyethyl methacrylate based cryogels, that is, 2-hydroxyethyl methacrylate (HEMA)–lactate–dextran have been studied under different bioreactor regimes (Bölgen et al., 2008) for bone tissue engineering applications. These cryogels were analysed for the cell ingrowth and extracellular matrix (ECM) synthesis of MG63 osteoblast-like cell line under static and dynamic (bioreactor) culture. Dynamic culture conditions include perfusion or compression-enhanced ECM synthesis and ingrowth of cells.

## 8.5.2 POLYMERIC-CERAMIC/BIOACTIVE GLASS COMPOSITE CRYOGELS FOR BONE TISSUE ENGINEERING

To the best of our knowledge, our group pioneered these cryogels for bone tissue engineering in 2011 when we introduced novel bone tissue engineered cryogels using a combination of organic along with an inorganic component. These cryogels were termed as polyvinyl alcohol-tetraethylorthosilicate-alginate-calcium oxide (PTAC) biocomposite cryogels (Mishra and Kumar, 2011). The organic components were alginate and polyvinyl alcohol, whereas tetraethylorthosilicate and calcium oxide formed the inorganic bioactive glass component. Due to the presence of a bioactive glass component, these cryogels were bioactive in nature and developed a hydroxyapatite-like layer on their surface on incubation with simulated body fluid (SBF). They also showed favourable pore size, porosity (Figure 8.1) and mechanical properties.

# Cryogel Biomaterials for Musculoskeletal Tissue Engineering

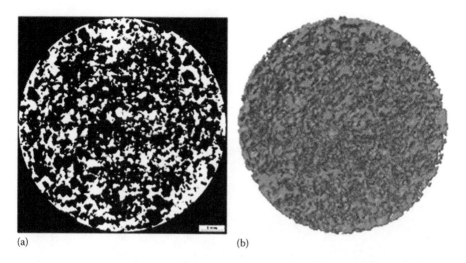

**FIGURE 8.1** Two-dimensional (a) and three-dimensional (b) images of heat-treated polyvinyl alcohol-tetraethylorthosilicate-alginate-calcium oxide (PTAC) biocomposite cryogels via micro X-ray computer tomographic scan, wherein black areas represent porous/void space while white regions represent cryogel scaffold material (scale bar = 1 mm). (From Mishra, R. and Kumar, A., *Journal of Biomaterial Science, Polymer Edition* 22(16), 2106–2126. Copyright (2011), VSP. With permission.)

Further, their *in vitro* biocompatibity was analysed and human osteoblasts showed good adherence and proliferation on the surface of these cryogels, along with high extracellular matrix production as well as mineralization at later time points (Figure 8.2).

In addition, our group has studied the *in vivo* biocompatibility, degradation and bone regeneration potential of these cryogels. The *in vivo* biocompatibility and degradation were determined in a C57Bl/10.Q inbred mice model and the results demonstrated that these cryogels did not show any local or systemic toxicity on subcutaneous implantation (Shakya et al., 2014). The cryogels were also analysed for their bone regeneration potential in a critical sized cranial defect model generated in rat skull. The bone healing of cryogel treated and nontreated rats was compared over a period of 4 weeks (Figure 8.3). The results at the macro-, micro- and gene-level demonstrated that the bone healing was much higher in the cryogel treated rats in comparison to nontreated rats, which showed negligible healing (Mishra et al., 2014).

Recently, in another study (Mishra and Kumar, 2014), we developed a new biocomposite cryogel using agarose and polyvinyl alcohol as the organic component and tetraethylorthosilicate and calcium chloride as the inorganic component. We termed these cryogels as polyvinyl alcohol-tetraethylorthosilicate-agarose-calcium chloride (PTAgC) biocomposite cryogels (Figure 8.4). These cryogels exhibit an interconnected porosity of 77 ± 0.16% and pore size of 190 ± 0.78 µm and showed biocompatibility with Saos-2 human osteoblasts as determined by MTT assay, SEM and alkaline phosphatase activity. Another interesting observation was that these cryogels exhibited the property of osteoinductivity, which is a desirable property for

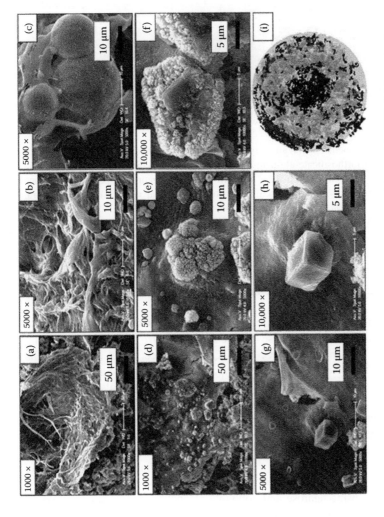

**FIGURE 8.2** The SEM images of polyvinyl alcohol–tetraethylorthosilicate–alginate–calcium oxide (PTAC) biocomposite cryogels after human osteoblast culture for 13 days (a–c) showing extracellular matrix production along with some calcium phosphate like mineral deposits, which increase in number after 24 days of culture (d–h). Micro X-ray computer tomographic (micro CT) image of cryogel after 24 days of culture showing the arrangement and density of cell sheet (blackish gray) on PTAC biocomposite cryogel (light gray) region (i). (From Mishra, R. and Kumar, A., *Journal of Biomaterial Science, Polymer Edition* 22(16), 2106–2126. Copyright (2011), VSP. With permission.)

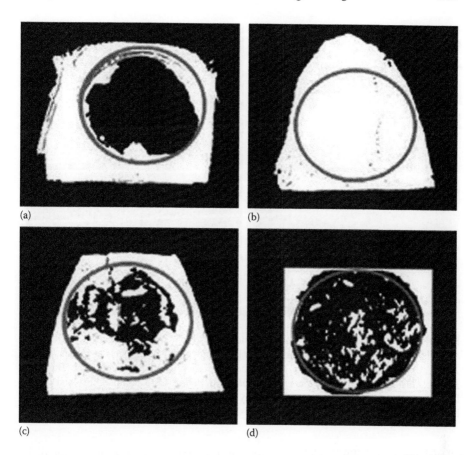

**FIGURE 8.3** The representative images (out of three sets of experiments) of nontreated (a), normal (b) and treated (c) at 4 weeks' time showing the difference in healing of defect (dark gray circle) in nontreated and treated rats obtained via microcomputed tomography (micro CT) analysis. The bone formation/healing could also be seen at the inner regions as represented by the cross-sectional view (d). (From Mishra, R., Goel, S.K., Gupta, K.C., Kumar, A., *Tissue Engineering Part A*, 20, 751–762. Copyright (2014), Mary Ann Liebert Inc. With permission.)

bone biomaterials. Their osteoinductive potential was determined via the osteoblastic differentiation of C2C12 myoblast progenitors seeded onto these cryogels (Figure 8.5).

Other composite cryogels that have been studied for bone tissue engineering are collagen-nanohydroxyapatite biocomposite cryogels (Rodrigues et al., 2013) and gelatin/hydroxyapatite cryogels (Ozturk et al., 2013). Collagen-nanohydroxyapatite biocomposite cryogels were studied at the *in vitro* level for the physical properties and the response of osteoblasts. The results demonstrated that the formation of a composite via hydroxyapatite incorporation in comparison to the use of only collagen scaffolds was favourable and provided better properties for bone tissue engineering

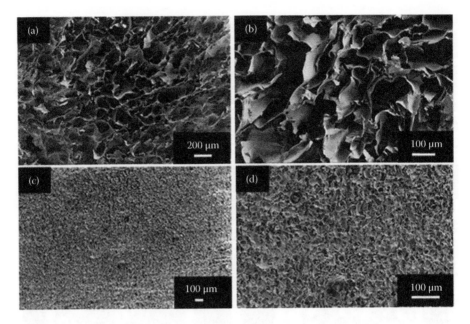

**FIGURE 8.4** The low and high magnification images via scanning electron microscopy analysis of pore size and overall porosity of PTAgC biocomposite cryogels at 16% (a, b) and 24% (c, d) monomer concentrations. (With kind permission from Springer Science+Business Media: *Journal of Material Science: Materials in Medicine*, Osteocompatibility and osteoinductive potential of supermacroporous polyvinyl alcohol-TEOS-agarose-Cacl2 (PTAgC) biocomposite cryogels. 25, 2014, 1327–37, Mishra, R. and Kumar, A.)

(Rodrigues et al., 2013). Gelatin/hydroxyapatite cryogels were analysed for their bone regeneration ability at the *in vivo* level, wherein, cryogels loaded with vascular endothelial growth factor (VEGF) were implanted in a rabbit critical sized tibial defect model and superior healing of defects was observed in cryogel implanted defects in comparison to the control groups (Ozturk et al., 2013).

## 8.6  CRYOGELS IN CARTILAGE TISSUE ENGINEERING

A fundamental prerequisite of a scaffold used for cartilage tissue engineering is the presence of a large and interconnected porous network together with an optimized rate of degradation. Cryogelation has emerged as a recent technology for the fabrication of scaffolds, which can have applications in the repair of cartilage defects. Researchers have used cryogels fabricated from either natural or synthetic polymers for their applicability in cartilage tissue engineering. Furthermore, the degradation rate of the cryogels can be modulated by introducing variations with respect to polymer or monomer concentration or with the percentage of cross-linkers. Furthermore, to date, different types of cryogels that have been fabricated using natural and synthetic polymers have shown application in the area of cartilage tissue engineering.

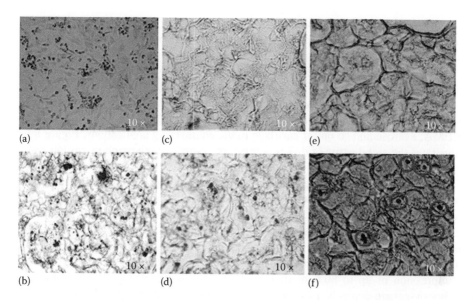

**FIGURE 8.5** Alkaline phosphatase enzyme represented by diffused blue coloration developed via BCIP-NBT staining of 200-µm thick sections of two-dimensional positive control (a), C2C12 seeded cryogel after day 3 of culture (b), negative control (without cells) (c) and C2C12 seeded cryogel after 20 days of culture (d). While blackish dots representing calcium-containing crystals can be seen via Von Kossa staining of 200-µm thick sections of negative control (without cells) (e) and C2C12 seeded cryogel after 20 days of culture (f). (With kind permission from Springer Science+Business Media: *Journal of Material Science: Materials in Medicine*, Osteocompatibility and osteoinductive potential of supermacroporous polyvinyl alcohol-TEOS-agarose-Cacl2 (PTAgC) biocomposite cryogels, 25, 2014. 1327–37, Mishra, R. and Kumar, A.)

### 8.6.1 Cryogels Fabricated Using Natural Polymers for Cartilage Tissue Engineering

Different types of natural polymers like gelatin, chitosan, dextran, lactate and so on, are being used either alone or in combination with other natural or synthetic polymers. The potential of one such hybrid cryogel matrix composed of biodegradable HEMA–lactate–dextran (HEMA–LLA–D) was analysed for cartilage tissue engineering. In this study, bovine articular chondrocytes were seeded onto cylindrical cryogels and cultured. Seeded chondrocytes proliferated rapidly and fully covered the scaffold surface; the chondrocytes also secreted a significant amount of extracellular matrix in the scaffolds, which was facilitated by the presence of interconnective porous morphology (Bolgen et al., 2011). Concepts of gene delivery have also been used to modulate the chondrocytes' properties and their efficient growth on the cryogel surface. In this study, primary rabbit chondrocytes were genetically modified with plasmid encoding bone morphogenetic protein-7 (BMP-7). Hence, transformed chondrocytes synthesized BMP-7 *in vitro* in the same way as they produced *in vivo*. Genetically altered chondrocytes were then seeded into gelatin–dextran scaffolds. Cartilage tissue formation was examined *in vivo* by the implantation of seeded scaffolds in auricular cartilage defects created in New Zealand (NZ) white rabbits for a period of 4 months. Histological analysis and other results demonstrated significantly better cartilage healing with BMP-7-transfected chondrocytes than in the nonmodified group as well as the control (Odabas et al., 2013).

Our group is working toward the development of novel and proficient cryogel matrices mainly from the natural and synthetic polymers like chitosan, gelatin, HEMA, agarose and so on for cartilage tissue engineering applications. Figure 8.6 illustrates the architecture of different types of cryogels synthesized in our lab analysed by SEM, micro CT and fluorescence microscopy.

Initial study was conducted on chitosan–gelatin cryogels using glutaraldehyde as a cross-linker. Pore diameter of chitosan–gelatin cryogels investigated by SEM was found to be in the range of 30–100 µm. Porosity, which is the void volume of chitosan–gelatin cryogels, was found to be greater than 90% using Archimedes's principle. Unconfined compression tests demonstrated that chitosan-gelatin cryogels are elastic and could maintain their physical integrity after subjecting them to the compression of up to 80% of their original length. Percentage degradation of chitosan–gelatin cryogels under *in vitro* conditions was found up to 13.58 ± 1.52% at 37°C after 8 weeks of incubation under sterile conditions. In addition to the physical characterizations, *in vitro* biocompatability of these matrices was confirmed by efficient cell adherence, proliferation and synthesis of extracellular matrix (ECM) by a model cell line, that is, fibroblast (Cos-7). The previous characterizations indicated the potential of chitosan–gelatin hybrid cryogels for cartilage tissue engineering (Kathuria et al., 2009). After the initial success with the fabrication of an ideal scaffold for cartilage tissue engineering, our group tried to fabricate a tailor made scaffold that could mimic the native architecture of the cartilage tissue. The major emphasis of this research work was the fabrication of a macroporous scaffold with controlled porosity to generate a gradient porous structure that could mimic the native architecture of the cartilage. Two natural polymers, that is, agarose and gelatin, were employed for the fabrication of gradient cryogel scaffolds.

Agarose–gelatin cryogels were synthesized in two different solvent systems (i.e., water and 0.1% acetic acid) at subzero temperature for the generation of an interconnected porous network. Out of the two solvent systems used, cryogels fabricated in water solvent system (WSS) showed a gradient porosity with an average pore diameter in the range of 76 to 187 µm. Synthesized cryogels exhibited a good swelling capacity and a tensile modulus of 380.23 ± 63.97 kPa. *In vitro* biocompatibility of the synthesized matrices was analysed by the growth and proliferation of fibroblast (Cos-7) with the aid of SEM and MTT assay. The native cartilage exhibits a gradient architecture so these matrices have a colossal applicability in the area of cartilage tissue engineering, as they closely resemble the native pore layout of the cartilage tissue (Tripathi et al., 2009). In another study done by our group, we fabricated matrices using the combination of HEMA and gelatin. Synthesized matrices supported the growth and proliferation of primary goat chondrocytes. Proliferation and metabolic activity was analysed by a continuous increase in glycosaminoglycan and collagen content with time. The results suggest the potential of HEMA-gelatin cryogel scaffold as a matrix for chondrocyte attachment and proliferation in a 3D

# Cryogel Biomaterials for Musculoskeletal Tissue Engineering

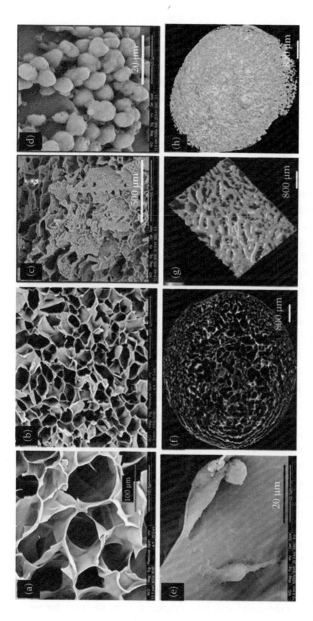

**FIGURE 8.6** SEMs of chitosan–gelatin and agarose–gelatin cryogels (a, b), SEM images showing growth of chondrocytes on HEMA–gelatin cryogels (c, d), SEM image showing the proliferation of primary goat chondrocytes of CAG cryogels (e), binarized micro-CT image showing the internal architecture of CAG cryogel (f), three-dimensional fluorescent microscopy images showing chondrocyte penetration through the scaffold (h). (Reprinted from *Acta Biomaterialia*. 5, Kathuria, N. et al., Synthesis and characterization of elastic and macroporous chitosan–gelatin cryogels for tissue engineering, 406–18, Copyright 2009, with permission from Elsevier; Tripathi, A. et al.: Elastic and macroporous agarose–gelatin cryogels with isotropic and anisotropic porosity for tissue engineering. *Journal of Biomedical Material Research*. 2009. 90 A. 680–94. Copyright Wiley-VCH Verlag GmbH & Co. KGaA. Reproduced with permission; Singh, D. et al., *Journal of Biomaterial Science: Polymer Edition* 22: 1733–51, Copyright (2011); Bhat, S. et al., *Journal of The Royal Society Interface*, 8: 540–54, Copyright (2011), Royal Society; Bhat, S. et al.: *In vitro* neo-cartilage formation on a three-dimensional composite polymeric cryogel matrix. *Macromolecular Bioscience*. 2013. 13: 827–37. Copyright Wiley-VCH Verlag GmbH & Co. KGaA. Reproduced with permission.)

environment and as a delivery system in cartilage-tissue engineering (Singh et al., 2011). In yet another study by our group, a hybrid scaffold fabricated from chitosan, agarose, gelatin and CAG exhibited great potential for cartilage tissue engineering. Microstructure analysis done by SEM revealed that CAG matrices possessed large and interconnected porous networks. The matrices were mechanically very stable, which makes them ideal as materials for cartilage replacement. Furthermore, fatigue tests revealed that the scaffolds did not develop any cracks when subjected to continuous cyclic strain, further validating their applicability as surrogate cartilage materials. To further confirm the mechanical stability, the cryogels were subjected to high frequency (5Hz) with 30% compression of their original length for $1 \times 10^5$ cycles. The CAG cryogels did not demonstrate any changes in their mass and dimensions during this run. Further, primary goat chondrocytes, isolated from 6- to 8-month-old female goat by collagen digestion, were grown on these scaffolds. SEM examination revealed that the CAG cryogels could support the growth and proliferation of these primary goat chondrocytes. In addition, CAG matrices were examined for their *in vitro* and *in vivo* biocompatibility. MTT assay results indicated that cryogel matrices did not inflict any type of toxic effect on the seeded chondrocytes. *In vivo* biocompatibility of the scaffolds was checked by the implantation of the sterile scaffolds in Wister rats. Histological analysis and SEM examination revealed the scaffold's biocompatibility when implanted subcutaneously with a faster degradation as compared to the *in vitro* conditions (Figure 8.7).

Results of this study fortified our idea that CAG cryogels can have potential as good three-dimensional scaffolds for cartilage tissue engineering (Bhat et al., 2011). Therefore, we furthered our studies with CAG scaffolds for the generation of neo-cartilage, which can be used as a sealant for treatment of lesions generated during the course of osteoarthritis. CAG matrices possess many fundamental properties like elasticity and polymer composition, which maintains chondrocytes phenotype and their property to synthesize large amounts of ECM. Such chondrocyte-seeded scaffolds when incubated under favourable conditions result in the generation of large amounts of ECM, which accumulates over and on the surface of the cryogel in the form of neo-cartilage (Figure 8.8).

Neo-cartilage generated on the surface of cryogels was unique in the sense that it could be detached from the associated matrix. Free neo-cartilage when compared with native cartilage exhibited histological and mechanical similarity. Biochemical components in neo-cartilage, that is, collagen and glycosaminoglycan, were on par with the native cartilage (i.e., 0.16 mg and 0.06 mg of dry weight of tissue, respectively). Animal studies with balb/c mice showed that neo-cartilage could integrate well with the host tissue, which justifies its biocompatibility and use as a sealant for localized lesions (Bhat et al., 2013). After the initial testing of the CAG cryogels, we tried to check the efficiency of these matrices in a more real system. Therefore, we created a cartilage defect model in rabbits and analysed the potential of CAG in the repair of these defects. Histological results showed that with time cartilage regeneration takes place. After the first week postsurgery, no sign of cartilage regeneration can be observed. However, by the eighth week of scaffold implantation, regeneration of cartilage was observed. Abundance of proteoglycans deposition was also observed in the regenerated tissue. Results illustrated that the

# Cryogel Biomaterials for Musculoskeletal Tissue Engineering 241

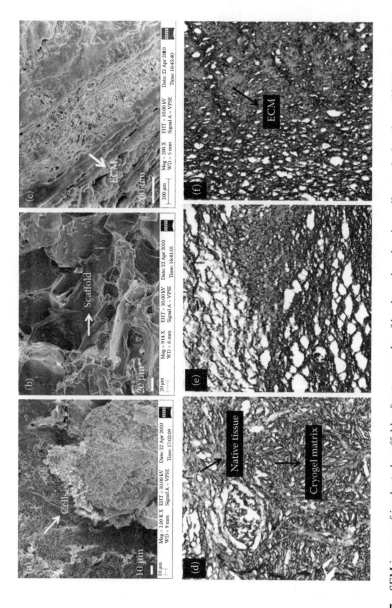

**FIGURE 8.7** SEM images of implanted scaffolds after two weeks of implantation showing cells attached to the scaffold surface (a), after four weeks of implantation (b) and after six weeks of implantation showing degraded cryogel matrix and deposition of ECM (c). Histological examination of the implanted constructs stained by H&E, after two weeks showing the integration of scaffolds with the native tissue and exhibiting the process of neovascularisation (d), after four weeks showing the disintegration of the cryogel matrix (e), after six weeks showing the deposition of the ECM (f). (From Bhat, S. et al., *Journal of The Royal Society Interface*. 8: 540–54. Copyright (2011), Royal Society. With permission.)

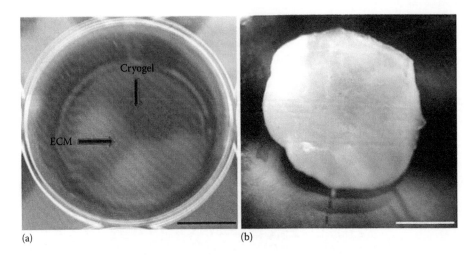

**FIGURE 8.8** Neo-cartilage generation in 3D cryogels. Digital images show the sequential development of neo-cartilage. Cells were initially seeded in cryogel matrix under appropriate culture conditions, which leads to the accumulation of ECM component in the CAG cryogel matrix (a), which results in the formation of neo-cartilage on the cryogel. The developed neo-cartilage is further disjoined physically from the partially degraded cryogel matrix (b). Scale bar is 5 mm. (Bhat, S. et al.: *In vitro* neo-cartilage formation on a three-dimensional composite polymeric cryogel matrix. *Macromolecular Bioscience*. 2013. 13: 827–37. Copyright Wiley-VCH Verlag GmbH & Co. KGaA. Reproduced with permission.)

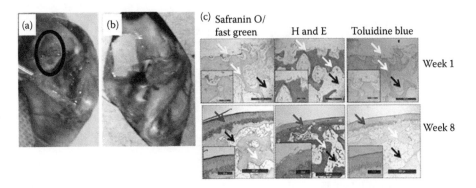

**FIGURE 8.9** Deep cartilage defect was created in New Zealand white rabbits, encircled area shows the defect site (a), CAG cryogel scaffold implanted over the defect site (b), histological examination of the defect after the first and eighth weeks (c), yellow arrow: absence of cartilage, red arrow: regenerated cartilage, white arrow: bone, black arrow: bone marrow. (Reproduced from Gupta, A., Bhat, S., Jagdale, P. R., Chaudhari, B. P., Lidgren, L., Gupta, K. C. and Kumar, A., *Tissue Engineering Part A* (doi: 10.1089/ten.TEA.2013.0702), Copyright (2014), Mary Ann Liebert Inc. With permission.)

healing process was much faster in test animals (animals implanted with CAG cryogels) than in control (animals without cryogels) (Gupta et al., 2014) (Figure 8.9).

Our group is also working toward the development of delivery systems for bioactive molecules on cryogel matrices. In this study, we have fabricated novel three-dimensional (3D) matrices using natural polymers like chitosan and gelatin. Scaffolds were modified by the addition of chondroitin sulphate. These matrices were characterized by rheology, scanning electron microscopy (SEM) and mechanical assay. Scaffolds exhibited compression modulus of 50 KPa and could support the growth and proliferation of primary goat chondrocytes indicating their potential in cartilage tissue engineering (Dwivedi et al., 2014).

### 8.6.2 CRYOGELS FABRICATED USING SYNTHETIC POLYMERS FOR CARTILAGE TISSUE ENGINEERING

In addition to natural polymers, synthetic polymers are widely explored for their applicability in the area of cartilage tissue engineering. In a study conducted by Hwang et al. (2010), cryogels fabricated from PEG were investigated for cartilage tissue engineering. Primary bovine chondrocytes were grown on the surface of cryogels and constructs were analysed for the production of cartilage-specific extracellular matrix. Biochemical quantification of DNA suggested that cells proliferate continuously under *in vitro* culture conditions. Synthesis of ECM by chondrocytes on and inside the cryogel scaffolds was further confirmed by SEM, histological and immunohistological analysis. Results indicated that macroporous PEG cryogels support the growth of chondrocytes and helped in the retention of biosynthetic activity of seeded chondrocytes (Hwang et al., 2010).

## 8.7 CONCLUSIONS

Based on the information provided in this chapter, we can conclude that cryogels have already been established as suitable biomaterials for tissue engineering applications in musculoskeletal tissues such as bone and cartilage. Further, due to their favourable properties such as the ability to provide supermacroporous structures with interconnected porosity, ease of fabrication without any involvement of production of any toxic intermediates, cryogels might soon be able to become important biomaterials for other musculoskeletal tissues like ligaments, tendons and muscles.

## ACKNOWLEDGEMENTS

The authors acknowledge the Department of Science and Technology (DST), Government of India and Department of Biotechnology (DBT), Government of India.

## LIST OF ABBREVIATIONS

| | |
|---|---|
| ACL | Anterior cruciate ligament |
| CAG | Chitosan, agarose, gelatin |
| ECM | Extracellular matrix |
| EGDE | Ethylene glycol diglycidyl ether |
| MACI | Matrix Associated Chondrocyte Implantation |
| MTT | 3-(4,5-dimethylthiazol-2-yl)-2,5-diphenyltetrazolium bromide) |
| PA | Peptide amphiphile |
| PCL | Polycaprolactone |
| PCL/nHA/Col | Polycaprolactone/nanohydroxyapatite/collagen |
| PEG | Polyethylene glycol |
| PEGDA | Poly (ethylene glycol) diacrylate |
| PGA | Polyglycolic acid |
| PLA | Polylactic acid |
| PLGA | Poly lactide-co-glycolide |
| poly(NiPAAm) | Poly(N-isopropylacrylamide) |
| PPF | Polypropylene fumarate |
| PTAC | Polyvinyl alcohol-tetraethylorthosilicate-alginate-calcium oxide |
| PTAgC | Polyvinyl alcohol-tetraethylorthosilicate-agarose-calcium chloride |
| PVA | Polyvinyl alcohol |
| SBF | Simulated body fluid |
| SDF 1-α | Stromal derived factors 1-α |
| TGF-β3 | Transforming growth factors-β3 |
| VEGF | Vascular endothelial growth factor |
| WSS | Water solvent system |

## REFERENCES

Ak, F., Oztoprak, Z., Karakutuk, I., Okay, O. (2013). Macroporous silk fibroin cryogels. *Biomacromolecules* 14: 719–27.

Albright, J.C., Carpenter, J.E., Graf, B.K., Richmond, J.C. (1999). Knee and leg: Soft-tissue trauma. *Orthopaedic Knowledge* Update 6, Beaty, J.H. (Ed), American Academy of Orthopaedic Surgeons, Rosemont, p. 533.

Alsberg, E., Anderson, K.W., Albeiruti, A., Franceschi, R.T., Mooney, D.J. (2001). Cell-interactive alginate hydrogels for bone tissue engineering. *J. Dent. Res* 80: 2025–29.

Amaral, M., Costa, M.A., Lopes, M.A., Silva, R.F., Santos, J.D., Fernandes, M.H. (2002). Si3N4—Bioglass composites stimulate the proliferation of MG63 osteoblast-like cells and support the osteogenic differentiation of human bone marrow cells. *Biomaterials* 23: 4897–906.

Aragona, J., Parsons, J.R., Alexander, H., Weiss, A.B. (1981). Soft tissue attachment of a filamentous carbon-absorbable polymer tendon and ligament replacement. *Clin. Orthop. Relat. Res* 160: 268–78.

Awad, H.A. (2012). Prospects of tendon tissue engineering in sports medicine. *Dtsch. Z. Sportmed.* 63: 132–5.

Badley, E.M., Rasooly, I., Webster, G.K. (1994). Relative importance of musculoskeletal disorders as a cause of chronic health problems, disability, and health care utilization: Findings from the 1990 Ontario Health Survey. *J. Rheumatol.* 21: 505–14.

Badylak, S.F., Freytes, D.O., Gilbert, T.W. (2009). Extracellular matrix as a biological scaffold material: Structure and function. *Acta Biomaterialia.* 5: 1–13.
Banwart, J.C., Asher, M.A., Hassanein, R.S. (1995). Iliac crest bone graft harvest donor site morbidity. A statistical evaluation. *Spine.* 20: 1055–60.
Bartlett, W., Skinner, J.A., Gooding, C.R., Carrington, R.W., Flanagan, A.M., Briggs, T. (2005). Autologous chondrocyte implantation versus matrix-induced autologous chondrocyte implantation for osteochondral defects of the knee: A prospective, randomised study. *J. Bone Jt. Surg. (British).* 87: 640–5.
Behravesh, E., Mikos, A.G. (2003). Three dimensional culture of differentiating marrow stromal osteoblasts in biomimetic poly (propylene fumarate-co-ethylene glycol)-based macroporous hydrogels. *J. Biomed. Mat. Res. A.* 66: 698–706.
Beynnon, B.D., Fleming, B.C. (1998). Anterior cruciate ligament strain *in vivo*: A review of previous work. *J. Biomech.* 31: 519–25.
Bhat, S., Lidgren, L., Kumar, A. (2013). *In vitro* neo-cartilage formation on a three-dimensional composite polymeric cryogel matrix. *Macromol. Biosci.* 13: 827–37.
Bhat, S., Tripathi, A., Kumar, A. (2011). Supermacroporous chitosan-agarose-gelatin cryogels: *In vitro* characterization and *in vivo* assessment for cartilage tissue engineering. *J. R. Soc. Interface.* 8: 540–54.
Blom, A. (2007). (V) Which scaffold for which application? *Curr. Orthopaed.* 21: 280–7.
Bolgen, N., Yang, Y., Korkusuz, P., Guzel, E., El Haj A.J., Piskin, E. (2011). 3D Ingrowth of bovine articular chondrocytes in biodegradable cryogel scaffolds for cartilage tissue engineering. *Tissue Eng. Regen. Med.* 5: 770–779.
Bölgen, N., Yang, Y., Korkusuz, P., Güzel, E., El Haj, A.J., Pişkin, E. (2008). Three-dimensional ingrowth of bone cells within biodegradable cryogel scaffolds in bioreactors at different regimes. *Tissue Eng. Part A.* 14: 1743–50.
Bosetti, M., Zanardi, L., Hench, L., Cannas, M. (2003). Type I collagen production by osteoblast-like cells cultured in contact with different bioactive glasses. *J. Biomed. Mater. Res.* 64: 189–95.
Bryant, S.J., Anseth, K.S. (2003). Controlling the spatial distribution of ECM components in degradable PEG hydrogels for tissue engineering cartilage. *J Biomed. Mater. Res A.* 64:70–79.
Butler, D.L., Juncosa-Melvin, N., Boivin, G.P., Galloway, M.T., Shearn, J.T., Gooch, C., Awad, H. (2008). Functional tissue engineering for tendon repair: A multidisciplinary strategy using mesenchymal stem cells, bioscaffolds, and mechanical stimulation. *J. Orthop. Res.* 26: 1–9.
Buckwalter, J.A. (2002). Articular cartilage injuries. *Clin. Orthop. Relat. Res.* 402: 21–37.
Cetin, D., Kahraman, A.S., Gümüşderelioğlu, M. (2011). Novel scaffolds based on poly(2-hydroxyethyl methacrylate) superporous hydrogels for bone tissue engineering. *J. Biomater. Sci. Polym. Ed.* 22: 1157–78.
Chen, G., Ushida, T., Tateishi, T. (2000). Hybrid biomaterials for tissue engineering: A preparative method for PLA or PLGA–collagen hybrid sponges. *Adv. Mater.* 12: 455–7.
Cheng, C.W., Solorio, L.D., Alsberg, E. (2014). Decellularized tissue and cell-derived extracellular matrices as scaffolds for orthopaedic tissue engineering. *Biotechnol. Adv.* 2: 462–84.
Choi, J.S., Lee, S.J., Christ, G.J., Atala, A., Yoo, J.J. (2008). The influence of electrospun aligned poly(epsiloncaprolactone)/collagen nanofiber meshes on the formation of self-aligned skeletal muscle myotubes. *Biomaterials* 29: 2899–906.
Chu, C.R., Szczodry, M., Bruno, S. (2010). Animal models for cartilage regeneration and repair. *Tissue Eng. Part B Rev.* 16: 105–15.
Costa, V.C., Costa, H.S., Vasconcelos, W.L., Pereira, M.M., Oréfice, R.L., Mansur, H.S. (2007). Preparation of hybrid biomaterials for bone tissue engineering. *Mat. Res.* 10: 21–6.

Dean, D., Topham, N., Meneghetti, C., Wolfe, M., Jepsen, K., He, S., Chen, K., Fisher, J., Cooke, M., Rimnac, C., Mikos, A.G. (2003). Poly(propylene fumarate) and poly(dllactic-co-glycolic acid) as scaffold materials for solid and foam-coated composite tissue-engineered constructs for cranial reconstruction. *Tissue Eng.* 9: 495–504.

Dorozhkin, S.V., Epple, M. (2002). Biological and medical significance of calcium phosphates. *Angew. Chem. Int. Ed. Engl.* 41: 3130–46.

De Diego, M.A., Coleman, N.J., Hench, L.L. (2000). Tensile properties of bioactive fibers for tissue engineering applications. *J. Biomed. Mater. Res.* 53: 199–203.

Drosse, I., Volkmer, E., Capanna, R., Biase, P.D., Mutschler, W., Schieker, M. (2008). Tissue engineering for bone defect healing: An update of a multi-component approach. *Injury* 39: S9–S20.

Dwivedi, P., Bhat, S., Nayak, V., Kumar, A. (2014). Microspheres and cryogel scaffold for application in cartilage tissue engineering. *Inter. J. Poly. Mater.* 63(16): 859–872.

Engelhardt, E.M., Micol, L.A., Houis, S., Wurm, F.M., Hilborn, J., Hubbell, J.A., Frey, P. (2011). A collagen-poly(lactic acid-co-varepsilon-caprolactone) hybrid scaffold for bladder tissue regeneration. *Biomaterials.* 32: 3969–76.

Fassina, L., Saino, E., Van Vlierberghe, S., Maliardi, V., Cusella De Angelis, M.G., Visai, L. (2010a). Electromagnetic vs. ultrasound stimulus of gelatin-based cryogels for bone tissue engineering. *J. Appl. Biomater. Biomech.* 8: 112.

Fassina, L., Saino, E., Visai, L., Avanzini, M.A., Cusella De Angelis, M.G., Benazzo, F., Van Vlierberghe, S., Dubruel, P., Magenes, G. (2010b). Use of a gelatin cryogel as biomaterial scaffold in the differentiation process of human bone marrow stromal cells. *Conf. Proc. IEEE Eng. Med. Biol. Soc.* 1: 247–50.

Friedman, C.D., Costantino, P.D., Takagi, S., Chow, L.C. (1998). Bone source hydroxyapatite cement: A novel biomaterial for craniofacial skeletal tissue engineering and reconstruction. *J. Biomed. Mater. Res. A.* 43: 428–32.

Freed, L.E., Marquis, J.C., Langer, R., Vunjak-Novakovic, G. (1994). Kinetics of chondrocyte growth in cell-polymer implants. *Biotechnol. Bioeng.* 43: 597–604.

Getgood, A., Brooks, R., Fortier, L., Rushton, N. (2009). Articular cartilage tissue engineering: Today's research, tomorrow's practice? *J. Bone Joint Surg. Br.* 91: 565–76.

Gough, J.E., Jones, J.R., Hench, L.L. (2004). Nodule formation and mineralization of human primary osteoblasts cultured of a porous bioactive glass scaffold. *Biomaterials* 25: 2039–46.

Gupta, A., Bhat, S., Jagdale, P.R., Chaudhari, B.P., Lidgren, L., Gupta, K.C., Kumar, A. (2014). Evaluation of three-dimensional chitosan-agarose-gelatin cryogel scaffold for the repair of subchondral cartilage defects: An *in vivo* study in a rabbit model. *Tissue Engineering Part A* 20(23–24):3101–11, doi: 10.1089/ten.TEA.2013.0702.

Henson, F., Getgood, A. (2011). The use of scaffolds in musculoskeletal tissue engineering. *Open Orthop. J.* 5: 261–6.

Hollinger, J.O., Battistone, G.C. (1986). Biodegradable bone repair materials. *Clin. Orthop. Relat. Res.* 207: 290–305.

Hutmacher, D.W., Schantz, T., Zein, I., Ngkw, Teoh, S.H., Tan, K.C. (2001). Mechanical properties and cell cultural response of polycaprolactone scaffolds designed and fabricated via fused deposition modelling. *J. Biomed. Mater. Res.* 55: 203–16.

Huang, A.H., Stein, A., Tuan, R.S., Mauck, R.L. (2009). Transient exposure to transforming growth factor beta 3 improves the mechanical properties of mesenchymal stem cell laden cartilage constructs in a density-dependent manner. *Tissue Eng. Part A* 15: 3461–72.

Hwang, Y., Sangaj, N., Varghese, S. (2010). Interconnected macroporous poly(ethylene glycol) cryogels as a cell scaffold for cartilage tissue engineering. *Tissue Eng. Part A* 16, 3033–41.

Irion, V.H., Flanigan, D.C. (2013). New and emerging techniques in cartilage repair: Other scaffold-based cartilage treatment options. *Oper. Tech. Sports Med.* 21: 125–37.

Iwasa, J., Ochi, M., Uchio, Y., Katsube, K., Adachi, N., Kawasaki, K. (2003). Effects of cell density on proliferation and matrix synthesis of chondrocytes embedded in atelocollagen gel. *Artif. Organs* 27: 249–55.

Katsen-Globa, A., Meiser, I., Petrenko, Y.A., Ivanov, R.V., Lozinsky, V.I., Zimmermann, H., Petrenko, A.Y. (2014). Towards ready-to-use 3-D scaffolds for regenerative medicine: Adhesion-based cryopreservation of human mesenchymal stem cells attached and spread within alginate-gelatin cryogel scaffolds. *J. Mater. Sci. Mater. Med.* 25: 857–71.

Kathuria, N., Tripathi, A., Kar, K.K., Kumar, A. (2009). Synthesis and characterization of elastic and macroporous chitosan–gelatin cryogels for tissue engineering. *Acta Biomaterialia* 5: 406–18.

Kim, H.W., Li, L.H., Lee, E.J., Lee, S.H., Kim, H.E. (2005). Fibrillar assembly and stability of collagen coating of titanium for improved osteoblast responses. *J. Biomed. Mater. Res. A.* 75: 629–38.

Kim, M.S., Jun, I., Shin, Y.M., Jang, W., Kim, S.I., Shin, H. (2010). The development of genipin-crosslinked poly(caprolactone) (PCL)/Gelatin nanofibers for tissue engineering applications. *Macromol. Biosci.* 10: 91–100.

Kin, S., Hagiwara, A., Nakase, Y., Kuriu, Y., Nakashima, S., Yoshikawa, T., Sakakura, C., Otsuji, E., Nakamura, T., Yamagishi, H. (2007). Regeneration of skeletal muscle using *in situ* tissue engineering on an acellular collagen sponge scaffold in a rabbit model. *ASAIO J.* 53: 506–13.

Knabe, C., Driessens, F.C.M., Planell, J.A., Gildenhaar, R., Berger, G., Reif, D., Fitzner, R., Radlanski, R.J., Gross, U. (2000). Evaluation of calcium phosphates and experimental calcium phosphate bone cements using osteogenic cultures. *J. Biomed. Mater. Res.* 52: 498–508.

Kneser, U., Schaefer, D.J., Polykandriotis, E., Horch, R.E. (2006). Tissue engineering of bone: The reconstructive surgeon's point of view. *J. Cell. Mol. Med.* 10: 7–19.

Kuo, C.K., Marturano, J.E., Tuan R.S. (2010). Novel strategies in tendon and ligament tissue engineering: Advanced biomaterials and regeneration motifs. *Sports Med. Arthrosc. Rehabil. Ther. Technol.* 2: 20.

Lai, W., Garino, J., Flaitz, C., Ducheyne, P. (2005). Excretion of resorption products from bioactive glass implanted in rabbit muscle. *J. Biomed. Mater. Res. A.* 75: 398–407.

Laurencin, C.T., Freeman, J.W. (2005). Ligament tissue engineering: An evolutionary materials science approach. *Biomaterials.* 26: 7530–6.

Lee C.H., Cook, J.L., Mendelson, A., Moioli, E.K., Yao, H., Mao, J.J. (2010). Regeneration of the articular surface of the rabbit synovial joint by cell homing: A proof of concept study. *Lancet* 376: 440–448.

Li, Z., Ramay, H.R., Hauch, K.D., Xiao, D., Zhang, M. (2005). Chitosan–alginate hybrid scaffolds for bone tissue engineering. *Biomaterials* 26: 3919–28.

Lin, H.R., Yeh, Y.J. (2004). Alginate/hydroxyapatite composite scaffolds for bone tissue engineering: Preparation, characterization, and *in vitro* studies. *J. Biomed. Mater. Res. B Appl. Biomater.* 71: 52–65.

Longo, U.G., Lamberti, A., Petrillo, S., Maffulli, N., Denaro, V. (2012). Scaffolds in tendon tissue engineering. *Stem Cells Int.* doi: 10.1155/2012/517165.

Lu, H.H., Cooper, J.A. Jr., Manuel, S., Freeman, J.W., Attawia, M.A., Ko, F.K., Laurencin, C.T. (2005). Anterior cruciate ligament regeneration using braided biodegradable scaffolds: *In vitro* optimization studies. *Biomaterials* 26: 4805–16.

Mishra, R., Basu, B., Kumar, A. (2009). Physical and cytocompatibility properties of bioactive glass-polyvinyl alcohol-sodium alginate biocomposite foams prepared via sol–gel processing for trabecular bone regeneration. *J. Mater. Sci. Mater. Med.* 20: 2493–500.

Mishra, R., Goel, S.K., Gupta, K.C., Kumar, A. (2014). Biocomposite cryogels as tissue-engineered biomaterials for regeneration of critical-sized cranial bone defects. *Tissue Eng. A.* 20: 751–62.

Mishra, R., Kumar, A. (2011). Inorganic-organic biocomposite cryogels for regeneration of bony tissues. *J. Biomater. Sci. Polym. Ed.* 22: 2107–26.

Mishra, R., Kumar, A. (2014). Osteocompatibility and osteoinductive potential of supermacroporous polyvinyl alcohol-TEOS-agarose-Cacl2 (Ptagc) biocompositecryogels. *J. Mater. Sci. Mater. Med* 25:1327–37. doi:10.1007/s10856-014-5166-8.

Molloy, T., Wang, Y., Murrell, G.A.C. (2003). The roles of growth factors in tendon and ligament healing. *Sports Med.* 33: 381–94.

Nair, M.B., Varma, H.K., Kumary, T.V., Babu, S.S., John, A. (2006). Cell interaction studies with novel bioglass coated hydroxyapatite porous blocks. *Trends Biomater. Artif. Organs.* 19: 108–14.

Netti, P.A., Shelton, J.C., Revel, P.A., Pirie, G., Smith S., Ambrosio, L., Nicolais, L., Bonfield, W. (1993). Hydrogels as an interface between bone and an implant. *Biomaterials.* 14: 1098–104.

O'Driscoll, S.W. (1998). The healing and regeneration of articular cartilage. *J. Bone Jt. Surg. Br.* 1795–812.

Odabas, S., Feichtinger, G.A., Korkusuz, P., Inci, I., Bilgic, E., Yar, A.S., Cavusoglu, T., Menevse, S., Vargel, I., and E. Piskin, E. (2013). Auricular cartilage repair using cryogel scaffolds loaded with BMP-7-expressing primary chondrocytes. *J. Tissue Eng. Regen. Med.* 7: 831–840.

Oseni, A.O., Crowley, C., Boland, M.Z., Butler, P.E., Seifalian, A.M. (2011). Cartilage tissue engineering: The application of nanomaterials and stem cell technology. In: *Tissue Engineering for Tissue and Organ Regeneration.* InTech, Croatia—European Union, doi: 10.5772/22453.

Ouyang, H.W., Goh, J.C.H., Mo, X.M., Teoh, S.H., Lee, E.H. (2002). Characterization of anterior cruciate ligament cells and bone marrow stromal cells on various biodegradable polymeric films. *Mater. Sci. Eng. C.* 20: 63–9.

Ozturk, B.Y., Inci, I., Egri, S., Ozturk, A.M., Yetkin, H., Goktas, G., Elmas, C., Piskin, E., Erdogan, D. (2013). The treatment of segmental bone defects in rabbit tibiae with vascular endothelial growth factor (VEGF)-loaded gelatin/hydroxyapatite 'cryogel' scaffold. *Eur. J. Orthop. Surg. Traumatol.* 7: 767–74.

Parrish, F.F., Murray, J.A., Urquhart, B.A. (1978). The use of polyethylene mesh (marlex) as an adjunct in reconstructive surgery of the extremities. *Clin. Orthop. Relat. Res.* 137: 276–86.

Paiva A.O., Duarte, M.G., Fernandes, M.H.V., Gil, M.H., Costa, N.G. (2006). In vitro studies of bioactive glass/polyhydroxybutyrate composites. *Mat. Res.* 9: 417–23.

Pelled, G., Atyelet, B.A., Hock, C. et al. (2010). Direct gene therapy for bone regeneration: Gene delivery, animal models, and outcome measures. *Tissue Eng. Part B Rev.* 16: 13–20.

Pennisi, E. (2002). Tending tender tendons. *Science* 295: 1011.

Phadke, A., Hwang, Y., Kim, S.H., Kim S.H., Yamaguchi, T., Masuda, K., Varghese. S. (2013). Effect of scaffold microarchitecture on osteogenic differentiation of human mesenchymal stem cells. *Eur. Cell. Mater.* 25: 114–28.

Place, E.S., Evans, N.D., Stevens, M.M. (2009). Complexity in biomaterials for tissue engineering. *Nature Materials.* 8: 457–70.

Reis, E.M., Vasconcelos, W.L., Mansur, H.S., Pereira, M.M. (2007). Synthesis and characterization of silica-chitosan porous hybrids for tissue engineering. *Key Eng. Mat.* 361–363: 967–70.

Ren, L., Tsuru, K., Hayakawa, S., Osaka, A. (2002). Novel approach to fabricate porous gelatin–siloxane hybrids for bone tissue engineering. *Biomaterials* 23: 4765–73.

Ringe, J., Kaps, C., Burmester, G.R., Sittinger, M. (2002). Stem cells for regenerative medicine: Advances in the engineering of tissues and organs. *Naturwissenschaften* 89: 338–51.

Rodrigues, S.C., Salgado, C.L., Sahu, A., Garcia, M.P., Fernandes, M.H., Monteiro, F.J. (2013). Preparation and Characterization of collagen-nanohydroxyapatite biocomposite scaffolds by cryogelation method for bone tissue engineering applications. *J. Biomed. Mater. Res. Part A.* 101: 1080–94.

Robinson, D., Efrat, M., Mendes, D.G., Halperin, N., Nevo, Z. (1993). Implants composed of carbon fiber mesh and bone-marrow-derived, chondrocyte-enriched cultures for joint surface reconstruction. *Bull Hosp. Jt. Dis.* 53: 75–82.

Saino, E., Fassina, L., Van Vlierberghe, S., Avanzini, M.A., Dubruel, P., Magenes, G., Visai, L., Benazzo, F. (2011). Effects of electromagnetic stimulation on osteogenic differentiation of human mesenchymal stromal cells seeded onto gelatin cryogel. *Int. J. Immunopathol. Pharmacol.* 24: 1–6.

Salgado, A.J., Coutinho, O.P., Reis, R.L. (2004). Bone tissue engineering: State of the art and future trends. *Macromol. Biosci.* 4: 743–65.

Schmitz, J.P., Hollinger, J.O. (1986). The critical size defect as an experimental model for craniomandibulofacial nonunions. *Clin. Orthop. Relat. Res.* 205: 299–308.

Shakya, A.K., Holmdahl, R., Nandakumar, K.S., Kumar, A. (2014). Polymeric cryogels are biocompatible and their biodegradation is independent of oxidative radicals. *J. Biomed. Mater. Res. A.* 102: 3409–18.

Shearn, J.T., Kinneberg, K.R., Dyment, N.A., Galloway, M.T., Kenter, K., Wylie, C., Butler, D.L. (2011). Tendon tissue engineering: Progress, challenges, and translation to the clinic. *J. Musculoskelet. Neuronal. Interact.* 11: 163–73.

Shive, M.S., Hoemann, C.D., Restrepo, A., Hurtig, M.B., Nicolas, D., Ranger, P., Stanish, W., Buschmann, M.D. (2006). BST-CarGel: *In situ* chondroinduction for cartilage repair. *Tech. Orthop. Surg.* 16: 271–8.

Shen, W., Chen, X., Chen, J., Yin, Z., Heng, B.C., Chen, W., Ouyang, H.W. (2010). The effect of incorporation of exogenous stromal cell-derived factor-1 alpha within a knitted silk-collagen sponge scaffold on tendon regeneration. *Biomaterials* 31: 7239–49.

Shah, R.N., Shah, N.A., Del Rosario Lim, M.M., Hsieh, C., Nuber, G., Stupp, S.I. (2010). Supramolecular design of self assembling nanofibers for cartilage regeneration. *PNAS.* 107: 3293–8.

Singh, D., Tripathi, A., Nayak, V., Kumar, A. (2011). Proliferation of chondrocytes on a 3-D modelled macroporous poly(hydroxyethyl methacrylate)–gelatin cryogel. *J. Biomater. Sci. Poly. Ed.* 22: 1733–51.

Tan, F., Xu, X., Deng, T., Yin, M., Zhang, X., Wang, J. (2012). Fabrication of positively charged poly(ethylene glycol)-diacrylate hydrogel as a bone tissue engineering scaffold. *Biomed. Mater.* 7: 055009 doi:10.1088/1748-6041/7/5/055009.

Tellnes, G., Bjerkedal, T. (1989). Epidemiology of sickness certification—A methodological approach based on a study from buskerud county in Norway. *Scandinavian J. Soc. Med.* 17: 245–51.

Thevenot, P.T., Nair, A.M., Shen, J., Lotfi, P., Ko, C., Tang, L. (2010). The effect of incorporation of SDF-1a into PLGA scaffolds on stem cell recruitment and the inflammatory response. *Biomaterials* 31: 3997–4008.

Tripathi, A., Kathuria, N., Kumar, A. (2009). Elastic and macroporous agarose–gelatin cryogels with isotropic and anisotropic porosity for tissue engineering. *J. Biomed. Mater. Res.* 90A: 680–94.

Venugopal, J., Vadgama, P., Sampath Kumar T.S., Ramakrishna S. (2007). Biocomposite nanofibres and osteoblasts for bone tissue engineering. *Nanotechnology* 18: 1–8.

Vogel, M., Voigt, C., Gross, U.M., Muller-Mai, C.M. (2001). *In vivo* comparison of bioactive glass particles in rabbits. *Biomaterials* 22: 357–62.

Wiria, F.E., Chua, C.K., Leong, K.F., Quah, Z.Y., Chandrasekaran, M., Lee, M.W. (2008). Improved biocomposite development of poly(vinyl alcohol) and hydroxyapatite for tissue engineering scaffold fabrication using selective laser sintering. *J. Mater. Sci. Mater. Med.* 19: 989–96.

Xynos, I.D., Edgar, A.J., Buttery, L.D., Hench, L.L., Polak, J.M. (2000a). Ionic products of bioactive glass dissolution increase proliferation of human osteoblasts and induce insulin growth factor II mRNA expression and protein synthesis. *Biochem. Biophys. Res. Commun.* 276: 461–5.

Xynos, I.D., Hukkanen, M.V.J., Batten, J.J., Buttery, L.D., Hench, L.L., Polak, J.M. (2000b). Bioglasss 45S5 stimulates osteoblast turnover and enhances bone formation *in vitro*: Implications and applications for bone tissue engineering. *Calcif. Tissue Int.* 67: 321–9.

Yang, Z., Yuan, H., Tong, W., Zou, P., Chen, W., Zhang, X. (1996). Osteogenesisin extraskeletally implanted porous calcium phosphate ceramics: Variability among different kinds of animals. *Biomaterials* 17: 2131–7.

Yoshikawa, H., Myoui, A. (2005). Bone tissue engineering with porous hydroxyapatite ceramics. *J. Artif. Organs.* 8: 131–6.

Younger, E.M., Chapman, M.W. (1989). Morbidity at bone graft donor sites. *J. Orthop. Trauma.* 3: 192–5.

Yuan, H., Fernades, H., Habibovic, P., de Boer, J., Barradas, A.M.C., de Ruiter, A., Walsh, W.R., van Blitterswijk, C.A., de Bruijn, J.D. (2010). Osteoconductive ceramics as a synthetic alternative to autologous bone grafting. *PNAS.* 107: 13614–9.

# 9 Cryogels for Neural Tissue Engineering

*Tanushree Vishnoi and Ashok Kumar**

## CONTENTS

9.1 Introduction ............................................................................................. 252
9.2 Nervous System ...................................................................................... 253
9.3 Cell Therapy for Neural Regeneration.................................................... 254
9.4 Conventional Strategies .......................................................................... 255
9.5 Fabrication of Scaffolds.......................................................................... 258
    9.5.1 Delivery of Biomolecules at Injury Site .................................... 258
    9.5.2 Electrospun Fibres for Neural Regeneration ............................. 258
    9.5.3 Hydrogels for Neural Recovery ................................................. 258
9.6 Nerve Conduits ....................................................................................... 259
9.7 Conducting Polymer ............................................................................... 261
9.8 Cryogels .................................................................................................. 262
9.9 Conducting Cryogel ................................................................................ 267
9.10 Conclusion .............................................................................................. 268
List of Abbreviations......................................................................................... 268
Acknowledgements ........................................................................................... 268
References......................................................................................................... 268

## ABSTRACT

Polymeric three-dimensional (3D) implants for tissue regeneration have gained momentum over the last years and have shown promising results. These scaffolds have been synthesized by various methods in order to attain optimum properties for neural regeneration. Cell transplantation either directly or encapsulated in the polymeric micro/nanospheres at the site of injury has shown improved results. However, there still seems to be a huge gap between the available strategies and their real application. One of the existing techniques that synthesizes interconnected porous gels at subzero temperature by using ice as a porogen is cryogelation. The properties of these gels can be modified as per the application by changing the polymer concentration, temperature and cross-linker concentration, which allows both soft and hard tissue cells to grow and proliferate on them. Recently, these cryogels have been incorporated with conducting polymer either during or after their synthesis in order to incorporate conducting properties. These conducting cyogels would allow electrical stimulation of the cells seeded

---

* Corresponding author: E-mail: ashokkum@iitk.ac.in; Tel: +91 512 2594051.

on them, which would change the membrane potential of these cells leading to altered cell proliferation and signaling. Moreover, synthesis of these scaffolds with natural polymers results in bioactive scaffolds that enable the native cells to migrate to the implanted scaffolds and thus result in faster regeneration.

**KEYWORDS**

Conducting polymer, micro/nanospheres, nerve regeneration, polymeric cryogels

## 9.1 INTRODUCTION

Neural cells, well recognized as excitable cells of the human body, are known for receiving the signals from the environment to the brain via afferent neurons whereas efferent neurons transmit information away from the brain (Schulz et al., 2006; Takashima et al., 2007). This transmission of impulses across the nerve is responsible for various biological functions. However, injuries due to accidents, trauma and so on, result in loss of functional neurons and thus inhibit these signals leading to loss of senses, paralysis, loss of memory and so on. The impact of these injuries in the day-to-day living of an individual is enormous and there is a constant need to fill this void. Many strategies have been employed to overcome this but with limited success. However, in the last decade neural tissue engineering has gained momentum and shown improvement in neural recovery (Huang and Huang, 2006; Melanie et al., 2013). Implantation of polymeric scaffolds either with or without growth factors and cells has been researched for better and viable results (Babensee et al., 2000; Richardson et al., 2001; Chen et al., 2003; Nicodemus et al., 2008; Smith et al., 2009; Shoichet, 2010; Lee et al., 2011). Initially, the cells were injected directly at the site of injury for recovery. Although these grafted cells showed improvement, they did not localize at the site of injury and further underwent apoptosis, resulting in varied results. Stem cells formed the major source of cell transplantation because of their ability to undergo differentiation in myriad lineages (Riess et al., 2007; Gaillard et al., 2011; Willerth, 2011; Nicolas et al., 2012; Antonic et al., 2013; Kannol, 2013; Nakamura and Okano, 2013; Piltti et al., 2013; Van Velthoven et al., 2013). However, in order to overcome the mentioned limitations, 3D scaffolds have been synthesized by several methods. Cells could be either encapsulated in these scaffolds or seeded on the synthesized scaffolds for transplantation at the site of injury. Entrapped cells are well protected from the microenvironment as well as localized to the injury site. Moreover, attaching certain ligands further allows the localization of these entrapped cells to the desired location. In the other case, scaffolds apart from providing protection to the cells also endowed them with mechanical strength and directionality. These scaffolds have been synthesized using either natural or synthetic polymers. Natural polymers used like chitosan and alginate are biodegradable and biocompatible and mostly resemble the glycosaminoglycans of the ECM (Freed et al., 1994; Mano et al., 2007) whereas synthetic polymers like polyethylene glycol (PEG), polyvinyl alcohol (PVA) and so on allow their mechanical properties and biodegradation to be tuned as per the requirement (Guanttilake and Adhikari, 2003; Li et al., 2005; Place et al., 2009). Different techniques like solvent extraction, salt leaching, electrospinning and so on

(Ho et al., 2004; Guan et al., 2005; Weigel et al., 2006; Yao et al., 2012; Riano-Velez and Echeverri-Cuartas, 2013) can be adapted for the synthesis of these 3D scaffolds in order to attain optimum properties. In the past few years, synthesis of macroporous gels, by cryogelation technique, and their application in the field of tissue engineering has been very convincing (Lozinsky et al., 2003; Kathuria et al., 2009). This might be because of their ease of synthesis, interconnected porous nature, lack of use of solvents and high temperature during synthesis, high mechanical strength, tunable properties and so on.

Thus, in this chapter our focus is to discuss the trends in neural tissue engineering and the application of cryogel scaffolds for improving nerve damage.

## 9.2 NERVOUS SYSTEM

The nervous system can be divided into the central (CNS) and peripheral nervous system (PNS). Neurons and glial cells are primarily the major cells present in the nervous system. Oligodendrocytes and astrocytes comprise the glial cells of the CNS whereas Schwann cells represent the PNS. The repair mechanism, after injury, is different for both the CNS and the PNS. However, it is much more complex in the former system (Reichert et al., 2008; Eric et al., 2009; Müller et al., 2012). The brain and spinal cord form the CNS wherein the outer layer of the brain consists of gray matter and the inner layer of white matter. The converse is present in the spinal cord. Depending on the extent of injury, the cell bodies (gray matter) or the axons (white matter) or both of them are damaged. Reactive microglia and astrocytes migrate to the injury site resulting in the formation of glial scar, which, along with the presence of proteoglycans, chondroitin sulfate, Nogo and so on, forms an inhibitory microenvironment retarding the repair and regeneration process. Further, it also limits the diffusion of growth-promoting biomolecules. Chondriotin sulfate and proteoglycans constitute the major inhibitory molecules of the CNS (Ray et al., 2002; Schmidt et al., 2003; Carmichael, 2006) (Figure 9.1).

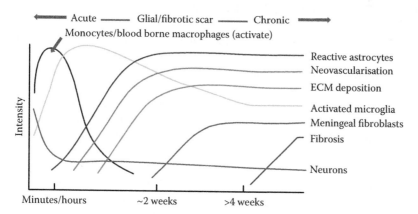

**FIGURE 9.1** Inflammation and wound healing response to implanted biomaterials in the CNS represented by temporal sequence. (Reproduced from, He, W. and Bellamkonda, R.V., In: *Indwelling Neural Implants: Strategies for Contending with the In Vivo Environment*. Reichert, W.M., Ed. Boca Raton, FL: CRC Press, Taylor & Francis Group, pp. 151–176. Copyright (2008), Taylor & Francis Group, LLC. With permission.)

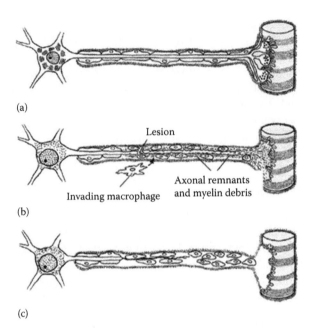

**FIGURE 9.2** Wallerian degeneration (a) cellular organization of neuron, Schwann cells and muscle fiber. (b) Following nerve transaction, the distal end undergoes Wallerian degeneration wherein the distal axon stump degenerates. Schwann cells dedifferentiate and proliferate in response and align to form bands of bunger for guiding the proximal end. The macrophages and Schwann cells are responsible for cleaning the debris. The cell body undergoes chromatolysis. (c) The proximal stump regenerates and forms connections with muscle fiber. (From Gillingwater, T.H. and Ribchester, R.R.: Compartmental neurodegeneration and synaptic plasticity in the Wld$^s$ mutant mouse. *J. Phys.* 2001. 534. 627–639. Copyright Wiley-VCH Verlag GmbH & Co. KGaA. Reproduced with permission.)

In the PNS after an injury, the distal end of the nerve undergoes 'Wallerian degeneration.' The myelin sheath degrades resulting in ovoid formation. The debris clearance is mediated by the macrophages, which secrete cytokines and chemokines. In response to these factors, Schwann cells align in the form of 'bands of bunger' thus providing directionality to the severed nerve axons whereas the proximal end of the injured nerve undergoes sprouting resulting in the formation of the growth cone, which migrates to the distal side due to the above-mentioned chemotactic factors as well as haptotactic cues as shown (Figure 9.2) (Gillingwater and Ribchester, 2001; Hall, 2005; Ribeiro-Resende et al., 2009; Rupprecht et al., 2010; Kim et al., 2013; Scheib and Hoke, 2013; Svenningsen and Dahlin, 2013).

## 9.3 CELL THERAPY FOR NEURAL REGENERATION

Cell therapy has been employed as a less invasive methodology in order to recover damaged neurons functionally. However, the recovery depends on the viability and localization of the transplanted cells at the site of injury. Neural progenitor cells, bone marrow, umbilical cord, adipose tissue, peripheral blood cells, oligodendrocytes,

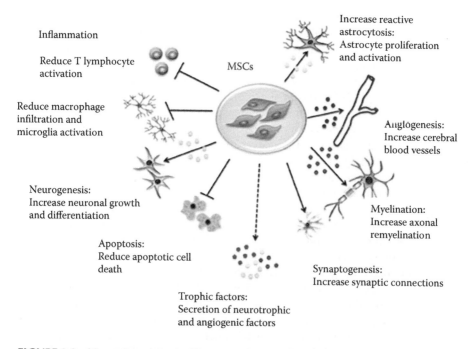

**FIGURE 9.3** Mesenchymal stem cells as potential candidates for neuroprotective and neurorestorative effect. (Reproduced from Castillo-Melendez, M., Yawno, T., Jenkin, G. and Miller, S., *Front. Neurosci.* 7: 1–14. Copyright (2013), With permission.)

Schwann cells and human amnion epithelial cells are a few examples of the kinds of transplanted cells (Chu et al., 2004; Shibashi et al., 2004). The improvement by this method can be attributed mainly to the release of diffusible factors; extracellular matrix formation and replacement of lost cells at the site of injury further enhance the functionality of the tissue (Tonya et al., 2007; Castillo-Melendez et al., 2013) (Figure 9.3). However, cell replacement therapy in the case of the CNS is more complex due to the hostile microenvironment present after injury. The presence of the blood–brain barrier, complex anatomy and delivery at the specific target site are the major challenges (Potts et al., 2013). It has been shown that the neuronal circuitry could be reestablished after transplantation of the neurons, differentiated from neural stem cells, in the transected spinal cord after 7 days of implantation in mice (Abematsu et al., 2010). Similarly, attempts have been made to restore the damaged cells in the cases of a neurodegenerative disease like Parkinson or a stroke, which leads to permanent loss of cells and thus functionality (Ma et al., 2010).

## 9.4 CONVENTIONAL STRATEGIES

The type of strategy used to repair the injured nerves depends on the kind and the extent of injury. If the nerve gap is not too large, then suturing the proximal and distal ends is the most successful approach used to date. Direct repair or end-to-end repair involves suturing the epineural covering of proximal and distal end as shown in Figure 9.4 (Wolford et al., 2003; Haninec et al., 2007). This is the most commonly used method in peripheral nerve regeneration.

Fascicular and group fascicular repair pertains to suturing of fascicules or a group of fascicules together. Although this surgery helps in the formation of proper connections required for functionality, it results in scars, fibrosis and inflammatory reactions (Spector et al., 2000; Pfister et al., 2011). Therefore, 'suture less nerve repair' has shown promising results. It not only makes the procedure less tedious but also reduces the other inflammatory reactions associated with microsutures. Recently, the use of fibrin glue during end-to-end repair was compared with traditional surgery and the results clearly showed that fibrin sealant gave better results with respect to not only reduced fibrosis and inflammation but also better nerve fiber alignment and axonal regeneration (Ornelas et al., 2006). However, larger nerve gaps cannot be repaired by direct suturing, as it would lead to compression and thus tension of the nerves affecting the regeneration, conduction velocity and the vascular supply at the site of injury. Therefore, an alternate option for these limitations is the use of autografts. Autografts do not suffer from limitations of immunogenic response and are still considered the gold standard method for nerve regeneration. It has successfully been employed to fill the gaps of 5 cm (Daly et al., 2008), although recently it has been reported that sural nerve autografts can repair nerve damage of 9 cm or more (Lee et al., 2008). It was also tested if the use of autologous nerve grafts gave better results when used either immediately after damage or as a delayed treatment. The observations showed that the latter gave better results (Spector et al., 2000). The size of the graft needs to be 10–20% longer than the nerve gap generated in

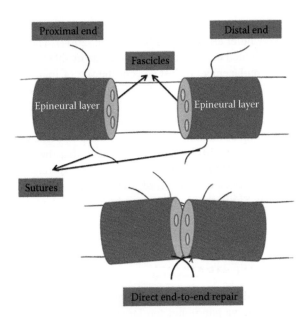

**FIGURE 9.4** Direct end-to-end repair. Suturing of the epineural covering of the proximal and distal nerves for nerve regeneration.

order to avoid fibrosis formation (Gordon et al., 2012). Useful motor recovery was observed in a series of experiments of autologous nerve grafting wherein the median, radial and ulnar nerve grafts were used (Millesi et al., 1976). However, the use of autologous nerve grafts leads to donor site morbidity, neuroma pain and difference in the dimensions of the implanted nerve from the native nerve (Slutsky et al., 2005). Therefore, nonnerve grafts have been sought as an alternative to the above. Muscle and vein grafts have shown better results compared to the others (Chiu et al., 1982; Glasby et al., 1991). Decellularization of these grafts with detergents leaves the cytoskeletal framework of these grafts intact, reducing the immune response (Hall et al., 1997; Szynkaruk et al., 2013; Zhao et al., 2014). The use of the femoral vein as grafts for the repair of a 1-cm gap generated in the sciatic nerve showed the nerve fibers could cross the vein and reach their distal end in 2 months (Chiu et al., 1982). An alternative to autograft is the use of allografts and xenografts as temporary 3D supports. Nonetheless, these have their own disadvantages of immune response, which makes the administration of immunosuppressant a necessity to the patient (Kvist et al., 2011; Lin et al., 2013). Moreover, very few clinical reports are available with little success. Considering the limitations associated with all the types mentioned above, there is always a need to find new methods and technologies for improved and better recovery. Tissue engineering and regenerative medicine, in the last 10 years, has shown promising results and thus can be an alternative remedy in the field of neural tissue engineering (O'Halloran and Pandit, 2007; Dionigi and Fauza, 2008; Galler and D'Souza, 2011; Salgado et al., 2013). Tissue engineering combines the principle of biology and engineering for designing scaffolds for repair and regeneration mechanisms. Scaffolds, constructs or matrices primarily function as supporting framework, which provides guidance and thus allows regeneration (Chan and Leong, 2008). The designing of these biomaterials should be such that it allows cell adherence, uniform distribution, proper morphology and exchange of nutrients and wastes. Meshes, foams, sponges and monoliths are various formats in which biomaterials have been synthesized (Chen et al., 2002). Polymers have been extensively used in the field of pharmaceutical and food industry and thus prove to be a potent candidate in the field of tissue engineering (Christina et al., 2011). Both natural and synthetic polymers have been used for the purpose. Natural polymers like gelatin, chitosan and so on resemble the extracellular matrix of the native environment. These polymers, apart from being biocompatible, are biodegradable which allows the cell adhesion properties, which is an added advantage for any tissue engineering approach. Synthetic polymers have also paved their way into the field of tissue engineering as their mechanical and degradation properties can be modulated as per the requirement. The mechanical properties of the scaffolds play an integral part in stem cell differentiation as scaffolds with less mechanical strength direct the stem cell differentiation toward soft tissues like neural cells (Goh et al., 2013; Lee et al., 2013) whereas scaffolds with high mechanical strength prefer chondrogenic, osteogenic differentiation (Park et al., 2011). Degradation of the implant is an important parameter and therefore it should maintain balance between the regeneration of the native tissue and degradation of the implanted scaffold. Synthetic polymers can also be modified with certain peptides to enhance cell adhesion. Neural tissue (including both the CNS and PNS) is a soft tissue and therefore requires that the implanted material should be such that it does not harm the neighboring tissues and thus these properties permit the use of polymers in the field of neural tissue engineering (Nair and Laurencin, 2006; Dhandayuthapani et al., 2011; Aurand et al., 2012).

## 9.5 FABRICATION OF SCAFFOLDS

### 9.5.1 Delivery of Biomolecules at Injury Site

Based on the molecular weight of the polymer, microspheres can be designed for long- or short-term release. Polymers with high molecular weight can be used for longer delivery compared to ones with lower molecular weight. Most of the growth factors whose shelf life periods are short and reside at the target site for limited duration can be delivered by microspheres/nanospheres (Siegel et al., 2000; Taylor et al., 2004; Liu et al., 2013). These vehicles allow them to be released in a controlled manner for a longer duration. Apart from this, various drugs and anti-inflammatory biomolecules can also be delivered (Nicodemius et al., 2008). It is observed that by varying the physical properties of the material, controlled release of the drugs can be enabled.

### 9.5.2 Electrospun Fibres for Neural Regeneration

Electrospun nanofibres have been flourishing as the new tissue engineered biomaterial because of their resemblance to the architecture of the *in vivo* protein fibres of the ECM (Cui et al., 2010). These fibres can be synthesized in the nanometer range unlike the conventional fibres (in micron range) and thus provide a high surface area for cell contact and adherence (Liu et al., 2013). The fibres' properties can also be modulated by varying the polymer concentration, type of polymers and electrospinning conditions (Xie et al., 2010; Lee and Arinzeh, 2011; Baiguera et al., 2014). Poly(l-lactic acid)-co-poly-(3-caprolactone)/collagen (PLCL/Coll) nanofibrous scaffolds showed potency when evaluated for human bone marrow mesenchymal stem cells differentiation into neuronal lineage. The seeded stem cells when analysed for their gene expression showed the presence of early and late neural marker like nestin and neurofilament (Prabhakaran et al., 2009). Orientation of these fibres has been shown to affect the cell alignment wherein the aligned fibres provided directionality to the seeded cells compared to the randomly synthesized fibres (Masaeli et al., 2013) (Figure 9.5). Moreover, it has been reported that Schwann cells seeded on aligned PCL fibres showed better cytoskeleton alignment, as well as increased expression of myelin specific gene, a marker for mature Schwann cells (Chew et al., 2008; Wang et al., 2009).

### 9.5.3 Hydrogels for Neural Recovery

Designing of scaffolds depends on the site of implantation, type of cell seeded and the severity of the injury. Hydrogels have been extensively used in the field of neural tissue engineering because of its capacity to absorb large amounts of water and optimum mechanical properties. These are physically and chemically cross-linked in nature (Bryant et al., 2002; Fromstein et al., 2002; Vanderhooft et al., 2009; Sung et

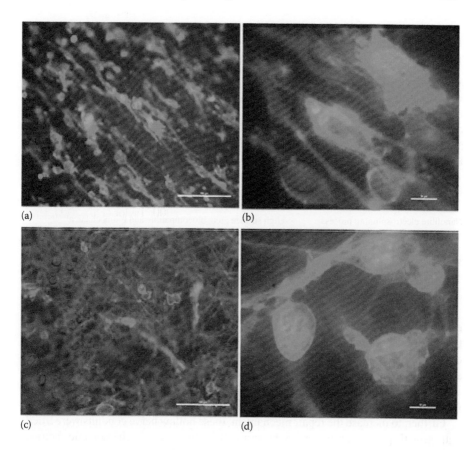

**FIGURE 9.5** Overlapped immunostain image for identification of adult rat stem cells seeded on poly(hydroxy alkanoate) composite nanofibrous scaffolds by p75LNGFR. Nucleus counterstained with DAPI. (Reproduced from Masaeli E. et al., *PLoS One.* 8: e57157. Copyright (2013). With permission.)

al., 2010). Traumatic brain injury leading to loss of brain tissue and thus formation of cavity has no determined shape. Therefore, the implant should be such that it does not mechanically damage the nearby tissue. Therefore, injectable hydrogels have been used that would eventually take the shape of the cavity in the brain evading the above limitations. Stimuli responsive polymers like poly(NIPAAm) and chitosan have been used for such damages (Van Tomme et al., 2008). There are various methods to synthesize these hydrogels as mentioned in Table 9.1 (Dhandayuthapani et al., 2011).

## 9.6 NERVE CONDUITS

Whenever there is a nerve injury leading to the generation of a gap in the proximal and distal end, there is a requirement to rebuild the lost connections for proper transmission of nerve impulses and thus functionality (Deal et al., 2012). In the earlier strategies, nondegradable silicone tubes were used to fill the nerve gaps. However, it

## TABLE 9.1
### Techniques for Scaffold Fabrication in Tissue Engineering

| Method | Unique Factors |
| --- | --- |
| Solvent casting/salt leaching method | Biodegradable controlled porous scaffolds |
| Ice particle leaching method | Control of pore structure and production of thicker scaffolds |
| Gas foaming/salt leaching method | Controlled porosity and pore structure sponge |
| Solvent evaporation | High density cell culture, due to high surface area |
| Freeze-drying method | 3D porous structure, durable and flexible |
| Thermally induced phase separation | Highly porous scaffold for cellular transplantation |
| Hydrogel bases injectable scaffolds | Biomimetically exhibit biocompatibility and cause minimum inflammatory response, thrombosis and damage |
| Nanofibre electrospinning process | High surface area, biocompatible and biomechanical |
| Microfiber wet spinning process | Biocompatible fibres with good mechanical properties |

*Source:* Reproduced from Dhandayuthapani, B., Yoshida, Y., Maekawa, T. and Kumar, D.S. (2011). *Int. J. of Polym. Sci.* 2011: 1–20. Copyright (2011). With permission.

eventually leads to compression of the nerve and further involved foreign body reaction in the long term (Moore et al., 2009). Then there was a shift from the first generation implant to degradable implants wherein the nerve conduits would degrade after providing support for the proximal end of the nerves to migrate to the distal end (Wang et al., 2009). These nerve conduits provided guidance and directionality to the nerve from the proximal to the distal end (Figure 9.6).

Further, to increase the repair mechanism, these hollow nerve conduits were filled with growth promoting factors, Schwann cells, which secrete chemotactic factors,

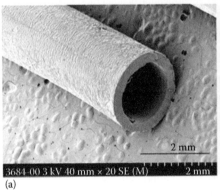
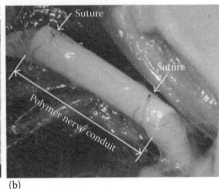

**FIGURE 9.6** Cross-linked poly (caprolactone) fumarate polymer nerve conduits for nerve regeneration in rat sciatic nerve (a) before and (b) after implantation. (Reprinted from *Acta Biomater.* 5, Wang, S., Yaszemski, M.J., Knight, A.M., Gruetzmacher, J.A., Windebank, A.J. and Lu, L., Photo-cross-linked poly(ε-caprolactone fumarate) networks for guided peripheral nerve regeneration: material properties and preliminary biological evaluations, 1531–1542. Copyright 2009, with permission from Elsevier.)

and so on (Belkas et al., 2004). Moreover, the architecture has also been modified by incorporating multiple channels in the gap as these would guide the sprouting axons to the distal end (de Ruiter et al., 2009). Recently, genipin cross-linked casein channels have shown better recovery in the 10-mm nerve gap generated in the rat sciatic nerve (Hsiang et al., 2011).

## 9.7 CONDUCTING POLYMER

It has long been reported that electrical stimulation to the neural cells is an important cue as these cells function in transmitting electrochemical signals. Apart from neural cells, these electrical cues have also been shown to enhance the proliferation and differentiation of other cell types (Zhao et al., 1999; Rivers et al., 2002; Dust et al., 2006; Shi et al., 2008). Electrical stimulation of nerve cells is capable of eliciting an action potential, which in return is responsible for various cellular responses. Further endogenous electric fields have been shown to polarize the nervous system during development. Therefore, various methods have been employed to incorporate this feature in the scaffolds synthesized for the neural regeneration. *In vitro* electric fields have been shown to direct the neurite growth of the nerve cells apart from initiation and growth of neurite. It has also been shown that electrical stimulation affects the cell–cell signalling and gene expression of neural cells. PC12 cells when electrically stimulated showed enhanced activation of the NGF gene (Kimura et al., 1998). Conducting polymers have long been in use since nineteenth century. However, recently they have emerged as a probable candidate in the field of tissue engineering (Bendrea et al., 2011; Laleh et al., 2011). These polymers are conducting in nature due to the presence of conjugate backbone. Doping of these polymers by various ions like $Cl^-$, $SO_3^-$ and so on, provide charges to these molecules that are otherwise neutral in nature (Hussain and Kumar, 2003; Rimbu et al., 2006). Doping facilitates these molecules with charges carriers and thus the property of electrical conductivity (Guimard et al., 2007; Arshak et al., 2009). In the field of tissue engineering, various biomolecules like extracellular matrix, peptides and growth factors have been used as dopants (Sanghvi et al., 2005; Stauffer et al., 2006; Thompson et al., 2006; Green et al., 2010). Apart from this, growth factors like NGF and BDNF have been entrapped or incorporated during synthesis of these conducting polymers resulting in enhanced neural regeneration (Lee et al., 2006; Thompson et al., 2006, 2010). Although doping with these biomolecules reduces their conductivity, the cellular properties like adherence, differentiation and morphology are enhanced. Moreover, if the dopant is attached noncovalently to the polymer backbone, it leaches out and is released in the microenvironment. This property is used in drug delivery wherein the drug can be released from the polymer backbone on electrical stimulation as it would bring a change in the oxidation/reduction state of the conducting polymer and hence would be released in the biological milieu (Fonner et al., 2008; Svirskis et al., 2010). Conducting polymers can be synthesized either by an electrochemical or a chemical method. Electrochemical synthesis results in film formation on a substrate on which cells can be grown and electrically stimulated. These can thus be used as electrodes for the growth and implantation of cells and provide a more compatible biological interface compared to the metals. Chemical synthesis is comparatively cheap and results in bulk synthesis of the polymer (Cho et al., 2010). It has also been shown to be compatible with the growth and stimulation of different cell types like fibroblast, keratinocytes and neural and cardiac cells (Ateh et al., 2006; Mihardja et al., 2008; Collazos-Castro et al., 2010; Onoda et al., 2010). Although conducting polymers have shown improved results in *in vitro* conditions, their usage in clinical trials is still very limited because of their inheritable nondegradable nature. However, it has been shown that this limitation can be overcome by blending poly(pyrrole) with other degradable polymers like chitosan, gelatin, agarose and so on. *In vivo* studies in mice have shown that chitosan polypyrrole blends implanted in rats did not cause much toxicity to the cells provided the poly(pyrrole) content was less than 6% because until then it can be removed from the body (George et al., 2005; Wan et al., 2005; Kai et al., 2011).

## 9.8 CRYOGELS

Hydrogels synthesized at subzero temperature are referred to as 'cryogels' and exhibit features like interconnected porous network, high mechanical strength, convective transport of nutrients and waste and large surface area (Lozinsky et al., 2003; Kumar, 2008; Bhat and Kumar, 2012; Zhou et al., 2012). These properties make cryogel a potential candidate in the field of tissue engineering. Over the last few years, cryogels have gained momentum in this field and shown improved results when used for cartilage, bone, skin and so on, tissue engineering (Dainiak et al., 2010; Fassina et al., 2010; Singh et al., 2010; Bhat et al., 2013; Jain and Kumar, 2013). Moreover, it is possible to vary the properties of these cryogels by changing the concentrations of the polymers, cross-linkers, temperature used and so on. Therefore, they can be synthesized as per the cell type/tissue, which needs to be engineered. Rheological studies have shown that mechanical moduli of soft tissues should range between 0.1 and 40 kPa. Neural tissue being a soft tissue grows and proliferates best when grown on scaffold stiffness of 0.5–1 kPa, which resembles the native brain tissue. Neural progenitor stem cells are capable of differentiating into both glial and neural cells on such substrates (Hang et al., 2012). Moreover, the mentioned properties are imperative to be present in soft tissue scaffolds so that they do not damage the nearby tissues. Gelatin-laminin cryogels were synthesized using varied concentration of cross-linkers (0.1%, 0.3% and 0.5%) in order to study the effect of cross-linking on scaffold degradation, cell adherence, pore size and mechanical strength. It is already well known that the mechanical strength of the gel can be varied by changing the cross-linker concentration, as a high concentration of the cross-linkers would covalently bind more numbers of polymer chains resulting in higher mechanical strength, compared to the scaffolds with their lower concentration (Jurga et al., 2011). It was observed that the cells responded better when seeded on 0.1 and 0.3% cross-linked cryogels. However, 0.1% cross-linker resulted in fragile cryogels and when implanted *in vivo* degraded within 24 h. This fast release of degraded products—gelatin and laminin—resulted in toxicity to the nearby cells as well as it was not in coordination with the regeneration of the injured brain tissue. Comparatively, 0.3% cryogels were mechanically better and did not result in any such kind of toxicity. Further, in order to deduce the optimum pore size for neural regeneration, OPLA cryogels with mean pore size of 100–200 μ, unlike the gelatin-laminin cryogels which had a mean pore size of 80–120 µ, were synthesized. Both the scaffolds allowed proliferation and differentiation of stem cells into neuroblasts. However, gelatin-laminin pore size allowed better cell expansion, network formation and contact to the scaffold as well as the adjacent cells compared to OPLA scaffolds. Gelatin-laminin cryogel was found to be better tolerated in the *in vivo* conditions when transplanted in mice as well as when cocultured with rat oragnotypic hippocampal slice cultures. It resulted in minimum microglial invasion at the centre, which was found to be mainly limited to the edges of the scaffold. Further, these scaffolds did not result in any scar formation due to noninvasion of astrocytes. Moreover, neuroblasts of the active tissue migrated to the implanted GL scaffold for integration. Thus, integration of the native tissue with the implanted scaffold is a prerequisite for the success in the field of tissue engineering as it enables stress distribution and weight bearing as well as inhibits tissue degeneration. Moreover, the migrated cells across the integration site, both from the implant to the native tissue and vice versa, allow for the compensation of the cells at the injury site (Hunziker et al., 2002; Jurga et al., 2011; Theodoropolous et al., 2011) (Figure 9.7).

It is well known that the greatest differentiation can only be achieved when the above engineered scaffold is seeded with an optimum kind of cell. It has been

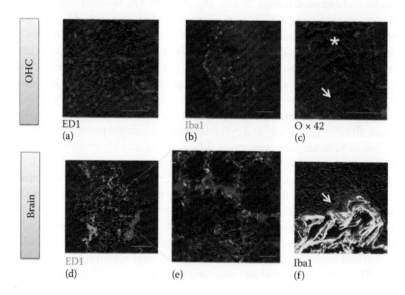

**FIGURE 9.7** Immunostaining of gelatin-laminin scaffold after transplantation into (a–c) organotypic hippocampal slices (OHC) and (d–e) rat brain to study the inflammatory response. Biodegradation of the scaffold started after 7 days and was limited to the edge of the cryogel (c–arrow). The scaffold also lacked microglia cell invasion at the core. (c–asterisk). The rate of biodegradation enhanced in the *in vivo* microenvironment, which was observed by a sufficient amount of scaffold decomposition (d–e). Both in OHC culture (a–c) and rat brain (d, e–arrow) scaffolds inflammatory cells were present only at the edges and not the centre of the scaffold (e–yellow colabelling) edges. Scaffold autofluorescence could be seen in images a–c in blue, d in red and e in yellow. Scale bar 100 microns. (Reprinted from *Biomaterials* 32, Jurga, M. et al., The performance of laminin-containing cryogel scaffolds in neural tissue regeneration, 3423–3434. Copyright 2011, with permission from Elsevier.)

revealed that cord blood stem cells (CBSCs), when seeded immediately after isolation, showed maximum differentiation when compared with the results of CBSCs transplanted after their differentiation into nestin positive neuroblasts or fully matured neural cells. Unlike the former type of cells, which showed reduced cell adherence and survival, fully matured cells did not survive in 3D scaffolds. Further, the same group showed that encapsulation of CBSCs in the gelatin-laminin scaffolds protected the cells from the *in vivo* microenvironment and provided a shield against the inflammatory cells (Sarnowska et al., 2013). After being seeded on the synthesized scaffolds, the cells migrated to the centre of the scaffold through the large pores, adhered and formed a niche in the small pores. The mean pore size of the scaffold was thus in accordance with the seeded CBSCs (20–160 µ). It was observed that after implantation the immune cells like microglia were limited to the boundary of the graft and did not migrate into the centre of the graft. Microglia and astrocytes are the key mediators of host response to the implant along with the plethora of other molecules like cytokines, growth factors and reactive oxygen species (He and Bellamkonda et al., 2008), thus allowing the stem cells a niche at the centre to proliferate and differentiate whereas the CBSCs that were transplanted without GL scaffold died within 72 h of implantation. GL scaffold with and without CBSCs were immunologically compatible with the host although the graft with the implanted cells was relatively better as MSCs also showed anti-inflammatory effect. Anti-inflammatory effects of stem cells have also been studied after implantation in TBI rat models. It was observed that these cells reduced the immigration of leukocyte cells as well as the presence of cells such as microglia/macrophages were also inhibited. This lead to low amounts of released cytokines, which are responsible for inflammation (Ichim et al., 2010; Zhang et al., 2013). Thus, the combination of bioengineered cryogel scaffold and cell transplantation could prove to be an alternative therapy in neurodegenerative cases. Moreover, it was observed that activity of MMP2/9 also increased when CBSC-seeded GL scaffolds were implanted. MMPs are mainly involved in ECM remodelling and cell proliferation. By altering the ECM, it exposes or hides the adhesion sites, thus modulating the cell adhesion and its downstream signalling (Liezhen et al., 2001; Papadimitriou et al., 2001). It also played a role in the fast degradation of the scaffold at the implantation site although it did not lead to any toxic effect. Further, the implanted cells remained viable after the GL scaffold degradation with no immune response reactivation (Sarnowska et al., 2013). Thus, the gelatin scaffold provided a physical barrier against the influx of immune cells (Figure 9.8).

Apart from using stem cells for neural regeneration, efforts have been made to isolate neural cells from brain as well as peripheral regions, for cell as well as cell seeded scaffold transplantation. However, the cell number achieved never meets the tissue-engineering requirement and therefore efforts have been made to increase their number by adding various growth promoting molecules (Silva et al., 2009). Recently, a lot of work has been done on alpha-ketoglutarate ($\alpha$-KG) as it has shown to increase proliferation rates of cells by reducing the toxic effect of ammonia (Vishnoi et al., 2013). Reports have shown that adding $\alpha$-KG exogenously in the media both in the free form as well as encapsulated in chitosan/gelatin microspheres leads to enhanced proliferation of neuro 2a cells seeded on cryogels as compared to their

# Cryogels for Neural Tissue Engineering

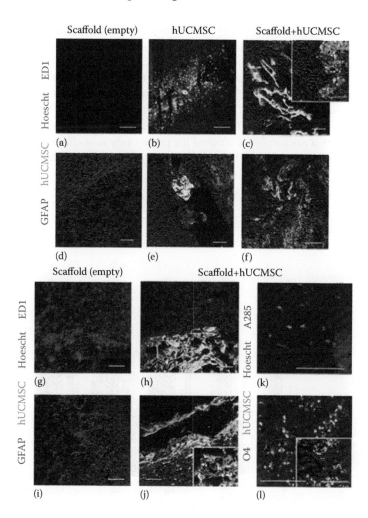

**FIGURE 9.8** Activation of microglia and reactive astrocytes in response to transplantation of empty gelatin laminin scaffold as well as cell-encapsulated scaffold in the rat brain. Empty scaffold (a,d) hUCMSCs (labelled with 5-chloromethyl-fluorescein-diacetate [CMFDA]), cell tracker in suspension (b,e) or 3D aggregates (c,f). Significant immune response was observed when hUCMSCs were injected in suspension labelled with CMFDA and stained with NuMa, green. Immune response ED1 and GFAP (glia) resulted in cell death (yellow; b,e). Transplantation of hUCMSCs encapsulated in GL scaffold reduced this response (c,f). The transplanted cells were also observed to migrate to the neighbouring host tissue. In the lower panel, the microglia and astrocyte activation in rat brain in response to empty scaffold (g,i), hUCMSCs (stained green with NuMa) and hUCMSCs in GL scaffold (h,j) is shown. hUCMSC-scaffold transplantation was observed to show reduced number of ED1 and GFAP–positive cells (red) when compared to the response of empty scaffold. In addition, GL scaffold not only provided support to the transplanted hUCMSCs but also directed their differentiation to microglia: astrocyte lineage (j) and oligodendrocytes (k,l). Cell nuclei were counter-stained with Hoechst (blue). Scale bar 100 um. (Reproduced from Sarnowska, A. et al., *Cell Transplant.* 22: 67–82. Copyright (2013), Cognizant Communication Corporation. With permission.)

control. Further analysis of their spent media showed reduced levels of ammonia, as α-KG is a potent scavenger for ammonia (Nedergaard et al., 2002; Park et al., 2005; Ketie et al., 2010). In the presence of ammonia, α-KG results in the formation of glutamate and glutamine. Both are involved in protein synthesis whereas glutamate is also a neurotransmitter (Shank and Campbell, 1982). However, when α-KG loaded microspheres were incorporated in the cryogel during synthesis, their efficiency reduced and thus the cell proliferation, as measured by MTT assay, declined. It was concluded that microspheres blocked the pores of these scaffolds, which inhibited transport of nutrients as well as the cell adherence (Vishnoi and Kumar, 2013) (Figure 9.9).

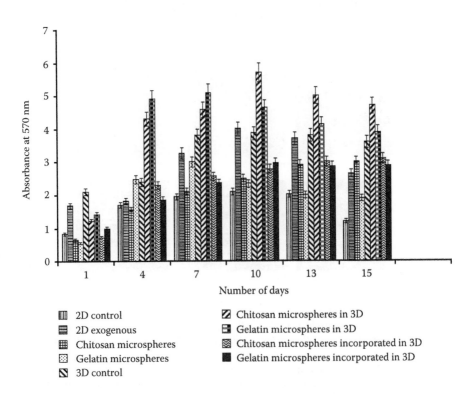

**FIGURE 9.9** MTT assay was used to study cellular metabolic activity/proliferation in the presence and absence of alpha ketoglutarate (α-KG). Neuro 2a cell proliferation was analysed both on 2D and 3D control (α-KG absent) as well as in samples containing α-KG. There were three different modes to administer α-KG: (1) exogenously in the media in freeform in 2D, (2) α-KG containing chitosan or gelatin microspheres in both 2D and 3D systems and (3) synthesized chitosan-gelatin polypyrrole cryogel incorporated with α-KG containing chitosan or gelatin microspheres. All the experiments were performed in triplicates, and the Student's $t$ test was performed to obtain the $P$ value. $P < 0.05$. (Reproduced from, Vishnoi, T. and Kumar, A., *BioMed. Res. Int.* 2013: 1–11. Copyright (2013), Tanushree Vishnoi and Ashok Kumar.)

## 9.9 CONDUCTING CRYOGEL

The role of bioelectricity and its importance in neural tissue was already discussed in an earlier section of this chapter. Moreover, few electrical devices have already been approved for stimulation like pacemakers, stimulators for spinal cord, vagal nerve and electrodes for stimulating brain (Hardy et al., 2013). However, in order to create a more biological interface with the native tissue, scaffolds with conducting properties have been synthesized. Various composites of conducting polymers with natural/synthetic polymers have been synthesized by different methods. These scaffolds, apart from providing the 3D microenvironment, also allow the localized stimulation of the cells seeded on them (Moghadam et al., 2010). Recently, conducting cryogels have been synthesized at subzero temperature (–12°C) using chitosan, gelatin and polypyrrole as the constituents. Polypyrrole being brittle in nature attains optimum mechanical properties when blended with chitosan and gelatin. It was observed that synthesis via the cryogelation process allowed the use of comparatively lesser amounts of polypyrrole, due to the process of cryoconcentration, for the synthesis of conducting cryogel (Vishnoi and Kumar, 2013) (Figure 9.10).

These scaffolds were characterized for their mechanical, biological and physical characteristics and were further used to stimulate, neuro 2a and C2C12 cells, which are mice neuroblastoma and myoblast cell line, respectively. The seeded cells were allowed to adhere on the scaffolds for a period of 24–72. Later an electrical stimulus of varying voltages for different durations were applied to the seeded cells which was observed to enhance the proliferation of the cells seeded on conducting cryogel as compared to their control. The synthesis of such conducting cryogels allowed controlled electrical stimulation to the cells. Fluorescent and SEM images revealed proper morphology of the seeded cells along with their homogenous distribution on the synthesized scaffolds (Vishnoi and Kumar, 2013). Another group has

**FIGURE 9.10** Digital image of wet (a) chitosan-gelatin cryogel, (b) chitosan gelatin-polypyrrole-conducting cryogel and (c) SEM of conducting cryogel. Polypyrrole could be seen uniformly distributed as a thin film in the synthesized cryogel (magnification ×500). (With kind permission from Springer Science+Business Media: *J. Mater. Sci. Mater. Med.*, Conducting cryogel scaffold as a potential biomaterial for cell stimulation and proliferation, 24, 2013, 447–459, Vishnoi, T. and Kumar, A.)

also reported synthesis of macroporous conducting gels by a freeze-drying method. PEDOT-PSS macroporous gels were synthesized using the ISIA method wherein ice segregation-induced selfassembly of PEDOT-PSS occurred when its precursors were unidirectionally frozen in liquid nitrogen. These gels were further characterized for their physical, chemical and morphological characteristics, which revealed that their synthesis process induced certain additional features like unidirectional longitudinal porous structure in them unlike their hydrogel counterparts. However, their synthesis did not alter the conductivity of the samples (Zhang et al., 2011; Pana et al., 2012).

## 9.10 CONCLUSION

It can be concluded that although various additional techniques have been employed for improving conditions of neural degeneration, there still seems to be a huge void that needs to be fulfilled. Because of their properties, cryogels have proved to be a potential candidate in the field of both soft and hard tissue engineering and have shown recovery both in the *in vitro* and *in vivo* conditions. Incorporating conducting polymers makes them suitable for stimulating excitable cells as well as provides a biological interface to the cells. However, complete recovery still remains a challenge and several attempts have to be made to achieve enhanced regeneration.

## LIST OF ABBREVIATIONS

| | |
|---|---|
| α-KG | alpha-ketoglutarate |
| 3D | Three-dimensional |
| CBSCs | Chord blood stem cells |
| CNS | Central nervous system |
| ECM | Extra-cellular matrix |
| GL | Gelatin-laminin |
| MSCs | Mesenchymal stem cells |
| NGF | Nerve growth factor |
| PEG | Poly(ethylene) glycol |
| PNS | Peripheral nervous system |
| PVA | Poly(vinyl) alcohol |

## ACKNOWLEDGEMENTS

T.V. acknowledges CSIR for her SRF.

## REFERENCES

Abematsu, M., Tsujimura, K., Yamano, M., Saito, M., Kohno, K., Kohyama, J., Namihira, M., Komiya, S., and Nakashima, K. (2010). Neurons derived from transplanted neural stem cells restore disrupted neuronal circuitry in a mouse model of spinal cord injury. *J. Clin. Invest.* 120: 3255–3266.

Antonic, A., Sena, E., Lees, J.S., Wills, T.E., Skeers, P., Batchelor, P.E., Macleod, M.R., and Howells, D. (2013). Stem cell transplantation in traumatic spinal cord injury: A systematic review and meta-analysis of animal studies. *Plos Biol.* 17: 1–14.

Arshak, K., Velusamy, V., Korostynska, O., Oliwa-Stasiak, K., and Adley, C. (2009). Conducting polymers and their applications to biosensors: Emphasizing on foodborne pathogen detection. *IEEE Sensor J.* 9: 1942–1951.

Ateh, D.D., Vadgama, P., and Navsaria, H.A. (2006). Culture of human keratinocytes on polypyrrole-based conducting polymers. *Tissue Eng.* 12: 645–655.

Aurand, E.R., Lampe, K.J., and Bjugstad, K.B. (2012). Defining and designing polymers and hydrogels for neural tissue engineering. *Neurosci. Res.* 72: 199–213.

Babensee, J.E., McIntire, L.V., and Mikos, A.G. (2000). Growth factor delivery for tissue engineering. *Pharm. Res.* 17: 497–504.

Baiguera, S., Gaudio, C.D., Lucatelli, E., Kuevda, E., Boieri, M., Mazzanti, B., Bianco, A., and Macchiarini, P. (2014). Electrospun gelatin scaffolds incorporating rat decellularized brain extracellular matrix for neural tissue engineering. *Biomater.* 35: 1205–1214.

Belkas, J.S., Shoichet, M.S., and Midha, R. (2004). Axonal guidance channels in peripheral nerve regeneration. *Oper. Tech. Orthop.* 14: 190–198.

Bendrea, A.D., Cianga, L., and Cianga, I. (2011). Review paper: Progress in the field of conducting polymers for tissue engineering applications. *J. Biomater. App.* 26: 605–615.

Bhat, S., and Kumar, A. (2012). Biomaterials in regenerative medicine. *J. Postgrad. Med. Edu. Res.* 46: 81–89.

Bhat, S., Lidgren, L., and Kumar, A. (2013). In vitro neo-cartilage formation on a three-dimensional composite polymeric cryogel matrix. *Macromol. Biosci.* 13: 827–837.

Bryant, S.J., and Anseth, K.S. (2002). Hydrogel properties influence ECM production by chondrocytes photoencapsulated in poly(ethyleneglycol) hydrogels. *J. Biomed. Mater. Res.* 59: 63–72.

Carmichael, S.T. (2006). Cellular and molecular mechanisms of neural repair after stroke: Making waves. *Ann. Neurol.* 59: 735–742.

Castillo-Melendez, M., Yawno, T., Jenkin, G., and Miller, S. (2013). Stem cell therapy to protect and repair the developing brain: A review of mechanisms of action of cord blood and amnion epithelial derived cells. *Front. Neurosci.* 7: 1–14.

Chan, B.P., and Leong, K.W. (2008). Scaffolding in tissue engineering: General approaches and tissue-specific considerations. *Eur. Spine J.* 17: 467–479.

Chen, G., Ushida, T., and Tateishi, T. (2002). Scaffold design for tissue engineering. *Macromol. Biosci.* 2: 67–77.

Chen, R.R., and Mooney, D.J. (2003). Polymeric growth factor delivery strategies for tissue engineering. *Pharm. Res.* 20: 1103–1112.

Chew, S.Y., Mi, R., Hoke, A., and Leong, K.W. (2008). The effect of the alignment of electrospun fibrous scaffolds on Schwann cell maturation. *Biomater.* 29: 653–661.

Chiu, D.T., Janecka, I., Krizek, T.J., Wolff, M., and Lovelace, R.E. (1982). Autogenous vein graft as a conduit for nerve regeneration. *Surgery.* 91: 226–233.

Christina, H., Tim, B., and Thomas, S. (2011). Polymers for neural implants. *J. Polym. Sci. Part B: Pol. Phy.* 49: 18–33.

Cho, Y., Pyo, M., and Zong, K. (2010). Chemical and electrochemical synthesis of highly conductive and processable poly pro DOP-alkyl derivatives. *J. Korean Electrochem. Soci.* 13: 57–62.

Chu, K., Kim, M., Park, K.I., Jeong, S.W., Park, H.K., Jung, K.H., Lee, S.T., Kang, L., Lee, K., Park, D.K., Kim, S.U., and Roh, J.K. (2004). Human neural stem cells improve sensorimotor deficits in the adult rat brain with experimental focal ischemia. *Brain Res.* 1016: 145–153.

Collazos-Castro, J.E., Polo, J.L., Hernández-Labrado, G.R., Padial-Cañete, V., and García-Rama. (2010). Bioelectrochemical control of neural cell development on conducting polymers. *Biomater.* 31: 9244–9255.

Cui, W., Zhou, Y., and Chang, J. (2010). Electrospun nanofibrous materials for tissue engineering and drug delivery. *Sci. Technol. Adv. Mater.* 11: 014108.

Dainiak, M.B., Allan, I.U., Savina, I.N., Cornelio, L., James, E.S., James, S.L., Mikhalovsky, S.V., Jungvid, H., and Galaev, I.Y. (2010). Gelatin-fibrinogen cryogel dermal matrices for wound repair: Preparation, optimisation and *in vitro* study. *Biomater.* 31: 67–76.

Daly, W., Yao, L., Zeugolis, D., Windebank, A., and Pandit, A. (2008). A biomaterials approach to peripheral nerve regeneration: Bridging the peripheral nerve gap and enhancing functional recovery. *J. R. Soc. Interface.* 9: 202–221.

deRuiter, G.C., Malessy, M.J.A., Michael, J.Y., Anthony, J.W., and Robert, J.S. (2009). Designing ideal conduits for peripheral nerve repair. *Neurosurg. Focus.* 26: 1–15.

Deal, D.N., Griffin, J.W., and Hogan, M.V. (2012). Nerve conduits for nerve repair or reconstruction. *J. Am. Acad. Orthop. Surg.* 20: 63–68.

Dhandayuthapani, B., Yoshida, Y., Maekawa, T., and Kumar, D.S. (2011). Polymeric scaffolds in tissue engineering application: A review. *Int. J. of Polym. Sci.* 2011: 1–20.

Dionigi, B., and Fauza, D.O. (2012). *Autologous Approaches to Tissue Engineering*, StemBook, Ed. The Stem Cell Research Community, Stem Book, doi/10.3824/stembook.1.90.1, http://www.stembook.org.

Dust, C., and Bawornluck, O. (2006). A study of electrical cell repellent from conductive polymer for biomedical application. *Tech. Meeting on Med. and Biol. Eng., IEE Japan; MBE.* 6: 25–28.

Eric, A.H., and Stephen, M.S. (2009). Axon regeneration in the peripheral and central nervous systems. *Results Probl. Cell. Differ.* 48: 339–351.

Fassina, L., Saino, E., Visai, L., Avanzini, M.A., Cusella De Angelis, M.G., Benazzo, F., Van Vlierberghe, S., Dubruel, P., and Magenes, G. (2010). Use of a gelatin cryogel as biomaterial scaffold in the differentiation process of human bone marrow stromal cells. *Conf. Proc. IEEE Eng. Med. Biol. Soc.* 2010: 247–250.

Fonner, J.M., Forciniti, L., Nguyen, H., Byrne, J.D., Kou, Y.F., Syeda, N.J., and Schmidt, C.E. (2008). Biocompatibility implications of polypyrrole synthesis techniques. *Biomed. Mater.* 3: 1–25.

Freed, I.E., Vunjak-Novakovic, G., Biron, R.J., Eagles, D.B., Lesnoy, D.C., Barlow, S.K., and Langer, R. (1994). Biodegradable polymer scaffolds for tissue engineering. *Nat. Biotechnol.* 12: 689–693.

Fromstein, J.D., and Woodhouse, K.A. (2002). Elastomeric biodegradable polyurethane blends for soft tissue applications. *J. Biomater. Sci. Polym.* 13: 391–406.

Gaillard, A., and Jaber, M. (2011). Rewiring the brain with cell transplantation in Parkinson's disease. *Trends Neurosci.* 34: 124–133.

Galler, K.M., and D'Souza, R.N. (2011). Tissue engineering approaches for regenerative dentistry. *Regen. Med.* 6: 111–124.

George, P.M., Lyckman, A.V., LaVan, D.A., Hegde, A., Leung, Y., Avasare, R., Testa, C., Alexander, P.M., Langer, R., and Sur, M. (2005). Fabrication and biocompatibility of polypyrrole implants suitable for neural prosthetics. *Biomater.* 26: 3511–3519.

Gillingwater, T.H., and Ribchester, R.R. (2001). Compartmental neurodegeneration and synaptic plasticity in the Wld$^S$ mutant mouse. *J. Phys.* 534: 627–639.

Glasby, M.A. (1991). Interposed muscle grafts in nerve repair in the hand: An experimental basis for future clinical use. *World J. of Sci.* 15: 501–510.

Goh, J.H., Hsi-Chin, W., Ming-Hong, C., Ming-Yi, C., Shun-Chih, C., and Tzu-Wei, W. (2013). Control of three-dimensional substrate stiffness to manipulate mesenchymal stem cell fate toward neuronal or glial lineages. *Acta Biomater.* 9: 5170–5180.

Gordon, M.C., and Khan, W. (2012). Recent advances and developments in neural repair and regeneration for hand surgery. *Open Orthop. J.* 6: 103–107.

Green, R.A., Lovell, N.H., and Poole, W.L.A. (2010). Impact of co-incorporating laminin peptide dopants and neurotrophic growth factors on conducting polymer properties. *Acta Biomater.* 6: 63–71.

Guanttilake, P.A., and Adhikari, R. (2003). Biodegradable synthetic polymers for tissue engineering. *Eur. Cells and Mater.* 5: 1–16.

Guan, J., Fujimoto, K.L., Sacks, M.S., and Wagner, W.R. (2005). Preparation and characterization of highly porous, biodegradable polyurethane scaffolds for soft tissue applications. *Biomater.* 26: 3961–3971.

Guimard, N.K., Gomezb, N., and Schmidt, C.E. (2007). Conducting polymers in biomedical engineering. *Prog. Polym. Sci.* 32: 876–921.

Hall, S. (1997). Axonal regeneration through acellular muscle grafts. *J. Anat. Jan.* 190. 57–71.

Hall, S. (2005). The response to injury in the peripheral nervous system. 87: 1309–1319.

Hang, L., Asanka, W., and Leipzig, N.D. (2012). 3D Differentiation of neural stem cells in macroporous: Photopolymerizable hydrogel scaffolds. *PLoS One.* 7: e48824.

Haninec, P., Sámal, F., Tomás, R., Houstava, L., and Dubovvý, P. (2007). Direct repair (nerve grafting), neurotization, and end-to-side neurorrhaphy in the treatment of brachial plexus injury *J. Neurosurg.* 106: 391–399.

Hardy, J.H., Lee, J.Y., and Schmidt, C.E. (2013). Biomimetic conducting polymer-based tissue scaffolds. *Curr. Opin. in Biotech.* 24: 847–854.

He, W., and Bellamkonda, R.V. (2008). A molecular perspective on understanding and modulating the performance of chronic central nervous system recording electrodes. In: *Indwelling Neural Implants: Strategies for Contending with the In Vivo Environment*, Reichert, W.M., Ed. Boca Raton, FL: CRC Press, Taylor & Francis Group, pp. 151–176.

Ho, M.H., Kuo, P.Y., Hsieh, H.J., Hsien, T.Y., Hou, L.T., Lai, J.Y., and Wang, D.M. (2004). Preparation of porous scaffolds by using freeze-extraction and freeze-gelation methods. *Biomater.* 25: 129–138.

Hsiang, S.W., Tsai, C.C., Tsai, F.J., Ho, T.Y., Yao, C.H., and Chen, Y.S. (2011). Novel use of biodegradable casein conduits for guided peripheral nerve regeneration. *J. R. Soc. Interface.* 8: 1622–1634.

Huang, Y.C., and Huang, Y.Y. (2006). Tissue engineering for nerve repair. *Biomed. Eng. Appl. Basis Commun.* 18: 100–110.

Hunziker, E.B. (2002). Articular cartilage repair: Basic science and clinical progress. A review of the current status and prospects. *Osteoarthr. Cartilage.* 10: 432–463.

Hussain, A.M.P., and Kumar, A. (2003). Electrochemical synthesis and characterization of chloride doped polyaniline. *Mater. Sci.* 26: 329–334.

Ichim, T.E., Alexandrescu, D.T., Solano, F., Lara, F., Campion, R.N., Paris, E., Woods, E.J., Murphy, M.P., Dasanu, C.A., Patel, A.N., Marleau, A.M., Leal, A., and Riordan, N.H. (2010). Mesenchymal stem cells as anti-inflammatories: Implications for treatment of Duchenne muscular dystrophy. *Cell Immunol.* 260: 75–82.

Jain, E., and Kumar, A. (2013). Disposable polymeric cryogel bioreactor matrix for therapeutic protein production. *Nat. Protocols.* 8: 821–835.

Jurga, M., Dainiak, M.B., Sarnowska, A., Jablonska, A., Tripathi, A., Plieva, F.M., Savina, I.N., Strojek, L., Jungvid, H., Kumar, A., Lukomska, B., Domanska-Janik, K., Forraz, N., and McGuckin, C.P. (2011). The performance of laminin-containing cryogel scaffolds in neural tissue regeneration. *Biomater.* 32: 3423–3434.

Kai, D., Prabhakaran, M.P., Jin, G., and Ramakrishna, S. (2011). Polypyrrole-contained electrospun conductive nanofibrous membranes for cardiac tissue engineering. *J. Biomed. Mater. Res. A.* 99: 376–385.

Kannol, H. (2013). Regenerative therapy for neuronal diseases with transplantation of somatic stem cells. *World J Stem Cells.* 5: 163–171.

Kathuria, N., Tripathi, A., Kar, K.K., and Kumar, A. (2009). Synthesis and characterization of elastic and macroporous chitosan–gelatin cryogels for tissue engineering. *Acta Biomater.* 5: 406–418.

Ketie, S., Leo, H.K., and Menno, L.W.K. (2010). Polymeric microspheres for medical applications. *Materials.* 3: 3537–3564.

Kim, H.A., Mindos, T., and Parkinson, D.B. (2013). Plastic fantastic: Schwann cells and repair of the peripheral nervous system. *Stem Cells Transl. Med.* 2: 553–557.

Kimura, K., Yanagida, Y., Haruyama, T., Kobatake, E., and Aizawa, M. (1998). Gene expression in the electrically stimulated differentiation of PC12 cells. *J. Biotechnol.* 63: 55–65.

Kumar, A. (2008). Designing new supermacroporous cryogel materials for bioengineering applications. *Eur. Cells and Mater.* 16: 11.

Kvist, M., Sondell, M., Kanje, M., and Dahlin, L.B. (2011). Regeneration in, and properties of, extracted peripheral nerve allografts and xenografts. *J. Plast. Surg. Hand Surg.* 45: 122–128.

Laleh, G.M., Molamma, P.P., Mohammad, M., Mohammad, H.N.E., Hossein, B., Sahar, K., Salem, S.D., and Seeram, R. (2011). Application of conductive polymers, scaffolds and electrical stimulation for nerve tissue engineering. *J. Tissue Eng. Regen. Med.* 5: e17–e35.

Lee, J.W., Serna, F., and Schmidt, C.E. (2006). Carboxy-endcapped conductive polypyrrole: Biomimetic conducting polymer for cell scaffolds and electrodes. *Langmuir.* 22: 9816–9819.

Lee, Y.H., Chung, M.S., Gong, H.S., Chung, J.Y., Park, J.H., and Baek, G.H. (2008). Sural nerve autografts for high radial nerve injury with nine centimeter or greater defects. *J. Hand Surg. Am.* 33: 83–86.

Lee, K., Silva, E.A., and Mooney, D.J. (2011). Growth factor delivery-based tissue engineering: General approaches and a review of recent developments. *J. R. Soc. Interface.* 8: 153–170.

Lee, Y.S., and Treena, L.A. (2011). Electrospun nanofibrous materials for neural tissue engineering. *Polymers.* 3: 413–426.

Lee, J., Abdeen, A.A., Zhang, D., and Kilian, K.A. (2013). Directing stem cell fate on hydrogel substrates by controlling cell geometry, matrix mechanics and adhesion ligand composition. *Biomaterials.* 34: 8140–8148.

Lee, Y.S., and T.L. Arinzeh. (2011). Electrospun nanofibrous materials for neural tissue engineering. *Polymers.* 3: 413–426.

Li, M., Mondrinos, M.J., Chen, X., and Lelkes, P.I. (2005). Electrospun blends of natural and synthetic polymers as scaffolds for tissue engineering. *Eng. Med. Biol. Soc.* 6: 5858–5861.

Liezhen, Fu., Takashi, H., Atsuko Ishizuya, O., and Yun-BoShi, E. (2007). Roles of matrix metalloproteinases and ecm remodeling during thyroid hormone-dependent intestinal metamorphosis in *Xenopuslaevis*. *Organogenesis.* 8: 181–194.

Lin, M.Y., Manzano, G., and Gupta, R. (2013). Nerve allografts and conduits in peripheral nerve repair. *Hand Clin.* 29: 331–348.

Liu, H., Wen, W., Hu, M., Bi, W., Chen, L., Liu, S., Chen, P., and Tan, X. (2013). Chitosan conduits combined with nerve growth factor microspheres repair facial nerve defects. *Neural Regen. Res.* 8: 3139–3147.

Lozinsky, V.I., Galaev, I.Y., Plieva, F.M., Savina, I.N., Jungvid, H., and Mattiasson, B. (2003). Polymeric cryogels as promising materials of biotechnological interest. *Trends Biotechnol.* 21: 445–451.

Ma, Y., Tang, C., Chaly, T., Greene, P., Breeze, R., Fahn, S., Freed, C., Dhawan, V., and Eidelberg, D. (2010). Dopamine cell implantation in Parkinson's disease: Long-term clinical and 18F-FDOPA PET outcomes. *J. Nucl. Med.* 51: 7–15.

Mano, J.F., Silva, G.A., Azevedo, H.S., Malafaya, P.B., Sousa, R.A., Silva, S.S., Boesel, L.F., Oliveira, J.M., Santos, T.C., Marques, A.P., Neves, N.M., and Reis, R.L. (2007). Natural origin biodegradable systems in tissue engineering and regenerative medicine: Present status and some moving trends. *J. R. Soc. Interface.* 4: 999–1030.

Masaeli, E., Morshed, M., Nasr-Esfahani, M.H., Sadri,S., Hilderink, J., Apeldoorn, A.V., Blitterswijk, C.A., and Moroni, L. (2013). Fabrication, characterization and cellular compatibility of poly(hydroxy alkanoate) composite nanofibrous scaffolds for nerve tissue engineering. *PLoS One.* 8: e57157.

Melanie, G., Stephen, C.J.B., Heather, A.D., Alison, J.L., Jonathan, P.G., and James, B.P. (2013). Engineered neural tissue for peripheral nerve repair. *Biomater.* 34: 7335–7343.

Mihardja, S., Sievers, R., and Lee, R.J. (2008). The effect of polypyrrole on arteriogenesis in an acute rat infarct model. *Biomater.* 29: 4205–4210.

Millesi, H., Meissl, G., and Berger. A. (1976). Further experience with interfascicular grafting of the median, ulnar, and radial nerves. *J. Bone Joint Surg. Am.* 58: 209–218.

Moghadam, P.N., and Zareh, E.N. (2010). Synthesis of conductive nanocomposites based on polyaniline/poly(styrene-alt-maleic anhydride)/polystyrene. *e-Polymers.* 10: 588–596.

Moore, A.M,. Kasukurthi, R., Magill, C.K., Farhadi, H.F. Borschel, G.H., and Mackinnnon, S.E. (2009). Limitations of conduits in peripheral nerve repairs. *Hand.* 4: 180–186.

Müller, H.W., Sendtner, M., and Bähr, M. (2012). Molecular basis of neural repair mechanisms. *Cell Tissue Res.* 349: 1–4.

Nair, L.S., and Laurencin, C.T. (2006). Polymers as biomaterials for tissue engineering and controlled drug delivery. *Adv. Biochem. Eng. Biotechnol.* 102: 47–90.

Nakamura, M., and Okano, H. (2013). Cell transplantation therapies for spinal cord injury focusing on induced pluripotent stem cells. *Cell Res.* 23: 70–80.

Nedergaard, M., Takano, T., and Hansen, A.J. (2002). Beyond the role of glutamate as a neurotransmitter. *Nat. Rev. Neurosci.* 3: 748–755.

Nicodemus, G.D., and Bryant, S.J. (2008). Cell encapsulation in biodegradable hydrogels for tissue engineering applications. *Tissue Eng. Part B Rev.* 14: 149–165.

Nicolas, G., Helen, B., Robin, J.M.F., and Nick, D.J. (2012). Autologous olfactory mucosal cell transplantsin clinical spinal cord injury: A randomizeddouble-blinded trial in a canine translational model. *Brain.* 135: 3227–3237.

O'Halloran, D.M., and Pandit, A.S. (2007). Tissue-engineering approach to regenerating the intervertebral disc. *Tissue Eng.* 13: 1927–1954.

Onoda, M., Abe, Y., and Tada, K. (2010). Experimental study of culture for mouse fibroblast used conductive polymer films. *Thin Solid Films.* 519: 1230–1234.

Ornelas, L., Padilla, L., Di Silvio, M., Schalch, P., Esperante, S., Infante, R.L., Bustamante, J.C., Avalos, P., Varela, D., and López, M. (2006). Fibrin glue: An alternative technique for nerve coaptation—Part II. Nerve regeneration and histomorphometric assessment. *J. Reconstr. Microsurg.* 22: 123–128.

Pana, L., Yu, G., Zhaia, D., Lee, H.R., Zhao, W., Liu, N., Wang, H., Tee, B., Shi, Y., Cui, Y., and Bao, Z. (2012). Hierarchical nanostructured conducting polymer hydrogel with high electrochemical activity. *Proc. Natl. Acad. Sci. USA.* 109: 9287–9292.

Papadimitriou, E., Waters, C.R., Manolopoulos, V.G., Unsworth, B.R., Maragoudakis, M.E., and Lelkes, P.L. (2001). Regulation of extracellular matrix remodeling and MMP-2 activation in cultured rat adrenal medullary endothelial cells. *Endothelium.* 8: 181–194.

Park, J.H., Ye, M., and Park, K. (2005). Biodegradable polymers for microencapsulation of drugs. *Molecules.* 10: 146–161.

Park, J.S., Chu, J.S., Tsou, A.D., Diop, R., Tang, Z., and Wang, A.L. (2011). The effect of matrix stiffness on the differentiation of mesenchymal stem cells in response to TGF-β. *Biomater.* 39: 3921–3930.

Pfister, B.J., Tessa, G., Joseph, R.L., Arshneel, S.K., Susan, E.M., and Cullen, D.K. (2011). Biomedical engineering strategies for peripheral nerve repair: Surgical applications, state of the art, and future challenges. *Crit. Rev. Biomed. Eng.* 39: 81–124.

Piltti, K.M., Salazar, D.L., Uchida, N., Cummings, B.J., and Anderson, A. (2013). Safety of human neural stem cell transplantation in chronic spinal cord injury. *J. Stem Cells Transl. Med.* 2: 961–974.

Place, E.S., George, J.H., Williams, C.K., and Stevens, M.M. (2009). Synthetic polymer scaffolds for tissue engineering. *Chem. Soc. Rev.* 38: 1139–1151.

Potts, M.B., Silvestrini, M.T., and Lim, D.A. (2013). Devices for cell transplantation into the central nervous system: Design considerations and emerging technologies. *Surg. Neurol. Int.* 4: 22–30.

Prabhakaran, M.P., Venugopal, J.R., and Ramakrishna, S. (2009). Mesenchymal stem cell differentiation to neuronal cells on electrospun nanofibrous substrates for nerve tissue engineering. *Biomater.* 30: 4996–5003.

Ray, S.K., Dixon, C.E., and Banik, N.L. (2002). Molecular mechanisms in the pathogenesis of traumatic brain injury. *Histol. Histopathol.* 17: 1137–1152.

Riano-Velez, J.A., and Echeverri-Cuartas, C.E. (2011). PLLA scaffold fabrication using salt leaching/gas foaming method. *Health Care Exchanges (PAHCE).* 2013: 2327–8161.

Ribeiro-Resende, V.T., Brigitte, K., Susanne, N., Sven, O., and Burkhard, S. (2009). Strategies for inducing the formation of bands of Büngner in peripheral nerve regeneration. *Biomater.* 30: 5251–5259.

Richardson, T.P., Peters, M.C., Ennett, A.B., and Mooney, D.J. (2001). Polymeric system for dual factor delivery. *Nat. Biotechnol.* 19: 1029–1034.

Riess, P., Molcanyi, M., Bentz, K., Maegele, M., Simanski, C., Carlitscheck, C., Schneider, A., Hescheler, J., Bouillon, B., Schäfer, U., and Neugebauer, E. (2007). Embryonic stem cell transplantation after experimental traumatic brain injury dramatically improves neurological outcome, but may cause tumors. *J. Neurotrauma.* 24: 216–225.

Rimbu, G.A., Stamatin, I., Jackson, C.L., and Scott, K. (2006). The morphology control of polyaniline as conducting polymer in fuel cell technology. *J. Optoelectron. Adv. M.* 8: 670–674.

Rivers, T.J., Hudson, T.W., and Schmidt, C.E. (2002). Synthesis of a novel biodegradable electrically conducting polymer for biomedical applications. *Adv. Funct. Mater.* 12: 33–37.

Rupprecht, R., Papadopoulos, V., Rammes, G., Baghai, T.C., Fan, J., Akula, N., Groyer, G., Adams, D., and Schumacher, M. (2010). Translocator protein (18 kDa) (TSPO) as a therapeutic target for neurological and psychiatric disorders. *Nat. Rev. Drug Discov.* 9: 971–988.

Salgado, A.J., Oliveira, J.M., Martins, A., Teixeira, F.G., Silva, N.A., Neves, N.M., Sousa, N., and Reis, R.L. (2013). Tissue engineering and regenerative medicine: Past, present, and future. *Intl. Rev. Neurobiol.* 108: 1–33.

Sanghvi, A.B., Miller, K.P., Belcher, A.M., and Schmidt, C.E. (2005). Functionalization using a novel peptide that selectively binds to a conducting polymer biomaterials. *Nat. Mater.* 4: 496–502.

Sarnowska, A., Jablonska, A., Jurga, M., Dainiak, M., Strojek, L., Drela, K., Wright, K., Tripathi, A., Kumar, A., Jungvid, H., Lukomska, B., Forraz, N., McGuckin, C., and Domanska-Janik, K. (2013). Encapsulation of mesenchymal stem cells by bioscaffoldsprotects cell survival and attenuates neuroinflammatory reaction in injured brain tissue after transplantation. *Cell Transplant.* 22: 67–82.

Scheib, J., and Hoke, A. (2013). Advances in peripheral nerve regeneration. *Nat. Rev. Neurol.* 9: 668–676.

Schmidt, C.E., and Leach J.B. (2003). Neural tissue engineering: Strategies for repair and regeneration. *Annu. Rev. Biomed. Eng.* 5: 293–347.

Shoichet, M.S. (2010). Polymer scaffolds for biomaterials applications. *Macromolecules.* 43: 581–591.

Schulz, D.J., Baines, R.R., Hempel, C.M., Li, L., Liss, B., and Misonou, H. (2006). Cellular excitability and the regulation of functional neuronal identity: From gene expression to neuromodulation. *J. Neurosci.* 26: 10362–10367.

Shank, R.P., and Campbell, G. (1982). Glutamine and alphaketoglutarate uptake and metabolism by nerve terminal enriched material from mouse cerebellum. *Neurochem. Res.* 7: 601–616.

Shi, G., Zhang, Z., and Rouabhia, M. (2008). The regulation of cell functions electrically using biodegradable polypyrrole–polylactide conductors. *Biomater.* 29: 3792–3798.

Shibashi, S., Sakaguchi, M., Kuroiwa, T., Yamasaki, M., Kanemura, Y., Shizuko, I., Shimazaki, T., Onodera, M., Okano, H., and Mizusawa, H. (2004). Human neural stem/progenitor cells, expanded in long-term neurosphere culturepromote functional recovery after focal ischemia in mongolian gerbils. *J. Neurosci. Res.* 78: 215–223.

Siegel, G.J., and Chauhan, N.B. (2000). Neurotrophic factors in Alzheimer's and Parkinson's disease brain. *Brain Res.Rev.* 33: 199–227.

Silva, A.K.A., Richard, C., Bessodes, M., Scherman, D., and Merten, O.W. (2009). Growth factor delivery approaches in hydrogels. *Biomacromolecules.* 10: 9–18.

Singh, D., Nayak, V., and Kumar, A. (2010). Proliferation of myoblast skeletal cells on three-dimensional super-macroporous cryogels. *Int. J. Biol. Sci.* 6: 371–381.

Slutsky, D.J. (2005). A practical approach to nerve grafting in the upper extremity. *Atlas Hand Clin.* 10: 73–92.

Smith, I.O., Liu, X.H., Smith, L.A., and Ma, P.X. (2009). Nanostructured polymer scaffolds for tissue engineering and regenerative medicine. *Wiley Interdiscip. Rev. Nanomed. Nanobiotechnol.* 1: 226–236.

Spector, J.G., Lee, P., and Derby, A. (2000). Rabbit facial nerve regeneration in autologous nerve grafts after antecedent injury. *Laryngoscope.* 110: 660–667.

Stauffer, W.R., and Cui, X.T. (2006). Polypyrrole doped with two peptide sequences from laminin. *Biomater.* 27: 2405–2413.

Stephanie, M.W. (2011). Neural tissue engineering using embryonic and induced pluripotent stem cells. *Stem Cell Res. Ther.* 2: 1–17.

Sung, J.H., Hwang, M.R., Kim, J.O., Lee, J.H., Kim, Y.I., Kim, J.H., Chang, S.W., Jin, S.G., Kim, J.A., Lyoo, W.S., Han, S.S., Ku, S.K., Yong, C.S., and Choi, H.G. (2010). Gel characterization and *in vivo* evaluation of minocycline-loaded wound dressing with enhanced wound healing using polyvinyl alcohol and chitosan. *Int. J. Pharm.* 392: 232–240.

Svirskis, D., Travas-Sejdic, J., Rodgers, A., and Garg, S. (2010). Electrochemically controlled drug delivery based on intrinsically conducting polymers. *J. Control Release.* 146: 6–15.

Svenningsen, A., and Dahlin, L. (2013). Repair of the peripheral nerve—Remyelination that works. *Brain Sci.* 3: 1182–1197.

Szynkaruk, M., Kemp, S.W., Wood, M.D., Gordon, T., and Borschel, G.H. (2013). Experimental and clinical evidence for use of decellularized nerve allografts in peripheral nerve gap reconstruction. *Tissue Eng. Part B Rev.* 19: 83–96.

Takashima, Y., Daniels, R.L., Knowlton, W., Teng, J., Liman, E.R., and McKemy, D.D. (2007). Diversity in the neural circuitry of cold sensing revealed by genetic axonal labeling of transient receptor potential melastatin 8 neurons. *J.Neurosci.* 27: 14147–14157.

Taylor, S.J., McDonald, J.W., and Sakiyama-Elbert S.E. (2004). Controlled release of neurotrophin-3 from fibrin gels for spinal cord injury. *J. Control Release.* 98: 281–294.

Theodoropoulos, J.S., De Croos, J.N., Park, S.S., Pilliar, R., and Kandel, R.A. (2011). Integration of tissue-engineered cartilage with host cartilage: An *in vitro* model. *Clin. Orthop. Relat. Res.* 469: 2785–2795.

Thompson, B.C., Moulton, S.E., Ding, J., Richardson, R., Cameron, A., O'Leary, S., Wallace, G.G., and Clark, G.M. (2006). Optimising the incorporation and release of a neurotrophic factor using conducting polypyrrole. *J. Control Release.* 116: 285–294.

Thompson, B.C., Richardson, R.T., Moulton, S.E., Evans, A.J., O'Leary, S., Clark, G.M., and Wallace, G.G. (2010). Conducting polymers, dual neurotrophins and pulsed electrical stimulation–dramatic effects on neurite out-growth. *J. Control Release.* 141: 161–167.

Tonya, B., Raphael, G., Marcel, D., and Gary, K.S. (2007). Cell transplantation therapy for stroke. *Stroke.* 38: 817–826.

Van, T.S.R., Storm, G., and Hennink, W.E. (2008). *In situ* gelling hydrogels for pharmaceutical and biomedical applications. *Int. J. Pharm.* 355: 1–18.

Van Velthoven, C.T., Sheldon, R.A., Kavelaars, A., Derugin, N., Vexler, Z.S., Willemen, H.L., Maas, M., Heijnen, C.J., and Ferriero, D.M. (2013). Mesenchymal stem cell transplantation attenuates brain injury after neonatal stroke. *Stroke.* 44: 1426–1432.

Vanderhooft, J.L., Alcoutlabi, M., Magda, J.J., and Prestwich, G.D. (2009). Rheological properties of cross-linked hyaluronan-gelatin hydrogels for tissue engineering. *Macromol. Biosci.* 9: 20–28.

Vishnoi, T., and Kumar, A. (2013). Comparative study of various delivery methods for the supply of alpha-ketoglutarate to the neural cells for tissue engineering. *BioMed. Res. Int.* 2013: 1–11.

Vishnoi, T., and Kumar, A. (2013). Conducting cryogel scaffold as a potential biomaterial for cell stimulation and proliferation. *J. Mater. Sci. Mater. Med.* 24: 447–459.

Walton, R.L., Brown, R.E., Matory, W.E. Jr., Borah, G.L., and Dolph, J.L. (1989). Autogenous vein graft repair of digital nerve defects in the finger: A retrospective clinical study. *Plast. Reconstr. Surg.* 84: 944–949.

Wan, Y., Yu, A., Wu, H., Wang, Z., and Wen, D. (2005). Porous-conductive chitosan scaffolds for tissue engineering II. in vitro and in vivo degradation. *J. Mater. Sci. Mater. Med.* 16: 1017–1028.

Wang, S., Yaszemski, M.J., Knight, A.M., Gruetzmacher, J.A., Windebank, A.J., and Lu, L. (2009). Photo-cross-linked poly($\varepsilon$-caprolactone fumarate) networks for guided peripheral nerve regeneration: Material properties and preliminary biological evaluations. *Acta Biomater.* 5: 1531–1542.

Weigel, T., Schinkel, G., and Lendlein, A. (2006). Design and preparation of polymeric scaffolds for tissue engineering. *Expert Rev. Med. Devices.* 3: 835–851.

Willerth, S.M. (2011). Neural tissue engineering using embryonic and induced pluripotent stem cells. *Stem Cell Res. Ther.* 2: 1–17.

Wolford, L.M., and Stevao, E.L.L. (2003). Considerations in nerve repair. *Proc. (Bayl. Univ. Med. Cent.).* 16: 152–156.

Xie, J., MacEwan, M.R., Schwartz, A.G., and Xia, Y. (2010). Electrospun nanofibers for neural tissue engineering. *Nanoscale.* 2: 35–44.

Yao, D., Dong, S., Lu, Q., Hu, X., Kaplan, D.L., Zhang, B., and Zhu, H. (2012). Salt-leached silk scaffolds with tunable mechanical properties. *Biomacromolecules* 13: 3723–3729.

Zhang, X., Li, C., and Luo, Y. (2011). Aligned/unaligned conducting polymer cryogels with three-dimensional macroporous architectures from ice-segregation-induced self-assembly of PEDOT-PSS. *Langmuir.* 27: 1915–1923.

Zhang, R., Liu, Y., Yan, K., Chen, L., Chen, X.R., Li, Chen, F.F., and Jiang, X.D. (2013). Anti-inflammatory and immunomodulatory mechanisms of mesenchymal stem cell transplantation in experimental traumatic brain injury. *J. Neuroinflam.* 10: 1–12.

Zhao, M., Forrester, J.V., and McCaig, C.D. (1999). A small, physiological electric field orients cell division. *Proc. Natl. Acad. Sci.* 96: 4942–4946.

Zhao, Z., Wang, Y., Peng, J., Ren, Z., Zhang, L., Guo, Q., Xu, W., and Lu, S. (2014). Improvement in nerve regeneration through a decellularized nerve graft by supplementation with bone marrow stromal cells in fibrin. *Cell Transplant.* 23: 97–110.

Zhou, D., Shen, S., Yun, J., Yao, K., and Lin, D.Q. (2012). Cryo-copolymerization preparation of dextran-hyaluronate based supermacroporous cryogel scaffolds for tissue engineering applications. *Front. Chem. Sci. Eng.* 6: 339–347.

# 10 Supermacroporous Cryogels as Scaffolds for Pancreatic Islet Transplantation

Y. Petrenko, A. Petrenko, P. Vardi and K. Bloch*

## CONTENTS

10.1 Introduction ................................................................................................278
10.2 Cryogels as Scaffolds for Implanted Pancreatic Islets................................280
      10.2.1 Induction of Neovascularisation in the Diabetic Environment ........280
      10.2.2 Making 3D-House for Implanted Insulin-Producing Cells..............282
10.3 The Development of Ready-to-Use 3D Cryogel/Stem Cell Constructs
      Supporting Pancreatic Islet Function............................................................286
      10.3.1 Mesenchymal Stem/Stromal Cells (MSCs) for Pancreatic Islet
            Support...............................................................................................286
      10.3.2 Alginate-Based Macroporous Cryogel Scaffolds
            for the Development of 3D Pancreatic Islet Support System............289
      10.3.3 Cryopreservation of MSCs Seeded Macroporous Alginate-
            Based Cryogel Scaffolds....................................................................294
10.4 Concluding Remarks ...................................................................................295
Acknowledgements................................................................................................296
List of Abbreviations..............................................................................................296
References..............................................................................................................296

## ABSTRACT

Clinical **islet transplantation** is a promising approach in treating Type 1 diabetes and a subgroup of patients with Type 2 diabetes requiring insulin administration. Currently, there are several limitations for long-term engraftment of functional islets: immune rejection, inadequate oxygen and nutrient supply, as well as inefficient metabolic waste removal from grafted endocrine tissue. To overcome such limitations, a tissue engineering approach based on utilization of three-dimensional (3D) supermacroporous scaffolds was developed. The main aim of using 3D scaffolds is to supply the grafted islets with a matrix that can promote local neovascularisation and support long-term survival and

---

* Corresponding author: Email: kbloch@post.tau.ac.il.

functional activity of the cells. Among the many approaches to construct such a bioengineered tissue construct, alginate- and agarose-based supermacroporous cryogels, manufactured by cryotropic gelation, showed several advantages. The unique interconnected pore structure of these biocompatible cryogels, in combination with their osmotic, chemical and mechanical stability, renders them as attractive matrices for the immobilization of biomolecules, hence facilitating cell attachment and enhancement of neovascularisation. In addition, cryogels are able to serve as a scaffold to support cell populations, such as mesenchymal stem/stromal cells (MSCs). These cell populations are of crucial importance to the field of tissue transplantation as they have been shown to attenuate immune rejection and secrete angiogenic, anti-inflammatory and anti-apoptotic factors. In addition, the unique properties of the supermacroporous cryogels enable the use of cryopreservation technologies to be used as an 'off-the-shelf' ready-to-use bio-artificial supporting system for further pancreatic islet transplantations.

## KEYWORDS

Cryogel, diabetes, islet transplantation, pancreas

## 10.1 INTRODUCTION

Transplantation of pancreatic islets is a promising approach for reversal of insulin dependency in Type 1 diabetes patients and in a subgroup of patients with Type 2 diabetes treated with insulin. However, there are several requirements for long-term successful engraftment of pancreatic islets: protection from immune rejection, adequate oxygen and nutrient supply, as well as efficient metabolic waste removal from grafted cells (Pepper et al., 2013; Zinger and Leibowitz, 2014). In addition, transplantation of pancreatic islets in several ectopic sites (e.g. subcutaneous, intramuscular, intraomental) requires a 3D scaffold to enable physical cell anchoring, protection, neovascularisation and retrievability of grafted islets.

Both allogeneic and xenogeneic implanted islets must be protected from immune rejection either by administration of general immunosuppressive drugs or by selective and local immunoisolation of the cells as in the bio-artificial pancreas (BAP) device (Colton, 1995). The second alternative, based on physical isolation of the grafted cells from the host defence system, is preferable as it does not cause severe systemic side effects typically induced by immunosuppressive drugs. The concept of BAP is to allow the long-term survival and functional activity of implanted pancreatic islets using various semipermeable materials such as hydrogels or membranes that prevent the penetration of antibodies and the immune system cells into the polymeric barrier. In addition, such materials should provide a supportive physical skeleton for the implanted cells as well as permeability for glucose and insulin, indispensable for an adequate metabolic response. Such a structure also should promote neovascularisation at the implantation site, permitting adequate oxygen and nutrient supply, as well as efficient metabolic waste removal (O'Sullivan et al., 2011).

One of the first prototypes of BAP composed of pancreatic islets microencapsulated in alginate hydrogel coated with poly-l-lysine was developed more than 30 years ago. These hydrogels are impermeable for immunoglobulins and immunocompetent cells, but allow the free diffusion of glucose and insulin. A single intraperitoneal implantation of alginate microencapsulated islets into diabetic rats induced normoglycemia for 2 to 3 weeks (Lim and Sun, 1980). Figure 10.1 shows a representative image of isolated pancreatic islets microencapsulated in an alginate bead.

Over the years, various designs and configurations of BAP have been developed by many research teams worldwide (Colton, 1995; de Vos et al., 2010; O'Sullivan et al., 2011; Lazard et al., 2012). Today, alginate, an anionic polysaccharide produced from seaweed and algae, still remains one of the most popular biopolymers for development of BAP due to its biocompatibility, as well as its mechanical and chemical properties (Vaithilingam and Tuch, 2011; de Vos et al., 2013). Apart from alginate, agarose—another polysaccharide polymer material generally extracted from seaweed—has also been widely used for islet encapsulation studies (de Vos et al., 2010, 2013). The immunoisolation procedure itself impedes the transplanted islets from revascularising, reenervating and remodelling the surrounding tissue and extracellular matrix (ECM). These conditions render the islets chronically dependent on the diffusion of oxygen, nutrition and metabolic waste across a relatively large barrier (Gibly et al., 2013). Recently, it has been shown that transplantation of islets immobilized on porous materials, together with encapsulation technology, has emerged as a promising strategy for the long-term functioning of a 3D islet tissue construct (Blomeier et al., 2006; Salvay et al., 2008; Kheradmand et al., 2011).

The extremely high sensitivity of pancreatic islets to oxygen deficiency leads to islet cell dysfunction and death, believed to constitute the main cause of islet

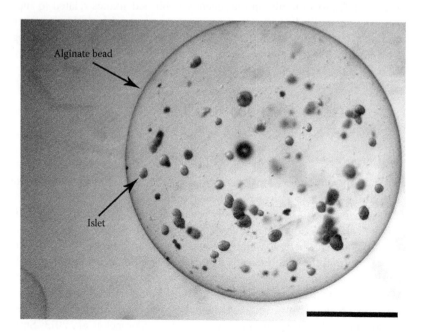

**FIGURE 10.1** Stereomicroscopic image of isolated rat pancreatic islets microencapsulated in alginate bead. Scale bars: 1 mm.

transplantation failure (Dionne et al., 1993). Such failure is induced by several factors: interruption of islet blood vessel network during isolation procedure, insufficient neovascularisation around grafted islets (Pepper et al., 2013), high level of islet oxygen consumption (Cornolti et al., 2009) and markedly decreased oxygen tension in transplanted pancreatic islets irrespective of the implantation site (Carlsson et al., 2001). These data indicate that adequate neovascularisation of the transplanted site is crucial for the survival of both encapsulated and nonencapsulated islets. Such conditions can be more critical in the case of BAP as the delivery of oxygen and nutrients is often limited by diffusion processes that can supply cells at distances only less than 200 μm from the nearest capillary (Rouwkema et al., 2008). A promising approach to overcome such mass transport limitations is to use 3D supermacroporous scaffolds produced from biocompatible polymers grafted with extracellular matrix molecules (Gibly et al., 2013; Pedraza et al., 2013).

Among many approaches trying to provide scaffolds with angiogenic and cytoprotective properties, the immobilization of functional mesenchymal stem/stromal cells (MSCs) on polymeric scaffolds was found to be very promising due to their ability to attenuate immune rejection (De Miguel et al., 2012) as well as to secrete angiogenic and anti-inflammatory factors enabling the long-term survival of engrafted islets (Borg et al., 2014; Scuteri et al., 2014). In addition, a direct role of MSCs on the protection of pancreatic islets from injury induced by hypoxia was recently reported (Lu et al., 2010). The authors showed that MSCs stimulated the expression of cytoprotective genes and enhanced viability and function of islets in the hypoxic environment.

This chapter focuses mainly on the recently published studies related to supermacroporous alginate- and agarose-based cryogels as potential scaffolds for pancreatic islet transplantation. Special attention is given to modifications of the cryogel surface by grafting of extracellular matrix molecules and MSCs to enhance neovascularisation in transplantation site and cytoprotection of implanted islets.

## 10.2 CRYOGELS AS SCAFFOLDS FOR IMPLANTED PANCREATIC ISLETS

### 10.2.1 Induction of Neovascularisation in the Diabetic Environment

Although the pancreas is the islets native home offering a favourable microenvironment with a high oxygen supply and a physiological sensitivity to basic nutrients (in particular, glucose), the pancreas has rarely been considered a potential implantation organ in clinical practice (Merani et al., 2008). Surgical interventions in the pancreas are difficult, and there is a high risk of acute complications due to the leakage of enzymes from the exocrine cells that cause tissue damage and inflammation (Carlsson, 2011). For these reasons, ectopic sites for islet transplantation in diabetic animals such as the liver, omental pouch, subcapsular kidney space, spleen, peritoneal cavity, subcutis, bone marrow, muscles, thymus, spinal fluids and brain have been explored (Rajab, 2010; Cantarelli and Piemonti, 2011; Coronel et al., 2013; Smink et al., 2013; Vériter et al., 2013). For clinical islet transplantation, the liver is usually used as a site for transplantation. However, islet transplantations in this site are hampered by insufficient oxygenation, cell rejection and procedure-related complications. Moreover, irrespective of the implantation sites, insufficient vascularisation of grafted islets is one of the most serious disadvantages in many ectopic sites (Carlsson et al., 2001).

Over the past decades, several approaches have been designed to overcome the severe limitations of ectopic transplantation sites including the deficient oxygen and nutrients supply as well as inadequate metabolic waste removal (Vaithilingam and Tuch, 2011; Lazard et al., 2012). One of the most promising approaches is to employ an angiogenesis-promoting highly porous polymeric scaffold facilitating the immobilization of islets throughout the device. Another important characteristic of such a scaffold is the existence of interconnected pores, which permit vascular infiltration into the interior area of implanted scaffolds and even allow intraislet vascularisation (Pedraza et al., 2013). In addition, the polymeric scaffolds can also present extracellular matrix (ECM) proteins, locally deliver trophic factors or encoding genes or serve as a vehicle for cell cotransplantation to improve islet survival and function (Chen et al., 2007; Salvay et al., 2008; Aviles et al., 2010; Gibly et al., 2013). In some cases, improved survival of immunoisolated islets was achieved by intramuscular scaffold implantation for the preparation of a prevascularised site for the subsequent implantation of a bioartificial pancreas (Balamurugan et al., 2003).

Among various scaffold types for cell transplantation, supermacroporous polymeric cryogels can become the superior material to create the above-outlined optimal microenvironment of an ectopic islet transplantation site. The three-dimensional sponge-like cryogels possess a wide pore morphology with interconnected large pores from tens to hundreds of microns (Lozinsky et al., 2003). A most practical and technological advantage is the ability to use a broad spectrum of various biocompatible biodegradable and nonbiodegradable polymers for cryogel preparation. In our studies, we employed supermacroporous (spongy) agarose-based cryogels prepared by a two-step freezing procedure (freezing at −30°C followed by incubation at a warmer subzero temperature) and subsequent thawing (Lozinsky et al., 2001). These agarose-based scaffolds provide a large surface area that can support intensive vascularisation and a high number of seeded islet cells. Extremely high porosity, with pore diameter of about several hundreds of microns, enables the improved mass transfer characteristics, which are crucial for cell nutrition and oxygenation. The unique structure of cryogels, in combination with their osmotic, chemical and mechanical stability, makes them attractive matrices for the immobilization of biomolecules of ECM, facilitating cell attachment to the surface of large pores (Lozinsky et al., 2003; Bloch et al., 2005; Tripathi et al., 2009). Additional advantages of the cryogel scaffolds made from agarose are their reported biocompatibility and their biodegradable characteristics, which are independent of the activity of reactive oxygen species, a property that could be useful for tissue engineering applications (Shakya et al., 2013).

The intensity of islet vascularisation depends not only on the tissue-specific transplantation site and scaffold characteristics, but also on the metabolic changes in the body induced by diabetes. Numerous aspects of the vascularisation are known to be defective in diabetes (Brem and Tomic-Canic, 2007). In this context, the level of glycaemia is the primary factor to be considered, and analysis of neovascularisation of polymeric implanted scaffolds in various ectopic sites of diabetic animals should be evaluated. In order to clarify this issue, we investigated the effect of glycaemia on neovascularisation of subcutaneously (SC) implanted wide pore agarose cryogels grafted with gelatin, which is known to facilitate cell adhesion (Bloch et al., 2005). Subcutaneous implantation of polymer scaffolds initiated a sequence of events similar to a foreign body reaction, starting with an acute inflammatory response leading, in some cases, to a chronic inflammatory response, fibrous capsule formation and neovascularisation (Anderson et al., 2008). In our study, we demonstrated the formation of a vascularised fibrous capsule surrounding the SC implanted cryogel scaffolds three weeks after transplantation. At this time point, we found strong expression of von Willebrand Factor (vWF) in endothelial cells of newly formed blood vessels and alpha-smooth muscle actin (alpha-SMA), a marker reflecting the mature state of blood vessels invading the scaffolds. Figure 10.2 shows representative images of blood vessel formation around SC implanted cryogels and immunohistochemical staining of blood vessels invading the scaffolds.

Comparative analysis of implanted scaffolds showed similar thicknesses of the fibrous capsules around the scaffolds in both diabetic and nondiabetic animals (Bloch et al., 2010). This observation is important, as a fibrous capsule surrounding the implanted device may act as a barrier to nutrient and oxygen diffusion (Wood et al., 1995). However, our data indicate that cryogel scaffolds made from agarose with grafted gelatin induced the formation of vascular structures that may facilitate mass transfers through the fibrous capsule. In this study, no differences were found in scaffold vascularisation between diabetic animals and controls. In addition, we showed the formation of mature blood vessels with pericytes in the fibrous capsule as well as in the tissue invading the scaffolds in both diabetic and intact mice using alpha-SMA staining (Bloch et al., 2010). Nevertheless, histological analysis of neovascularisation is insufficient to estimate the effect of oxygen and nutrient delivery to the implant if not supported by functional tests, especially in **diabetes**, which is a disease associated with secondary micro and macrovascular complications evolving from defective and excessive blood vessel formation (e.g. retinopathy and glomerular nephropathy) (Cooper et al., 2001). For that reason, functional evaluation is a complementary tool in the estimation of blood supply to SC implanted scaffolds. Part of such an evaluation is the ability of immobilized pancreatic islets in implanted cryogel to achieve normoglycaemia in diabetic animals. Carrying out such research will facilitate the estimation of the expected outcomes of subcutaneously implanted cryogel scaffolds for future human islet transplantation.

### 10.2.2 Making 3D-House for Implanted Insulin-Producing Cells

In healthy individuals, insulin-producing beta cells occupy about 70–80% of the total volume of pancreatic islets. These highly vascularised endocrine organs are also composed of alpha- and delta-cells that produce glucagon and somatostatin, respectively. With a diameter of about 50–250 µm, the islets constitute approximately 1–2% of the total mass of the intact pancreas. In a hypoxic environment, insulin-producing beta cells are almost totally destroyed, while alpha cells survive, but lose the stimulus-specific glucagon response (Bloch et al., 2012). Figure 10.3 shows representative images of intact pancreatic islets and islets exposed to hypoxia.

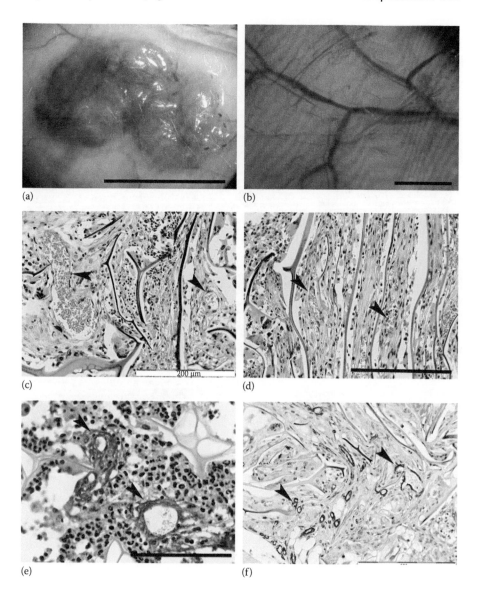

**FIGURE 10.2** Representative images of agarose-gelatin cryogel scaffolds implanted subcutaneously. Stereophotography of implanted cryogel (a) and neovascularisation surrounding implantation side (b). Hematoxylin and eosin staining of tissue invasion into the scaffolds (c and d). vWF and alpha-SMA staining of blood vessels in tissue invading scaffolds. Arrows highlight some stained vessels. Three weeks postimplantation. Scale bars: 1 cm (a); 50 μm (b); 200 μm (c–f).

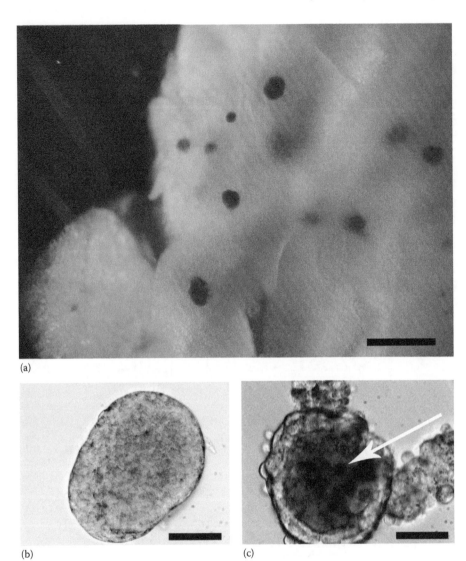

**FIGURE 10.3** Representative images of dithizone-stained islets in intact rat pancreas (a). The isolated rat pancreatic islets cultured in normoxia (b) or in hypoxia (c). Arrow shows sign of necrotic cell death in the central islet area. Scale bars: 300 µm (a); 50 µm (b and c).

Interestingly, transplantation of whole islets or purified beta-cells is known to provide a long-term normoglycaemia in diabetic animals; thus, it is likely that islet non-beta-cells are not essential for successful transplantation (King et al., 2007). Hence, both islets and purified populations of beta-cells isolated from pancreatic islets or derived from stem cells can be utilized for cell therapy of diabetes. Therefore, the scaffolds used for islet transplantation should provide suitable housing for either large cell aggregates or individual cells as well as support cells growing in monolayers.

In addition, such polymeric scaffolds should prevent islet clumping, which leads to functional abnormalities and cell death, and should maintain mechanical stability as well as allowing easy retrieval when cell replacement is necessary.

In our studies we aimed to evaluate a sponge-like morphology of agarose-based cryogels (Figure 10.4a) and to test the feasibility of these supermacroporous gel materials as 3D scaffolds for *in vitro* culturing of islet-like insulinoma cell aggregates (Lozinsky et al., 2008). It was shown that such cryogel sponges prepared from agarose alone displayed very low adhesion properties for anchorage-dependent growth of insulinoma cells. Instead of the typical monolayer architecture (Figure 10.4b), the cells cultured in the 3D cryogel sponges formed clusters, morphologically resembling pancreatic islets. The formation of islet-like structures or pseudo-islets is an inherent property of both purified pancreatic islet cells and insulinoma cells, resulting in homotypic beta-cell communications required for appropriate insulin secretory responses (Beger et al., 1998; Hauge-Evans et al., 1999; Luther et al., 2006). Our results suggest that agarose-based cryogels induce close cell-to-cell contact, improving the functional responsiveness of beta-cells. Thus, pseudo-islets formation may

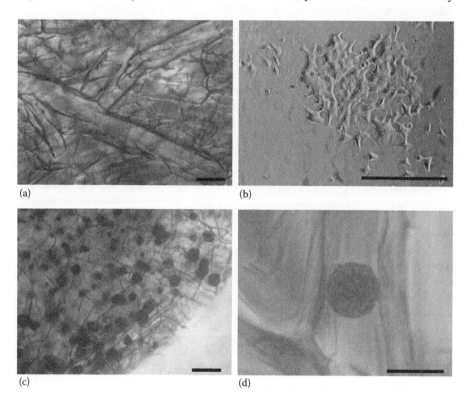

**FIGURE 10.4** Insulin producing insulinoma INS-1 cells and pancreatic islets cultured in agarose-based supermacroporous cryogels. Stereomicroscopic images of agarose-based cryogel scaffold (a). Insulinoma cells grown as a monolayer on surface of agarose-gelatin cryogel (b). Insulinoma cells grown as pseudo-islets in agarose cryogel (c). Rat pancreatic islet cultured in agarose cryogel (d). Two weeks in culture. Scale bars: 200 μm (a, c, d); 50 μm (b).

be a useful research model in the functional study of beta-cells as well as possible tissue source for islet transplantation. Indeed, when the insulinoma cells were seeded in agarose-based cryogels, the cells did not adhere to the agarose surface but, during overnight incubation, formed 3D, multicellular, spherical aggregates (Figure 10.4c).

The viable insulin producing pseudo-islets were about 50–100 μm in diameter, and they did not increase significantly in size during the 10 days in culture, suggesting a cryogel-dependent inhibiting proliferative activity and advanced cell differentiation. In fact, pseudo-islets in the 3D cryogels demonstrated enhanced potassium-dependent insulin secretion in contrast to cells cultured as a monolayer. Interestingly, the stimulation index for insulin response to glucose was significantly higher in pseudo-islets compared to cell monolayer. These results suggest that the formation of pseudo-islets in agarose-based cryogels does not harm cell integrity and improves the functional responsiveness of beta-cells. The large inner surface area of spongy cryogels can support the survival and differentiation of beta-cells cultivated at a high density, probably due to improved mass-transfer characteristics. An additional advantage of the cryogels is their transparency, which enables the microscopic evaluation of the cell morphology and viability (Lozinsky et al., 2008).

In order to compare further the effect of the cryogel scaffold on cell viability and the functioning of insulin producing cells in different respiratory oxygen demands, we used INS-1E, a highly differentiated insulinoma cell line, which has a low oxygen consumption rate, and pancreatic islets, which have a significantly raised respiratory activity. Microscopic analysis showed that in contrast to the agarose cryogel, the cryogel spongy construct prepared from agarose with grafted gelatin was suitable for *in vitro* cultivation of insulinoma cells growing as a monolayer. Insulinoma cell monolayers cultivated in cryogel scaffold for 2 weeks displayed a normal growth rate, insulin secretion and content. In contrast, 2 weeks of *in vitro* cultivation of adult pancreatic islets in the cryogel sponge (Figure 10.4d) resulted in impaired insulin response to glucose and decreased intracellular insulin content, compared to islets cultured in plastic dishes as free-floating cell aggregates (Bloch et al., 2005). Thus, it can be deduced that transport limitations related to oxygenation, nutrient delivery and waste removal within the 3D constructs were responsible for the abnormal functional activity of islets cultured in such *in vitro* static conditions. This obstacle is common for *in vitro* static cultures of cells with a high level of oxygen consumption in 3D scaffolds. In order to overcome such transport limitations *in vitro*, continuous medium perfusion was applied to the 3D tissue construct (Cartmell et al., 2003). *In vivo*, continuous perfusion is achieved by the vascular blood supply to cells immobilized in implanted 3D constructs.

## 10.3 THE DEVELOPMENT OF READY-TO-USE 3D CRYOGEL/STEM CELL CONSTRUCTS SUPPORTING PANCREATIC ISLET FUNCTION

### 10.3.1 Mesenchymal Stem/Stromal Cells (MSCs) for Pancreatic Islet Support

Successful engraftment of allogeneic pancreatic islets or insulin-producing cells may be achieved using an effective immunomodulation strategy along with adequate oxygen and blood supply to the grafted islets (Abdi et al., 2008; Busch et al., 2011; Ellis et al., 2013; Hematti et al., 2013). The search for methods that can provide these conditions is one of the most topical issues in the field of pancreatic islet transplantation. MSCs may be a promising solution to the above-mentioned problems (Busch et al., 2011; Hematti et al., 2013). MSCs are self-renewing, multipotent progenitor cells that can be isolated from many different tissues and have the capacity to differentiate into various lineages (Pittenger et al., 1999; Dominici et al., 2006; Vija et al., 2009). Furthermore, the successful differentiation of MSCs toward endothelial (Planat-Benard et al., 2004; Cao et al., 2005; Zhang et al., 2009), hepatic (Lee et al., 2004; Shu et al., 2004), neuronal (Zhang et al., 2011) and myogenic (Di Rocco et al., 2006) lineages as well as into insulin-producing cells (Xie et al., 2009; Kim et al., 2012) has been shown. The unique multilineage differentiation potential of MSCs opens possibilities for their widespread application in regenerative medicine, especially for the treatment of different degenerative and immune disorders (Parekkadan and Mildwid, 2010; Shi et al., 2010).

MSC transplantation is currently considered a promising approach for therapeutic use in humans (see reviews Parekkadan and Milwid, 2010; Lee et al., 2011; Patel et al., 2013). Two main therapeutic mechanisms of transplanted MSCs are being considered: (1) The differentiation and replacement of damaged cells and (2) the secretion of trophic factors.

Despite intensive study of the localization and distribution of MSCs after systemic administration, in most situations, functional improvements were observed without replacing damaged cells (Iso et al., 2007; Lee et al., 2009; Prockop et al., 2010). This evidence suggests that most of the beneficial effects of MSCs transplantation could be explained by the secretion of trophic factors (Prockop et al., 2010).

Trophic factors secreted by MSCs have multiple effects including (a) the modulation of immune reactions and inflammation, (b) the protection from cell death and (c) the stimulation of tissue repair. It is often difficult to distinguish the difference between these effects. For example, the authors have shown that MSCs in culture secrete a wide range of proinflammatory and anti-inflammatory cytokines and chemokines and inhibit the proliferation of lymphocytes in mixed lymphocyte culture as well as the protection of beta-cells from autoimmune attack, promoting the temporary restoration of glucose regulation (Fiorina et al., 2009).

Initial indications regarding the immunomodulatory effect of MSCs were first reported in the context of MSC transplantation studies in animals and humans. It has been shown that autologous and allogeneic MSCs can be transplanted without immune rejection (Lazarus et al., 1995; Horwitz et al., 1999). Further preclinical studies have demonstrated that MSCs reduce the course of a variety of immune-mediated diseases, including graft rejection, graft-vs.-host disease, collagen-induced arthritis and myelin oligodendrocyte glycoprotein-induced experimental autoimmune encephalomyelitis (Bartholomew et al., 2002; Ryan et al., 2005; Zappia et al., 2005; Augello et al., 2007). The immunomodulatory effect of MSCs is explained by their potential to suppress both lymphocyte proliferation and activation in response to allogeneic antigens. MSCs can induce the development of $CD^{8+}$ regulatory T ($T_{reg}$) cells that in turn can successfully suppress allogeneic lymphocyte responses (Djouad et al., 2003) and prohibit the differentiation of monocytes and

CD34-positive progenitors into antigen presenting dendritic cells (Djouad et al., 2007). MSCs stimulated the T cells arrest in the $G_1$ phase because of cyclin D2 downregulation (Glennie et al., 2005). MSCs are also able to inhibit the proliferation of IL-2 or IL-15 stimulated NK cells (Spaggiari et al., 2006). Besides, MSCs can alter B cell proliferation, activity and chemotactic behaviours as well as reduce the expression of major histocompatibility complex class II (MHC-II), CD40 and CD86 on dendritic cells following maturation induction (Petrie and Tuan, 2010). Among the factors that MSCs produce to suppress immune reactions, prostaglandin E2 (PGE2), indoleamine 2,3-dioxygenase (IDO), nitric oxide, TGF-β1 and human leukocyte antigen G (HLA-G) play important roled. The immunoregulatory abilities of MSCs may be considered a natural function of these cells to maintain the homeostasis of their local microenvironment. It is well known that in bone marrow MSCs participate in haematopoietic stem cells niche organization by secreting specific cytokines and growth factors to support haematopoiesis. MSCs derived from both bone marrow and adipose tissue secrete a large number of cytokines under normal culture conditions (Caplan, 2009; Blaber et al., 2012). At the expansion stage, MSCs secrete various biologically active factors such as G-CSF, M-CSF, IL-6, IL-7, IL-10, IL-11, IL-12, IL-13, SCF, IFN-γ and VEGF. Hsiao et al. (2012) performed a comparative analysis of paracrine factor profiles of MSCs derived from bone marrow, adipose and dermal tissues, and showed that VEGF-A and VEGF-D contribute to the proangiogenic paracrine effect of adipose tissue derived MSCs. The trophic factors secreted by MSCs may not only protect the host cells, but also prolong the therapeutic effects of MSCs supporting survival of different cells or tissues in the pre- or post-transplantation period.

It is important to note that under an external stimulus, paracrine factor secretion is activated and changes its profile. Under normal culture condition, MSCs do not secrete inflammatory factors. Of special interest is TNF-alpha induced protein 6 (TSG-6), which manifests multipotent anti-inflammatory effects: (1) it inhibits the inflammatory network of proteases, (2) it binds to fragments of hyaluronan and thereby abrogates their pro-inflammatory effects and (3) it suppresses neutrophil infiltration into sites of inflammation. Under inflammatory signals (TNF-α, IL-1b or LPS) or environmental stress such as aggregation or hypoxia, TSG-6 secretion is rapidly activated (Lee et al., 2011). Considering that excessive inflammatory responses contribute to pathological changes in many diseases, the anti-inflammatory effects of TSG-6, IL-1RA and cytokines, secreted by MSCs at the initial phase of acute inflammation, can make a significant contribution to the therapeutic effects of MSCs.

The above-mentioned paracrine effects of MSCs could be efficiently applied for the support of pancreatic islets during *in vitro* processing and following transplantation leading to the improved survival of grafted tissue. It has recently been shown that the preculturing of islets with MSCs using a direct contact configuration maintains functional beta-cell mass *in vitro* and the capacity of cultured islets to reverse hyperglycaemia in diabetic mice (Rackham et al., 2013). Jung et al. showed that contact coculture of MSCs and islets resulted in sustained survival and retention of glucose-induced insulin secretion (Jung et al., 2011). Yeung et al. showed that culture of MSCs with islets prevented beta-cell apoptosis after cytokine treatment and improved the glucose-stimulated insulin secretion *in vitro* (Yeung et al., 2012). Other

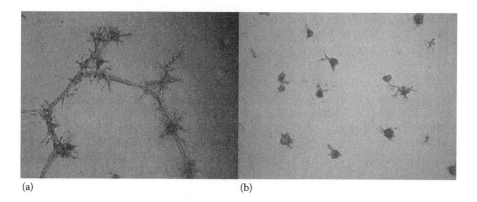

(a)  (b)

**FIGURE 10.5** Formation of capillary-like structures by adipose tissue stromal cells within Matrigel: (a) cells from primary cultures; (b) MSCs from fourth passage without endothelial induction. (With kind permission from Springer Science+Business Media: *Cytology and Genetics*, Phenotypical properties and ability to multilineage differentiation of adipose tissue stromal cells during subculturing, 46(1), 2012. 36–40, Petrenko YA and Petrenko AY.)

authors have indicated that islets, cocultured with MSCs, secreted an increased level of insulin after 14 days, whereas noncultured islets gradually deteriorated and cell death occurred proving the cytoprotective, anti-inflammatory and anti-apoptotic effects of MSCs as a supporting cell type (Karaoz et al., 2010). Furthermore, it was shown that culture of islets with MSCs protected the islets from hypoxia/reoxygenation-induced injury by decreasing the apoptotic cell ratio and increasing HIF-1α, HO-1 and COX-2 mRNA expression (Lu et al., 2010).

Despite the fact that secreting large quantities of trophic factors is one of the primary and substantial functions of MSCs responsible for their therapeutic or supporting activity, the role of MSCs in the artificial pancreatic niche is not limited only by the paracrine mechanisms. We cannot exclude that some portion of MSCs in response to external stimulus *in vivo* will be able to differentiate toward endothelial or insulin-producing cell lineages. We have previously demonstrated that primary cultures of human adipose tissue stromal cells contain a number of FLK+ endothelial cells, which could form capillary-like structures in Matrigel *in vitro* (Figure 10.5). Furthermore, after several passage expansions and subsequent inductions, adipose tissue MSCs could differentiate into endothelial-like cells (Petrenko and Petrenko, 2012) and insulin-producing cells *in vitro* (Petrenko et al., 2011c).

These results prove the suitability and prospects of MSC application for generating a microenvironment favourable for the repair and longevity of pancreatic islets in the coculture/cotransplantation system.

### 10.3.2 Alginate-Based Macroporous Cryogel Scaffolds for the Development of 3D Pancreatic Islet Support System

The development of artificial multicellular pancreatic substitutes and efficient MSCs-based support systems for the improvement of islet function *in vitro* and post-transplantation demands the translation of two-dimensional studies into 3D environments, which will more fully mimic the conditions that exist *in vivo*. Such an environment could be created by the application of 3D scaffolds, based on different biomaterials. The scaffold should both retain the cells within the defect site and promote tissue ingrowth and vascularisation. An ideal scaffold should be biocompatible, nonimmunogenic and (potentially) biodegradable; it should also provide optimal attachment, proliferation and differentiation of cells (Shoichet et al., 2010).

The main problems occurring when fabricating scaffolds for pancreatic islets are the pore sizes to facilitate islets entry, and the interconnectivity of pores, necessary for the sufficient nutrient and oxygen flow in the pre- or post-transplantation periods (Daoud et al., 2010). Various technologies have been applied for the preparation of porous scaffolds, such as salt leaching (Mikos et al., 1994; Murphy et al., 2002), freeze-drying (Lawson et al., 2004), cryotropic gelation (Lozinsky et al., 2002; 2008) and electrospinning (Li et al., 2005; Venugopal et al., 2008). The cryogenic methods, freeze-drying and cryotropic gelation utilize ice crystals as porogens. Both of these cryogenic approaches are noted for their high potential to induce interconnected gross porosity in the resulting polymeric matrices (Dvir et al., 2005) and are most commonly used techniques for the fabrication of macroporous matrices. We have shown that the use of cryotropic gelation for the fabrication of agarose- or alginate-based scaffolds enables the interconnection of macroporous 3D structures with average pore sizes of more than 100 µm (Figure 10.6), which is wide enough for free penetration of MSCs and islets on their seeding into the scaffold.

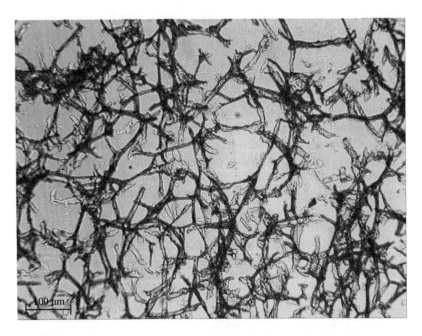

**FIGURE 10.6** The macroporous structure of alginate-gelatin scaffold. (From Petrenko YA, Ivanov RV, Petrenko AY, and Lozinsky VI. (2011). *Journal of Materials Science: Materials in Medicine* 22(6): 1529–1540. Copyright (2011), Springer. With permission.)

Alginate has been safely used to deliver proteins and cells to patients, which makes the alginate-based scaffolds a highly attractive tool for clinical applications. Alginate does not contain bonds that are hydrolysable by the human organism, but they are often used as resorbable biomaterials. This is achieved by the *in vitro* preformation of water-insoluble gels through ionic cross-linking of alginate chains with divalent cations such as $Ca^{2+}$, while the subsequent dissolution of these gels *in vivo* occurs via the ion exchange of alginate-bound $Ca^{2+}$ for monovalent cations, such as $Na^+$, $K^+$ and $NH_4^+$. The rate of the dissolution process can be considerably reduced by using additional chemical cross-linking of ionotropic alginate gel. Furthermore, water-soluble biopolymers can be attached to the alginate matrices by covalent coupling using glutaraldehyde, carbodiimide or carbonyldiimidazole (Jagur-Grodzinski et al., 2006). Simple alginate capsules were shown to increase canine islet survival *in vitro* for up to 3 weeks (Korbutt et al., 2004; Daoud et al., 2010). Furthermore, this alginate-encapsulated islet exhibited improved functional properties post-transplantation. Qi et al. showed that human islets maintained the viability and *in vitro* function after encapsulation and the alginate microbeads enable long-term *in vivo* human islet graft function (Qi et al., 2008).

Another key point in the development of two- or multicellular 3D support systems for pancreatic islets using alginate-based scaffolds is the ability of scaffold to provide optimal functional properties for the supporting cell types. When using MSCs as a supporting cell type for islets in a 3D environment, the ability of the scaffold to provide attachment, proliferation and multilineage differentiation of MSCs during culture is a critical point.

Seeding MSCs into macroporous alginate scaffolds (AS) resulted in the formation of viable multicellular aggregates with no signs of attachment and spreading of cells on the pore surfaces of the scaffold. These results confirm the data reporting the absence of cellular recognition proteins on alginate matrices, which limits cell attachment to the natural polymer (Alsberg et al., 2001; Lawson et al., 2004). Many studies improved the attachment and proliferation of different animal and human cell lines within AS either by grafting peptide sequences onto alginate materials (Alsberg et al., 2001) or by the incorporation of other substances such as hyaluronic acid (Miralles et al., 2001), tricalcium phosphate (Lawson et al., 2004), gelatin (Yang et al., 2009), chitosan (Tan et al., 2003), or even by preparing multicomponent scaffolds, such as hydroxyapatite-alginate-gelatin (Bernhardt et al., 2009) or tricalcium phosphate-alginate-gelatin (Eslaminejad et al., 2007).

We have assessed two AS modification protocols (Petrenko et al., 2011a). The first protocol included a mechanical incorporation of gelatin (type A) into the bulk of scaffolds (AGS/i). The second protocol included four main steps: (1) freezing the initial Na-alginate solution, (2) ice sublimation from the frozen sample, (3) curing the freeze-dried matter with calcium ions in polar organic medium with (4) subsequent chemical coupling of gelatin to the wide pore matrix (AGS/c). The efficiency of the modified AS was assessed by Alamar blue assay and fluorescein diacetate (FDA) staining. Figure 10.7 shows that the mechanical incorporation of gelatin at different concentrations (AGS/i group) could not promote proliferation of MSCs within the scaffolds. However, chemical activation of the polymer matrix followed by the attachment of gelatin molecules to the pore inner surface (AGS/c group) resulted in

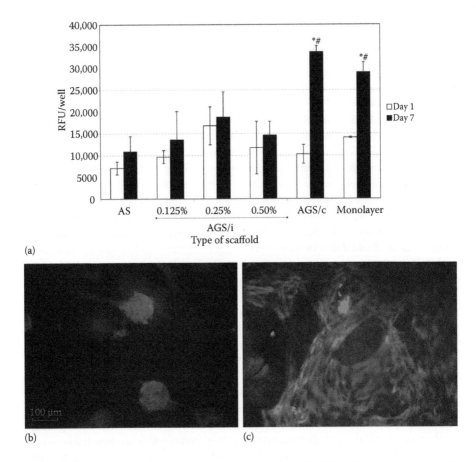

**FIGURE 10.7** Adhesion, morphology and proliferation of MSCs within macroporous AS, AGS/i and AGS/c: (a) Alamar blue assay of MSCs seeded into macroporous alginate scaffolds modified by different methods; (b, c) Morphology and distribution of MSCs within matrices AS (b) and AGS/c (c) on the seventh day of culture. *, Values are significantly higher versus that of AS and AGS/i groups (P\0.05). #, Values are significantly higher versus that measured on day 1 in the same groups (P\0.05). (With kind permission from Springer Science+Business Media: *Bulletin of Experimental Biology and Medicine*, Comparison of the methods for seeding human bone marrow mesenchymal stem cells to macroporous alginate cryogel carriers, 150(4), 2011, 543–6, Petrenko YA, Ivanov RV, Lozinsky VI, and Petrenko AY.)

the 3.5-fold increase in Alamar blue fluorescence on day 7 compared to day 1, indicating cell proliferation (Figure 10.7a).

The obtained results were confirmed by the analysis of viability and morphological properties of MSCs within the AGS by applying FDA staining. It was revealed that after 7 days of MSCs culture within the AS and AGS/i, cells preserved their viability, but did not adhere and flatten on the pore surfaces. In this case, the formation of cell aggregates was observed (Figure 10.7b). In contrast, during culture of MSCs within the AGS/c type of scaffolds, cells showed typical fibroblast-like morphology and actively populated the porous carrier (Figure 10.7c).

The further induction of MSCs into adipogenic differentiation resulted in the accumulation of intracellular lipids, positively stained by Nile red. Osteogenic induction led to alkaline phosphatase expression, confirming osteogenic differentiation of cells. In chondrogenic media, MSCs accumulated extracellular matrix, which was positively stained by alcian blue.

Therefore, although both technical approaches resulted in the fabrication of macroporous scaffolds from alginate solutions, the functional tests revealed a marked superiority of the second protocol. The scaffolds obtained using this methodology possessed the desired porous morphology and ability to adhere stromal cells. The scaffold preparation technique based on covalent binding of gelatin not only promoted attachment and growth of MSCs within the scaffold, but also helped preserve the main functional parameters of the cells, for example, the capacity of multilineage differentiation (Petrenko et al., 2011b).

A significant issue in 3D islet culture is seeding efficiency. Optimally, cells and islets must be immobilized in scaffolds in a uniform manner throughout the 3D support system. Technically, simple static seeding is often used for the majority of cell types and carriers, but it has some drawbacks: unequal cell distribution, low seeding efficiency, cell condensation and so on (Wendt et al., 2003). Dynamic seeding with the use of perfusion, centrifugation, negative pressure and so on is more effective (Roh et al., 2007; Petrenko et al., 2008). It provides higher efficiency of cell seeding and distribution within the scaffold.

Recently, we have described a simple perfusion method, which ensures rapid and equal MSCs distribution within macroporous AGS (Petrenko et al., 2011a). Figure 10.8 represents two applications of seeding techniques.

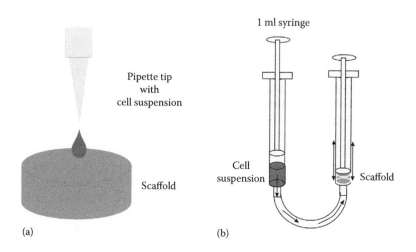

**FIGURE 10.8** Schematic presentation of the methods for seeding cells to macroporous alginate cryogel scaffolds: (a) static method; (b) perfusion method.

The static method (Figure 10.8a) consists of an application of a minimal volume (20 µl) of concentrated cell suspension ($3 \times 10^5$ cells/ml) on the surface of a 3D scaffold using an automated pipette. The scaffold with cells was incubated for 3 h at 37°C and then transferred into wells of a 24-well plate containing 1 ml medium. For seeding by the perfusion method (Figure 10.8b), two 1-ml syringes connected with an elastic plastic tube were used. Porous AS was placed in one syringe (its diameter corresponded to the inner diameter of the syringe), 100-µl cell suspension ($3 \times 10^5$ cells/ml) was placed in another syringe, and the scaffold was gradually saturated with cells by gently moving the syringe piston back and forth. The scaffold saturated with cells was incubated in a syringe for 3 h at 37°C and then transferred into wells of a 24-well plate containing 1 ml culture medium. The results of seeding the macroporous scaffolds by applying the cell suspension to the surface of the alginate sponge or by gentle perfusion were different. After static seeding, most cells were located in the surface area of the scaffold and only a few cells were submerged. The use of the perfusion method ensured more equal distribution of cells over the 3D structure of the alginate cryogel: MSCs were located not only on the surface, but also over the entire volume of the scaffold, a feature necessary for the development of the uniformly organized system.

### 10.3.3 Cryopreservation of MSCs Seeded Macroporous Alginate-Based Cryogel Scaffolds

Application of cell-scaffold constructs in regenerative medicine implies a linear workflow from cell seeding in appropriate scaffolds through proliferation *in vitro* to transplantation *in vivo*, a procedure that does not permit the pause or even the storing of the tissue constructs in biobanks for future supply. Cryopreservation of MSCs/scaffold constructs with maintained cell viability and functionality is a desirable approach (Umemura et al., 2011; Popa et al., 2013; Pravdyuk et al., 2013) to overcome shortage in supply and facilitate immediate application of the constructs by their ready-to-use character. In spite of several decades of research, it is still very difficult to cryopreserve adherent cells. The cells with cell–cell and cell–substrate contacts are much more sensitive to freeze–thaw injury than single cells in suspension and their spacious plasma membrane and cytoskeleton is affected by mechanical ruptures, followed by cell detachment and death (Acker et al., 1999; Ebertz et al., 2004; Xu et al., 2014). These contacts, mediated by cytoskeleton proteins, are involved in the anchorage, spreading and motility of adherent cells (Gumbiner, 1996). However, the cryopreservation effect on cell spreading using such properties has not yet been proven. Attachment and spreading processes depend on the duration of cultivation and can already be detected after a few hours (Anselme et al., 2006). To enhance cryopreservation success, the strained cytoskeleton of adherent cells has to be protected against injury caused by freezing and thawing procedures. Since it is well known that water molecules are the main cause of cryo-injury (solution effects, mechanical damage by ice crystals) (Muldrew et al., 1994), hydrogel scaffolds seem to have beneficial effects for cryopreservation procedures. Recently, the possibility for cryopreservation of MSCs within macroporous alginate-based cryogel scaffold has been demonstrated (Katsen-Globa et al., 2014). The authors demonstrated that the cryopreservation outcome depends on the adhesion and spreading degree of MSCs within the scaffolds prior to the cryopreservation procedure. The obtained result could become a basis for the development of off-the-shelf available 3D tissue engineered supporting system for further pancreatic islet transplantation studies.

## 10.4 CONCLUDING REMARKS

Summarizing this chapter, we propose a schematic presentation of the MSCs-based 3D supporting system for pancreatic islet cell culture *in vitro* and islet transplantation (Figure 10.9). For the synthesis of core forming matrix, we suggest using alginate- or agarose-based supermacroporous cryogel scaffold with grafted gelatin. Such a scaffold provides a beneficial microenvironment for grafted islets promoting local neovascularisation, adequate nutrition, oxygen supply and efficient removal of metabolic waste. In addition, interconnected pores with grafted molecules of ECM offer a superior environment for attachment, proliferation and differentiation of MSCs.

Secreting anti-inflammatory, immunomodulatory and angiogenic factors, MSCs seeded into 3D scaffolds create the macrosupport needed for the improved engraftment of pancreatic islets after transplantation. Moreover, MSCs secreting anti-apoptotic factors within 3D scaffolds provide the necessary cytoprotective effects during pre- and postimplantation periods. Finally, the unique properties of the supermacroporous cryogels facilitate the application of cryopreservation technologies for the cryogel-based construct and create a basis for the development of off-the-shelf bioartificial support system for further pancreatic islet transplantations.

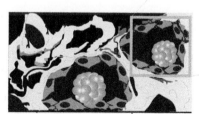
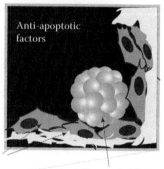

**FIGURE 10.9** Schematic presentation of the MSCs-based 3D supporting system for pancreatic islet culture and transplantation.

## ACKNOWLEDGEMENTS

The authors gratefully acknowledge the financial support of Israel–Ukraine joint research program (# 1244512) from the Israel Ministry of Science and Technology, and Ministry of Science and Education of Ukraine. The authors thank Sara Dominitz for her editorial assistance.

## LIST OF ABBREVIATIONS

| | |
|---|---|
| AGS | Alginate scaffold with either incorporated or chemically coupled gelatin |
| AGS/c | Alginate scaffolds with chemically coupled gelatin |
| AGS/I | Alginate scaffolds with incorporated gelatin |
| AS | Alginate scaffold |
| FDA | Fluorescein diacetate |
| MSCs | Mesenchymal stem/stromal cells |

## REFERENCES

Abdi R, Fiorina P, Adra C, Atkinson M, and Sayegh MH. (2008). Immunomodulation by mesenchymal stem cells a potential therapeutic strategy for type 1 diabetes. *Stem Cells* 57(7): 1759–1767.

Acker JP, Larese A, Yang H, Petrenko A, and McGann LE. (1999). Intracellular ice formation is affected by cell interactions. *Cryobiology* 38(4): 363–371.

Alsberg E, Anderson KW, Albeiruti A, Franceschi RT, and Mooney DJ. (2001). Cell-interactive alginate hydrogels for bone tissue engineering. *Journal of Dental Research* 80(11): 2025–2029.

Anderson JM, Rodriguez A, and Chang DT. (2008). Foreign body reaction to biomaterials. *Seminars in Immunology* 20(2): 86–100.

Anselme K, and Bigerelle M. (2006). Modelling approach in cell/material interactions studies. *Biomaterials* 27(8): 1187–1199.

Augello A, Tasso R, Negrini SM, Cancedda R, and Pennesi G. (2007). Cell therapy using allogeneic bone marrow mesenchymal stem cells prevents tissue damage in collagen-induced arthritis. *Arthritis and Rheumatism* 56(4): 1175–1186.

Aviles MO, Lin CH, Zelivyanskaya M et al. (2010). The contribution of plasmid design and release to *in vivo* gene expression following delivery from cationic polymer modified scaffolds. *Biomaterials.* 31(6): 1140–1147.

Balamurugan AN, Gu Y, Tabata Y et al. (2003). Bioartificial pancreas transplantation at prevascularized intermuscular space: Effect of angiogenesis induction on islet survival. *Pancreas* 26(3): 279–285.

Bartholomew A, Sturgeon C, Siatskas M, Ferrer K, McIntosh K, Patil S et al. (2002). Mesenchymal stem cells suppress lymphocyte proliferation *in vitro* and prolong skin graft survival *in vivo*. *Experimental Hematology* 30(1): 42–48.

Beger C, Cirulli V, Vajkoczy P, Halban PA, and Menger MD. (1998). Vascularisation of purified pancreatic islet-like cell aggregates (pseudoislets) after syngeneic transplantation. *Diabetes.* 47(4):559–565.

Bernhardt A, Despang F, Lode A, Demmler A, Hanke T, and Gelinsky M. (2009). Proliferation and osteogenic differentiation of human bone marrow stromal cells on alginate-gelatine-hydroxyapatite scaffolds with anisotropic pore structure. *Journal of Tissue Engineering and Regenerative Medicine* 3(1): 54–62.

Blaber SP, Webster A, Hill CJ, Breen EJ, Kuah D, Vesey G, and Herbert BR. (2012). Analysis of *in vitro* secretion profiles from adipose-derived cell populations. *Journal of Translational Medicine* 10: 172.

Bloch K, Lozinsky VI, Galaev IY, Yavriyanz K, Vorobeychik M, Azarov D et al. (2005). Functional activity of insulinoma cells (INS-1E) and pancreatic islets cultured in agarose cryogel sponges. *Journal of Biomedical Materials Research. Part A* 75(4): 802–809.

Bloch K, Vanichkin A, Damshkaln LG, Lozinsky VI, and Vardi P. (2010). Vascularisation of wide pore agarose-gelatin cryogel scaffolds implanted subcutaneously in diabetic and non-diabetic mice. *Acta Biomaterialia.* 6(3): 1200–1205.

Bloch K, Vennäng J, Lazard D, and Vardi P. (2012). Different susceptibility of rat pancreatic alpha and beta cells to hypoxia. *Histochemistry and Cell Biology* 137(6): 801–810.

Blomeier H, Zhang X, Rives C et al. (2006). Polymer scaffolds as synthetic microenvironments for extrahepatic islet transplantation. *Transplantation.* 82(4): 452–459.

Borg DJ, Weigelt M, Wilhelm C et al. (2014). Mesenchymal stromal cells improve transplanted islet survival and islet function in a syngeneic mouse model. *Diabetologia.* 57(3): 522–531.

Brem H, and Tomic-Canic M. (2007). Cellular and molecular basis of wound healing in diabetes. *Journal of Clinical Investigation* 117(5): 1219–1222.

Busch SA, van Crutchen ST, Deans RJ, and Ting AE. (2011). Mesenchymal stromal cells as a therapeutic strategy to support islet transplantation in type 1 diabetes mellitus. *Cell Medicine* 2(2): 43–53.

Cantarelli E, and Piemonti L. (2011). Alternative transplantation sites for pancreatic islet grafts. *Current Diabetes Reports.* 11(5): 364–374.

Cao, Y, Sun Z, Liao L, Meng Y, Han Q, and Chunhua Zhao R. (2005). Human adipose tissue-derived stem cells differentiate into endothelial cells *in vitro* and improve postnatal neovascularization *in vivo*. *Biochemical and Biophysical Research Communications* 332(2): 370–379.

Caplan AI. (2009). Why are MSCs therapeutic? New data: New insight. *The Journal of Pathology* 217(2): 318–324.

Carlsson PO, Palm F, Andersson A, and Liss P. (2001). Markedly decreased oxygen tension in transplanted rat pancreatic islets irrespective of the implantation site. *Diabetes* 50(3): 489–495.

Carlsson PO. (2011). Influence of microenvironment on engraftment of transplanted beta-cells. *Upsala Journal of Medical Sciences* 116(1): 1–7.

Cartmell SH, Porter BD, García AJ, and Guldberg RE. (2003). Effects of medium perfusion rate on cell-seeded three-dimensional bone constructs *in vitro*. *Tissue Engineering* 9(6): 1197–1203.

Chen RR, Silva EA, Yuen WW, and Mooney DJ. (2007). Spatio-temporal VEGF and PDGF delivery patterns blood vessel formation and maturation. *Pharmaceutical Research* 24(2): 258–264.

Colton CK. (1995). Implantable biohybrid artificial organs. *Cell Transplantation* 4(4): 415–436.

Cooper ME, Bonnet F, Oldfield M, and Jandeleit-Dahm K. (2001). Mechanisms of diabetic vasculopathy: An overview. *American Journal of Hypertension* 14(5 Pt 1): 475–486.

Cornolti R, Figliuzzi M, and Remuzzi A. (2009). Effect of micro- and macroencapsulation on oxygen consumption by pancreatic islets. *Cell Transplantation* 18(2): 195–201.

Coronel MM, and Stabler CL. (2013). Engineering a local microenvironment for pancreatic islet replacement. *Current Opinion in Biotechnology* 24(5): 900–908.

Daoud J, Rosenberg L, and Tabrizian M. (2010). Pancreatic islet culture and preservation strategies: Advances, challenges, and future outlook. *Cell Transplantation* 19(12): 1523–1535.

De Miguel MP, Fuentes-Julián S, Blázquez-Martínez A et al. (2012). Immunosuppressive properties of mesenchymal stem cells: Advances and applications. *Current Molecular Medicine* 12(5): 574–591.

De Vos P, Lazarjani HA, Poncelet D, and Faas MM. (2013). Polymers in cell encapsulation from an enveloped cell perspective. *Advanced Drug Delivery Reviews* doi: 10.1016/j.addr.2013.11.005. [Epub ahead of print]

De Vos P, Spasojevic M, and Faas M. (2010). Treatment of diabetes with encapsulated islets. *Advances in Experimental Medicine and Biology* 670: 38–53.

Di Rocco G, Iachininoto MG, Tritarelli A, Straino S, Zacheo A, Germani A et al. (2006). Myogenic potential of adipose-tissue-derived cells. *Journal of Cell Science* 119(Pt 14): 2945–2952.

Dionne KE, Colton CK, and Yarmush ML. (1993). Effect of hypoxia on insulin secretion by isolated rat and canine islets of Langerhans. *Diabetes*. 42(1): 12–21.

Djouad F, Charbonnier LM, Bouffi C, Louis-Plence P, Bony C, Apparailly F et al. (2007). Mesenchymal stem cells inhibit the differentiation of dendritic cells through an interleukin-6-dependent mechanism. *Stem Cells (Dayton, Ohio)* 25(8): 2025–2032.

Djouad F, Plence P, Bony C, Tropel P, Apparailly F, Sany J et al. (2003). Immunosuppressive effect of mesenchymal stem cells favors tumor growth in allogeneic animals. *Blood* 102(10): 3837–3844.

Dominici M, Le Blanc K, Mueller I, Slaper-Cortenbach I, Marini F, Krause D et al. (2006). Minimal criteria for defining multipotent mesenchymal stromal cells. The International Society for Cellular Therapy position statement. *Cytotherapy* 8(4): 315–317.

Dvir T, Tsur-Gang O, and Cohen S. (2005). Designer scaffolds for tissue engineering and regeneration. *Israel Journal of Chemistry* 45(4): 487–494.

Ebertz SL, and Mcgann LE. (2004). Cryoinjury in endothelial cell monolayers. *Cryobiology* 49(1): 37–44.

Ellis CE, Suuronen E, Yeung T, Seeberger K, and Korbutt GS. (2013). Bioengineering a highly vascularized matrix for the ectopic transplantation of islets. *Islets* 5(5): 216–225.

Eslaminejad MB, Mirzadeh H, Mohamadi Y, and Nickmahzar A. (2007). Bone differentiation of marrow-derived mesenchymal stem cells using β-tricalcium phosphate – alginate – gelatin hybrid scaffolds. *Tissue Engineering* 1(6): 417–424.

Fiorina P, Jurewicz M, Augello A, Vergani A, Dada S, La Rosa S et al. (2009). Immunomodulatory function of bone marrow-derived mesenchymal stem cells in experimental autoimmune type 1 diabetes. *Journal of Immunology* 183(2): 993–1004.

Gibly RF, Zhang X, Lowe WL Jr, and Shea LD. (2013). Porous scaffolds support extrahepatic human islet transplantation, engraftment, and function in mice. *Cell Transplantation*. 22(5): 811–819.

Glennie S, Soeiro I, Dyson PJ, Lam EW, and Dazzi F. (2005). Bone marrow mesenchymal stem cells induce division arrest energy of activated T cells. *Blood* 105(7): 2821–2827.

Gumbiner BM. (1996). Cell adhesion: The molecular basis of tissue architecture and morphogenesis. *Cell* 84(3): 345–357.

Hauge-Evans AC, Squires PE, Persaud SJ, and Jones PM. (1999). Pancreatic beta-cell-to-beta-cell interactions are required for integrated responses to nutrient stimuli: enhanced $Ca^{2+}$ and insulin secretory responses of MIN6 pseudoislets. *Diabetes* 48(7): 1402–1408.

Hematti P, Kim J, Stein AP, and Kaufman D. (2013). Potential role of mesenchymal stromal cells in pancreatic islet transplantation. *Transplantation Reviews (Orlando)* 27(1): 21–29.

Horwitz EM, Prockop DJ, Fitzpatrick DJ, Koo WW, Gordon PL, Neel M et al. (1999). Transplantability and therapeutic effects of bone marrow-derived mesenchymal cells in children with osteogenesis imperfecta. *Nature Medicine* 5(3): 309–313.

Hsiao STF, Asgari A, Lokmic Z, Sinclair R, Dusting GJ, Lim SY et al. (2012). Comparative analysis of paracrine factor expression in human adult mesenchymal stem cells derived from bone marrow, adipose, and dermal tissue. *Stem Cells and Development* 21(12): 2189–2203.

Iso Y, Spees JL, Serrano C, Bakondi B, Pochampally R, Song YH et al. (2007). Multipotent human stromal cells improve cardiac function after myocardial infarction in mice without long-term engraftment. *Biochemical and Biophysical Research Communications* 354(3): 700–706.

Jagur-Grodzinski J. (2006). Polymers for tissue engineering, medical devices, and regenerative medicine. Concise general review of recent studies. *Polymers for Advanced Technologies* 17(6): 395–418.

Jung EJ, Kim SC, Wee YM, Kim YH, Choi MY, Jeong SH et al. (2011). Bone marrow-derived mesenchymal stromal cells support rat pancreatic islet survival and insulin secretory function *in vitro*. *Cytotherapy* 13(1): 19–29.

Karaoz E, Genç ZC, Demircan PC, Aksoy A, and Duruksu G. (2010). Protection of rat pancreatic islet function and viability by coculture with rat bone marrow-derived mesenchymal stem cells. *Cell Death & Disease* 1: e36.

Katsen-Globa A, Meiser I, Petrenko YA, Ivanov RV, Lozinsky VI, Zimmermann H, and Petrenko AY. (2014). Towards ready-to-use 3-D scaffolds for regenerative medicine: Adhesion-based cryopreservation of human mesenchymal stem cells attached and spread within alginate–gelatin cryogel scaffolds. *Journal of Materials Science: Materials in Medicine* 25(3): 857–871.

Kheradmand T, Wang S, Gibly RF et al. (2011). Permanent protection of PLG scaffold transplanted allogeneic islet grafts in diabetic mice treated with ECDI-fixed donor splenocyte infusions. *Biomaterials*. 32(20): 4517–4524.

Kim SJ, Choi YS, Ko ES, Lim SM, Lee CW, and Kim DI. (2012). Glucose-stimulated insulin secretion of various mesenchymal stem cells after insulin-producing cell differentiation. *Journal of Bioscience and Bioengineering* 113(6): 771–777.

King AJ, Fernandes JR, Hollister-Lock J, Nienaber CE, Bonner-Weir S, and Weir GC. (2007). Normal relationship of beta- and non-beta-cells not needed for successful islet transplantation. *Diabetes* 56(9): 2312–2318.

Korbutt GS, Mallett AG, Ao Z, Flashner M, and Rajotte RV. (2004). Improved survival of microencapsulated islets during *in vitro* culture and enhanced metabolic function following transplantation. *Diabetologia* 47(10): 1810–1818.

Lawson MA, Barralet JE, Wang L, Shelton RM, and Triffitt JT. (2004). Adhesion and growth of bone marrow stromal cells on modified alginate hydrogels. *Tissue Engineering* 10: 9–10: 1480–1491.

Lazard D, Vardi P, and Bloch K. (2012). Induction of beta-cell resistance to hypoxia and technologies for oxygen delivery to transplanted pancreatic islets. *Diabetes/Metabolism Research and Reviews* 28:6: 475–484.

Lazarus HM, Haynesworth SE, Gerson SL, Rosenthal NS, and Caplan AI. (1995). Ex vivo expansion and subsequent infusion of human bone marrow-derived stromal progenitor cells (mesenchymal progenitor cells): Implications for therapeutic use. *Bone Marrow Transplantation* 16:4: 557–564.

Lee KD, Kuo TKC, Whang-Peng J, Chung YF, Lin CT, Chou SH et al. (2004). In vitro hepatic differentiation of human mesenchymal stem cells. *Hepatology* 40:6: 1275–1284.

Lee RH, Pulin AA, Seo MJ, Kota DJ, Ylostalo J, Larson BL et al. (2009). Intravenous hMSCs improve myocardial infarction in mice because cells embolized in lung are activated to secrete the anti-inflammatory protein TSG-6. *Cell Stem Cell* 5:1: 54–63.

Lee JW, Fang X, Krasnodembskaya A, Howard JP, and Matthay MA. (2011). Concise review: Mesenchymal stem cells for acute lung injury: Role of paracrine soluble factors. *Stem Cells* 29(6): 913–919.

Li WJ, Tuli R, Huang X, Laquerriere P, and Tuan RS. (2005). Multilineage differentiation of human mesenchymal stem cells in a three-dimensional nanofibrous scaffold. *Biomaterials* 26(25): 5158–5166.

Lim F, and Sun AM. (1980). Microencapsulated islets as bioartificial endocrine pancreas. *Science* 210(4472): 908–910.

Lozinsky VI. (2002). Cryogels on the basis of natural and synthetic polymers: Preparation, properties and application. *Russian Chemical Reviews* 71(6): 489–511.

Lozinsky VI, Damshkaln LG, Bloch KO, Vardi P, Grinberg NV, Burova TV, and Grinberg VYa. (2008). Cryostructuring of polymer systems. XXIX. Preparation and characterization of supermacroporous (spongy) agarose-based cryogels used as 3D-scaffolds for culturing insulin-producing cell aggregates. *Journal of Applied Polymer Science* 108(5): 3046–3062.

Lozinsky VI, Damshkaln LG, Plieva FM, Galaev IYu, and Mattiasson B. (2001). The polymeric composition for the preparation of macroporous agarose gel and the method for the gel producing. Russian Patent No. 2, 220, 987.

Lozinsky VI, Galaev IYu, Plieva F, Savina IN, Jungvid H, and Mattiasson B. (2003). Polymeric cryogels as promising materials of biotechnological interest. *Trends in Biotechnology* 21(10): 445–451.

Lu Y, Jin X, Chen Y, Li S, Yuan Y, Mai G, Tian B et al. (2010). Mesenchymal stem cells protect islets from hypoxia/reoxygenation-induced injury. *Cell Biochemistry and Function* 28(8): 637–643.

Luther M J, Hauge-Evans A, Souza K LA, Jörns A, Lenzen S, Persaud SJ, and Jones PM. (2006). MIN6 beta-cell-beta-cell interactions influence insulin secretory responses to nutrients and non-nutrients. *Biochemical and Biophysical Research Communications* 343(1): 99–104.

Merani S, Toso C, Emamaullee J, and Shapiro AM. (2008). Optimal implantation site for pancreatic islet transplantation. *British Journal of Surgery* 95:12: 1449–1461.

Mikos AG, Thorsen AJ, Czerwonka LA, Bao Y, and Langer R. (1994). Preparation and characterization of poly(L-lactic acid) foams. *Polymer* 35:5: 1068–1077.

Miralles G, Baudoin R, Dumas D, Baptiste D, Hubert P, Stoltz JF et al. (2001). Sodium alginate sponges with or without sodium hyaluronate: *In vitro* engineering of cartilage. *Journal of Biomedical Materials Research* 57:2: 268–278.

Muldrew, K, and McGann LE. (1994). The osmotic rupture hypothesis of intracellular freezing injury. *Biophysical Journal* 66(2 Pt 1): 532–541.

Murphy WL, Dennis RG, Kileny JL, and Mooney DJ. (2002). Salt fusion: An approach to improve pore interconnectivity within tissue engineering scaffolds. *Tissue Engineering* 8:1: 43–52.

O'Sullivan ES, Vegas A, Anderson DG, and Weir GC. (2011). Islets transplanted in immunoisolation devices: A review of the progress and the challenges that remain. *Endocrine Reviews* 32(6): 827–844.

Parekkadan B, and Milwid JM. (2010). Mesenchymal stem cells as therapeutics. *Annual Review of Biomedical Engineering* 12: 87–117.

Patel DM, Shah J, and Srivastava AS. (2013). Therapeutic potential of mesenchymal stem cells in regenerative medicine. *Stem Cells International* 496218. doi: 10.1155/2013/496218.

Pedraza E, Brady AC, Fraker CA Molano RD, Sukert S, Berman DM et al. (2013). Macroporous three-dimensional PDMS scaffolds for extrahepatic islet transplantation. *Cell Transplantation* 22(7): 1123–1135.

Pepper AR, Gala-Lopez B, Ziff O, and Shapiro AM. (2013). Revascularization of transplanted pancreatic islets and role of the transplantation site. *Clinical and Developmental Immunology* doi: 10.1155/2013/352315.

Petrenko YA, Ivanov RV, Lozinsky VI, and Petrenko AY. (2011a). Comparison of the methods for seeding human bone marrow mesenchymal stem cells to macroporous alginate cryogel carriers. *Bulletin of Experimental Biology and Medicine* 150(4): 543–546.

Petrenko YA, Ivanov RV, Petrenko AY, and Lozinsky VI. (2011b). Coupling of gelatin to inner surfaces of pore walls in spongy alginate-based scaffolds facilitates the adhesion, growth and differentiation of human bone marrow mesenchymal stromal cells. *Journal of Materials Science. Materials in Medicine* 22(6): 1529–1540.

Petrenko YA, Mazur SP, Grischuk VP et al. (2011c). The choice of induction factors for differentiation of multipotent mesenchymal stromal cells derived from human adipose tissue to insulin-producing cells *in vitro*. *Cellular Transplantation and Tissue Engineering* 6(1): 73–79.

Petrenko YA, Volkova NA, Zhulikova EP, Damshkaln LG, Lozinsky VI, and Petrenko AY. (2008). Choice of conditions of human bone marrow stromal cells seeding into polymer macroporus sponges. *Biopolymers and Cell* 24(5): 399–405.

Petrenko YA., and Petrenko AY. (2012). Phenotypical properties and ability to multilineage differentiation of adipose tissue stromal cells during subculturing. *Cytology and Genetics* 46(1): 36–40.

Petrie A, and Tuan RS. (2010). Therapeutic potential of the immunomodulatory activities of adult mesenchymal stem cells. *Birth Defects Research Part C: Embryo Today: Reviews* 90(1): 67–74.

Pittenger MF, Mackay AM, Beck SC, Jaiswal RK, Douglas R, Mosca JD et al. (1999). Multilineage potential of adult human mesenchymal stem cells. *Science* 284(5411): 143–147.

Planat-Benard V, Silvestre JS, Cousin B et al. (2004). Plasticity of human adipose lineage cells toward endothelial cells: Physiological and therapeutic perspectives. *Circulation* 109(5): 656–663.

Popa EG, Rodrigues MT, Coutinho DF et al. (2013). Cryopreservation of cell laden natural origin hydrogels for cartilage regeneration strategies. *Soft Matter* 9(3): 875.

Pravdyuk AI, Petrenko YA, Fuller BJ, and Petrenko AY. (2013). Cryopreservation of alginate encapsulated mesenchymal stromal cells. *Cryobiology* 66(3): 215–222.

Prockop DJ, Kota DJ, Bazhanov N, and Reger RL. (2010). Evolving paradigms for repair of tissues by adult stem/progenitor cells (MSCs). *Journal of Cellular and Molecular Medicine* 14(9): 2190–2199.

Qi M, Strand BL, Mørch Y, Lacík I et al. (2008). Encapsulation of human islets in novel inhomogeneous alginate-ca2+/ba2+ microbeads: *In vitro* and *in vivo* function. *Artificial Cells, Blood Substitutes, and Immobilization Biotechnology* 36(5): 403–420.

Rackham CL, Dhadda PK, Chagastelles PC et al. (2013). Pre-culturing islets with mesenchymal stromal cells using a direct contact configuration is beneficial for transplantation outcome in diabetic mice. *Cytotherapy* 15(4): 449–459.

Rajab A. (2010). Islet transplantation: Alternative sites. *Current Diabetes Reports* 10(5):332–337.

Roh JD, Nelson GN, Udelsman BV et al. (2007). Centrifugal seeding increases seeding efficiency and cellular distribution of bone marrow stromal cells in porous biodegradable scaffolds. *Tissue Engineering* 13(11): 2743–2749.

Rouwkema J, Rivron NC, and Van Blitterswijk CA. (2008). Vascularisation in tissue engineering. *Trends Biotechnol.* 26(8): 434–441.

Ryan JM, Barry FP, Murphy JM, and Mahon BP. (2005). Mesenchymal stem cells avoid allogeneic rejection. *Journal of Inflammation (London)* 2: 8.

Salvay DM, Rives CB, Zhang X et al. (2008). Extracellular matrix protein-coated scaffolds promote the reversal of diabetes after extrahepatic islet transplantation. *Transplantation.* 85(10): 1456–1464.

Scuteri A, Donzelli E, Rodriguez-Menendez V, Ravasi M, Monfrini M, Bonandrini B et al. (2014). A double mechanism for the mesenchymal stem cells' positive effect on pancreatic islets. *PLoS One* 9(1):e84309.

Shakya AK, Holmdahl R, Nandakumar KS, and Kumar A. (2013). Polymeric cryogels are biocompatible and their biodegradation is independent of oxidative radicals. *J Biomed Mater Res A.* doi: 10.1002/jbm.a.35013. [Epub ahead of print]

Shi Y, Hu G, Su J, Li W, Chen Q, Shou P, Xu C et al. (2010). Mesenchymal stem cells: A new strategy for immunosuppression and tissue repair. *Cell Research* 20(5): 510–518.

Shoichet SM. (2010). Polymer scaffolds for biomaterials applications. *Macromolecules* 43(2): 581–591.

Shu SN, Wei L, Wang JH, Zhan YT, Chen HS, and Wang Y. (2004). Hepatic differentiation capability of rat bone marrow-derived mesenchymal stem cells and hematopoietic stem cells. *Journal of Gastroenterology* 10(19): 2818–2822.

Smink AM, Faas MM, and de Vos P. (2013). Toward engineering a novel transplantation site for human pancreatic islets. *Diabetes.* 62(5): 1357–1364.

Spaggiari GM, Capobianco A, Becchetti S, Mingari MC, and Moretta L. (2006). Mesenchymal stem cell-natural killer cell interactions: Evidence that activated NK cells are capable of killing MSCs, whereas MSCs can inhibit IL-2-induced NK-cell proliferation. *Blood* 107(4): 1484–1490.

Tan TW, Hu B, Jin XH, and Zhang M. (2003). Release behavior of ketoprofen from chitosan/alginate microcapsules. *Journal of Bioactive and Compatible Polymers* 18(3): 207–218.

Tripathi A, Kathuria N, and Kumar A. (2009). Elastic and macroporous agarose-gelatin cryogels with isotropic and anisotropic porosity for tissue engineering. *J Biomed Mater Res A.* 90(3): 680–694.

Umemura E, Yamada Y, Nakamura S, Ito K, Hara K, and Ueda M. (2011). Viable cryopreserving tissue-engineered cell-biomaterial for cell banking therapy in an effective cryoprotectant. *Tissue engineering. Part C, Methods* 17(8): 799–807.

Vaithilingam V, Quayum N, Joglekar MV et al. (2011). Effect of alginate encapsulation on the cellular transcriptome of human islets. *Biomaterials* 32(33): 8416–8425.

Vaithilingam V, and Tuch BE. (2011). Islet transplantation and encapsulation: An update on recent developments. *Rev Diabet Stud.* 8(1): 51–67.

Venugopal J, Low S, Choon AT, and Ramakrishna S. (2008). Interaction of cells and nanofiber scaffolds in tissue engineering. *Journal of Biomedical Materials Research. Part B, Applied Biomaterials* 84(1): 34–48.

Vériter S, Gianello P, and Dufrane D. (2013). Bioengineered sites for islet cell transplantation. *Curr Diab Rep.* 13(5): 745–755.

Vija L, Farge D, Gautier J-F, Vexiau P, Dumitrache C, Bourgarit A et al. (2009). Mesenchymal stem cells: Stem cell therapy perspectives for type 1 diabetes. *Diabetes & Metabolism* 35(2): 85–93.

Wendt D, Marsano A, Jakob M, Heberer M, and Martin I. (2003). Oscillating perfusion of cell suspensions through three-dimensional scaffolds enhances cell seeding efficiency and uniformity. *Biotechnology and Bioengineering* 84(2): 205–214.

Wood RC, LeCluyse EL, and Fix JA. (1995). Assessment of a model for measuring drug diffusion through implant-generated fibrous capsule membranes. *Biomaterials* 16(12): 957–959.

Xie QP, Huang H, Xu B, Dong X, Gao SL, Zhang B, and Wu YL. (2009). Human bone marrow mesenchymal stem cells differentiate into insulin-producing cells upon microenvironmental manipulation *in vitro*. *Differentiation; Research in Biological Diversity* 77(5): 483–491.

Xu X, Liu Y, and Cui ZF. (2014). Effects of cryopreservation on human mesenchymal stem cells attached to different substrates. *Journal of Tissue Engineering and Regenerative Medicine* 8(8): 664–672.

Yang C, Frei H, Rossi FM, and Burt HM. (2009). The differential *in vitro* and *in vivo* responses of bone marrow stromal cells on novel porous gelatin-alginate scaffolds. *Journal of Tissue Engineering and Regenerative Medicine* 3(8): 601–614.

Yeung TY, Seeberger KL, Kin T et al. (2012). Human mesenchymal stem cells protect human islets from pro-inflammatory cytokines. *PloS One* 7(5): e38189.

Zappia E, Casazza S, Pedemonte E et al. (2005). Mesenchymal stem cells ameliorate experimental autoimmune encephalomyelitis inducing T-cell anergy. *Blood* 106(5): 1755–1761.

Zhang HT, Chen H, Zhao H, Dai YW, and Xu RX. (2011). Neural stem cells differentiation ability of human umbilical cord mesenchymal stromal cells is not altered by cryopreservation. *Neuroscience Letters* 487(1): 118–122.

Zhang P, Baxter J, Vinod K, Tulenko TN, and Di Muzio PJ. (2009). Endothelial differentiation of amniotic fluid-derived stem cells: Synergism of biochemical and shear force stimuli. *Stem Cells and Development* 18(9): 1299–308.

Zinger A, and Leibowitz G. (2014). Islet transplantation in type 1 diabetes: Hype, hope and reality—A clinician's perspective. *Diabetes Metab Res Rev.* 30(2): 83–87.

# Section III

*Application of Supermacroporous Cryogels in Biotechnology*

# 11 Enzymatic Biocatalysts Immobilized on/in the Cryogel-Type Carriers

*Elena N. Efremenko,\* Ilya V. Lyagin and Vladimir I. Lozinsky*

## CONTENTS

11.1 Introduction ................................................................................................308
11.2 Protein Immobilization by Covalent Attachment to Cryogel Matrix...........309
11.3 Protein Immobilization via Entrapment Technique ....................................315
11.4 Protein Immobilization by Ionic Binding....................................................320
11.5 Protein Immobilization by Affine Binding .................................................320
       11.5.1 Protein Immobilization via Antigen-Antibody Binding..................321
       11.5.2 Protein Immobilization via Lectin Affinity Binding.......................322
       11.5.3 Protein Immobilization via Metal Ion Chelating Binding...............322
11.6 Protein Immobilization by Combined Methods ..........................................325
11.7 Conclusion ...................................................................................................325
List of Abbreviations............................................................................................326
References............................................................................................................326

### ABSTRACT

This chapter aims to present the latest advances in the field of the development of immobilized enzymatic biocatalysts and protein products using various cryogel-type carriers. The key approaches to the production of immobilized forms of functionally and catalytically active products with high stability in aqueous media and in media with organic solvents were analysed. The diversity of enzymatic and protein systems whose characteristics can be substantially improved by immobilization in polymeric cryogel matrices with a **macroporous** or **wide-porous structure** was performed.

### KEYWORDS

Cryogel, enzyme, immobilization, macroporous structure, wide-porous structure

---

\* Corresponding author: E-mail: elena_efremenko@list.ru; Tel: +74959393170.

## 11.1 INTRODUCTION

Enzymatic and protein-based products for biomedicine and biotechnology are becoming increasingly widespread. However, due to the natural characteristics of macromolecules used in these products, additional treatment or specific production conditions are required in order to enhance their stability during storage and everyday use. Immobilization of proteins/enzymes on/in various carriers has been successfully applied for these purposes (Sheldon and van Pelt, 2013); in addition to the fact that immobilization enhances the durability of catalytic and functional properties, it adds the possibility of reusing stabilized forms of enzymes/proteins in various biotechnological processes, which improves their economic attractiveness. A variety of techniques employing various carriers is used to immobilize enzymes/proteins; among these, cryogel-type polymeric matrices are particularly notable. Processes of cryotropic gelation of polymeric systems occur in the nondeep freezing, storage in the frozen state and thawing of the solutions or colloidal dispersions containing monomeric or polymeric precursors potentially capable of producing gels. Polymeric materials formed under these conditions are termed cryogels and they possess some specific features as compared to conventional gels formed at temperatures higher than the crystallization point of the solvent (Lozinsky, 2008). Thawing of the frozen sample produces a cryogel containing cavities filled with the liquid that formed in the sites of melted crystals. Thus, polycrystals of the frozen solvent act as a porogen during cryogel formation. Depending on the properties and initial concentration of the precursors and conditions of cryogenic processing, it is possible to produce macroporous matrices with the pores from tenth fractions of micrometers to ~10 μm in the cross-section and wide-porous (spongy) systems with pores of tens and hundreds of micrometers. Therefore, there are two main types of cryogel carriers with wide-porous and macroporous structure, respectively.

Polymeric cryogels have been shown to demonstrate improved performance, including chemical stability and mechanical strength, which is greater than that of many other types of carriers. In addition, a cryogel polymer matrix readily lends itself to modification with various spacers and ligands required for immobilizing enzymes/proteins, which substantially expands the scope of potential applications of such carriers for the production of various new biocatalysts and their use in catalytic reactions. The high chemical stability of polymeric cryogel carriers makes them applicable both in aqueous media and in media containing organic solvents (Filippova et al., 2001; Lysogorskaya et al., 2008). Taken together, these properties allow for solving of nontrivial problems of biotechnology and biomedicine using proteins and, in particular, enzymes immobilized in/on such matrices.

Analysis of articles and patents (data obtained from scholar.google.com) published in the field of development of immobilized enzyme forms employing macroporous cryogels as carriers (Figure 11.1) indicated exponential growth of interest in this problem over the past 10 years. Given this information, one can make predictions about future trends in developments in this field of study. Thus, the total number of reported study results in the field of development of immobilized enzyme products based on macroporous cryogels can be expected to double by 2018 relative to the number of reports already published by 2014. By 2021, the number of these

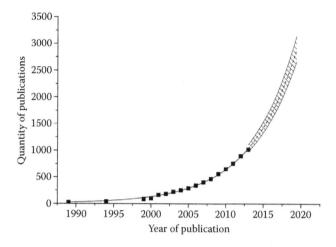

**FIGURE 11.1** Publication activity (articles and patents) in the development of enzyme products immobilized on cryogel-type carriers (based on the data of scholar.google.com).

works is likely to increase by a factor of at least 2.6, if the current rate of growth of the number of such studies remains stable.

Worldwide, the key line of research in this field concentrates on the development of new, novel methods for immobilizing various enzymes on polymeric cryogel carriers. Owing to this, the review focuses on state-of-the-art immobilized biocatalytic systems that are created using the potential of various enzymes and cryogel-type carriers.

## 11.2 PROTEIN IMMOBILIZATION BY COVALENT ATTACHMENT TO CRYOGEL MATRIX

Immobilization by strong covalent attachment of enzymes to the carrier is a rather actively used and thoroughly characterized technique for producing stabilized forms of broad-spectrum enzymes. The simplicity of this technique lies in the fact that the stage of carrier production and activation is temporally and spatially separated from the immobilization procedure. This circumvents limitations associated with the production of the carrier itself, allowing the use of reagents or conditions that partially or completely can inactivate the enzymes to be immobilized. This is also important in the case of cryogels, as they are obtained at subzero temperatures, while freezing is known to render many enzymes largely inactive (Feller, 2010). On the other hand, the immobilization technique discussed here enables optimization of multiple process parameters, which ultimately affects efficacy in a positive way.

It follows from Table 11.1 that the development of new biocatalytic systems involved testing of a rather wide range of enzymes from a variety of classes as moieties for covalent immobilization on cryogel-type carriers. The selection of individual enzymes in this study was based on practical interest in the immobilized catalyst under development and in its catalytic functions.

## TABLE 11.1
## Known Types of Covalent Immobilization on Cryogel-Type Carriers

| Enzyme | Polymer Used for the Carrier Preparation[a] | Activating Agent[a] | Biocatalyst Application | Ref. |
|---|---|---|---|---|
| Thermolysin | PVA | GA, epichlorohydrin | Peptide synthesis | Filippova et al. (2001); Belyaeva et al. (2008) |
| Subtilisin | PVA | GA, divinylsulfone, epichlorohydrin | Peptide synthesis | Bacheva et al. (2001, 2003, 2005); Filippova et al. (2001); Belyaeva et al. (2005) |
| α-Chymotrypsin, subtilisin | PVA | GA | Peptide synthesis | Semashko et al. (2008) |
| α-Chymotrypsin | PVA | GA | Peptide synthesis | Markvicheva et al. (2005) |
| Trypsin | PVA | GA, divinylsulfone, terephthalic dialdehyde, succinic dialdehyde | Peptide synthesis and hydrolysis | Lysogorskaya et al. (2008) |
| Lipase | PVA | GA | Amino acid synthesis | Plieva et al. (2000) |
| Organophosphate hydrolase | PVA | GA | Organophosphorus compounds hydrolysis | Efremenko et al. (2002) |
| Glucose oxidase/horseradish peroxidase, savinase/esperase | Mixture of albumin and chitosan | GA | Glucose analysis, protein hydrolysis | Hedström et al. (2008) |
| Laccase | PVA | GA | Oxidation of aromatic substrates | Stanescu et al. (2010, 2012) |
| Lysozyme | PAAm | GA | Capture of bacterial endotoxins | Hanora et al. (2005) |

[a] GA – glutaraldehyde; PAAm – poly(acryl amide); PVA – poly(vinyl alcohol).

Notably, a variety of proteases capable of hydrolysing proteins and peptides, when covalently immobilized in polyvinyl alcohol (PVA) cryogels, can be efficiently employed for biosynthetic reactions in water-organic media largely consisting of organic solvents that inactivate the soluble forms of the same enzymes. Thus, trypsin immobilized in PVA cryogel could efficiently catalyse the synthesis of N-carbobenzoxy-L-Phe-L-Arg-L-Leu p-nitroanilide from N-carbobenzoxy-L-Phe-L-Arg methyl ester (or N-carbobenzoxy-L-Phe-L-Arg) and L-Leu p-nitroanilide, as well as the formation of N-carbobenzoxy-L-Ala-L-Ala-L-Arg-L-Phe p-nitroanilide from N-carbobenzoxy-L-Ala-L-Ala-L-Arg and L-Phe p-nitroanilide in organic solvents mixture (dimethylformamide and acetonitrile, DMF-MeCN) (Lysogorskaya et al., 2008). The chemistry of the activator employed for introducing reactive groups in the cryogel carrier matrix is important for the production of a highly active catalyst. Thus, among the activating agents tested, which included dialdehydes (terephthalic, succinic or glutaric) or divinyl sulfone, glutaric aldehyde yielded maximum carrier capacity in terms of the protein. However, the highest specific activity relative to 1 mg of protein was achieved, with somewhat lesser carrier capacity in terms of the protein and greater activity of the immobilized biocatalyst, when the carrier matrix was activated using divinyl sulphone. Owing to the macroporous cryogel structure, peptide synthesis carried out using this immobilized enzyme produced the target product with a 65% yield of over 24 h.

Covalent immobilization of subtilisin and thermolysin in a matrix based on PVA cryogel carried out by linking the enzymes to a carrier modified with glutaraldehyde or epichlorohydrin (Filippova et al., 2001) allowed developing a biocatalyst capable of synthesizing various N-acylated p-nitroanilides of tetrapeptides with the general formula Z–Ala–Ala–Xaa–Yaa–pNA (where Z is benzyloxycarbonyl; Xaa is Leu, Lys or Glu; Yaa is Phe or Asp; and pNA is p-nitroanilide) in anhydrous media. This type of immobilization was demonstrated to ensure high stability of enzymes in a DMF-MeCN medium and a greater than 90% yield of target peptides over just 2 h.

The use of a similar approach to the production of immobilized biocatalysts based on thermolysin covalently linked with PVA cryogel using glutaraldehyde (Belyaeva et al., 2008) allowed optimizing synthetic conditions for tetrapeptides of the general formula Z–Ala–Ala–Xaa–pNA, where Xaa is Leu, Ile, Phe, Val or Ala. The following characteristics were especially important for optimizing the application of such immobilized biocatalyst in the biosynthetic process carried out in the organic solvent: total enzymatic activity introduced into the reaction medium, substrate concentration, water content and organic solvent ratio. Efficient work of the biocatalyst itself under empirically selected conditions of the biosynthetic reaction allowed achieving almost quantitative yields of the target products.

Another series of studies, where subtilisin 72 was the key study subject, demonstrated that the biocatalyst obtained by covalent immobilization of this enzyme in PVA cryogel can be reused (at least three times) for the synthesis of tetrapeptides Z-Ala-Ala-Xaa-Phe-pNA (Xaa is Leu, Glu or Lys) with only a slight decrease in the yield of the target product in media with organic solvents mixture (DMF-MeCN) (Bacheva et al., 2001).

Direct comparison of the catalytic activity of native subtilisin 72, its noncovalently linked enzyme–polyelectrolyte complex with polyacrylic acid, and subtilisin

72 covalently immobilized in PVA cryogel (Bacheva et al., 2003) in the synthesis of tetrapeptides of the general formula Z-Ala-Ala-Xaa-Yaa-pNA (where Xaa is Leu, Lys or Glu; Yaa is Phe or Asp) demonstrated a clear advantage of PVA cryogel in media with high levels of organic solvents (>80% DMF).

Tests of various activating agents (epichlorohydrin, divinylsulfone or glutaraldehyde) used for the covalent immobilization of subtilisin 72 in PVA cryogel matrix (Bacheva et al., 2005) established that the maximum content of immobilized enzyme in the biocatalyst is achieved with the use of divinyl sulphone and glutaraldehyde. At the same time, increases in enzyme levels in immobilized biocatalyst were invariably associated with a decline in specific activity per 1 mg of protein introduced. Thus, it was demonstrated that covalently immobilized enzyme products are characterized by an optimum protein load; exceeding it causes suboptimum enzyme function in the resulting product. From this standpoint, biocatalyst samples obtained using a carrier activated with glutaraldehyde were the best performers. An immobilized biocatalyst obtained using this technique based on subtilisin 72, when used for the synthesis of the tetrapeptide Z-Ala-Ala-Leu-Phe-pNA, was stable not only in media with high DMF concentration, which was due to aqueous microenvironment of the enzyme inside the cryogel matrix, but retained over 80% of initial activity in a 5 M solution of urea, an extremely strong chaotropic agent causing protein denaturation (Almarza et al., 2009).

Biocatalysts based on subtilisin 72 covalently immobilized in PVA cryogel were also shown to be usable for stereoselective enzymatic synthesis of dipeptides and tripeptides (Belyaeva et al., 2005). Yields of the end product were as high as 90%. It should be noted that there are reports (Anikina et al., 2008) in which cryogels were used as carriers for solid-phase matrix enzymatic synthesis of longer oligopeptides.

According to the studies performed, α-chymotrypsin that is covalently immobilized in PVA cryogel carriers and capable of hydrolysing proteins can also be used for synthesizing peptides *de novo*.

Comparison of the efficacy of action of two proteases (α-chymotrypsin and subtilisin) covalently immobilized in PVA cryogel in similar reactions of the synthesis of fluorogenic substrates for cysteine proteases of the papain family, Abz-Phe-Ala-pNA and Glp-Phe-Ala-Amc (Abz, pNA, Glp and Amc are o-aminobenzoyl, p-nitroanilide, pyroglutamyl and 4-amino-7-methylcoumaride, respectively), carried out in DMF-MeCN medium (Semashko et al., 2008), demonstrated that covalently immobilized α-chymotrypsin ensures a 100% yield of the product over a shorter time span (24 h) than covalently immobilized subtilisin, which produces the maximum attainable yield (56.2%) in 96 h. Both biocatalysts were consistently active in media with organic solvents.

Enzymes covalently immobilized on PVA cryogel-type matrices were demonstrated not only to produce high yields of catalytic reaction products, but also to preserve enantioselectivity, which is typical of the soluble form of the enzyme. Thus, enantioselective hydrolysis of a Schiff's base derived from *p*-chlorobenzaldehyde and ethyl ester of *DL*-phenylalanine was demonstrated in a water-acetonitrile medium using hog pancreas lipase attached to PVA cryogel with glutaraldehyde (Plieva et al., 2000). The main product of the reaction was *L*-phenylalanine, with an enantiomeric excess of 83% achieved in 144 h.

It should be noted that the properties of immobilized biocatalyst are affected not only by the method of enzyme immobilization, but also by the enzyme's own characteristics. Thus, when a common approach was used to produce samples of enzymes covalently attached to PVA cryogel modified with glutaraldehyde, the resulting biocatalyst based on α-chymotrypsin (Markvicheva et al., 2005) had a lower enantioselectivity than a similar immobilized biocatalyst based on lipase (Plieva et al., 2000), despite the high yield of the target product (80–90%).

Analysis of results published by a number of researchers revealed that immobilized enzymes could be rendered more stable using additional techniques and stabilizers. Thus, immobilized enzyme in PVA cryogel could be additionally stabilized by immobilizing it in the form of a presynthesized enzyme-polyelectrolyte complex. The efficiency of this technique was demonstrated in experiments with organophosphate hydrolase (OPH) attached to PVA cryogel modified by glutaraldehyde (Efremenko et al., 2002) (Figure 11.2). Thus, if immobilization was carried out using a polyelectrolyte complex consisting of the enzyme and 1,5-dimethyl-1,5-diazaundecamethylene polymethobromide (PB), the activity of the resulting biocatalyst in 50% ethanol was twice as high as the activity of the biocatalyst produced by immobilizing nonpolyelectrolyte complex-stabilized enzyme. In addition, this immobilized enzymatic biocatalyst was significantly more stable during storage and use.

Additional enzyme stabilization by enzyme–polyelectrolyte complexes synthesized prior to covalent immobilization in cryogel-type carriers was also successfully employed in work with lysozyme (Hanora et al, 2005). Thus, lysozyme was initially transformed into a complex with polyethylenimine, which was additionally doped with the antibiotic polymyxin B; this version of stabilized enzyme was subsequently immobilized by covalently attaching it to polyacrylamide (PAAm) cryogel via a glutaraldehyde linkage. This biocatalyst was capable of efficiently capturing bacterial endotoxins of *Escherichia coli* even from media containing ballast proteins. In addition, owing to the use of a wide-porous sponge-like cryogel, removal of cells and cell debris from the endotoxin-containing culture medium was not necessary.

Another approach to the development of highly stable biocatalysts based on covalently immobilized proteins involved investigation of a wide range of enzymes (glucose oxidase, horseradish peroxidase, savinase, esperase), which were introduced into macroporous composite cryogel-type carriers obtained based on hen egg albumin and chitosan cross-linked by glutaraldehyde (Hedström et al., 2008). A biocatalyst based on glucose oxidase/horseradish peroxidase was subsequently used for optical glucose flow assay. However, due to the low efficiency of the spectrophotometric detection system, the limit of detection of this biosensor was rather unsatisfactory at 0.5 mM. Investigation of the properties of a second biocatalyst based on savinase/esperase confirmed that, in theory, it could be used for protein hydrolysis in the flow systems. This possibility specifically depends on the use of macroporous cryogels. Nevertheless, further studies of the biocatalysts developed are obviously required, as the simplicity and originality of the implemented approach to enzyme immobilization provoke a vivid interest.

High efficacy of a method for producing immobilized biocatalysts with laccase activity that involves the covalent attachment of this enzyme to PVA cryogel was demonstrated in a series of works (Stanescu et al., 2010, 2012). The authors developed and optimized a procedure for producing immobilized biocatalysts, which

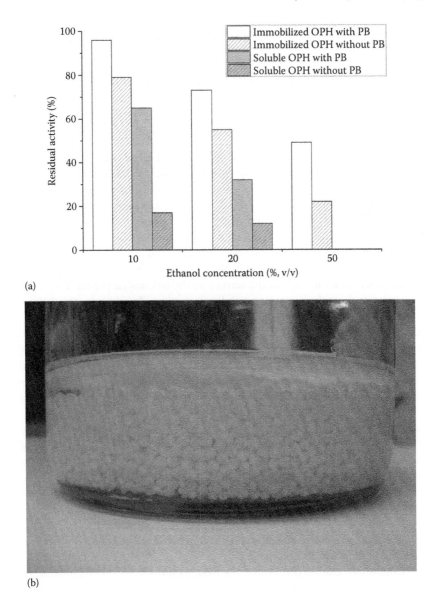

**FIGURE 11.2** Influence of ethanol concentration on the enzymatic activity of different organophosphate hydrolase preparations (a) and photo of the granules of immobilized biocatalyst obtained with polybrene (PB) addition (b).

involved testing three different approaches to laccase immobilization. According to one approach, cryogel-type carrier activated by glutaraldehyde was directly contacted with laccase solution. In another technique, PVA cryogel was initially modified by glutaraldehyde, then contacted with N-(3-dimethylaminopropyl)-N′-ethyl-carbodiimide hydrochloride, and afterward, with laccase solution. In a third method, PVA cryogel was activated by glutaraldehyde and then contacted with aqueous β-alanine solution, whereupon it was treated with N-(3-dimethylaminopropyl)-N'-ethyl-carbodiimide hydrochloride; only then laccase immobilization was carried out.

These experiments demonstrated that increase in spacer length leads to a reduction in the quantity of immobilized enzyme, and maximum enzyme capacity can be achieved with a carrier activated only with glutaraldehyde, omitting any further tricks. This immobilization allowed improvement of the stability of laccase during storage by an order of magnitude, as well as use of the developed laccase biocatalyst for oxidising various phenolic compounds in apple juice (Stanescu et al., 2012). In principle, the cryogel-type matrix employed should not create diffusion or mass exchange hindrances for the substrate. However, judging by the catalytic constants of the immobilized enzyme, its activity was substantially reduced, and the catalytic activity was less than that of the soluble form by a factor of 100 or more. As a result, treatment of natural apple juice samples with the developed biocatalyst resulted in a greater total residual concentration of phenolic compounds than in the case of free enzyme. However, it should be noted that, in theory, this biocatalyst could be used in a number of very different fields; the most important of these is probably wastewater treatment, including elimination of stable polyphenolic compounds. In this case, the benefits of the developed biocatalyst are expressed to the maximum extent—most importantly, enhanced stability and separability allowing for reuse, as well as lack of mass exchange hindrances, which are typical of all cryogel-based carriers.

Owing to the development of immunoassays for antigen detection, special interest is drawn to works in which polyclonal antibodies, rather than enzymes, are covalently immobilized on cryogel-type carriers. Of interest are studies in which polyclonal antibodies capable of efficient capture of viral particles from the solution are immobilized in PVA cryogel activated by glutaraldehyde (Plieva et al., 1998). The resulting product could be employed to efficiently identify and isolate virions from solutions. Moreover, due to the use of PVA cryogel, such a process was comparable in efficiency to traditionally used sorbents in this field or in the case of large virus particles, surpassed them.

To summarize the information presented in this section on enzyme immobilization by chemical attachment to cryogels, it must be noted that in all cases this type of immobilization represents a good approach to the production of efficient biocatalysts that are resistant to various inactivating factors (temperature, pH, presence of organic solvents, etc.) and exhibit high stability during storage and use. The greatest efficiency is usually observed when glutaraldehyde is used to modify the surface of cryogel-type carriers followed by covalent immobilization of enzymes on such carriers. A positive effect is also achieved by introducing additional stabilizing agents, including other polymers that form enzyme-polyelectrolyte complexes with the enzyme to be immobilized, with beneficial effects for a variety of biocatalyst characteristics, predominantly for catalytic performance.

## 11.3 PROTEIN IMMOBILIZATION VIA ENTRAPMENT TECHNIQUE

The second most important, although not the second most popular method, for immobilizing enzymes in cryogels is the entrapment technique. Entrapment of the enzyme inside a carrier matrix can be achieved in a number of ways (Table 11.2).

## TABLE 11.2
## Some Samples of Enzyme Immobilization by Its Entrapment

| Enzyme | Polymer Used for Cryogel Production | Gel-Formation Technique | Biocatalyst Application | Ref. |
|---|---|---|---|---|
| **1. Wide Porous Cryogel Carriers** | | | | |
| Peroxidase-labeled anti-rabbit IgG goat IgG, anti-peroxidase, peroxidase | PHEMA,[a] poly(neopentylglycol dimethacrylate), poly(trimethylolpropane triacrylate) | Radical polymerization induced by γ-rays | ELISA | Kumakura et al. (1983) |
| Urease | Poly(N-isopropylacrylamide) | Absorption, freezing–thawing | Urea degradation | Petrov et al. (2011) |
| Urease | PHEMA covered by PEG | Absorption by inner layer | Analysis of heavy metals, urea degradation | Petrov et al. (2012) |
| **2. Macroporous Cryogel Carriers** | | | | |
| Glucoamylase | PVA | Iterative freezing–thawing | Starch conversion | Kokufuta and Jinbo (1992) |
| PEGylated glucose oxidase | PVA | Iterative freezing–thawing | Analysis of glucose | Doretti et al. (1998) |
| PEGylated ribonuclease A, superoxide dismutase or catalase | PVA | Iterative freezing–thawing | Various biotechnological and therapeutic applications | Veronese et al. (1999) |
| AChE, PEGylated choline oxidase | PVA | Iterative freezing–thawing | Analysis of acetylcholine and choline | Doretti et al. (2000) |
| PEGylated lipase | PVA | Iterative freezing–thawing | Hydrolysis of acetoxycoumarins | Veronese et al. (2001) |
| Keratinase | Composite of PVA and pectin | Iterative freezing–thawing | Controlled release of antimicrobial preparation | Martínez et al. (2013) |
| Naringinase | PVA | Freezing–thawing | Hydrolysis of naringin | Busto et al. (2007) |
| Benzaldehyde lyase, mixture of carbonyl reductase with formate dehydrogenase | Mixture of PVA with PEG | Freezing–thawing | Stereoselective organic synthesis | Hischer et al. (2006) |

[a] PAAm – poly(acryl amide); PEG – poly(ethylene glycol); PHEMA – poly(hydroxyethyl methacrylate); PVA – poly(vinyl alcohol).

One of the first known techniques for entrapping enzymes in cryogel-type matrices is radical polymerization of frozen solutions of hydroxyethyl methacrylate, neopentylglycol dimethacrylate or trimethylolpropane triacrylate driven by γ-radiation, which was tested for the immobilization of peroxidase-labelled anti-rabbit IgG goat IgG, antiperoxidase or peroxidase (Kumakura et al., 1983). These proteins, which were introduced into the monomer mixture before freezing and polymerization, were embedded in a matrix of immobilized biocatalyst. An increase in the concentration of hydrophilic or hydrophobic monomer in the mixture caused a decrease or increase in the biocatalyst's enzymatic activity, respectively. Solid-phase sandwich-ELISA was one possible application of the developed biocatalyst. Immobilized form of the enzymes allowed reusing them as biosensitive elements, while the cryogel performed the function of a matrix with an extensive internal structure.

Another method for enzyme immobilization by entrapment in a cryogel-type carrier was demonstrated in a work where PVA cryogel was produced using a glucoamylase-containing solution of the polymer (Kokufuta and Jinbo, 1992). Concanavalin A was used as an additional component for boosting the biocatalyst's enzymatic activity, which could not be achieved by adding human serum albumin. Introduction of concanavalin A had almost no effect on carrier capacity in terms of glucoamylase; at the same time, the biocatalyst's enzymatic activity in the presence of concanavalin A grew 2.7 times relative to the biocatalyst without added concanavalin A. The resulting immobilized biocatalyst could be reused to carry out efficient hydrolysis of starch to glucose.

Other polymers can also be introduced in the enzyme mixture as additional components; for instance, polyethylene glycol (PEG) can be used as a cryoprotecting agent for the biocomponent (Hischer et al., 2006). Biocatalysts based on benzaldehyde lyase, as well as a mixture of carbonyl reductase with formate dehydrogenase, immobilized in PEG-containing PVA cryogel were efficiently used for organic synthesis in a two-phase reaction medium. In this reaction, an aqueous medium inside the cryogel-type matrix acted as one of the phases, with poorly soluble substrates diffusing into it from the organic solvent (hexane). Thus, the authors successfully used the benefits afforded by cryogel-type carriers to create a biocatalytic system.

Another approach to efficacy enhancement of protein immobilization in a carrier matrix lies in increasing the molecular weight of the biological product prior to entrapment in the polymer by means of a targeted chemical modification of the enzyme before immobilization. For instance, pegylation of glucose oxidase allowed for a more than twofold increase in the enzymatic activity of a biocatalyst obtained by the entrapment of the enzyme in a PVA matrix that was produced by multiple freeze–thaw cycles of the source polymer solution (Doretti et al., 1998). This immobilized biocatalyst was suitable as the biosensitive element of a biosensor for blood glucose assay.

In the same way, that is, using a technique aimed at increasing the molecular weight of the protein by pegylation followed by entrapment of the modified bioproduct in a wide-porous cryogel-type carrier, multiple enzymes can be modified simultaneously. For instance, pegylated choline oxidase was co-immobilized with acetylcholinesterase in PVA cryogel (Doretti et al., 2000). The resulting immobilized biocatalyst was suitable for determining choline and acetylcholine.

The influence of the molecular weight of PEG used for modification of the enzyme surface, as well as the molecular weight of PVA used for preparing polymer solutions, which are subsequently subjected to cryostructuring, on the efficacy of enzyme immobilization and retention in the carrier matrix has been thoroughly studied. One study covered a variety of enzymes (ribonuclease A, superoxide dismutase, catalase), which are used for various biotechnological and therapeutic applications (Veronese et al., 1999). The authors established a nonlinear relationship between the amount of enzyme eluted from the cryogel carrier after immobilization and the number of introduced PEG molecules with a molecular weight of 5 kDa. As a result, as the number of PEG monomer units increases from 9 to 67, the difference between residual activities of pegylated and nonpegylated immobilized enzyme grows from 20% for ribonuclease (molecular weight 14 kDa) to 50% for catalase (molecular weight 250 kDa). The authors concluded that the molecular weight of such pegylated complexes is the key factor for the preservation of enzymatic activity in a biocatalyst matrix. Nevertheless, this effect apparently can be produced by a number of factors acting together, most notably by interpolymeric complexes that may form between PEG residues introduced into the protein structure and the polymer matrix (PVA). As a result, the efficacy of enzyme retention in the carrier must depend on the number of PEG molecules—an effect that was also demonstrated. In addition, an increase in the molecular weight of PVA from 31–50 kDa to 124–186 kDa was also demonstrated to entail a 1.2–2.5-fold decrease in activity eluted from the carrier.

Veronese et al. (2001) investigated the possibility of using such pegylated enzymes entrapped in a cryogel-type carrier matrix in media with organic solvents. In particular, pegylated lipase was immobilized in PVA cryogel and subsequently used for deacylation of acetoxycoumarins. It was demonstrated that despite substantially reduced elution of the enzyme from the carrier, the resulting immobilized biocatalyst was not stable in media with organic solvents (n-hexane/isopropanol), losing up to 60% of its initial activity over six cycles of use for the hydrolysis of acetoxycoumarins. Thus, these studies identified substantial limitations in the use of this technique of enzyme immobilization in cases when organic solvent-based reaction media had to be used, despite the fact that an aqueous phase was present inside the cryogel-type matrix.

A separate line of research encompassed the influence of the concentration of PVA solution used for immobilization and the concentration of enzyme introduced in the polymer solution prior to cryostructuring on the efficacy of enzyme immobilization in a cryogel-type matrix (Busto et al., 2007). The example of naringinase, an immobilized product of which was developed for efficient hydrolysis of naringin in simulated juice, helped to establish that the maximum efficacy of enzyme immobilization could be attained with the maximum PVA concentration used (8%). Enzyme concentration in the investigated range (0.32–1.08 U/$g_{carrier}$) had negligible effect on the resulting efficacy of immobilization and activity of the immobilized biocatalyst obtained. Thus, cryostuctured polymer and, in particular, its concentration used for obtaining the carrier was demonstrated to be the main factor determining the efficacy of enzyme immobilization.

Another known technique for introducing enzymes into cryogel-type matrices is based on physical absorption of the enzyme by the bulk of presynthesized carrier, which is an attractive option by virtue of the simplicity of its implementation. One example is the immobilization of urease in preliminary prepared poly(N-isopropylacrylamide) cryogel (Petrov et al., 2011). Nevertheless, in comparison with entrapment, when the protein is frozen in a solution with the polymer or monomer mixture undergoing polymerization, this approach proved to be less fruitful. This is because immobilized enzyme could be freely eluted from the carrier matrix, and residual activity fell to 70% of the initial value in as little as 24 h. Meanwhile, a biocatalyst obtained using immobilization by entrapment could be used to decompose urea in a flow mode for six work cycles. No substantial elution of the enzyme from the cryogel matrix or reduction of its catalytic activity was detected in the process.

A similar study, where urease was immobilized in another preformed cryogel of poly(2-hydroxyethyl methacrylate), demonstrated that enzyme elution from the cryogel matrix could be prevented as follows: after the enzyme is absorbed by the carrier, the latter is coated with a mixture of PEG and low molecular weight poly(ethylene glycol) diacrylate, which is then polymerized using the photosensitive agent ((4-benzoylbenzyl) trimethylammonium chloride) (Petrov et al., 2012). Due to the presence of macropores, the substrate can freely penetrate into the cryogel-type matrix, while elution of the enzyme from the carrier is prevented by the additional coating. Immobilized biocatalyst produced in this way was used to determine the concentration of Cu(II), which inhibits urease; it was also employed for the decomposition of urea in a batch process.

The possibility of solving a problem requiring controlled slow release of enzyme, rather than simple retention of enzyme in a cryogel-type matrix, was demonstrated in a work with keratinase (Martínez et al., 2013). According to the authors, keratinase was co-immobilized with enrofloxacin in a composite cryogel-type carrier produced based on PVA and pectin and intended for use as an antiseptic cutaneous product. The amount of enzyme released varied depending on the content of pectin in the composite and on the degree of its esterification, temperature and NaCl concentration in the medium, as well as incubation time. However, in *in vitro* studies under nearly physiological conditions, cutaneous application of this antiseptic drug resulted in a 6.9% release of the second component (enrofloxacin) in the presence of keratinase, in contrast to the expected 15% release. Nevertheless, the first step in the development of controlled slow release systems based on enzyme-containing cryostructured materials can be considered successful, and this field has potential for development in this field of research.

In summary, one can conclude that enzyme immobilization by entrapment in polymeric cryogel matrix has a number of advantages and disadvantages. Disadvantages include low efficacy of immobilization and decline of biocatalyst stability during use, which is due to the elution of enzyme from wide-porous cryogel-type matrix. Correction of this situation necessitates additional measures, such as chemical modification of immobilized enzyme, or optimization of carrier, or postproduction modification of the biocatalyst (additional treatment with cross-linking agents, utilization of additional coatings) and so on. It must be noted that this field is developing more rapidly and yields more interesting methods of implementation than the classical enzyme immobilization by chemical attachment. The most attractive feature is the possibility of controlled release of immobilized products produced based on cryogel-type carriers using this technique.

## 11.4 PROTEIN IMMOBILIZATION BY IONIC BINDING

Protein ionic binding has been used in the laboratory for a long time, predominantly for analytical and preparative chromatography. Chromatography-related peculiarities of this approach are described in detail elsewhere (Di Palma et al., 2012). Advantages of cryogels, in turn, can be efficiently utilized for enzyme immobilization by ionic binding.

A team who developed a biocatalyst based on hen egg-white lysozyme (Bibi et al., 2011) used a composite carrier produced by embedding EXPRESS-ION™ Exchanger Q particles in a polymerization cryogel based on poly(hydroxyethyl methacrylate) (PHEMA) in order to immobilize the enzyme. This biocatalyst was developed for breaking down bacterial cells by the action of lysozyme immobilized by ionic binding. However, despite the large capacity of ion exchange particles per se, the capacity of the composite material remained at almost the same level as in the case of PAAm-based cryogel without added particles. Moreover, grafting PAAm cryogel by sulpho groups increased its capacity in terms of the protein by a factor of 14.5. This opportunity resulted from the use of a cryogel carrier that freely admitted monomer solution into its pores; the latter then polymerized based on this matrix.

The use of a chemical initiator of graft polymerization, $(Ce(NH_4)_2(NO_3)_6)$, instead of the traditional γ-radiation, led to a 1.6-fold decline in the ionized molecule capacity (Bibi and Fernández-Lahore, 2013). At the same time, the dynamic binding capacity of carrier in terms of the target immobilized enzyme (lysozyme) grew by a factor of 3.3. The authors' explanation was that the biomolecules could easily access the tentacles or brush-type polymeric structure produced by grafting the inner surface of the cryogel. The tentacle-type polymer chains provide a three-dimensional space to biomolecules and thus high binding capacity is observed because of multipoint interaction. Additionally, the authors compared cation exchange and anion exchange moieties. The capacity of sulpho group-containing carriers was demonstrated to exceed that of cryogels with diethylamine groups sixfold.

Summarizing the available data, it must be noted that enzyme immobilization by ionic binding in cryogel-type carriers has found very limited application for the development of immobilized enzymatic biocatalysts. This is due to a number of objective causes. First, this method is not selective, and the immobilized enzyme can contain various protein impurities, which may improve or degrade the catalytic properties of the biocatalyst. Second, the stability issue of such products during storage and use is often omitted. Of course, ionic binding is a strong interaction; nevertheless, it can be degraded or changed by substances present in an aqueous solution. As a result, immobilized enzyme can be expected to leach from the carrier at a substantial rate, reducing the enzymatic activity of the biocatalyst.

## 11.5 PROTEIN IMMOBILIZATION BY AFFINE BINDING

Enzyme immobilization by affine binding is undergoing rather active development owing to the fact that this approach to the production of new biocatalysts has a number of crucial advantages.

First, high specificity and selectivity of binding can be considered the main advantage of this technique. On the one hand, this allows immobilizing enzymes without prior purification while ensuring specifically the immobilization of the target enzyme. On the other hand, it enhances the efficacy of immobilization and relaxes high capacity requirements for the carrier.

Second, high strength of enzyme binding by the carrier is an undisputed advantage of this immobilization technique. This produces favourable effects for the stability of such biocatalysts during storage and use. In addition, this allows regenerating the biocatalyst under controlled conditions by removing inactivated enzyme and depositing a new active product.

Many variants of biocatalysts have been produced using this approach to protein immobilization in/on cryogels by now. However, only a few of them will be considered below, namely, immobilization via antigen-antibody binding, lectin affinity immobilization and immobilization by metal chelation interactions.

### 11.5.1 Protein Immobilization via Antigen-Antibody Binding

Carrier modification with antigens is one of the most thoroughly understood approaches to protein (antibody) immobilization. Protein A is one of the simplest to use; at the same time, it has the widest specificity for various immunoglobulins. Its antibody binding strength is so great that it can hold human B-lymphocytes via conjugation with antibodies localized on the cell surface (Kumar et al., 2003).

Protein A is often affixed to a cryogel-type carrier using immobilization by covalent attachment, which, on the one hand, enables strong attachment to the matrix, while on the other hand, ensures that the carrier produces no detrimental effects on its antigenic activity. One example includes the use of PHEMA cryogel, to which protein A was attached via cyanogens bromide activation of matrix (Alkan et al., 2009). This allowed development of an efficient system for permanent and temporary immobilization, which was subsequently employed for multiple separations of immunoglobulin IgG from blood plasma. Cryogel-type carriers, due to the presence of wide pores, ensured high efficacy of this method, by allowing deposition of a complex protein mixture (blood plasma). This system also demonstrated high efficacy of IgM separation (Alkan et al., 2010).

Another possible approach used immobilization of an antibody or its fragment, rather than an antigen, on a cryogel-type carrier. For instance, anti-hIgG was immobilized in PHEMA cryogel activated by carbodiimide via additional binding of the anti-hIgG with an F(c) fragment of IgG, which was modified with a functional monomer (N-methacryloly-(L)-cysteine methylester) and built into the polymer matrix (so-called 'oriented binding') (Bereli et al., 2013). For comparison, anti-hIgG was also immobilized on a carrier containing no F(c) fragments of IgG ('random binding'). With oriented binding of anti-hIgG to the carrier, IgG separation efficacy was three times higher than with random binding, while the purity of IgG was 96.7%. Owing to the cryogel matrix, target antibodies could be separated directly from the blood plasma, which was stripped of blood cells by filtering through a membrane with a 3-μm pore size. This biosorbent could be used at least 10 times, with minimum decline in capacity.

Obviously, immunosorbents are considered in this text because antibodies can be conjugated with various immobilized reporter enzymes. Therefore, immobilization of enzymes on cryogel carriers can be realized through the antibody.

### 11.5.2 Protein Immobilization via Lectin Affinity Binding

Unlike systems based on antigen–antibody interactions, which have limited applicability (they are mostly used for separating immunoglobulins or for immunoassays), a wide selection of lectins can be used for immobilization as a solution for a variety of problems. High affinity for carbohydrates, binding strength and specificity of interaction are the key characteristics of lectins.

One representative of lectins is concanavalin A. A number of systems have been developed on its basis. Thus, concanavalin A was covalently attached to PAAm–allylglycidyl ether copolymer cryogel (Babac et al., 2006). This lectin-modified affine sorbent was subsequently used for multiple separations and purifications of human IgG, which was, evidently, glycosylated. It should be taken into account that proteins obtained because of such carrier interactions will be adulterated by other glycosylated proteins present in blood plasma, and further purification may be necessary.

Concanavalin A can also efficiently bind to nonglycosylated proteins. For instance, invertase was very efficiently immobilized on poly(ethylene glycol dimethacrylate) cryogel; concanavalin A was attached to an aminated derivative of this cryogel via glutaraldehyde (Uygun et al., 2012). Then conditions of biocatalyst preparation (medium pH, ionic strength, flow rate and enzyme concentration) were optimized. This approach to enzyme immobilization on concanavalin A-modified cryogel carrier allowed boosting carrier capacity in terms of the target enzyme by an order of magnitude and using the resulting biocatalyst for efficient hydrolysis of sucrose, yielding glucose-fructose syrup. Owing to poly(ethylene glycol dimethacrylate) cryogel having an extensive internal structure, highly concentrated viscous substrate solutions can be used for conversion.

A similar technique was used to solve a problem of immobilizing inulinase, which catalyses the hydrolysis of inulin to fructose (Altunbaş et al., 2013). The method for producing a biocatalyst based on inulinase was the same as in the previous case (Uygun et al., 2012). It should be noted that the optimum pH value for the immobilization of these two enzymes was approximately equal (pH 4 to 5). Apparently, these carbohydrate-converting enzymes (invertase and inulinase) have identical glycosylated binding sites for concanavalin A. Presumably, other carbohydrases could also be immobilized on cryogel-type carriers modified with concanavalin A using such interactions. This, undoubtedly, presents many new opportunities for the development of biocatalytic systems based on various bioproducts of considerable importance for biotechnology. The use of wide-porous cryogel carriers for the preparation of biocatalyst, in turn, can allow for the efficient conversion of carbohydrate-containing substrates, which often exist in the form of heterogeneous viscous media.

### 11.5.3 Protein Immobilization via Metal Ion Chelating Binding

Another kind of affine interaction is metal chelating binding, which is based on the formation of complexes between metal ions, which are firmly attached to the carrier, and various ligands present in the structure of immobilized enzymes. This approach is mostly used in a laboratory setting for separation and purification of various proteins; for instance, by means of complexation between carrier-bound metal ion and His-tag that was genetically introduced into a protein/enzyme molecule (Efremenko et al., 2006) or antibody (Dainiak et al., 2004).

Optimization of the conditions of modified carrier preparation, in addition to immobilization conditions, allows substantially enhancing the capacity of such a carrier in terms of the target protein, and, consequently, biocatalyst activity (Plieva et al., 2006).

Immobilization of organophosphate hydrolase (OPH) containing an N- or C-terminal polyhistidine tag in PAAm–allylglycidyl ether copolymer cryogel modified by iminodiacetic acid ligands and charged with Cu(II) or Co(II) ions allowed the development of a biocatalyst for decomposing organophosphorous compounds in a flow system (Efremenko et al., 2007, 2009). The fact that enzymes immobilized in this way could function under a wide range of conditions (Table 11.3), as well as their high stability during storage and use (for hexahistidine-tagged OPH ($His_6$-OPH), the activity half-life was over 420 days) justifies the conclusion that this approach to protein immobilization has substantial practical significance and can be readily applied in practice. In particular, 1 mL of $His_6$-OPH immobilized in this way was demonstrated to be capable of treating 10.8 m³ of a highly neurotoxic solution of the pesticide Paraoxon at the concentration of 27.5 ppm over one half-life. Due to the wide-porous structure of PAAm cryogel, medium flow rate through the bioreactor, at which the biocatalyst can function, could be substantially increased (up to two reactor volumes per minute). In addition, the wide pores of this carrier obviate the need to remove microscopic particulate matter from sewage waters containing hydrolysable pesticides, which also improves the operating characteristics of this immobilized product from the standpoint of practical use.

Metal chelating binding is less specific in the absence of specialized genetic modification of immobilized enzyme. This interaction can only be possible if amino

### TABLE 11.3
### pH Range of Maximal Activity (90–100%) Typical of Soluble and Immobilized Derivatives of Organophosphate Hydrolase in Hydrolysis of Organophosphorus Compounds

| Enzyme[a] | Biocatalyst Form | pH Range of Maximal Activity | Ref. |
|---|---|---|---|
| $His_6$-OPH | Soluble | 9.5–11.0 | Efremenko et al. (2008) |
| | Immobilized | 8.0–11.5 | Efremenko et al. (2009) |
| OPH-$His_6$ | Soluble | 9.7–11.2 | Efremenko et al. (2008) |
| | Immobilized | 8.0–11.5 | Efremenko et al. (2009) |
| $His_{12}$-OPH | Soluble | 10.2–11.7 | Efremenko et al. (2008) |
| | Immobilized | 8.0–11.5 | Efremenko et al. (2009) |

*Note:* Immobilization of enzymes was carried out via metal chelating binding to cryogel matrixes.

[a] $His_6$-OPH is organophosphate hydrolase (OPH) containing $His_6$-tag at the N-terminus, OPH-$His_6$ is OPH containing $His_6$-tag at the C-terminus, $His_{12}$-OPH is OPH containing $His_{12}$-tag at the N-terminus.

acid residues of histidine, cysteine, lysine and arginine are present in the native protein structure. At the same time, the use of cryogel carriers modified with various ligands, such as iminodiacetic acid residues, which can be charged with various metal ions, including Cu (II), allow not only to separate enzymes from media with a complex composition (Bansal et al., 2006; Kumar et al., 2006), but also to produce stable immobilized enzymatic biocatalysts (Stanescu et al., 2011). To wit, a biocatalyst capable of flow-system oxidation of anthraquinone derivatives (Acid Blue 62, bromaminic acid) was developed using laccase immobilized via this technique in a cryogel of poly(dimethylacryl amide) copolymer with allylglycidyl ether. In this case, the use of a cryogel carrier additionally improved the operating characteristics of the product that can be applied for the treatment of wastewater.

The use of a cryogel of PHEMA–allylglycidyl ether copolymer, which was also modified with iminodiacetic acid ligands, but was charged with Ni(II) ions, allowed producing a biocatalyst for removing urea from wastewaters (Uygun et al., 2013).

Ni(II), Cu(II) and Co(II) are not the only species that can be used as chelating metal ions. Thus, it was shown that cryogel of PAAm-allylglycidyl ether copolymer modified with the Cibacron Blue F3GA dye and charged with Fe (III) ions can be used to immobilize catalase (Tüzmen et al., 2012). Despite the fact that the modified carrier itself could adsorb substantial quantities of the enzyme, its loading with iron ions allowed boosting the capacity in terms of protein by an additional 25%. Of course, one could doubt whether immobilization in this case is indeed achieved owing to metal-chelate binding. Nevertheless, this variant of the method opens up new possibilities for immobilizing proteins using less toxic metal ions than $Cu^{2+}$, $Co^{2+}$ or $Ni^{2+}$.

An interesting consequence of the strong selective binding of biological products to metal-chelate carriers is the possibility of further manipulations with the immobilized protein itself. This non-trivial approach was used for the enzyme $His_6$-OPH (Gudkov and Efremenko, 2007). Immobilization of this enzyme was used to carry out its refolding on PAAm-allylglycidyl ether copolymer cryogel, since the initial enzyme was obtained in the form of inclusion bodies and solubilized in 6 M urea. Thus, this approach opens new possibilities for the production of proteins and, in particular enzymes, which can only be produced as inclusion bodies, rather than in the soluble form. Further practical development of this technique is to be expected, especially using macroporous cryogels, which allow direct deposition of bioparticles (inclusion bodies) in the form of a suspension.

Metal-chelate binding with cryogel-based carriers provides the most ample opportunities for the development of immobilized enzymatic biocatalysts, as it combines such features as strength of protein–carrier binding, comparable to that of immobilization by covalent attachment, and simplicity of production, as in the case of immobilization by entrapment. Meanwhile, enzymes do not undergo inactivation, which usually accompanies immobilization by covalent attachment; moreover, unlike with immobilization by entrapment, enzyme desorption from the carrier does not occur. At the same time, the carrier can be modified using simple chemical techniques using readily available and inexpensive reagents. However, under this approach, genetic modification of the original protein with introduction of an affine sequence is required for increasing the efficacy of immobilization.

Since the desirable features of this immobilization technique, particularly for recombinant proteins, has already led to metal-chelate binding on the base of cryogel-type carriers being actively used for the production of various bioproducts, continued active development of this field in connection with the design of a variety of new enzymatic biocatalysts can be expected.

## 11.6 PROTEIN IMMOBILIZATION BY COMBINED METHODS

A few developments in the field of cryogel-type carrier application for protein immobilization are difficult to assign to a particular type of immobilization, as they combine multiple approaches and, for this reason, merit special attention, as they potentially can be used for a range of enzymes.

Thus, the authors of several works used two different techniques for the immobilization of IgG or IgY immunoglobulins on cryogel-type carriers via an unspecified protein with the molecular weight of 33 kDa, which consisted of 306 amino acids and was produced in the form of inclusion bodies using biotechnological methods. According to the first immobilization technique, inclusion bodies were introduced by absorption into PAAm–allylglycidyl ether copolymer cryogel hydrophobically modified with sulfamethazine (Ahlqvist et al., 2006a) or phenol (Ahlqvist et al., 2006b). Polyclonal antibodies specific for this protein were then immobilized in the form of inclusion bodies.

According to the second immobilization technique, inclusion bodies were mixed with antibodies, yielding antigen–antibody complexes, followed by immobilization of the resulting conjugates in PAAm-allylglycidyl ether copolymer cryogel modified with protein A. These works aimed to create a prototype system for sandwich ELISA assay of bioparticles in the form of inclusion bodies.

Omitting the discussion of the advantages and disadvantages of this approach relative to popular immunoassay techniques (especially in view of the fact that immobilized complexes may undergo elution when the carrier is mechanically compressed [Dainiak et al., 2006]), one should mention that the immobilization of bioobjects on hydrophobically modified cryogel-type carriers is quite attractive. Apparently, this immobilization technique can be of interest for the case of large enzymatic complexes, as well as transmembrane enzymes with a large hydrophobic surface area.

## 11.7 CONCLUSION

The results obtained using enzymes and proteins immobilized on cryogel-type carriers confirm the existence of many opportunities for developing a variety of efficient biocatalytic systems that are required for solving a wide range of problems. Such biocatalysts often have serious advantages, as immobilization all by itself allows substantially enhancing their stability and reusing them in technological processes. Enzymes, on the other hand, allow for the use of biocatalysts under mild conditions and have a much higher catalytic activity even in comparison to conventional catalysts, which often helps significantly simplify complex processes. These advantages are combined with benefits afforded by wide-porous cryogel carriers, to wit, absence of mass transfer limitations and, consequently, opportunities for using higher flow rates, simplicity of production and scalability.

Many techniques for immobilization of enzymes in cryogel-type matrices have been proposed and tested. Of course, further promises of development in the field of enzymatic biocatalysts are closely associated with advances in other fields of knowledge. For instance, composite materials can be used to enhance the mechanical strength of biocatalysts; in particular, the introduction of $TiO_2$ nanoparticles in the carrier is known to substantially increase the strength of cryogels (Zhan et al., 2013) and improve sorption characteristics of such composite carriers (Yao et al., 2006). Moreover, additives can be introduced in cryogels not only in the form of nanoparticles, but also in the form of microparticles or submicroparticles (Ceylan and Odabaşı, 2013).

Even now, microstructured enzymatic drug delivery systems (Fejerskov et al., 2012) and controlled drug release systems have been created based on PVA hydrogels (Fejerskov et al., 2013). Similar drug delivery and controlled drug release systems based on cryogels can be expected to emerge in the near future.

Overall, we can conclude that the development and application of immobilized enzymatic biocatalysts produced using cryogel-type carriers are on the increase. Studies in this field open new vistas in the development and practical use of original biocatalysts with unique features that depend not only on new types of cryogels, but also on the properties of new enzymes and proteins, including recombinant species, which originally have different origins (viruses, animal and plant cells, microorganisms).

## LIST OF ABBREVIATIONS

| | |
|---|---|
| Abz | o-Aminobenzoic acid |
| Amc | 4-Amino-7-methylcoumarin |
| DMF | Dimethylformamide |
| ELISA | Enzyme-linked immunosorbent assay |
| Glp | Pyroglutamate |
| $His_6$-OPH | Hexahistidine-tagged OPH |
| IgG | Immunoglobulin G |
| IgM | Immunoglobulin M |
| IgY | Immunoglobulin Y |
| MeCN | Acetonitrile |
| OPH | Organophosphate hydrolase |
| PAAm | Polyacrylamide |
| PB | 1,5-Dimethyl-1,5-diazaundecamethylene polymethobromide |
| PEG | Polyethylene glycol |
| PHEMA | Poly(hydroxyethyl methacrylate) |
| pNA | p-Nitroanilide |
| PVA | Polyvinyl alcohol |

## REFERENCES

Ahlqvist, J., Kumar, A., Sundström, H. et al. (2006a). Affinity binding of inclusion bodies on supermacroporous monolithic cryogels using labeling with specific antibodies. *Journal of Biotechnology* 122 no. 2: 216–25.

Ahlqvist, J., Dainiak, M.B., Kumar, A., Hörnsten, E.G., Galaev, I.Y., and Mattiasson, B. (2006b). Monitoring the production of inclusion bodies during fermentation and enzyme-linked immunosorbent assay analysis of intact inclusion bodies using cryogel minicolumn plates. *Analytical Biochemistry* 354 no. 2: 229–37.

Alkan, H., Bereli, N., Baysal, Z., and Denizli, A. (2009). Antibody purification with protein A attached supermacroporous poly (hydroxyethyl methacrylate) cryogel. *Biochemical Engineering Journal* 45 no. 3: 201–8.

Alkan, H., Bereli, N., Baysal, Z., and Denizli, A. (2010). Selective removal of the autoantibodies from rheumatoid arthritis patient plasma using protein A carrying affinity cryogels. *Biochemical Engineering Journal* 51 no. 3: 153–59.

Almarza, J., Rincon, L., Bahsas, A., and Brito, F. (2009). Molecular mechanism for the denaturation of proteins by urea. *Biochemistry* 48 no. 32: 7608–13.

Altunbaş, C., Uygun, M., Uygun, D.A., Akgöl, S., and Denizli, A. (2013). Immobilization of inulinase on concanavalin A-attached super macroporous cryogel for production of high-fructose syrup. *Applied Biochemistry and Biotechnology* 170 no. 8: 1909–21.

Anikina, O.M., Lysogorskaya, E.N., Oksenoit, E.S., Losinskii, V.I., and Filippova, I.Yu. (2008). Subtilisin Carlsberg in complex with sodium dodecyl sulfate is an effective catalyst for the solid phase segment coupling of peptides on polyvinyl alcohol cryogel. *Russian Journal of Bioorganic Chemistry* 34 no. 3: 329–33.

Babac, C., Yavuz, H., Galaev, I.Yu., Pişkin, E., and Denizli, A. (2006). Binding of antibodies to concanavalin A-modified monolithic cryogel. *Reactive and Functional Polymers* 66 no. 11: 1263–71.

Bacheva, A.V., Baibak, O.V., Belyaeva, A.V. et al. (2003). Native and modified subtilisin 72 as a catalyst for peptide synthesis in media with a low water content. *Russian Journal of Bioorganic Chemistry* 29 no. 5: 502–8.

Bacheva, A.V., Belyaeva, A.V., Lysogorskaya, E.N., Oksenoit, E.S., Lozinsky, V.I., and Filippova, I.Yu. (2005). Biocatalytic properties of native and immobilized subtilisin 72 in aqueous-organic and low water media. *Journal of Molecular Catalysis B: Enzymatic* 32 no. 5–6: 253–60.

Bacheva, A.V., Plieva, F.M., Lysogorskaya, E.N., Filippova, I.Yu., and Lozinsky, V.I. (2001). Peptide synthesis in organic media with subtilisin 72 immobilized on poly (vinyl alcohol)-cryogel carrier. *Bioorganic & Medicinal Chemistry Letters* 11 no. 8: 1005–8.

Bansal, V., Roychoudhury, P.K., Mattiasson, B., and Kumar, A. (2006). Recovery of urokinase from integrated mammalian cell culture cryogel bioreactor and purification of the enzyme using p-aminobenzamidine affinity chromatography. *Journal of Molecular Recognition* 19 no. 4: 332–39.

Belyaeva, A.V., Bacheva, A.V., Oksenoit, E.S., Lysogorskaya, E.N., Lozinskii, V.I., and Filippova, I.Yu. (2005). Peptide synthesis in organic media with the use of subtilisin 72 immobilized on a poly (vinyl alcohol) cryogel. *Russian Journal of Bioorganic Chemistry* 31 no. 6: 529–34.

Belyaeva, A.V., Smirnova, Yu.A., Lysogorskaya, E.N. et al. (2008). Biocatalytic properties of thermolysin immobilized on polyvinyl alcohol cryogel. *Russian Journal of Bioorganic Chemistry* 34 no. 4: 435–41.

Bereli, N., Ertürk, G., Tümer, M.A., Say, R., and Denizli, A. (2013). Oriented immobilized anti-hIgG via F(c) fragment-imprinted PHEMA cryogel for IgG purification. *Biomedical Chromatography* 27 no. 5: 599–607.

Bibi, N.S., and Fernández-Lahore, M. (2013). Grafted megaporous materials as ion-exchangers for bioproduct adsorption. *Biotechnology Progress* 29 no. 2: 386–93.

Bibi, N.S., Gavara, P.R., Soto Espinosa, S.L., Grasselli, M., and Fernández-Lahore, M. (2011). Synthesis and performance of 3D-Megaporous structures for enzyme immobilization and protein capture. *Biotechnology Progress* 27 no. 5: 1329–38.

Busto, M.D., Meza, V., Ortega, N., and Perez-Mateos M. (2007). Immobilization of naringinase from *Aspergillus niger* CECT 2088 in poly (vinyl alcohol) cryogels for the debittering of juices. *Food Chemistry* 104 no. 3: 1177–82.

Ceylan, Ş., and Odabaşı, M. (2013). Novel adsorbent for DNA adsorption: $Fe^{3+}$-attached sporopollenin particles embedded composite cryogels. *Artificial Cells, Nanomedicine, and Biotechnology* 41 no. 6: 376–83.

Dainiak, M.B., Kumar, A., Galaev, I.Yu., and Mattiasson, B. (2006). Detachment of affinity-captured bioparticles by deformation of a macroporous hydrogel. *Proceedings of the National Academy of Sciences* 103 no. 4: 849–54.

Dainiak, M.B., Kumar, A., Plieva, F.M., Galaev, I.Yu., and Mattiasson, B. (2004). Integrated isolation of antibody fragments from microbial cell culture fluids using supermacroporous cryogels. *Journal of Chromatography A* 1045 no. 1–2: 93–98.

Di Palma, S., Hennrich, M.L., Heck, A.J., and Mohammed, S. (2012). Recent advances in peptide separation by multidimensional liquid chromatography for proteome analysis. *Journal of Proteomics* 75 no. 13: 3791–813.

Doretti, L., Ferrara, D., Gattolin, P., Lora, S., Schiavon, F., and Veronese, F.M. (1998). PEG-modified glucose oxidase immobilized on a PVA cryogel membrane for amperometric biosensor applications. *Talanta* 45 no. 5: 891–98.

Doretti, L., Ferrara, D., Lora, S., Schiavon, F., and Veronese, F.M. (2000). Acetylcholine biosensor involving entrapment of acetylcholinesterase and poly(ethylene glycol)-modified choline oxidase in a poly(vinyl alcohol) cryogel membrane. *Enzyme and Microbial Technology* 27 no. 3–5: 279–85.

Efremenko, E., Lyagin, I., Gudkov, D., and Varfolomeyev, S. (2007). Immobilized biocatalysts for detoxification of neurotoxic organophosphorus compounds. *Biocatalysis and Biotransformation* 25 no, 2–4: 359–64.

Efremenko, E.N., Lyagin, I.V., Plieva, F.M., Galaev, I.Y., and Mattiasson, B. (2009). Dried-reswollen immobilized biocatalysts for detoxification of organophosphorous compounds in the flow systems. *Applied Biochemistry and Biotechnology* 159: 251–60.

Efremenko, E.N., Lozinsky, V.I., Sergeeva, V.S., et al. (2002). Addition of Polybrene improves stability of organophosphate hydrolase immobilized in poly(vinyl alcohol) cryogel carrier. *Journal of Biochemical and Biophysical Methods* 51: 195–201.

Efremenko, E., Votchitseva, Y., Plieva, F., Galaev, I., and Mattiasson, B. (2006). Purification of $His_6$-organophosphate hydrolase using monolithic supermacroporous polyacrylamide cryogels developed for immobilized metal affinity chromatography. *Applied Microbiology and Biotechnology* 70 no. 5: 558–63.

Efremenko, E.N., Lyagin, I.V., Votchitseva, Y.A. et al. (2008). The influence of length and localization of polyhistidine tag in the molecule of organophosphorus hydrolase on the biosynthesis and behavior of fusion protein. In: *Biotechnology: State of the Art and Prospects for Development*. G.E. Zaikov, Ed. pp. 87–101. New York: Nova Science Publishers Inc.

Fejerskov, B., Jensen, B.E., Jensen, N.B., Chong, S.F., and Zelikin, A.N. (2012). Engineering surface adhered poly(vinyl alcohol) physical hydrogels as enzymatic microreactors. *ACS Applied Materials & Interfaces* 4 no. 9: 4981–90.

Fejerskov, B., Smith, A.A.A., Jensen, B.E.B., Hussmann, T., and Zelikin, A.N. (2013). Bioresorbable surface-adhered enzymatic microreactors based on physical hydrogels of poly(vinyl alcohol). *Langmuir* 29 no. 1: 344–54.

Feller, G. (2010). Protein stability and enzyme activity at extreme biological temperatures. *Journal of Physics: Condensed Matter*. 22 no.32: 323101 (17 p.).

Filippova, I.Yu., Bacheva, A.V., Baibak, O.V. et al. (2001). Proteinases immobilized on poly (vinyl alcohol) cryogel: Novel biocatalysts for peptide synthesis in organic media. *Russian Chemical Bulletin* 50 no. 10: 1896–902.

Gudkov, D.A., and Efremenko, E.N. (2007). Refolding of hexahistidine-tagged organophosphorous hydrolase from inclusion bodies. *Moscow University Chemistry Bulletin* 62 no. 6: 320–24.

Hanora, A., Plieva, F.M., Hedström, M., Galaev, I.Y., and Mattiasson, B. (2005). Capture of bacterial endotoxins using a supermacroporous monolithic matrix with immobilized polyethyleneimine, lysozyme or polymyxin B. *Journal of Biotechnology* 118 no. 4: 421–33.

Hedström, M., Plieva, F., Galaev, I.Yu., and Mattiasson, B. (2008). Monolithic macroporous albumin-chitosan cryogel structure: A new matrix for enzyme immobilization. *Analytical and Bioanalytical Chemistry* 390 no. 3: 907–12.

Hischer, T., Steinsiek, S., and Ansorge-Schumacher, M.B. (2006). Use of polyvinyl alcohol cryogels for the compartmentation of biocatalyzed reactions in non-aqueous media. *Biocatalysis and Biotransformation* 24 no. 6: 437–42.

Kokufuta, E., and Jinbo, E. (1992). A hydrogel capable of facilitating polymer diffusion through the gel porosity and its application in enzyme immobilization. *Macromolecules* 25: 3549–52.

Kumakura, M., Kaetsu, I., Suzuki, M., and Adachi, S. (1983). Immobilization of antibodies and enzyme-labeled antibodies by radiation polymerization. *Applied Biochemistry and Biotechnology* 8 no. 2: 87–96.

Kumar, A., Bansal, V., Nandakumar, K.S. et al. (2006). Integrated bioprocess for the production and isolation of urokinase from animal cell culture using supermacroporous cryogel matrices. *Biotechnology and Bioengineering* 93 no. 4: 636–46.

Kumar, A., Plieva, F.M., Galaev, I.Y., and Mattiasson, B. (2003). Affinity fractionation of lymphocytes using a monolithic cryogel. *Journal of Immunological Methods* 283 no. 1–2: 185–94.

Lozinsky, V.I. (2008). Polymeric cryogels as a new family of macroporous and supermacroporous materials for biotechnological purposes. *Russian Chemical Bulletin* 57 no. 5: 1015–32.

Lysogorskaya, E.N., Roslyakova, T.V., Belyaeva, A.V., Bacheva, A.V., Lozinskii, V.I., and Filippova, I.Yu. (2008). Preparation and catalytic properties of trypsin immobilized on cryogels of polyvinyl alcohol. *Applied Biochemistry and Microbiology* 44 no. 3: 241–46.

Markvicheva, E.A., Lozinsky, V.I., Plieva, F.M. et al. (2005). Gel-immobilized enzymes as promising biocatalysts: Results from Indo-Russian collaborative studies. *Pure and Applied Chemistry* 77 no. 1: 227–36.

Martínez, Y.N., Cavello, I., Hours, R., Cavalitto, S., and Castro, G.R. (2013). Immobilized keratinase and enrofloxacin loaded on pectin PVA cryogel patches for antimicrobial treatment. *Bioresource Technology* 145: 280–84.

Petrov, P., Jeleva, D., and Tsvetanov, C.B. (2012). Encapsulation of urease in double-layered hydrogels of macroporous poly(2-hydroxyethyl methacrylate) core and poly(ethylene oxide) outer layer: Fabrication and biosensing properties. *Polymer International* 61 no. 2: 235–39.

Petrov, P., Pavlova, S., Tsvetanov, C.B., Topalova, Y., and Dimkov, R. (2011). *In situ* entrapment of urease in cryogels of poly (N-isopropylacrylamide): An effective strategy for noncovalent immobilization of enzymes. *Journal of Applied Polymer Science* 122 no. 3: 1742–48.

Plieva, F., Bober, B., Dainiak, M., Galaev, I.Y., and Mattiasson, B. (2006). Macroporous polyacrylamide monolithic gels with immobilized metal affinity ligands: The effect of porous structure and ligand coupling chemistry on protein binding. *Journal of Molecular Recognition* 19 no. 4: 305–12.

Plieva, F.M., Kochetkov, K.A., Singh, I., Parmar, V.S., Belokon', Yu.N., and Lozinsky, V.I. (2000). Immobilization of hog pancreas lipase in macroporous poly (vinyl alcohol)-cryogel carrier for the biocatalysis in water-poor media. *Biotechnology Letters* 22 no. 7: 551–54.

Plieva, F.M., Lozinskii, V.I., and Isaeva, E.I. (1998). Use of polyvinyl alcohol cryogels in biotechnology. VI. Bioaffinity sorbents based on a supermacroporous carrier for work with viral particles. *Russian Biotechnology* 10: 12–17.

Semashko, T.A., Lysogorskaya, E.N., Oksenoit, E.S., Bacheva, A.V., and Filippova, I.Yu. (2008). Chemoenzymatic synthesis of new fluorogenous substrates for cysteine proteases of the papain family. *Russian Journal of Bioorganic Chemistry* 34 no. 3: 339–43.

Sheldon, R.A., and van Pelt, S. (2013). Enzyme immobilisation in biocatalysis: Why, what and how. *Chemical Society Reviews* 42: 6223–35.

Stanescu, M.D., Fogorasi, M., Shaskolskiy, B.L., Gavrilas, S., and Lozinsky, V.I. (2010). New potential biocatalysts by laccase immobilization in PVA cryogel type carrier. *Applied Biochemistry and Biotechnology* 160 no. 7: 1947–54.

Stanescu, M.D., Gavrilas, S., Ludwig, R., Haltrich, D., and Lozinsky, V.I. (2012). Preparation of immobilized *Trametes pubescens* laccase on a cryogel-type polymeric carrier and application of the biocatalyst to apple juice phenolic compounds oxidation. *European Food Research and Technology* 234 no. 4: 655–62.

Stanescu, M.D., Sanislav, A., Ivanov, R.V., Hirtopeanu, A., and Lozinsky, V.I. (2011). Immobilized laccase on a new cryogel carrier and kinetics of two anthraquinone derivatives oxidation. *Applied Biochemistry and Biotechnology* 165 no. 7–8: 1789–98.

Tüzmen, N., Kalburcu, T., and Denizli, A. (2012). Immobilization of catalase via adsorption onto metal-chelated affinity cryogels. *Process Biochemistry* 47 no. 1: 26–33.

Uygun, M., Akduman, B., Akgöl, S., and Denizli, A. (2013). A new metal-chelated cryogel for reversible immobilization of urease. *Applied Biochemistry and Biotechnology* 170 no. 8: 1815–26.

Uygun, M., Uygun, D.A., Özçalışkan, E., Akgöl, S., and Denizli, A. (2012). Concanavalin A immobilized poly (ethylene glycol dimethacrylate) based affinity cryogel matrix and usability of invertase immobilization. *Journal of Chromatography B: Biomedical Sciences and Applications* 887–888: 73–78.

Veronese, F.M., Mammucari, C., Caliceti, P., Schiavon, O., and Lora, S. (1999). Influence of PEGylation on the release of low and high molecular-weight proteins from PVA matrices. *Journal of Bioactive and Compatible Polymers* 14 no. 4: 315–30.

Veronese, F.M., Mammucari, C., Schiavon, F. et al. (2001). Pegylated enzyme entrapped in poly (vinyl alcohol) hydrogel for biocatalytic application. *Il Farmaco* 56 no. 8: 541–47.

Yao, K., Yun, J., Shen, S., Wang, L., He, X., and Yu, X. (2006). Characterization of a novel continuous supermacroporous monolithic cryogel embedded with nanoparticles for protein chromatography. *Journal of Chromatography A* 1109 no. 1: 103–10.

Zhan, X.-Y., Lu, D.-P., Lin, D.-Q., and Yao, S.-J. (2013). Preparation and characterization of supermacroporous polyacrylamide cryogel beads for biotechnological application. *Journal of Applied Polymer Science* 130 no. 5: 3082–89.

# 12 Application of Cryogels in Water and Wastewater Treatment

*Linda Önnby\**

## CONTENTS

12.1 Introduction .................................................................................................332
12.2 Cryogel-Based Adsorbents ..........................................................................333
    12.2.1 Functionalized Cryogels ...................................................................333
    12.2.2 Protected Cryogels............................................................................334
12.3 Composite Cryogels.....................................................................................335
    12.3.1 Metal-Oxide Composite Cryogels ....................................................336
    12.3.2 Other Types of Composites..............................................................340
    12.3.3 Cryogel Composites with Molecularly Imprinted Polymers ...........341
12.4 Pollutant Source Determines the Adsorbent ...............................................345
    12.4.1 Industrial and Municipal Wastewater...............................................345
    12.4.2 Point-of-Use Filters...........................................................................346
    12.4.3 Oil Removal—Taking Advantage of Cryogel Pore Space................347
    12.4.4 Metal Removal in Connection to Biogas Production ......................347
    12.4.5 Microbial Cryogel Systems...............................................................349
12.5 Future Challenges ........................................................................................349
    12.5.1 System Regeneration.........................................................................349
    12.5.2 Deposition .........................................................................................350
    12.5.3 Risk Assessment ...............................................................................351
    12.5.4 Costs ..................................................................................................352
12.6 Concluding Remarks ...................................................................................353
Acknowledgements.................................................................................................354
List of Abbreviations..............................................................................................354
References...............................................................................................................355

## ABSTRACT

Cryogels have potential for applications in water and wastewater treatment. However, maximum levels of exposure to both incorporated materials in cryogels (i.e., nanoparticles [NPs]) and monomers, as a result of leakage, must be established before cryogel composites can be implemented in the large-scale

---

\* Corresponding author: E-mail: linda.oennby@eawag.ch; Tel: +41 58 7655405.

treatment of water. Other factors, such as cost and reuse, must also be studied for each specific adsorbent. One of the greater advantages of cryogels is the flexibility of their preparation. The material can be prepared as monoliths, discs, beads or particles, to suit specific requirements. Results on a laboratory scale may be different from those obtained when treating real water and wastewater due to differences in scale and water chemistry. The social acceptance of the technology and its overall environmental impact must also be assessed. This chapter starts with describing how cryogel materials have been developed from their initial applications in water treatment to more recent developments. This is followed by current published results obtained from the treatment of water from different sources, including both lab-scale evaluations as well as real water samples. As the main application of cryogels within water treatment today is as adsorbent materials, these will be given the main focus for this chapter but still include more recent research results from applications regarding disinfection and oil absorption. The main pollutants of interest for this chapter are **inorganic pollutants** such as metals and metalloids (e.g. cadmium and arsenic), but organic contaminants are also mentioned. Future challenges in this area today, regarding cryogel applications, cover the need for system regeneration, the deposition of the pollutant and, finally, the total cost of the treatment.

**KEYWORDS**

Arsenic, cadmium, cryogels, inorganic pollutants, water treatment

## 12.1 INTRODUCTION

The supply of clean and safe water in all parts of the world is one of our greatest challenges. Today, more than 780 million people do not have access to safe, clean water or adequate sanitation (UN, 2013). Every hour, more than 80 children die somewhere in the world because of water-related diseases (UN, 2013; UNICEF, 2013). Water is becoming as important a resource as oil. Without water, life is impossible.

Environmental pollution poses a threat to the global water supply. Both old and new kinds of pollutants can be found in our water systems and in the environment in general. Examples of environmental pollutants are polychlorinated biphenyls, polycyclic aromatic hydrocarbons, heavy metals, pesticides, flame-retardants, herbicides, pharmaceutical residues and endocrine-disrupting compounds. Many new kinds of organic pollutants as well as heavy metals are potentially harmful, even at low concentrations. Environmental pollution may be the result of anthropogenic activities or natural events, and contaminated water can be found everywhere in the world, in both rich and poor countries.

Human exposure to environmental pollutants may be direct or indirect. Groundwater, which serves as drinking water in many parts of the world, is one direct source of exposure. Irrigation of crops for food production is an example of an indirect exposure to environmental pollutants. Toxic substances such as **cadmium** and **arsenic** have been detected, for example, in rice, threatening sustainable agriculture in many parts of Asia. Furthermore, in Western societies, findings of pharmaceuticals, endocrine disruptors and so on in surface waters, wastewaters and drinking water jeopardizes future sound and safe water supply (Ternes, 1998; Kolpin et al., 2002; Bendz et al., 2005; Schwarzenbach et al., 2006).

To prevent the contamination of water, or at least to reduce it, sustainable measures must be developed and implemented. In developing countries, 90% of all wastewater is not treated at all, and flows directly into rivers, lakes and coastal zones (UN, 2013). Apart from the effective implementation of existing water treatment methods, novel methods must be developed. One example of a separation technique that can be applied to treat wastewater is adsorption. Applying cryogels in water treatment has mainly been shown through the concept of cryogel-based adsorbents (Le Noir et al., 2009; Hajizadeh et al., 2010a,b; Kumar et al., 2013). However, other applications where cryogel materials function as supports for degrading organisms also exist (Kuyukina et al., 2009; Magri et al., 2012).

New water treatment methods must meet certain requirements. They should be low in cost and have little negative environmental impact over time, to meet the needs for future generations. The aim is to replace potential pollutants with greener alternatives or avoid them altogether.

## 12.2 CRYOGEL-BASED ADSORBENTS

The supermacroporosity of cryogels combined with their thin polymer walls the materials have low surface areas which thereby results in low adsorption capacities. As a result, cryogels have been prepared as composites, with added particles, to increase the adsorption capacity, which will be discussed later in this chapter. Another way to add particles to the cryogel matrix is to attach ligands, for example, functional groups, that are able to adsorb or chelate specific target ions, for example, metal ions.

### 12.2.1 FUNCTIONALIZED CRYOGELS

Capture of metal ions using cryogel has been exploited in ion metal affinity chromatography (IMAC), as a means to recover proteins, for example, histamine (HIS)-tagged proteins (Porath et al., 1975). Although cryogel materials do not compete as well due to their low capacities for these chromatographic applications, cryogel materials have shown other benefits such as the capability for processing particulate-containing media (Plieva et al., 2007). In the context of IMAC, copper ions are most frequently mentioned, as this metal ion shows high affinity for peptides and proteins (Porath, 1992). However, cryogel-oriented studies have also focused on using both $Zn^{2+}$ and $Fe^{2+}$ in the purification of, for example, lysozyme (Ergün, 2007; Derazshamshir et al., 2008). The capacity of the cryogel matrix for these kind of applications was also shown to be increased using embedded $Cu^{2+}$-attached sporopollenin particles with high surface area, which were embedded in poly(2-hydroxy ethyl methacrylate) (p(HEMA)) cryogels, which were further used for human serum albumin purification (Erzengin et al., 2011). As an extension to these applications, we modified a polyacrylamide-based cryogel (pAAm-cryogel) with attached iminodiacetic acid (resulting in an IDA-ligand), and studied it for metal ion removal; for example, $Cu^{2+}$, $Ni^{2+}$, $Zn^{2+}$ and $Cd^{2+}$ (Önnby et al., 2010). Cryogels with IDA-ligands have been studied both in pure water treatment processes and in connection with biogas production (Lehtomäki and Björnsson, 2006;

Selling et al., 2008; Önnby et al., 2010). In addition to the IDA-ligand, another ligand with higher carboxyl (–COOH) density was developed and compared with the IDA-ligand (Önnby et al., 2010). The new ligand, named TBA-ligand, was produced from a stepwise attachment using tris(2-aminoethyl)amine and bromoacetic acid. Besides attaching –COOH groups to cryogels, materials consisting of –SH groups and imidazole groups have also been evaluated for adsorption of a range of different metal ions, for example, $Cu^{2+}$, $Cd^{2+}$, $Pb^{2+}$ and $Zn^{2+}$ (Tekin et al., 2011; Önnby et al., 2012).

### 12.2.2 Protected Cryogels

An adsorbent material with a rational design was developed, inspired by two earlier studies (Plieva and Mattiasson, 2008; Selling et al., 2008). The TBA- and the IDA-cryogel, based on pAAm, were prepared inside a plastic carrier and were able to remove heavy metals in synthetic wastewater with high efficiency. This adsorbent could be used in a stirred reactor or in packed bed configurations (e.g. columns), in a continuous mode, without loss of shape or performance, as can be seen in Figure 12.1. It was also suggested that the cryogels protected by the plastic carriers could be used in open treatment basins or in rotating reactors.

Municipal wastewater treatment is primarily designed to remove organic matter, nitrogen and phosphorous using biological systems employing activated sludge. The removal of heavy metals from these systems can therefore be regarded as a further benefit. Occasionally, wastewater treatment plants (WWTPs) receive water with increased metal ion content. Treatment with adsorbents could be beneficial for these situations. The high inflow of water to WWTPs is a challenge, as it is not yet possible to produce cryogels in sufficiently large quantities. A municipal treatment plant in southern Sweden serving the city of Malmö receives approximately 1.6 $m^3/s$ (VA-Syd, 2009). In a study by Karvelas et al. (2003), at a WWTP in Greece, over 70% of the Cu and Mn were found in the solid fraction, while roughly 47–63% of the remaining (Pb, Cd, Ni, Zn and Fe) were found in the liquid (Karvelas et al., 2003). Hence, the removal of metals at lower concentrations should be carried out early in the treatment process before sedimentation has taken place. The high organic-matter content for wastewater at this point, however, can pose a problem, as it could both interact with the adsorbent material and with heavy metal ions (forming chelating complexes with humic acids). If metals were removed, sewage sludge could be of higher environmental quality. The use of sewage sludge as a fertilizer is currently the subject of debate, due to its relatively high levels of heavy metals (see more details next).

Prepared functionalized cryogels were protected, as they were prepared inside two kinds of plastic carriers made by Kaldnes (AnoxKaldnes, Sweden): K1 with a diameter of 10 mm and K2 with a diameter of 25 mm (Plieva and Mattiasson, 2008; Önnby et al., 2010). The carriers containing gel showed better stability after vigorous stirring for 48 h than a gel monolith without protection by Kaldnes carriers, as shown in Figure 12.2. Other applications of cryogels prepared in carriers are found in the discussed microbial cryogel systems.

Ligand attachment serves as a viable option in water treatment when specific pollutants should be targeted, as the choice of ligand will determine which targets are suitable. As an example, –SH groups show high affinity for both As and Hg.

# Application of Cryogels in Water and Wastewater Treatment

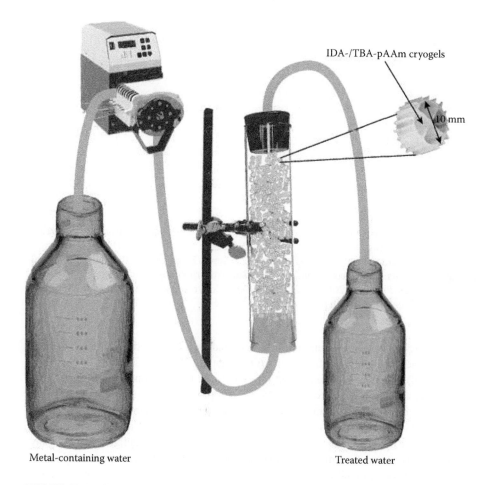

**FIGURE 12.1** Schematic illustration showing how water contaminated with metal ions can be treated in continous mode. A chelating cryogel was prepared in plastic carriers which were packed in a column. (Illustrated by Fia Persson.)

However, due to the low surface area of cryogels, the ligand density is restricted and affects the final adsorption capacity obtained from these materials.

## 12.3 COMPOSITE CRYOGELS

Adding high surface-area particles, which may provide a better way of preparing adsorbents for metal-ion removal, can increase the adsorption capacity of cryogels. In some of our work, composite cryogels were produced using adsorbent particles of metal oxides (e.g. NPs or microscaled particles) and molecularly imprinted polymers (MIPs, see Section 12.3.3. for more information), embedded in pAAm cryogels.

In contrast to the preparation of functionalized cryogels, the preparation of composite cryogels is simple and can be completed in one step. In our work, different particles were embedded in cryogels, for example, $Al_2O_3$ nanoparticles (AluNPs), Fe-Al

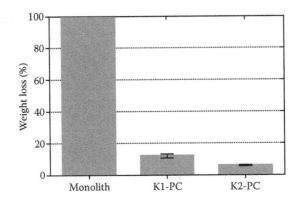

**FIGURE 12.2** Mechanical stability in terms of weight loss after 48 h of rigorous stirring of IDA-pAAm cryogel composites (K1-PC and K2-PC) and a piece of IDA-pAAm cryogel monolith. (Modified from Önnby, L., Giorgi, C., Plieva, F. M. and Mattiasson, B.: Removal of heavy metals from water effluents using supermacroporous metal chelating cryogels. *Biotechnology Progress.* 2010. 26(5). 1295–1302. Copyright Wiley-VCH Verlag GmbH & Co. KGaA. Reproduced with permission.)

hydrous oxides or titanate nanotubes (TNTs), and then evaluated with regard to their ability to adsorb metal ions from water (Önnby et al., 2010, 2012, 2015; Kumar et al., 2013, 2014) and these can be divided into two main groups, consisting of larger particles (including metal hydroxides and MIPs > 0.1 μm) and smaller particles (<100 nm). The cryogels holding smaller particles are often referred to as nanocomposites and are discussed later in this section. In principle, the cryogels were used as support for highly adsorbent particles, and the adsorption by the support itself was negligible. As with any other material used for immobilization, it should ideally not affect (reduce) the performance of the immobilized adsorbents (e.g. by blocking or decreasing the available surface area of the adsorbent particles). Moreover, particle-containing cryogels can be made with different types of particles. For this section, metal-oxides, MIPs and other types of composite cryogels will be discussed in more detail.

### 12.3.1 Metal-Oxide Composite Cryogels

For the metal-containing composites developed in our group, it was confirmed that the embedment of the particles did not result in lower adsorption capacities as compared to the adsorption obtained from particles applied in suspension (Önnby et al., 2012; Kumar et al., 2013). Cryogel composites with embedded AluNPs (Alu-cryo) showed a similar adsorption capacity for As(V) as AluNPs in suspension. Similarly, As(III) removal by Fe-Al cryogels (Fe-Al-cryo) was almost identical to that with Fe-Al hydrous oxide particles in suspension (Kumar et al., 2013). However, Savina et al. (2011) reported a reduction in adsorption when using cryogel composites. Only 33% of the adsorption of As(III) achieved with free NPs ($\alpha$-$Fe_2O_3$ and $\alpha$-$Fe_3O_4$) was exhibited by the respective cryogel composites (the cryogel-support was made out of p(HEMA)) (Savina et al., 2011).

The embedded particles are usually physically entrapped in the cryogel, placing a limit on the particle loading. Also, with this kind of immobilization of NPs, the reactivity of NPs can be compromised. We evaluated the optimal particle load of the Alu-cryo composite, testing 1, 2, 3, 4, 5 and 6% (w/v) of AluNPs embedded in pAAm cryogels (Önnby et al., 2014). Parameters such as the loss of NPs during the production and use of the composite, and As(V) removal were studied. The optimal mass of AluNPs was determined to be 4% (w/v) for the composite evaluated, as can be seen in Figure 12.3. At the initial concentration of As(V) of 5 mg/l, the degree of saturation of the adsorbent varied between the Alu-cryogels with various amounts of AluNPs. The uptake of As(V) by the 4-Alu-cryo was similar to that of free AluNPs in suspension. Both of these had similar loadings of 2 g AluNPs/l.

It was shown that high volumes of water could be filtered through the Alu-cryo material composite without any detectable loss of NPs, or significant reduction in pollutant removal (demonstrated for As(V)) (Önnby et al., 2014). The use of embedded metal oxides in cryogels to remove inorganic pollutants is an attractive solution. A water volume 8000 times greater than the volume of the Alu-cryo composite itself was able to pass through the filter material without particle leaching (see Section 12.5.3 about risk assessment for more details). However, the results presented in Önnby et al. (2014) indicated that acrylamide (AAm) leaked out of the polymer support, as has been observed in previous studies (Plieva et al., 2004). AAm leakage was also seen during the production phase. After washing the composite with 80 bed volumes of water, the concentration of AAm in the filtrate water was significantly less than the permissible limit of 0.5 μg/l (WHO, 2011). These results show that AAm should not be used in the production of cryogel supports

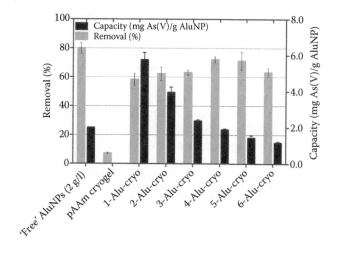

**FIGURE 12.3** Comparison of As(V) removal and adsorption capacity of free AluNPs in suspension at 2 g/l, unmodified pAAm cryogel and Alu-cryo composites, containing 1–6 and (w/w) AluNPs. The initial As(V) concentration was 5 mg/l, pH = 7 ± 0.5 and T = 22°C. (Reprinted from *Science of The Total Environment* 473–474(0), Onnby, L., Svensson, C., Mbundi, L., Busquets, R., Cundy, A. and Kirsebom, H., γ-Al2O3-based nanocomposite adsorbents for arsenic(V) removal: assessing performance, toxicity and particle leakage, 207–214. Copyright 2014, with permission from Elsevier.)

and must be replaced. Toxicity of AAm is very well documented (Fullerto and Barnes, 1966; Odland et al., 1994; Friedman, 2003). Thus, when applying adsorbent materials for environmental purposes, these aspects need to be considered as well.

Particles with higher density, for example, Fe-Al hydrous oxides ($\rho \approx 2.9$ g/cm$^3$) and TNTs ($\rho \approx 4.6$ g/cm$^3$), have a tendency to settle fast in the monomer + particle suspension prior to freezing polymerization. The radius of the particles also determines the settling velocity and is described by Stoke's law (McCabe, 2001). The particles in the resulting monolithic composite are thus heterogeneously distributed; more being found at the bottom of the monolith. Although cryogels can be used to immobilize NPs or other particles, in a stable form, the particles may still agglomerate. This could be avoided with heavier particles by using high-viscosity cryogel suspension, or by changing the charge distribution in the suspension, to avoid particle aggregation and sedimentation of the particles.

The actual loading for each composite embedded with metal oxides was determined experimentally using thermo gravimetric analysis and compared with the theoretical load. In one of the studies (Kumar et al., 2013), the Fe-Al hydrous oxide particle load was 67% (0.67 g/g dried gel) of the theoretical amount. For the smaller AluNPs, however ($\rho \approx 4$ g/cm$^3$), the particle yield was close to 90% of the theoretical (0.9 g/g dried gel) (Önnby et al., 2012), supporting the theory that particle incorporation is affected by particle density and size. Similar evaluations were done in the work of Savina et al. (2011), both using thermo gravimetric analysis, and by elution of particles using HCl for composite-particle cryogels containing iron NPs ($\alpha$-Fe$_2$O$_3$ and Fe$_3$O$_4$), indicating that the mass load varied between 0.4 and 0.6 g/g dried gel and there were minor differences between the two types of particles (Savina et al., 2011). From a production point of view, low incorporation rates will affect the overall cost using these adsorbents, and losses should thus be avoided. Additionally, low incorporation prepared in large cryogel matrix volumes will also take up space in the treatment process. In the case of the incorporation of TNTs in the pAAm cryogel, only 2% (w/v) particle loading could be achieved due to rapid settling of the particles (Önnby et al., 2015). By measuring the dry weight of the TNT-cryo, it was concluded that 82.5% of the loaded particles remained in the monolith after production, corresponding to 0.20 g TNTs/g dry polymer. The tendency of TNTs or high-density particles in general, to form aggregates also affects settling, as the aggregates will have a larger 'particle' radius than non-aggregated particles.

In a recently published study, we explored the possibility of coating Fe-Al hydrous oxide particles onto macroporous support (e.g. pAAm cryogels) using a reversible *in situ* precipitation method, developed in our group (Kumar et al., 2014). The procedure was simple and flexible and involved less toxic chemicals than in our previous study (Kumar et al., 2013). The mass loading of the metal hydroxide coated macroporous polymers (MHCMPs) could typically be adjusted by increasing the number of precipitation cycles or by altering the initial metal salt solutions. In Figure 12.4, an illustration shows how the precipitation and flow-through procedure was done. In addition to Fe-Al hydrous oxides, we also showed that the method was applicable to other metal hydroxides as well, for example, copper, cobalt and nickel hydroxides. Using this coating method, the specific surface area, determined using nitrogen isotherms for Brunauer-Emmet-Teller (BET) surface area, could be increased

# Application of Cryogels in Water and Wastewater Treatment

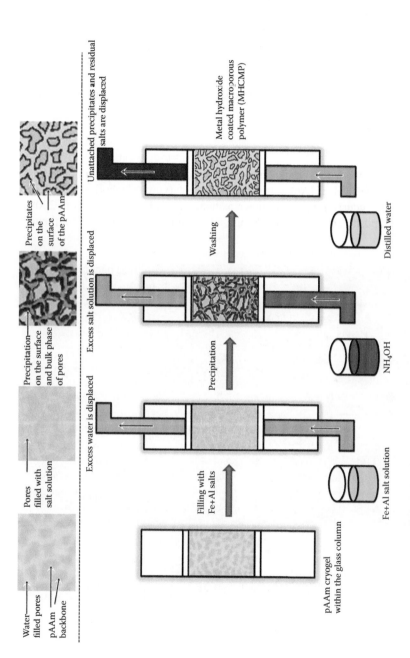

**FIGURE 12.4** Schematic showing a flow-through method developed for metal hydroxide precipitation on macroporous support, for the production of metal hydroxide coated macroporpus polymer (MHCMP). (From Kumar, P. S., Önnby, L. and Kirsebom, H. *Journal of Materials Chemistry A* 2(4): 1076–1084. Copyright (2014). Reproduced by permission of The Royal Society of Chemistry.)

effectively (from initial 18 m²/g for uncoated samples to 63 m² g⁻¹ for the highest mass incorporation of the studied MHCMPs).

The particles are incorporated into the thin but dense polymer walls, and their accessibility is key for successful adsorption. Scanning electron microscopy (SEM) and transmission electron microscopy were used to investigate how the composites with embedded particles were immobilized in the cryogel support. For the MHCMPs, SEM-coupled energy-dispersive X-ray (EDX) spectra was used and could well indicate the metal hydroxide coating on the macroporous support, as shown in Figure 12.5.

### 12.3.2 OTHER TYPES OF COMPOSITES

Min and co-workers, studied cryogel materials for adsorption of $Cd^{2+}$ using hydroxyapatite ($Ca_5(PO_4)_3(OH)$, HAp) immobilized in poly(vinyl alcohol) (PVA) cryogels, a naturally occurring mineral known as bone mineral (Wang and Min, 2008). PVA/HAp composite cryogels showed high adsorption capacity for $Cd^{2+}$ although

**FIGURE 12.5** SEM images capturing MHCMPs with Fe/Al produced using a single cycle (a and b) with a 1 M salt solution of Al and Fe, respectively, (c) with a 0.4 M salt concentration. The highlighted sections in (b) indicate the regions studied using EDX and in (d) obtained spectra from EDX-analysis. (From Kumar, P. S., Önnby, L. and Kirsebom, H., *Journal of Materials Chemistry A* 2(4): 1076–1084. Copyright (2014). Reproduced by permission of The Royal Society of Chemistry.)

evaluated using high adsorbent concentrations (40 g/l). Other examples were demonstrated in the work of Hajizadeh et al. (2010b), showing that carbon composite cryogels exhibited a markedly lower adsorption capacity due to blockage in the cryogel pores by reagents used in the production. By choosing a different strategy (e.g. cryostructured particle gel produced in one-step freezing procedure), the adsorption of phenol could be improved fivefold but was still half as compared to the pure carbon particles, demonstrated in aqueous solution (Hajizadeh et al., 2010b). Similar reduced adsorption behavior has been seen when employing cryogel material as carrier, as in the case of immobilized marine algal biomass in PVA cryogels, where the adsorption capacity was much lower as compared to freely suspended biomass, when tested for $Cu^{2+}$ adsorption (Sheng et al., 2008).

Another type of composite cryogel, that is, a semi-interpenetrating polymer network (IPN), is double-ionic cryogels prepared with at least one of the networks polymerized in the presence of the other (Myung et al., 2008). An IPN cryogel based on pAAm was modified with anionic potato starch and evaluated for heavy metal adsorption (Apopei et al., 2012). Recently, the same research group could show substantial adsorption (667.7 mg/g; details in Table 12.1) of the dye methylene blue (used as a model compound for textile dyes), by a modification of the previously studied IPN-cryogel material with potato starch (Dragan and Apopei Loghin, 2013). The combination of bio-based and synthetic material is indeed an interesting concept to further develop.

### 12.3.3 Cryogel Composites with Molecularly Imprinted Polymers

The use of MIPs for selective removal of specific pollutants, especially pharmaceuticals, antibiotics or endocrine disruptors, commonly referred to as micropollutants, has been evaluated in a number of studies for cryogel-supported adsorbent materials (Le Noir et al., 2007; Baggiani et al., 2010; Koç et al., 2011; Hajizadeh et al., 2013). There are two main categories for these studies: (1) MIP cryogel composites to be used in solid phase extraction (Baggiani et al., 2010; Hajizadeh et al., 2013), enabling preconcentration of an analyte prior to analysis, and (2) MIP-cryogel composites to be used for adsorption (Le Noir et al., 2007, 2009; Hajizadeh et al., 2010a).

In brief, MIPs are polymers produced with the target compound present during polymerization, giving the resulting polymer particle an adsorption site frequently compared with synthetic antibodies (Wulff, 1995; Chen et al., 1997, 2011). The adsorption site is given a specificity already in the production phase, as is shown in Figure 12.6, where the template (e.g. target ion or molecule) is present during the polymerization. After the polymerization, the template is removed, leaving behind an adsorption site with high selectivity for the template.

MIP adsorbent materials can play a key role in certain kinds of pollutant sources. For example, high levels of phosphate often coexist with arsenic and compete for adsorption sites. For these cases, a more selective adsorbent, such as the MIP-cryo developed in one previous study (Önnby et al., 2012), may provide a useful alternative. In our work, we found that the MIP-cryo exhibited higher selectivity for As(V), although the overall adsorption capacity was lower than that of Alu-cryo (Önnby et al., 2012) and Fe-Al-cryo (Kumar et al., 2013). Other than wastewater,

## TABLE 12.1
## Characteristics of Some Ligand-Based and Particle-Based (Metal Oxides and MIPs) Cryogel Composites Used as Adsorbents for Removal of Inorganic and Organic Pollutants

| Adsorbent (Functional Group) | Ligand Density ($\mu$mol/g) | Adsorbent Dose (g/l) | Contact Time (h) | pH | Adsorption Capacity (mg/g) | Ref. |
|---|---|---|---|---|---|---|
| | | Inorganic Pollutants | | | | |
| IDA-pAAm[b] (COOH) | 23 | | 24 | 5 | Cu = 3.65<br>Ni = 3.37<br>Zn = 3.76<br>Cd = 6.4 | Önnby et al., 2010 |
| TBA-pAAm (COOH) | 32 | | 24 | 5 | Cu = 5<br>Zn = 5.2 | Önnby et al., 2010 |
| p(HEMA-VIM)/p(HEMA) (imidiazole) | 45.3 | 5 | 24 | 5.5 | Pb = 7.62<br>Cd = 5.8<br>Zn = 4.34<br>Cu = 2.54 | Tekin et al., 2011 |
| PVA/HAp (HAp) | 1990 | 20 | 24 | na[c] | Cd = 53.3 | Wang and Min, 2008 |
| pAAm/PA[d] (polyanion, PA) | 12.3[e] | 0.5 | 24 | 4.7 | Cu = 40.7<br>Cd = 19.3<br>Ni = 9.3<br>Zn = 7.5 | Apopei et al., 2012 |
| pAAm-MIP-cryogel(MIP) | 20[f] | 0.1 | 10 | na | $BrO_3^-$ ($C_f$ = 30 $\mu$g/l) | Hajizadeh et al., 2010a |
| **Particle Composite Cryogels** | Specific Surface Area (m$^2$/g)[a] | | | | | |
| TNT-cryo (TNTs) | 260 | 1 | 24 | 7 | Cd = 156.3 | Önnby et al., 2015 |

*(Continued)*

## TABLE 12.1 (CONTINUED)
### Characteristics of Some Ligand-Based and Particle-Based (Metal Oxides and MIPs) Cryogel Composites Used as Adsorbents for Removal of Inorganic and Organic Pollutants

| Particle Composite Cryogels | Specific Surface Area (m²/g)[a] | Adsorbent Dose (g/l) | Contact Time (h) | pH | Adsorption Capacity (mg/g) | Ref. |
|---|---|---|---|---|---|---|
| Alu-cryo (AluNPs) | 40 | 2 | 24 | 7 | As(V) = 20.1 | Önnby et al., 2012 |
| Fe-Al-cryo (Fe-Al hydrous oxides) | 74 | 2 | 24 | 7 | As(III) = 24.1 | Kumar et al., 2013 |
| MHCMP (Fe-Al) | 63 | 0.7 | 24 | 7 | As(III) = 26.4 ($C_i$ = 23 mg/l) | Kumar et al., 2014 |
| $Fe_3O_4$-HEMA ($Fe_3O_4$) | 60 | 1 | 24 | 7 | As(III) = 3.1 ($C_i$ = 2 mg/l) | |

### Adsorbent (Functional Group)

**Organic Pollutants**

| | | | | | | |
|---|---|---|---|---|---|---|
| C-pAAm/PA60.5[g] (polyanion, PA) | na | 1 | 12 | na | Methylene Blue = 667.7 | Dragan and Apopei Loghin, 2013 |
| PHEMA/MIP (MIP) | 123.2 | na | 2 | na | E2 = 5.32 | Koç et al., 2011 |
| Carbon-cryostructured particle gel (carbon) | na | 1 | 24 | | Phenol = 1250 | Hajizadeh et al., 2010b |

[a] Particle-containing composites are characterized with surface area from the particle state if not mentioned otherwise.
[b] Two different ligands were evaluated in the same study, both based on COOH.
[c] na, not available.
[d] Semi-IPN cryogel based on pAAm and PA from hydrolysed potato starch (PS) with poly(acrylonitrile).
[e] Expressed in wt% of total adsorbent.
[f] %(w/w, for MIPs to total monomer weight).
[g] C-pAAm/PA60.5.

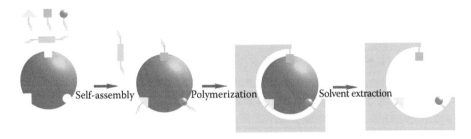

**FIGURE 12.6** A schematic illustration of the production of a molecularly imprinted polymer (MIP). From self-assembly between the template, surrounded by the functional groups, to polymerization, solvent extraction and final MIP. (Illustrated by Fia Persson, inspired by a figure in Haupt, K. and Mosbach, K., *Chemical Reviews* 100(7): 2495–2504.)

MIP adsorbents would be highly beneficial when extracting valuable precious metals from diluted waste streams.

One great challenge for MIP-technology is water compatibility, as these polymers are more efficient in organic solvents, which is the solvent used during the production of the polymer (Wulff, 1995). However, attempts are done to improve these characteristics and by incorporating MIPs in cryogels, the adsorption process has been demonstrated successful even in columns without blockage as shown for 17β-estradiol (E2) (Le Noir et al., 2007), which otherwise is the case when applying unsupported particles in columns. Another MIP-cryogel composite for E2 was evaluated in competitive solutions (cholesterol and stigmasterole) giving selectivity coefficients (k) of 7.6 and 85.8, respectively (Koç et al., 2011). The high selectivity for stigmasterole can be understood by comparing the molecular structure between the competitors and E2, where stigmasterole is expected to be less flexible (methylated alkene chain) as compared to the alkane chain in cholesterol.

Adsorption characteristics from a range of different cryogel materials with various functional groups or embedded particles are summarized in Table 12.1. Different cryogel materials demonstrated for adsorption of inorganic pollutants are dominantly indicating on similar adsorption capacities for heavy metal ions (e.g. Cu, Ni, Zn, etc.), but readily higher capacities are also seen for Cu and Cd (Apopei et al., 2012; Önnby et al., 2015).

Most of the developed MIP composites for organic pollutants have also been evaluated for desorption and demonstrate repeated adsorption-desorption cycles (Le Noir et al., 2007, 2009; Hajizadeh et al., 2010a; Koç et al., 2011), however, often demonstrated with (toxic) organic solvents (e.g. methanol). On the other hand, very few are tested on realistic water samples, suggesting limitations or further studies needed for this technique. However, there are examples demonstrating the possibility with treatment of realistic water samples (Le Noir et al., 2007; Önnby et al., 2012; Kumar et al., 2013). Embedding and thereby supporting particles in cryogels does not always result in easy and facilitated adsorption between the target ion and the adsorption site (supported particles can show reduced adsorption as compared to being nonsupported). To improve these capacities, (nano)composite cryogels should be prepared differently, for example, by immobilizing the adsorbent particle with maintained (individual) particle size (e.g. avoiding agglomeration) or with high accessibility to the adsorption surface (e.g. by coating or surface attachment). One example was recently published in our group (Kumar et al., 2014).

## 12.4 POLLUTANT SOURCE DETERMINES THE ADSORBENT

Depending on the wastewater source and pollutant to be removed, different adsorbent materials might be necessary for the process. Other factors determining material choice might be quantities of processed water (e.g. wastewater treatment plant or treating water from the tap) or the qualities of the water (e.g. water matrix in terms of coexisting ions/compounds, natural organic matter, etc.). Thus, to find a universal adsorbent for arsenic or phenol, for example, might not be possible. The following sections will introduce some different applications tested with cryogel materials, including comparisons with other types of adsorbent choices studied regarding a similar application area.

### 12.4.1 Industrial and Municipal Wastewater

Cryogel adsorbents can be applied as a polishing step to remove arsenic, both As(III) and As(V), without the need for pretreatment, from wastewater at a smelting plant (Kumar et al., 2013). The limit for arsenic release from industrial sources according to the European Commission is currently 0.150 mg/l (EU 2010). The use of an iron and aluminium-based cryogel adsorbent (e.g. Fe-Al-cryo) on real wastewater from a Swedish smelting plant showed that this limit could be achieved (Kumar et al., 2013). Adding adsorption as a polishing step would improve the overall treatment of an existing precipitation process. It would also reduce the amount of arsenic released from the smelting plant, thus improving the quality of the water released to the environment. It is noteworthy to mention that postprecipitated wastewater often is of high acidic/alkaline character, as a result from the treatment process. This, in turn, puts certain limitations on the metal-based adsorbent particles, for example, iron and aluminium are solubilized with low and high pH, respectively. Thus, although As(III and V) could be lowered to the set guidelines in the present study, aluminium ions were detected after the treatment, suggesting that the material had been solubilized in the highly alkaline wastewater. In addition, the cryogel matrix might also show limitations with regard to pH (see Section 12.4.2).

A synthetic wastewater was used in one study to evaluate the adsorption obtained for three heavy metal ions (at 0.5 mmol/l)—$Cd^{2+}$, $Hg^{2+}$ or $Pb^{2+}$. Besides the target ion, the synthetic wastewater contained $Ni^{2+}$, $Zn^{2+}$, $Fe^{2+}$, $Co^{2+}$, $Sn^{2+}$ and $Ag^+$, and added salt (NaCl, 700 mg/l). The studied bead-like adsorbent material—poly(methylmethacrylate methacryloyl amidoglutamicacid) or poly(MMA-MAGA)—indicated on little reduced adsorption capacity for the synthetic wastewater (between 80 and 75% reduction) as compared to a water solution only containing the target ion (Denizli et al., 2004). In one of our studies, municipal wastewater was selected in order to obtain a highly anion-rich matrix, which could be useful for the evaluation of selective or nonselective adsorption between a MIP-adsorbent material and

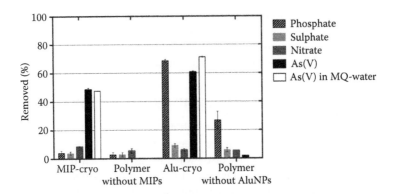

**FIGURE 12.7** Removal of anions from As(V)-spiked municipal wastewater (with initial concentrations (mg/l) of 17.7 ± 0.06 ($NO_3^-$), 56.2 ± 0.15 ($SO_4^{2-}$), 3.2 ± 0.07 ($PO_4^{3-}$) and As(V) = 5 mg/l)) using the polymer adsorbents Alu-cryo and MIP-cryo. (Control samples were the polymer support of each adsorbent.) Adsorption compared for As(V) present in ultra pure water (As(V) in MQ-water). pH = 7 ± 0.5, T = 22°C. (Reprinted from *Water Research* 46(13), Önnby, L., Pakade, V., Mattiasson, B. and Kirsebom, H., Polymer composite adsorbents using particles of molecularly imprinted polymers or aluminium oxide nanoparticles for treatment of arsenic contaminated waters, 4111–4120. Copyright 2012, with permission from Elsevier.)

a metal-oxide-adsorbent material. Figure 12.7 shows the adsorption of As(V) in the absence and presence of coexisting anions ($NO_3^-$, $SO_4^{2-}$, $PO_4^{3-}$) obtained from a spiked municipal wastewater sample by (1) an MIP-cryogel adsorbent (MIP-cryo) and (2) by Alu-cryo along with adsorption obtained from respective polymer support (pAAm and amine modified pAAm cryogels). Overall, it was shown that a gained selective adsorption of As(V) observed for MIP-cryo was obtained at the expense of a lowered specific adsorption (compare adsorption capacity between Alu-cryo and MIP-cryo). Thus, choice of adsorbent for a specific waste stream should ideally focus on the pollutant source and the extent of removal needed, for example, selectivity gain will imply loss in removal loads/efficiency.

### 12.4.2 Point-of-Use Filters

The application of composite cryogels as a POU filter is another interesting area. POU filters are necessary in areas where groundwater used for drinking water is contaminated (e.g. metal contamination). The challenge in this application, as drilled wells are often located in rural areas with limited communications and waste management, is to avoid hazardous waste generation from exhausted adsorbent materials (Leupin and Hug, 2005). In the case of arsenic contamination of groundwater, arsenic exists predominantly as As(III), and preoxidation would be necessary prior to adsorption with certain adsorbent materials. To avoid a two-stage treatment process, an iron-bearing adsorbent can be beneficial, as demonstrated with coprecipitated Fe-Al-cryo, which can remove As(III) in a single step (Kumar et al., 2013). Using cryogels as a support in POU filters results in porous membranes that potentially can be operated at low pressures. Ideally, a large range of contaminants should be removed by one single filter, for example, by adding more than one adsorbent in the supporting material, as this would increase their overall potential.

Other studies published on promising POU filters have involved metal oxide (nano) particles, as well as the immobilization of the particles on a support/matrix (Leupin and Hug, 2005; Leupin et al., 2005; Neumann et al., 2013; Sankar et al., 2013). The main advantage for most of these published materials, for example, sand filters with iron matrix or chitosan composites with silver NPs, in comparison to a cryogel-based adsorbent, is that the technology is simple and can be performed onsite. Sankar et al. (2013) produced a POU filter at ambient temperatures to be used in India. POU filters should also be low in cost (Sankar et al., 2013). For example, in the same study, it was shown that filters using AgNPs for water disinfection had a long lifetime, reducing the cost of treatment. Neumann et al. (2013) showed that As(III) removal is possible using iron-sand filters (SONO filters) based on locally available iron in Bangladesh. High amounts of natural organic matter and suspended solids would clog a sand filter, which is less likely for a cryogel filter. However, the amount of suspended solids in groundwater treated to provide drinking water is low.

The swelling ability of cryogels can also be used in situations when water access is scarce (e.g. emergencies, civil wars, etc.). To be able to disinfect a portion of water using cryogels has therefore been demonstrated using poly sodium acrylate cryogel materials decorated with silver NPs (Loo et al., 2013a). The same group of authors emphasizes the use of these materials in water emergencies and evaluate ways to alter morphology, pore size and mechanical strength to yield high swelling degrees and high swelling rates (Loo et al., 2013b), both vital properties for the intended application: provide potable water in emergencies.

### 12.4.3 Oil Removal—Taking Advantage of Cryogel Pore Space

The application of cryogels has been studied for other environmental applications, examples being the successful use of adsorbents for oil spills and polycyclic aromatic hydrocarbons (PAHs) from seawater (Ceylan et al., 2009; Hu et al., 2013). These applications are benefitted by the large pore space of the cryogel material, which can yield interesting solutions for the future. Apart from higher adsorption capacities than the commercially available polypropylene adsorbent, Okay demonstrated higher sorption rates and repetitive removal using butyl rubber as a sorbent (Ceylan et al., 2009). Similar results were achieved in the removal of organic liquids (e.g. benzene and toluene) as well as crude oils such as diesel oils. However, from an environmental perspective, the use of benzene in the production of the material is questionable (Karakutuk and Okay, 2010; Hu et al., 2013), but the crystals formed from benzene yielded larger pore sizes ($10^1$–$10^2$ μm) (Karakutuk and Okay, 2010), necessary for the high removal capacities.

### 12.4.4 Metal Removal in Connection to Biogas Production

Indirect effects on health arise from the use of contaminated water for irrigation, or fertilizers (solid or liquid) applied to agricultural land. In northern Europe, soil is naturally rich in cadmium; as a result, grown crops for food or biogas production contain this toxic heavy metal. In one study, we evaluated the possibility of removing $Cd^{2+}$ from biogas leachate using titanate nanotubes embedded in pAAm-cryogels (TNT-cryo) (Önnby et al., 2015). The nutrients can be recycled to the soil, and in the long term, better and safer crops can be produced. Further studies must be carried out to determine the best stage in connection to the anaerobic digestion process to apply TNT-cryo filters in order to achieve optimal mobilization of $Cd^{2+}$ and subsequent $Cd^{2+}$ removal. One approach could be to apply the TNT-cryo filter between hydrolysis and the anaerobic step, prior to the precipitation of metal sulphides, as has been demonstrated in previous studies using pAAm cryogel materials applied on other plant material (Lehtomäki and Björnsson, 2006; Selling et al., 2008). The results of the flow-through experiments with TNT-cryo, using real leachate with an initial (spiked) $Cd^{2+}$ concentration of 1 mg/l, are shown in Figure 12.8. After close to 2000 bed volumes of inflow water, no breakthrough was obtained and cadmium concentrations were kept low throughout the experiment (Önnby et al., 2015). Present regulations (for 2013) for bio-leachate are currently set to 35 mg Cd per kilogram added phosphor and in 2025, the regulation will be more stringent, requiring lower Cd additions (17 mg Cd/kg P, Figure 12.8).

Despite the suspended solids in all these studies (e.g. seaweed leachate contained total solids out of 3.4 wt% and total suspended solids of 7.8 mg $l^{-1}$) (Önnby et al., 2015), cryogels manage well and avoid getting clogged, owing to their macroporosity (Lehtomäki and Björnsson, 2006; Selling et al., 2008). Further studies, however, are needed for the reduction of $Cd^{2+}$ in marine biomass and in biomass in general. In addition to removing $Cd^{2+}$ from the liquid phase, it must be shown that $Cd^{2+}$ mobilization from the biomass can be achieved without jeopardizing potential products such as high-value nutrients and biogas processes with low overall

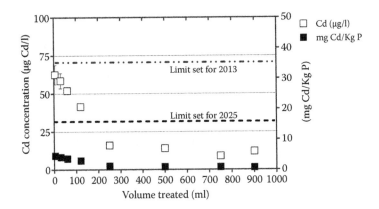

**FIGURE 12.8** Adsorption of $Cd^{2+}$ in real seaweed leachate, obtained in flow-through mode for 2000 bed volumes using TNT-cryo. White squares indicate $Cd^{2+}$ concentration (μg $l^{-1}$), filled squares indicate Cd amount expressed per kg P present in the leachate. Regulations set for bio-based fertilizer (2013 and 2025) are included in the graph. T = 20°C, pH = 7 ± 0.5. (Unpublished work.)

environmental impact. In addition, the goal should optimally be achieved at a reasonable cost.

### 12.4.5 MICROBIAL CRYOGEL SYSTEMS

Another interesting area of exploration for macroporous materials is to use them as support for biomaterial, for example, cells, enzymes and so on, and to further apply them in water treatment. These applications are exciting as the porous structure of cryogels is highly relevant for microbial growth and offer improved qualities as compared to other commercial hydrogels used for cell entrapment, for example, unhindered mass transfer of metabolites and substrates (Lozinsky and Plieva, 1998). Different pollutants have been studied, for example, degradation of PAHs in water by *Rhodococcus* cells as biocatalyst, demonstrated in fluidized-bed bioreactor with cryogel carriers (pAAm and PVA) as well as a sawdust-based carrier (Kuyukina et al., 2009). The better catalytic activity and hence oxidation of PAHs by sawdust-based carrier was attributed to the higher cell entrapment obtained for these materials.

PVA cryogels have frequently been used as carriers for bacteria, for example, *Thiobacillus denitrificans*, and demonstrated for denitrification processes, with the PVA entrapped cells achieving higher nitrate removal compared to the free cells (Zhang et al., 2009). In addition to denitrification, nitrification of ammonia-polluted seawater was studied using immobilized marine sludge in PVA carriers (Furukawa et al., 1993). Recently, deammonification was studied with immobilized anammox sludge in PVA cryogels, applied on real (swine) wastewater in stirred tank reactors. The study could show that high removal rates (>93%) were obtained during a time period of 5 months, and the immobilized bacteria were resistant to nitrate inhibition during the same time period (Magri et al., 2012).

## 12.5 FUTURE CHALLENGES

A range of water treatment applications has been studied for cryogel materials. The potential is high, provided that other aspects such as costs, regeneration and deposition can be fulfilled and follow set requirements by environmental law, for example. As such, environmental applications are challenging compared to biomedical applications where higher costs meets higher acceptance. Additionally, new technology faces great challenges in proving that it is environmentally friendly and giving improved qualities compared to the present solutions. Using adsorbent materials of toxic origin might therefore be problematic. Some of these matters will be discussed in the last sections of this chapter. The discussion will be given from a general perspective, including materials other than cryogels. Finally, a cost estimation will be presented with regard to some of the cryogel materials studied for metal removal in our group.

### 12.5.1 SYSTEM REGENERATION

The adsorption of pollutants from contaminated water provides purified water. However, the adsorbent will become saturated or exhausted with time. Regeneration of the adsorbent is preferable to disposal after use, as disposal is associated with higher operational costs and a greater burden on the environment. Many of the MIP-containing composites were regenerated using organic solvents, for metal saturated adsorbents with chelating cryogel matrix, both HCl and ethylenediaminetetraacetic acid (EDTA) were proven to be successful eluents (Dinu and Dragan, 2008; Önnby et al., 2010; Hajizadeh et al., 2013).

With the increased use of metal-based adsorbents, regeneration has recently been shown to be possible by altering the pH (Wang et al., 2013). The cost of treatment using an adsorbent that can be regenerated decreases with the number of times the adsorbent can be reused (Lu et al., 2007; Qu et al., 2013). There are some limitations on recycling of the materials based on cryogel supports. High pH will probably hydrolyse the pAAm backbone. The recovery of other heavy metals is more interesting than the recovery of arsenic or cadmium, which are both extremely toxic. Some heavy metals can be reused by industry, whereas the use of arsenic and cadmium has practically ceased in recent years (Bissen and Frimmel, 2003; Järup, 2003).

POU filters with long lifetimes are preferable to POU filters that need to be frequently replaced. SONO filters have been evaluated over an 8-year period, showing that the iron had been continuously oxidized (Neumann et al., 2013), and new adsorption sites are generated by iron oxidation and chemical reactions on the surface (Ahamed and Hussam, 2009). Apart from developing efficient POU filters, householders must be encouraged to use these filters, which has proven to be difficult (Neumann et al., 2013).

## 12.5.2 Deposition

Deposition of a highly toxic element, which in turn is not degradable, puts high requirements on the procedure. Taking arsenic as an example, which is highly toxic, precipitation, for example, with excess iron, can be used to remove the less soluble arsenic (Palfy et al., 1999). Due to the poor stability of iron arsenates, calcium can be added to ensure safe deposition of these arsenic compounds (Palfy et al., 1999). The solidified material can thereafter be disposed of in secure landfills (Leist et al., 2000). The lab test or procedure for evaluating arsenic waste is called the toxicity characteristic leaching procedure (TCLP) and was developed by the US Environmental Protection Agency (USEPA). This procedure is used to evaluate possible arsenic (or other toxic content) leaching from the solidified product (USEPA, 2002). The concentration of toxic substances being leached out of the solid should not exceed the guidelines for drinking water by more than 100 times. In practice, this means that no more than 1 mg/l can be leached out of the waste deposited in landfills (Ahamed and Hussam, 2009). If this cannot be guaranteed, the waste must be disposed of in a sanitary landfill (Leist et al., 2000). Conditions that can lead to the leaching of arsenic, for example, reducing conditions or low oxygen concentration, should be avoided at the deposition site. A TCLP evaluation should therefore be conducted for all sorbent materials developed (Ahamed and Hussam, 2009). Thus, apart from developing and producing efficient adsorbent materials for pollutant removal, their regeneration and final deposition must also be investigated.

Despite their long lifetimes, cryogel composites will have to be disposed of at some point. For the pAAm cryogel, this is risky in terms of environmental impact and safety. The polymer will probably be hydrolysed at high pH, leading to the leakage of acrylic monomers. AAm is also a chemical on the REACH list. (REACH denotes Registration, Evaluation, Authorization and Restriction of Chemicals and is used to regulate chemicals in the EU; it was adopted by the European Parliament in June 2007.). It contains information on the toxicity of chemicals and their future impact on the environment (Pesendorfer, 2006). The pAAm cryogel should therefore be replaced prior to the implementation of cryogel composites in commercial systems. Other bio-based materials should be studied and evaluated in further studies.

### 12.5.3 Risk Assessment

Nanotechnology, or engineered nanomaterials (ENMs), provides attractive opportunities for the development of novel water treatment methods. The characteristics of ENMs, such as their high specific surface area, antimicrobial activity, photosensitivity, catalytic activity and so on make their use of these materials interesting in many different applications in water and wastewater treatment (Hennebel et al., 2009; Khin et al., 2012; Qu et al., 2013). However, concerns regarding their possible environmental consequences remain. As the surface-to-volume ratio increases with decreasing particle size, ENMs or NPs are of great interest in adsorption-governed separation processes. In addition, both surface energy and surface structure are size dependent, and the nanoscale thus offers highly active adsorption sites (Auffan et al., 2008). Currently, there is more focus on results of treatment from novel ENM-based technology, and evidence of potential toxicity from the technology during use is often overlooked. Apart from potentially negative health effects on the environment, the cost effectiveness and social acceptability of these treatment technologies must also be investigated (Qu et al., 2013). The use of ENMs in water systems will inevitably lead to loss of them. The high production of ENMs has already resulted in these materials being found in water systems in our environment (Xu et al., 2012; Kaegi et al., 2013). It should be noted that it is not always the NPs themselves that are toxic, but the ions dissolved in them (Franklin et al., 2007). Identifying the possible risks associated with NPs is thus a challenge, likewise can be stated for the long-term ecotoxicological effects of NPs freely dispersed in the environment (Nel et al., 2006).

For the above-mentioned reasons, possible leakage of NPs from the nanocomposites with different AluNP loading (e.g. 1–6% w/v) was assessed in one of our studies (Önnby et al., 2014). In addition to the detection of traces of NPs during the production and use of the Alu-cryo, toxicity and the removal of As(V) were evaluated for the optimal Alu-cryogel (4-Alu-cryo). AluNP leakage into water filtered through the 4-Alu-cryo (0.5 ml cryogel) was measured using particle induced X-ray emission (PIXE) (Johansson and Johansson, 1984). Traces of AluNPs were investigated in different volumes of filtered water: 1000, 2000 and 4000 ml as well as a control sample (ultra pure water was used, sample denoted 0).

The samples of filtrated water were filtered through a filter paper and analysed regarding elemental Al using PIXE. A known amount of 30 mg AluNP/l was also included in the analysis (sample denoted 'known'). No relative increase of elemental

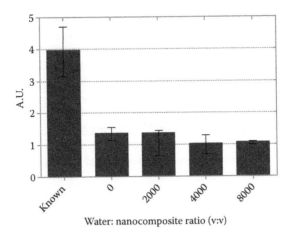

**FIGURE 12.9** The relative amount (A.U.) of elemental Al in water fractions during leakage studies from the Alu-cryo composite (4% AluNPs w/v) at different water-to-nanocomposite ratios (0, 2000, 4000 and 8000). The known amount of AluNPs was 30 mg/l. (Reprinted from *Science of The Total Environment* 473–474(0), Önnby, L., Svensson, C., Mbundi, L., Busquets, R., Cundy, A., and Kirsebom, H., γ-Al2O3-based nanocomposite adsorbents for arsenic(V) removal: assessing performance, toxicity and particle leakage, 207–214. Copyright 2014, with permission from Elsevier.)

Al was recognized as compared with the known concentration, as can be seen in Figure 12.9. This indicates that there was no detectable release of AluNPs into the water samples. Toxicity was partly measured as the reduction in the viability of epithelial cells. Based on the results of these measurements, the observed reduction in viability (≈40% for the first 20 bed volumes) was determined as being more likely to be the result of leakage of the monomer (AAm) than leakage of AluNPs (Önnby et al., 2014). This study showed how important it is to design a reliable toxicity test that includes all the components in the material. These findings suggest that the polymer used in the adsorbents, pAAm, should be replaced in future materials. The monomer AAm is already known to be a potent carcinogen (IARC, 1999).

### 12.5.4 Costs

To estimate the costs of the adsorbents studied in this work, calculations were performed based on price from suppliers offering high bulk volumes of the chemicals required. However, it is difficult to find prices for such large quantities of NPs. The prices of low quantities were consequently reduced by factors between 10 and 30. The costs were estimated for the materials Alu-cryo, Fe-Al-cryo and TNT-cryo. To facilitate comparison, all these materials were considered as POU filters. The obtained cost for three different POU filters with particles (TNTs, AluNPs and Fe-Al-hydrous oxides) was calculated for delivering a daily consumption of 10 l of water per day for a family of five for one year. As there are no specific cadmium filters available in Sweden, other than carbon filters, calculated costs were compared with the price for a commercial supplier of arsenic filters in Sweden. The calculations included

Application of Cryogels in Water and Wastewater Treatment 353

personnel costs (laboratory assistant) and production costs for a laboratory providing low volumes of cryogels.

The choice of particles significantly affects the total material cost, depending on how the particles were obtained. Three different routes were considered: (1) they were used as received from the supplier (AluNPs for Alu-cryo), (2) they were produced from metal salts (Fe-Al particles for Fe-Al-cryo) or (3) they were obtained via modification of purchased nanopowder ($TiO_2$-NPs for TNT-cryo). The loss of particles during production was also taken into account in these costs. It was assumed that the supporting material was pAAm cryogel. The calculations indicated that cryogel POU filters were more expensive as compared to the Swedish commercial filter supplier (containing an iron matrix). The market price of Swedish filter devices for the removal of arsenic consists of the cost of the holder/supporting material and the active filter material (consisting of an iron matrix). The market price also contains costs for labour, transport, marketing and profit (Johnsson, 2013). The iron matrix is usually exchanged every second year. The cost of the container was assumed to be 50% of the initial price of the filter device.

The lowest cost of a cryogel POU filter was obtained for TNT-cryo, as it has a higher adsorption capacity, and hence longer lifetime can be expected. The most cost-effective POU cryogel filter was shown to be 30 times more expensive than the Swedish commercial arsenic filter. The commercial Swedish filter device is designed for higher flow rates than those evaluated for the cryogel materials used in this example, and can thus produce higher amounts of pure water per day. It should also be noted that the regeneration of cryogel materials was not considered in these cost estimates. This will reduce the total cost by a factor equal to the number of times the material can be regenerated.

## 12.6 CONCLUDING REMARKS

Before cryogels can optimally compete with other adsorbent materials, there are certain areas that need to be addressed. Although cryogels demonstrate high potential as carriers for high surface-area adsorbent particles, chelating ligands, cells, bacteria and so on, in water treatment, few studies are addressed on real water and wastewater samples. In addition, it would be interesting to see how cryogel materials stand in comparison to commercially available adsorbents. Clearly, the use of toxic chemicals in the production of adsorbents for environmental applications should be abandoned or at least less considered. Synthetic chemicals, however, can withstand microbial degradation, otherwise possible when applied in wastewater, for example. Other aspects such as biofilm formation and its impact of adsorption have not been thoroughly addressed, which suggests that there is a fine balance between bio-based materials and synthetic materials when applied in bioactive media.

Considering the low volumes of cryogels currently produced, applications close to the tap or in small-scale treatment plants should be addressed as an application area before reaching the wastewater treatment plant as it exhibits very high flow rates ($>m^3/s$). In addition, different waste sources contain different matrix components, which need to be considered when tailoring the most appropriate material. This can be advantageous for cryogel materials as they are both flexible and show promise for many areas. To the best of my knowledge, applications using cryogels in, for example, dehalogenation reactions are still not investigated, which could be an interesting area to explore. To conclude, and regarding the relatively high estimated costs, these materials may be better suited for countries where labor and transport are available, than in rural areas where material availability and infrastructures are limited.

## ACKNOWLEDGEMENTS

This chapter is based on my PhD project, which resulted in a thesis in 2013 entitled 'Water Treatment Using Cryogel-Based Adsorbents: Targeting Environmental Polutants at Low Concentrations', which was financially supported by the Swedish Research Council, the EU-FP7 project Carbosorb and MISTRA—the foundation for strategic and environmental research. Region Scania (Region Skåne) in Sweden is also acknowledged for financial support during 2014. I also express my deepest gratitude toward Dr. Harald Kirsebom, one very valuable supervisor during my PhD studies. Also, thanks to Prof. Em. Bo Mattiasson for financial support during these years.

## LIST OF ABBREVIATIONS

| | |
|---|---|
| **AAm** | Acrylamide |
| **Alu-cryo** | AluNPs embedded in pAAm cryogels |
| **AluNP** | Aluminium oxide nanoparticles |
| **As(V)** | Arsenate |
| **BET** | Brunauer-Emmet-Teller |
| **–COOH** | Carboxyl |
| **E2** | 17β-estradiol |
| **ENM** | Engineered nanomaterials |
| **Fe-Al-cryo** | Fe-Al hydrous oxides embedded in pAAm cryogels |
| **HAp** | Hydroxyapatite |
| **HIS** | Histamine |
| **IDA** | Iminodiacetic acid |
| **IMAC** | Ion metal affinity chromatography |
| **IPN** | Inter penetrating network |
| **MHCMPs** | Metal hydroxide coated macroporous polymers |
| **MIP** | Molecularly imprinted polymers |
| **MIP-cryo** | MIPs embedded in pAAm cryogels |
| **NPs** | Nanoparticles |
| **p(HEMA)** | Poly(2-hydroxy ethyl methacrylate) |
| **pAAm** | Polyacrylamide |
| **PAH** | Polycylic aromatic hydrocarbon |
| **PIXE** | Particle induced X-ray emission |
| **POU** | Point-of-use filters |
| **PVA** | Poly(vinyl alcohol) |
| **REACH** | Registration, Evaluation, Authorisation and Restriction of Chemicals |
| **–SH** | Thiol |

SONO       Iron hydroxide/sand filters for arsenic removal
TCLP       Toxic characteristic leaching procedure
TNT-cryo   TNTs embedded in pAAm cryogels
TNTs       Titante nanotubes
WWTP       Wastewater treatment plant

## REFERENCES

Ahamed, S. and Hussam, A. (2009). Groundwater arsenic removal technologies based on sorbents: Field applications and sustainability. In: *Handbook of Water Purity and Quality*. S. Ahuja, Ed. Academic Press, Amsterdam, Netherlands.

Apopei, D. F., Dinu, M. V., Trochimczuk, A. W. and Dragan, E. S. (2012). Sorption isotherms of heavy metal ions onto semi-interpenetrating polymer network cryogels based on polyacrylamide and anionically modified potato starch. *Industrial & Engineering Chemistry Research* **51**(31): 10462–10471.

Auffan, M., Rose, J., Proux, O., Borschneck, D., Masion, A., Chaurand, P., Hazemann, J.-L., Chaneac, C., Jolivet, J.-P., Wiesner, M. R., Van Geen, A. and Bottero, J.-Y. (2008). Enhanced adsorption of arsenic onto maghemite nanoparticles: As(III) as a probe of the surface structure and heterogeneity. *Langmuir* **24**(7): 3215–3222.

Baggiani, C., Baravalle, P., Giovannoli, C., Anfossi, L. and Giraudi, G. (2010). Molecularly imprinted polymer/cryogel composites for solid-phase extraction of bisphenol A from river water and wine. *Analytical and Bioanalytical Chemistry* **397**(2): 815–822.

Bendz, D., Paxeus, N. A., Ginn, T. R. and Loge, F. J. (2005). Occurrence and fate of pharmaceutically active compounds in the environment, a case study: Hoje River in Sweden. *Journal of Hazardous Materials* **122**(3): 195–204.

Bissen, M. and Frimmel, F. H. (2003). Arsenic—A review. Part I: Occurrence, toxicity, speciation, mobility. *Acta Hydrochimica et Hydrobiologica* **31**(1): 9–18.

Ceylan, D., Dogu, S., Karacik, B., Yakan, S. D., Okay, O. S. and Okay, O. (2009). Evaluation of butyl rubber as sorbent material for the removal of oil and polycyclic aromatic hydrocarbons from seawater. *Environmental Science & Technology* **43**(10): 3846–3852.

Chen, H., Olmstead, M. M., Albright, R. L., Devenyi, J. and Fish, R. H. (1997). Metal-ion-templated polymers: Synthesis and structure of N-(4-vinylbenzyl)-1,4,7-triazacyclononanezinc(II) complexes, their copolymerization with divinylbenzene, and metal-ion selectivity studies of the demetalated resins—Evidence for a sandwich complex in the polymer matrix. *Angewandte Chemie International Edition in English* **36**(6): 642–645.

Chen, L., Xu, S. and Li, J. (2011). Recent advances in molecular imprinting technology: Current status, challenges and highlighted applications. *Chemical Society Reviews* **40**(5): 2922–2942.

Denizli, A., Sanli, N., Garipcan, B., Patir, S. and Alsancak, G. (2004). Methacryloylamidoglutamic acid incorporated porous poly(methyl methacrylate) beads for heavy-metal removal. *Industrial and Engineering Chemistry Research* **43**(19): 6095–6101.

Derazshamshir, A., Ergün, B., Peşint, G. and Odabaşı, M. (2008). Preparation of Zn2+-chelated poly(HEMA-MAH) cryogel for affinity purification of chicken egg lysozyme. *Journal of Applied Polymer Science* **109**(5): 2905–2913.

Dinu, M. V. and Dragan, E. S. (2008). Heavy metals adsorption on some iminodiacetate chelating resins as a function of the adsorption parameters. *Reactive & Functional Polymers* **68**(9): 1346–1354.

Dragan, E. S. and Apopei Loghin, D. F. (2013). Enhanced sorption of methylene blue from aqueous solutions by semi-IPN composite cryogels with anionically modified potato starch entrapped in PAAm matrix. *Chemical Engineering Journal* **234**: 211–222.

Ergün, B., Derazshamshir, A. and Odabaşı, M. (2007). Preparation of Fe(III)-chelated poly(HEMA-MAH) cryogel for lysozyme adsorption. *Hacettepe Journal of Biology and Chemistry* **35**(2): 143–148.

Erzengin, M., Unlu, N. and Odabasi, M. (2011). A novel adsorbent for protein chromatography: Supermacroporous monolithic cryogel embedded with Cu2+-attached sporopollenin particles. *Journal of Chromatography A* **1218**(3): 484–490.

EU. (2010). EU Directive (2010/75/EU) in industrial emissions. Brussels, Belgium, EU Commission.

Franklin, N. M., Rogers, N. J., Apte, S. C., Batley, G. E., Gadd, G. E. and Casey, P. S. (2007). Comparative toxicity of nanoparticulate ZnO, bulk ZnO, and ZnCl2 to a freshwater microalga (*Pseudokirchneriella subcapitata*): The importance of particle solubility. *Environmental Science & Technology* **41**(24): 8484–8490.

Friedman, M. (2003). Chemistry, biochemistry, and safety of acrylamide. A review. *Journal of Agricultural and Food Chemistry* **51**(16): 4504–4526.

Fullerto, P. and Barnes, J. M. (1966). Peripheral neuropathy in rats produced by acrylamide. *British Journal of Industrial Medicine* **23**(3): 210–221.

Furukawa, K., Ike, A., Ryu, S.-L. and Fujita, M. (1993). Nitrification of NH4-N polluted sea water by immobilized acclimated marine nitrifying sludge (AMNS). *Journal of Fermentation and Bioengineering* **76**(6): 515–520.

Hajizadeh, S., Kirsebom, H., Galaev, I. Y. and Mattiasson, B. (2010a). Evaluation of selective composite cryogel for bromate removal from drinking water. *Journal of Separation Science* **33**(12): 1752–1759.

Hajizadeh, S., Kirsebom, H. and Mattiasson, B. (2010b). Characterization of macroporous carbon-cryostructured particle gel, an adsorbent for small organic molecules. *Soft Matter* **6**(21): 5562–5569.

Hajizadeh, S., Xu, C., Kirsebom, H., Ye, L. and Mattiasson, B. (2013). Cryogelation of molecularly imprinted nanoparticles: A macroporous structure as affinity chromatography column for removal of beta-blockers from complex samples. *Journal of Chromatography A* **1274**: 6–12.

Haupt, K. and Mosbach, K. (2000). Molecularly imprinted polymers and their use in biomimetic sensors. *Chemical Reviews* **100**(7): 2495–2504.

Hennebel, T., De Gusseme, B., Boon, N. and Verstraete, W. (2009). Biogenic metals in advanced water treatment. *Trends in Biotechnology* **27**(2): 90–98.

Hu, Y., Liu, X., Zou, J., Gu, T., Chai, W. and Li, H. (2013). Graphite/Isobutylene-isoprene rubber highly porous cryogels as new sorbents for oil spills and organic liquids. *ACS Applied Materials & Interfaces* **5**(16): 7737–7742.

IARC (1999). Monographs on the evaluation of carcinogenic risks to humans: Some industrial chemicals. Volume 60. Lyon, France, World Health Organization.

Järup, L. (2003). Hazards of heavy metal contamination. *British Medical Bulletin* **68**(1): 167–182.

Johansson, E. M. and Johansson, S. A. E. (1984). PIXE analysis of water at the parts per trillion level. *Nuclear Instruments & Methods in Physics Research Section B-Beam Interactions with Materials and Atoms* **3**(1–3): 154–157.

Johnsson, K. (2013). Manager of Vattenteknik, Sweden. Personal communication.

Kaegi, R., Voegelin, A., Ort, C., Sinnet, B., Thalmann, B., Krismer, J., Hagendorfer, H., Elumelu, M. and Mueller, E. (2013). Fate and transformation of silver nanoparticles in urban wastewater systems. *Water Research* **47**(12):3866–3877.

Karakutuk, I. and Okay, O. (2010). Macroporous rubber gels as reusable sorbents for the removal of oil from surface waters. *Reactive & Functional Polymers* **70**(9): 585–595.

Karvelas, M., Katsoyiannis, A. and Samara, C. (2003). Occurrence and fate of heavy metals in the wastewater treatment process. *Chemosphere* **53**(10): 1201–1210.

Khin, M. M., Nair, A. S., Babu, V. J., Murugan, R. and Ramakrishna, S. (2012). A review on nanomaterials for environmental remediation. *Energy & Environmental Science* **5**(8): 8075–8109.

Koç, İ., Baydemir, G., Bayram, E., Yavuz, H. and Denizli, A. (2011). Selective removal of 17β-estradiol with molecularly imprinted particle-embedded cryogel systems. *Journal of Hazardous Materials* **192**(3): 1819–1826.

Kolpin, D. W., Furlong, E. T., Meyer, M. T., Thurman, E. M., Zaugg, S. D., Barber, L. B. and Buxton, H. T. (2002). Pharmaceuticals, hormones, and other organic wastewater contaminants in US streams, 1999–2000: A national reconnaissance. *Environmental Science & Technology* **36**(6): 1202–1211.

Kumar, P. S., Önnby, L. and Kirsebom, H. (2013). Arsenite adsorption on cryogels embedded with iron-aluminium double hydrous oxides: Possible polishing step for smelting wastewater? *Journal of Hazardous Materials* **250**: 469–476.

Kumar, P. S., Önnby, L. and Kirsebom, H. (2014). Reversible *in situ* precipitation: A flow-through approach for coating macroporous supports with metal hydroxides. *Journal of Materials Chemistry A* **2**(4): 1076–1084.

Kuyukina, M. S., Ivshina, I. B., Serebrennikova, M. K., Krivorutchko, A. B., Podorozhko, E. A., Ivanov, R. V. and Lozinsky, V. I. (2009). Petroleum-contaminated water treatment in a fluidized-bed bioreactor with immobilized Rhodococcus cells. *International Biodeterioration & Biodegradation* **63**(4): 427–432.

Le Noir, M., Plieva, F., Hey, T., Guieysse, B. and Mattiasson, B. (2007). Macroporous molecularly imprinted polymer/cryogel composite systems for the removal of endocrine disrupting trace contaminants. *Journal of Chromatography A* **1154**(1–2): 158–164.

Le Noir, M., Plieva, F. M. and Mattiasson, B. (2009). Removal of endocrine-disrupting compounds from water using macroporous molecularly imprinted cryogels in a moving-bed reactor. *Journal of Separation Science* **32**(9): 1471–1479.

Lehtomäki, A. and Björnsson, L. (2006). Two-stage anaerobic digestion of energy crops: Methane production, nitrogen mineralisation and heavy metal mobilisation. *Environmental Technology* **27**(2): 209–218.

Leist, M., Casey, R. J. and Caridi, D. (2000). The management of arsenic wastes: Problems and prospects. *Journal of Hazardous Materials* **76**(1): 125–138.

Leupin, O. X. and Hug, S. J. (2005). Oxidation and removal of arsenic (III) from aerated groundwater by filtration through sand and zero-valent iron. *Water Research* **39**(9): 1729–1740.

Leupin, O. X., Hug, S. J. and Badruzzaman, A. B. M. (2005). Arsenic removal from Bangladesh tube well water with filter columns containing zerovalent iron filings and sand. *Environmental Science & Technology* **39**(20): 8032–8037.

Loo, S.-L., Fane, A. G., Lim, T.-T., Krantz, W. B., Liang, Y.-N., Liu, X. and Hu, X. (2013a). Superabsorbent cryogels decorated with silver nanoparticles as a novel water technology for point-of-use disinfection. *Environmental Science & Technology* **47**(16): 9363–9371.

Loo, S.-L., Krantz, W. B., Lim, T.-T., Fane, A. G. and Hu, X. (2013b). Design and synthesis of ice-templated PSA cryogels for water purification: Towards tailored morphology and properties. *Soft Matter* **9**(1): 224–234.

Lozinsky, V. I. and Plieva, F. M. (1998). Poly(vinyl alcohol) cryogels employed as matrices for cell immobilization. 3. Overview of recent research and developments. *Enzyme and Microbial Technology* **23**(3–4): 227–242.

Lu, C., Chiu, H. and Bai, H. (2007). Comparisons of adsorbent cost for the removal of zinc (II) from aqueous solution by carbon nanotubes and activated carbon. *Journal of Nanoscience and Nanotechnology* **7**(4–5): 1647–1652.

Magri, A., Vanotti, M. B. and Szoegi, A. A. (2012). Anammox sludge immobilized in polyvinyl alcohol (PVA) cryogel carriers. *Bioresource Technology* **114**: 231–240.

McCabe, W., Smith, J., and Harriot, P. (2001). *Unit of Operations Chemical Engineering.* McGraw Hill, Singapore.

Myung, D., Waters, D., Wiseman, M., Duhamel, P.-E., Noolandi, J., Ta, C. N. and Frank, C. W. (2008). Progress in the development of interpenetrating polymer network hydrogels. *Polymers for Advanced Technologies* **19**(6): 647–657.

Nel, A., Xia, T., Madler, L. and Li, N. (2006). Toxic potential of materials at the nanolevel. *Science* **311**(5761): 622–627.

Neumann, A., Kaegi, R., Voegelin, A., Hussam, A., Munir, A. K. M. and Hug, S. J. (2013). Arsenic removal with composite iron matrix filters in Bangladesh: A field and laboratory study. *Environmental Science & Technology* **47**(9): 4544–4554.

Odland, L., Romert, L., Clemedson, C. and Walum, E. (1994). Glutathione content, glutathione transferase-activity and lipid-peroxidation in acrylamide-treated neuroblastoma N1E-115 cells. *Toxicology in Vitro* **8**(2): 263–267.

Önnby, L., Giorgi, C., Plieva, F. M. and Mattiasson, B. (2010). Removal of heavy metals from water effluents using supermacroporous metal chelating cryogels. *Biotechnology Progress* **26**(5): 1295–1302.

Önnby, L., Kirsebom, H. and Nges, I. A. (2015). Cryogel-supported titanate nanotubes for waste treatment: Impact on methane production and bio-fertilizer quality. *Journal of Biotechnology* **207**:58–66.

Önnby, L., Pakade, V., Mattiasson, B. and Kirsebom, H. (2012). Polymer composite adsorbents using particles of molecularly imprinted polymers or aluminium oxide nanoparticles for treatment of arsenic contaminated waters. *Water Research* **46**(13): 4111–4120.

Önnby, L., Svensson, C., Mbundi, L., Busquets, R., Cundy, A. and Kirsebom, H. (2014). γ-Al2O3-based nanocomposite adsorbents for arsenic(V) removal: Assessing performance, toxicity and particle leakage. *Science of the Total Environment* **473–474**(0): 207–214.

Palfy, P., Vircikova, E. and Molnar, L. (1999). Processing of arsenic waste by precipitation and solidification. *Waste Management* **19**(1): 55–59.

Pesendorfer, D. (2006). EU environmental policy under pressure: Chemicals policy change between antagonistic goals? *Environmental Politics* **15**(1): 95–114.

Plieva, F. M., Andersson, J., Galaev, I. Y. and Mattiasson, B. (2004). Characterization of polyacrylamide based monolithic columns. *Journal of Separation Science* **27**(10–11): 828–836.

Plieva, F. M., Galaev, I. Y. and Mattiasson, B. (2007). Macroporous gels prepared at subzero temperatures as novel materials for chromatography of particulate-containing fluids and cell culture applications. *Journal of Separation Science* **30**(11): 1657–1671.

Plieva, F. M. and Mattiasson, B. (2008). Macroporous gel particles as novel sorbent materials: Rational design. *Industrial & Engineering Chemistry Research* **47**(12): 4131–4141.

Porath, J. (1992). Immobilized metal-ion affinity-chromatography. *Protein Expression and Purification* **3**(4): 263–281.

Porath, J., Carlsson, J., Olsson, I. and Belfrage, G. (1975). Metal chelate affinity chromatography, a new approach to protein fractionation. *Nature* **258**(5536): 598–599.

Qu, X., Alvarez, P. J. J. and Li, Q. (2013). Applications of nanotechnology in water and wastewater treatment. *Water Research* **47**(12): 3931–3946.

Qu, X., Brame, J., Li, Q. and Alvarez, P. J. J. (2013). Nanotechnology for a safe and sustainable water supply: Enabling integrated water treatment and reuse. *Accounts of Chemical Research* **46**(3): 834–843.

Sankar, M. U., Aigal, S., Maliyekkal, S. M., Chaudhary, A., Anshup, Kumar, A. A., Chaudhari, K. and Pradeep, T. (2013). Biopolymer-reinforced synthetic granular nanocomposites for affordable point-of-use water purification. *Proceedings of the National Academy of Sciences of the United States of America* **110**(21): 8459–8464.

Savina, I. N., English, C. J., Whitby, R. L. D., Zheng, Y., Leistner, A., Mikhalovsky, S. V. and Cundy, A. B. (2011). High efficiency removal of dissolved As(III) using iron nanoparticle-embedded macroporous polymer composites. *Journal of Hazardous Materials* **192**(3): 1002–1008.

Schwarzenbach, R. P., Escher, B. I., Fenner, K., Hofstetter, T. B., Johnson, C. A., von Gunten, U. and Wehrli, B. (2006). The challenge of micropollutants in aquatic systems. *Science* **313**(5790): 1072–1077.

Selling, R., Hakansson, T. and Bjornsson, L. (2008). Two-stage anaerobic digestion enables heavy metal removal. *Water Science and Technology* **57**(4): 553–558.

Sheng, P. X., Wee, K. H., Ting, Y. P. and Chen, J. P. (2008). Biosorption of copper by immobilized marine algal biomass. *Chemical Engineering Journal* **136**(2–3): 156–163.

Tekin, K., Uzun, L., Sahin, C. A., Bektas, S. and Denizli, A. (2011). Preparation and characterization of composite cryogels containing imidazole group and use in heavy metal removal. *Reactive & Functional Polymers* **71**(10): 985–993.

Ternes, T. A. (1998). Occurrence of drugs in German sewage treatment plants and rivers. *Water Research* **32**(11): 3245–3260.

UN. (2013). UN Water. Retrieved October 17, 2013, from http://www.unwater.org/water-cooperation-2013/water-cooperation/facts-and-figures/.

UNICEF. (2013). Clean Drinking Water. Retrieved, October 17, 2013, from http://www.unicefusa.org/work/water/.

USEPA. (2002). Arsenic treatment technologies for soil, waste, and water. EPA-542-R-02-004. USEPA. Cincinnati, OH.

VA-Syd (2009). Sjölunda avloppsreningsverk. Malmö, Sweden.

Wang, T., Liu, W., Xu, N. and Ni, J. R. (2013). Adsorption and desorption of Cd(II) onto titanate nanotubes and efficient regeneration of tubular structures. *Journal of Hazardous Materials* **250**: 379–386.

Wang, X. and Min, B. G. (2008). Comparison of porous poly (vinyl alcohol)/hydroxyapatite composite cryogels and cryogels immobilized on poly (vinyl alcohol) and polyurethane foams for removal of cadmium. *Journal of Hazardous Materials* **156**(1–3): 381–386.

WHO. (2011). Guidelines for drinking-water quality. Geneva, Switzerland, World Health Organization.

Wulff, G. (1995). Molecular imprinting in cross-linked materials with the aid of molecular templates—A way towards artificial antibodies. *Angewandte Chemie-International Edition in English* **34**(17): 1812–1832.

Xu, P., Zeng, G. M., Huang, D. L., Feng, C. L., Hu, S., Zhao, M. H., Lai, C., Wei, Z., Huang, C., Xie, G. X. and Liu, Z. F. (2012). Use of iron oxide nanomaterials in wastewater treatment: A review. *Science of The Total Environment* **424**: 1–10.

Zhang, Z., Lei, Z., He, X., Zhang, Z., Yang, Y. and Sugiura, N. (2009). Nitrate removal by Thiobacillus denitrificans immobilized on poly(vinyl alcohol) carriers. *Journal of Hazardous Materials* **163**(2–3): 1090–1095.

# 13 Cryogels in High Throughput Processes

*Joyita Sarkar, Ankur Gupta and Ashok Kumar\**

## CONTENTS

13.1 Introduction ........................................................................................362
13.2 High Throughput Screening of Optimal Conditions for Cell Separation.....364
13.3 High Throughput Analysis of Proteins......................................................366
13.4 Other High Throughput Processes ...........................................................372
    13.4.1 Cell Surface Analysis ....................................................................374
    13.4.2 Cell-Based Assays .........................................................................375
    13.4.3 Panning of Phage Display Library................................................376
    13.4.4 Pollutant Analysis..........................................................................377
13.5 Conclusions...........................................................................................379
Acknowledgements..........................................................................................382
List of Abbreviations........................................................................................382
References........................................................................................................382

### ABSTRACT

High throughput processes have been able to improve the efficiency as well as minimize the cost, labour and time required in a process. These high throughput techniques have found there major applications in drug discovery, genomics, proteomics and toxicological studies. Although the high throughput platform has been miniaturized to microarrays, many research groups still prefer 96-well plates due to their better sensitivity and cheap infrastructure required by them. Cryogels, when inserted in an open-ended 96-well plate, provide an excellent leakage- and drainage-protected platform that can be used in a high throughput manner. Different cryogel minicolumns in 96-well plate format have been used for parallel screening of optimum cell separation conditions and peptide affinity tags, processing of particulate containing samples, analysis of cell surface, pollutants and cell-based cytotoxicity assays, monitoring inclusion body production during fermentation, enzyme-linked immunosorbent assay (ELISA) and panning of phage display library.

### KEYWORDS

High throughput screening, cryogels, 96-well plates, ELISA, cytotoxicity studies

---

\* Corresponding author: E-mail: ashokkum@iitk.ac.in; Tel: +91-512-2594051.

## 13.1 INTRODUCTION

Improving the efficiency of the process and meanwhile minimizing the cost has been the constant demand of both industry as well as researchers. High throughput processes have been a boon to such demands. These processes have revolutionized the pharmaceutical and biotechnological industries and have been leveraged by drug discovery processes. These processes are gaining popularity not only in industries but also among research laboratories due to their inherent advantages like fast and simultaneous processing of a large number of samples and miniaturized platform that leads to reduced processing time, amount of reagent and space required. These properties significantly reduce the processing cost and labour.

High throughput processes have found major applications in drug discovery, toxicology, genomics and protein biology. They play a major role in the initial stages of drug discovery where there is a large number of chemical compounds to be screened against a large number of biological targets, and a number of assays are performed to validate the compounds and targets (Martis et al., 2011; Mayr and Bojanic, 2009; White, 2000). High throughput processes have also been explored in toxicological studies to predict the toxicity of a particular compound (Szymański et al., 2012). In cases of genomics and protein biology, these processes have been used to determine nucleic acid–protein interactions, protein–protein interactions, single nucleotide polymorphism, genomic library screening, proteomics and so on. (Bulyk, 2006; Huang et al., 2001; Kircher et al., 2010; Sasakura et al., 2004). In the recent past, high throughput chromatographic processes have also been developed that help in screening and choosing for the right matrix, buffers, flow rates and so on simultaneously (Chhatre and Titchener-Hooker, 2009). Recently, these high throughput processes are also being used for screening or remediation of a plethora of environmental pollutants (Desmet et al., 2012; Gupta et al., 2013; Stenuit et al., 2008).

In the past decade, there has been an increasing need for further reduction in working volume, time and cost. This has driven high throughput processes to follow a trend towards miniaturization and the paradigm is shifting from 96-well plates to higher density plates (384- and 1536-well plates) and microarrays. These higher density plates require much lower volume and can assay a larger number of samples compared to 96-well plates. For example, the 1536-well plates can bring down the working volume to 2.5 µl and can perform 60,000 assays per day (Mayr and Bojanic, 2009). Although a large increase in throughput has been achieved by switching from 96-well plates to higher density plates and microarrays, a majority of research groups still rely on 96-well plates. The reason behind this is that higher density plates create hurdles in sensitivity of the assays, mixing issues and so forth, while microarrays require long design and implementation time and automated robotic operation, which make them very expensive and reduce their error recovery abilities (Szymański et al., 2012).

Matrixes developed using cryogelation technology, that is, cryogels, have been used in a variety of applications, for example, for chromatography, separation, analytical purposes, cell culture and bioreactor and so on (Ahlqvist et al., 2006; Dainiak et al., 2006a,b; Jain and Kumar, 2009; Jain et al., 2013; Kumar and Srivastava, 2010). When prepared in appropriate moulds, cryogels can be inserted in 96-well plates and can be used for high throughput purposes (Figure 13.1a). The cryogels fit tightly with the walls of the well of 96-well plate leaving no gap and making the system leakage protected. The cryogels also retain the applied liquid on them due to capillary forces owing to their large and interconnected pores, making the system drainage protected (Dainiak et al., 2006a,b; Galaev et al., 2005). Drop forming units are created at the open ends of the wells. Liquid coming out through the open ends of the wells can be collected using another 96-well plate kept below the open-ended plate. The latter acts as a collection reservoir, which can be directly analysed in a multiplate reader

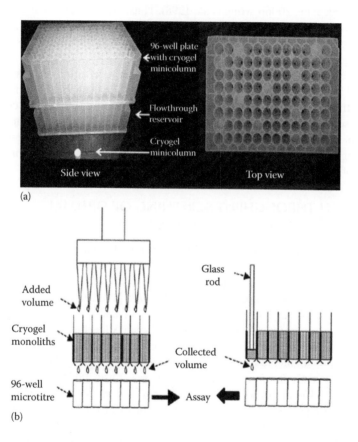

**FIGURE 13.1** (a) Digital image of high throughput platform with 0.5 ml poly(4-vinyl pyridine-*co*-divinyl benzene) cryogel minicolumns inserted in open-ended 96-well plate. (b) Schematic representation of high throughput procedures carried out in 96-well plate format. (Reprinted from *J. Chromatogr. A*. 1278, Gupta, A., Sarkar, J., and Kumar, A., High Throughput Analysis and Capture of Benzo[a]pyrene using Supermacroporous Poly(4-vinyl pyridine-*co*-divinyl benzene) Cryogel Matrix, 16–21. Copyright 2013, with permission from Elsevier; Reprinted from *Enzym. Microb. Technol.* 40, Dainiak, M.B., Galaev, I.Y., and Mattiasson, B., Macroporous Monolithic Hydrogels in a 96-Minicolumn Plate Format for Cell Surface-Analysis and Integrated Binding/Quantification of Cells, 688–695. Copyright 2007, with permission from Elsevier.)

or can be further processed (Dainiak et al., 2007b) (Figure 13.1b). In other cases, for example, while using fluorescent tags, the captured analytes can be assayed directly on the cryogel (Hanora et al., 2005). The properties of the cryogels, like porosity, volumetric and linear flow rates, microstructure and so on are quite reproducible. The use of a cryogel minicolumn in a multiwell format offers the unique methods of recovery of the captured analyte by mechanical deformation because of the inherent elastic property of cryogels (Figure 13.1b) (Dainiak et al., 2007a; Lozinsky et al., 2003). The commonly used flow-induced shear force leads to dilution of the eluted cells and there is also a chance of cells getting damaged (Cozens-Roberts et al., 1990; Dainiak et al., 2006b; Ming et al., 1998). Thus, use of mechanical compression helps to overcome these drawbacks. All these properties make cryogels suitable to be used in high throughput formats. These cryogel scaffolds can also be easily scaled-up for large-scale operations.

The chapter provides an overview of different cryogel minicolumns that have been used in 96-well plates for a variety of high throughput applications like screening of optimum conditions for cell separation, processing of particulate containing samples, screening of peptide affinity tags from peptide libraries, monitoring production of inclusion bodies during fermentation, ELISA, analysis of cell surface, optimizing cell-based cytotoxicity assays, panning of phage display libraries and analysis of pollutants.

## 13.2 HIGH THROUGHPUT SCREENING OF OPTIMAL CONDITIONS FOR CELL SEPARATION

An inexpensive and efficient process for separation of a particular cell type from a mixed population of cells is highly desirable for a number of biomedical and biotechnological applications. Chromatography provides many advantages over other conventional methods of separations like very pure products can be obtained from complex mixtures, large-scale operation is possible and a variety of methods can be applied for separation. A monolithic cryogel matrix has been used for chromatographic separation of many biological particles like plasmids, viruses, cell organelles and mammalian and bacterial cells (Dainiak et al., 2005; Hanora et al., 2006; Kumar et al., 2003; Teilum et al., 2006; Williams et al., 2005). The large pores of cryogels of 10–100 μm make it possible to separate particles of size up to 1–10 μm without clogging the columns. In addition, particles of this size range will not be mechanically entrapped in the columns and the cells are not disrupted as they do not experience much shear forces due to the interconnected pores that allow laminar flow of the fluid. Therefore, cryogels can be used successfully for an efficient matrix for cell separation (Dainiak et al., 2007a).

However, the surfaces of cells are highly complex and heterogeneous in nature, which leads to a variety of interactions of the cell surface and chromatographic matrix. Therefore, there are a large number of factors that may affect the binding and elution of cells in these matrices (Dainiak et al. 2006a). High throughput processes can provide a platform for parallel chromatography to screen for optimal conditions for cell separation. High throughput processes provide easy, fast and parallel approaches for analysing the optimal condition for binding and elution of cells depending on the multivalent interactions it experiences with the matrix. Indeed, concanavalin A-cryogel (Con A-cryogel) minicolumns inserted in open-ended 96-well microtitre plates have been used to optimize suitable conditions for separation of *Escherichia coli* (*E. coli*) and *Saccharomyces cerevisiae* (*S. cerevisiae*) cells. These minicolumns were also stacked over one another to scale-up the process and carry out chromatography.

Affinity (Con A-) and blank (native) cryogel monoliths of height 18.8 mm and diameter 7.1 mm were inserted in wells of 96-well plate. *E. coli* and *S. cerevisiae* cells were first loaded at 0.2 units of $OD_{600}$. In the absence of any incubation, no *E. coli* cells were retained on both ConA- and native cryogels. However, as the incubation time was increased, nonspecific retention of *E. coli* cells on both blank and Con A-cryogels increased and reached up to 70% binding in 30 min incubation. Therefore, *E. coli* cells cannot be separated from *S. cerevisiae* cells based on incubation time on the adsorbent due to nonspecific binding of *E. coli* cells. Therefore, the load of cells, that is, $OD_{600}$ units, were further optimized and it was found that at a cell load of 0.03 units of $OD_{600}$, which corresponds to about $2 \times 10^5$ cells, there was maximum binding of *S. cerevisiae* cells (93%) and negligible nonspecific retention of *E. coli* cells (Dainiak et al. 2006a) (Table 13.1).

### TABLE 13.1
### Screening for Optimal Conditions for Separation of *S. cerevisiae* and *E. coli* Cells on Con-A Cryogel Monoliths in 96-Well Plate

| Cell Type | Column | Cell Load, Units $OD_{600}$ | Incubation Time[a] (min) | Bound Cells[b] (%) |
|---|---|---|---|---|
| *E. coli* | Native | 0.2 | 30 | 69 |
| *E. coli* | ConA | 0.2 | 30 | 70 |
| *S. cerevisiae* | Native | 0.2 | 30 | 39 |
| *S. cerevisiae* | ConA | 0.2 | 30 | 89 |
| *E. coli* | Native | 0.2 | 0 | 0 |
| *E. coli* | ConA | 0.2 | 0 | 1 |
| *S. cerevisiae* | Native | 0.2 | 0 | 0 |
| *S. cerevisiae* | ConA | 0.2 | 0 | 35 |
| *E. coli* | Native | 0.1 | 0 | 3 |
| *E. coli* | ConA | 0.1 | 0 | 2 |
| *S. cerevisiae* | Native | 0.1 | 0 | 65 |
| *S. cerevisiae* | ConA | 0.1 | 0 | 1 |
| *E. coli* | ConA | 0.03 | 0 | 93 |
| *S. cerevisiae* | ConA | 0.03 | 0 | |

*Source:* Reprinted from *J. Chromagr. A.* 1123, Dainiak. M. B., Galaev, I. Y., and Mattiason, B., 145–50. Copyright 2006, with permission from Elsevier.

[a] Cells were incubated with the adsorbent prior to washing step.
[b] Amount of applied cells was assumed to be 100%.

Different cells were detached from affinity cryogel columns by elastic deformation, namely, yeast cells from Con A-cryogels, $His_6$-*E. coli* cells from Ni(II)-iminodiacetate-cryogels (Ni(II)-IDA-cryogels) and human lymphocytes from Protein A-cryogels. The effect of other parameters on cell release by mechanical deformation like rigidity of cryogel pore walls, nature and concentration of eluents, affinity and density of binding groups and so on were analysed simultaneously in open ended wells of 96-well plates. The effect of increasing concentration of ethylenediaminetetracetate (EDTA) on the elution of $His_6$-*E. coli* cells from Ni(II)-IDA-cryogels on 'dense' (6%) and 'soft' (5%) matrices with or without compression was studied in different wells (Figure 13.2a). It was concluded that only compression is enough to elute about 80% of the bound cells without affecting the immobilized affinity ligands on the cryogel matrix (Dainiak et al., 2006b).

Similarly, in another study, dense and soft Con A-cryogel minicolumns were explored in parallel experiments for efficient detachment conditions of yeast cells and the effect of concentration of different eluents. It was found that for dense (6%) cryogels, which are highly elastic (elastic modulus 0.065 MPa), deformation alone is enough to elute the cells whereas in case of soft (5%) cryogels with elastic moduli of 0.01 MPa, compression along with 0.5 M of α-D-manno-pyranoside or glucose in the elution buffer is required for maximum elution of yeast cells from the matrix (Galaev et al., 2007) (Figure 13.2b,c).

## 13.3 HIGH THROUGHPUT ANALYSIS OF PROTEINS

Using cryogels as a matrix in 96-well format has also shown the possibility where different processes like capture and clarification of the protein from nonclarified samples can be integrated with direct on column assay of captured protein (Arvidsson et al., 2003). The large and interconnected pores of the cryogel matrix allow unhindered passage of the cell debris and other impurities of the cell culture broth (Galaev et al., 2005). In 2005, Galaev and co-workers synthesized polydimethylacrylamide (pDMAA) using dimethylacrylamide and allyl glycidyl ether (AGE) as monomers and N,N'-Methylenebisacrylamide (MBAAm) as cross-linker. Immobilized metal affinity chromatography (IMAC) monolithic columns were prepared by coupling the synthesized cryogel with copper ions. These IMAC columns (0.5 ml) were inserted into wells of microtitre plate. This system was used for direct on column quantification of $His_6$-tagged lactate dehydrogenase ($His_6$-LDH) from crude homogenate. Different amounts of *E. coli* cells in nonclarified crude homogenate were directly applied on the minicolumns previously equilibrated with a buffer. After washing, a buffer having an enzyme substrate for the enzyme lactate dehydrogenase was added directly on the minicolumn and the coloured product formed was displaced using an elution buffer. Analysis was done by measuring absorbance at 340 nm. The result showed that as the amount of homogenate increases, the amount of the captured protein also increases (Figure 13.3a). In another experiment, monolith-to-monolith variation was monitored loading the same amount of crude homogenate on the different columns. The standard deviation of only 0.08 was observed among the columns (Figure 13.3b). Further, it was also shown that the optimal cell disrupting condition can also be screened (Table 13.2). Similar results were found when a $His_6$-tagged

# Cryogels in High Throughput Processes

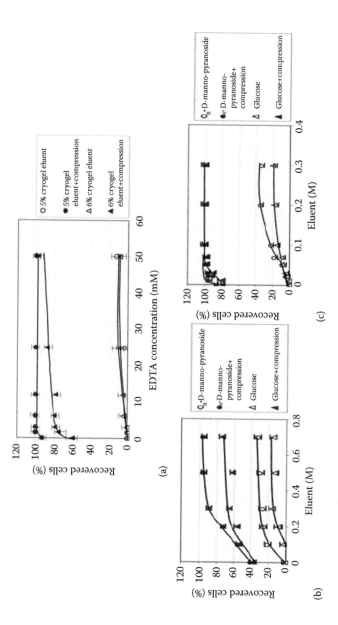

**FIGURE 13.2** Release of (a) His$_6$ *E. coli* captured on Ni(II)-IDA-cryogel columns; yeast cells bound to (b) 5% and (c) 6% Con-A-cryogels by conventional elution and compression of cryogels. Experimental conditions: Column was equilibrated in running buffer, 20 mM HEPES with 0.2 M NaCl, pH 7.0 for *E. coli* and Ni(II)- IDA-cryogel and 0.1 M Tris with 150 mM NaCl, 5 mM CaCl$_2$ and 5 mM MgCl$_2$, pH 7.4 for yeast cells and Con A-cryogels. *E. coli* and yeast cells were suspended in proper running buffer to an OD$_{450}$ of 0.727 and OD$_{600}$ of 0.855, respectively. After incubation for 15–20 min and washing with 3.5 ml running buffer, elution was done with different concentration of EDTA (for *E. coli* cells) and α-D-manno-pyranoside and glucose (for yeast cells) in running buffer. (Reproduced from Dainiak, M.B., Kumar, A., Galaev, I.Y., and Mattiasson, B., *Proc. Natl. Acad. Sci.* 103: 849–854. Copyright 2006. National Academy of Sciences, U.S.A.; Reprinted with permission from Galaev, I.Y., Dainiak, M.B., Plieva, F., and Mattiasson, B., *Langmuir.* 23: 35–40. Copyright 2007. American Chemical Society.)

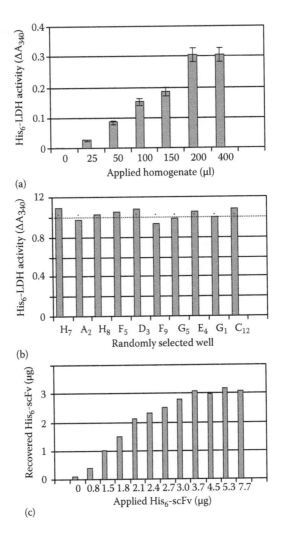

**FIGURE 13.3** Binding of $His_6$-LDH from (a) different amounts, (b) 100 μl of crude *E. coli* cell homogenate and (c) the dependence of recovered amount of His6-ScFv on the volume of *E. coli* cell culture fluid applied to 0.5 ml of Cu (II)-IDA-cryogel minicolumns inserted in wells of open-ended 96-well plates. Experimental conditions: (a and b) After samples were applied to the minicolumns, unbound cell debris and proteins were washed with 0.2 M Tris-HCl, pH 7.3 (buffer A). After incubation of substrate (2 mM pyruvate) and 0.45 mM NADH on the cryogel, the reaction mixture was displaced from the minicolumn by applying 0.5 ml of buffer A and its absorbance was measured at 340 nm. (c) Columns were equilibrated with 20 mM HEPES, 0.2 M NaCl, 2mM imidazole, pH 7.0. Different volumes of cell culture fluid (with 30 μg/ml of $His_6$-ScFv) were loaded on the column. Elution was done with 1 ml of 20 mM EDTA, 20 mM NaCl, pH 7.5. Quantification of $His_6$-ScFv was done by ELISA. (Reprinted from *J. Chromatogr. A*. 1065, Galaev, I.Y., Dainiak, M.B., Plieva, F.M., Hatti-Kaul, R., and Mattiasson, B., High Throughput Processing of Particulate-Containing Samples Using Supermacroporous Elastic Monoliths in Microtiter [Multiwell] Plate Format, 169–175. Copyright 2005, with permission from Elsevier.)

### TABLE 13.2
### Direct Assay of His$_6$-LDH Activity in Cell Homogenate of *E. coli* Using Cu (II)-IDA-Cryogel Columns, under Different Sonication Conditions

| Sonication Conditions | | His$_6$-LDH |
|---|---|---|
| Cells per Second | Amplitude (%)[a] | Activity ($\Delta A_{340}$)[b] |
| 0.5 | 20 | 0 |
| 0.7 | 20 | 0.025 |
| 0.5 | 20 40 | 0.142 |
| 0.7 | 20 40 | 0.160 |
| 0.5 | 20 40 60 | 0.175 |
| 0.7 | 20 40 60 | 0.183 |
| 0.5 | 20 40 60 60 | 0.187 |
| 0.7 | 20 40 60 60 | 0.186 |
| 0.5 | 20 40 60 60 60 60 | 0.229 |

*Source:* Reprinted from *J. Chromatogr. A.* 1065, Galaev, I.Y., Dainiak, M.B., Plieva, F.M., Hatti-Kaul, R., and Mattiasson, B., 169–175. Copyright 2005, with permission from Elsevier.

[a] Sonication was carried out for 1 min at each amplitude and 1 min of interruption was given between each treatment. His$_6$-LDH, from 150 µl of different samples of cell homogenate obtained under varying conditions of sonications, were isolated simultaneously.

[b] Enzyme reaction for LDH was carried out directly in 96-well plate filled with 0.5 ml of Cu (II)-IDA-cryogel minicolumns.

single chain Fv-antibody fragment (His$_6$-scFv) was captured directly from *E. coli* cell culture fluid. The amount of captured His$_6$-scFv fragment was quantified by ELISA. About 80–90% of the His$_6$-scFv loaded on the matrix can be recovered, and the recovered protein was directly proportional to the amount of His$_6$-scFv applied (Figure 13.3c).

Another study was published in the same year where Hanora et al. were able to develop a system to simultaneously screen metal binding peptide tags using different metal ions. The cryogel column was synthesized using pDMAA polymer. These minicolumns were coupled with different metal ions, namely, copper ($Cu^{2+}$), nickel ($Ni^{+2}$), zinc ($Zn^{+2}$), cobalt ($Co^{2+}$) and cadmium ($Cd^{2+}$). A peptide library of 23 tagged green fluorescent proteins (GFP) was generated by insertion of different amino acid sequences at the amino terminal of GFP. These tagged proteins were loaded on the cryogel minicolumns coupled with different metal ions to which they bind and enables simultaneous screening of a large number of peptides from the library (Figure 13.4; Table 13.3). This can also be used to screen the binding strength of the tagged protein with different metal ligands. The screening of bacterial peptide library by conventional methods involves individual selection, culturing and expression of the desired clones (Hanora et al., 2005).

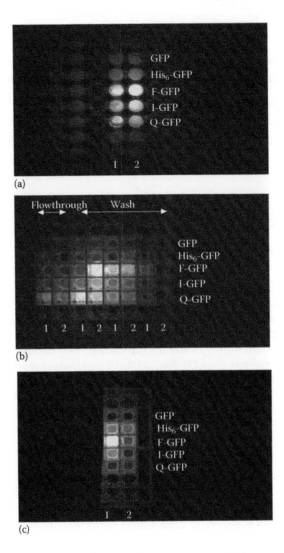

**FIGURE 13.4** (a) Image of open-ended 96-well plate filled with 0.5 ml pDMAA-cryogel minicolumns coupled to $Ni^{2+}$ ions, after protein extracts have been applied. Columns 1 and 2 represent cells treated with PopCulture reagent and sonicated lysate, respectively. (b) Image of 96-well plate containing flowthrough and wash effluents from the minicolumns shown in (a). (c) Image of 96-well plate containing elution fractions of the column shown in (a). Experimental conditions: 5 column volumes (CV) of 100 mM $NiCl_2$ was applied to the minicolumn and washed with 6 CV water, 6 CV imidazole (to remove the unbound and loosely bound $Ni^{2+}$ ions) and again with 6 CV water. After equilibration with 6 CV of 20 mM Tris-HCl, 0.5 M NaCl, pH 8.0 (buffer A), 250 μl of PopCulture reagent or sonicated lysate was applied to the column and washed with 6 CV buffer A. Stepwise gradient elution was done with 10, 20, 30, 40 and 200 mM imidazole. (Reprinted from *J. Chromatogr. A.* 1087, Hanora, A., Bernaudat, F., Plieva, F.M., Dainiak, M.B., Bülow, L., Galaev, I.Y., and Mattiasson, B., Screening of Peptide Affinity Tags Using Immobilised Metal Affinity Chromatography in 96-Well Plate Format, 38–44. Copyright 2005, with permission from Elsevier.)

## TABLE 13.3
## Binding of GFP Having Different Tag Sequences to Cryogel Minicolumns Coupled to Different Metal Ions ($Ni^{2+}$, $Zn^{2+}$, $Co^{2+}$ and $Cd^{2+}$)

| Name | Tag Sequence | $Ni^{2+}$-IDA | $Zn^{2+}$-IDA | $Co^{2+}$-IDA | $Cd^{2+}$-IDA |
|---|---|---|---|---|---|
| Native-GFP | MEFELGT | | | | |
| His$_6$-GFP | MGHHHHHHGT | + | + | + | + |
| A-GFP | MEFHVRLKH | + | | | |
| B-GFP | MEFHVCMHH | + | | | |
| C-GFP | MEFHQETEH | + | | | |
| D-GFP | MEFHPKLEH | + | | | |
| E-GFP | MEFHNWMDH | + | | | |
| F-GFP | MEFHFKSH | + | + | + | + |
| G-GFP | MEFHNAILH | + | | | |
| H-GFP | MEFHNRSRH | + | | | |
| I-GFP | MEFHANHMH | + | + | + | + |
| J-GFP | MEFHWRSRH | + | | | |
| K-GFP | MEFHNGSEH | + | | | |
| L-GFP | MEFHTRSGH | + | | | |
| M-GFP | MEFHNWMDH | + | | | |
| N-GFP | MEFHEIDVH | + | | | |
| O-GFP | MEFHWRARH | + | | | |
| P-GFP | MEFHWGYLH | + | | | |
| Q-GFP | MEFHTSMLH | + | + | + | |
| R-GFP | MEFHSRLSH | + | | | |
| S-GFP | MEFHQKVLH | + | | | |
| T-GFP | MEFHALRGH | + | | | |
| U-GFP | MEFHFQFDH | + | | | |
| V-GFP | MEFHRSLAH | + | | | |
| W-GFP | MEFHVWMRH | | | | |

*Source:* Reprinted from *J. Chromatogr. A*. 1087, Hanora, A., Bernaudat, F., Plieva, F.M., Dainiak, M.B., Bülow, L., Galaev, I.Y., and Mattiasson, B., 38–44. Copyright 2005, with permission from Elsevier.

Another study was published in the following year (2006) in which Ahlqvist et al. developed a high throughput system for monitoring inclusion body production during fermentation. Further, this system was used for ELISA of the intact inclusion body. In a bioprocess industry, constant monitoring of performance of bacteria and other microorganisms during fermentation is required. Parameters such as pH, oxygen and target product are generally monitored. In the recent past, there has been an increasing demand for an analysis system that is more rapid, cost-effective and safe to analyse the target product. This will help in reducing the cost of the product. During protein production, monitoring is important so that the effective downstream processes can be used for its isolation. Although it is highly desirable that the protein produced should be in a soluble and active form, it is commonly found that a high production of protein results in the formation of its aggregates, primarily due to the misfolding of polypeptides. On the other hand, there are conditions when target protein can only be produced in the form of inclusion body. Therefore, direct monitoring is necessary in the bioprocess (Ahlqvist et al., 2006; Villaverde et al., 2003). Techniques like flow cytometry and light scattering have been used for monitoring but they have many operational complexities (Wållberg, 2004; Wittrup et al., 1988). These limitations were overcome by designing a system using a cryogel matrix and 96-well plates. The cryogel matrix was synthesized at subzero temperature using acrylamide and AGE as monomers and MBAAm as a cross-linker. The epoxy-containing cryogel matrix was coupled with Protein A as an affinity ligand. The samples from the fermentation broth were collected at different time points. The inclusion bodies in the samples were coupled with antibodies and the samples were passed through minicolumns in the microtitre plate for the analysis. It was seen that the production of inclusion body and protein increases with time (Figure 13.5a).

**ELISA** is one of the most sensitive techniques for analysis of proteins. Conventional ELISA requires many steps such as cell disruption, isolation and purification of the antigen displayed on the bioparticle surface. However, some of the techniques have been developed in which protein assay can be done in a heterogeneous mixture (Halim et al., 2005; Treci et al., 2004). Nevertheless, these techniques are quite cumbersome as they require special sample treatments, and target protein is not specifically captured (Treci et al., 2004). On the other hand, a whole cell ELISA requires preculture and is suitable for cells only. Phenyl cryogel minicolumns were used to develop a system in which cryogel matrix was used as a solid phase for ELISA samples containing particulates. These cryogel mincolumns were fixed into open-ended 96-well plates. The inclusion bodies were immobilized onto the cryogels by loading a suspension having inclusion bodies. The inclusion body is evenly distributed over the whole surface of the cryogel wall (Figure 13.5b). Then on-column ELISA was performed, the microcolumn-to-microcolumn have standard deviation of only 0.007 (Figure 13.5c).

Cryogels have shown the potential to be used as a solid phase matrix for ELISA. It provides for stable binding to antibodies and antigens along with high binding capacity. Additionally, the immunological activity of the bound molecules does not get affected (Kemeny et al., 1988). Along with it, sensitivity of the reaction can be increased by increasing the amount applied of the sample and increasing the incubation time. The cryogel minocolumns and 96-well plate can be used to develop a high throughput screening system for analysing serum samples.

## 13.4 OTHER HIGH THROUGHPUT PROCESSES

High throughput processes have been leveraged for screening purposes. Advances in molecular biology techniques have provided a large number of libraries like complementary DNA library, genomic library, peptide library and so on, which has led to the need of fast and simple screening techniques for these libraries. Similar increases in new compounds, targets and assays are leading to development of high throughput cytotoxic assays that can provide specific information of effects of chemicals on the properties of cells. In addition, integrated approaches of chromatography and

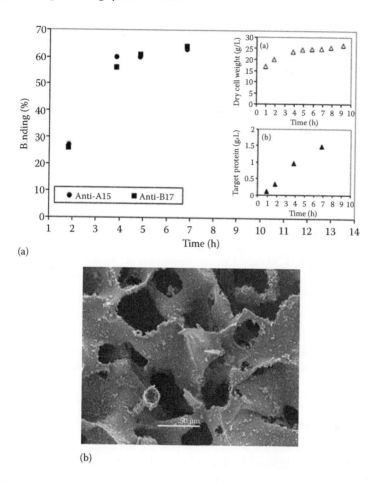

**FIGURE 13.5** (a) Monitoring fermentation process by analysis of binding of either anti-A15 IgG- (●) or anti-B17 IgG- (■) bound inclusion bodies to protein-A cryogel minicolumns. Inset A and inset B represent, respectively, amount of dry cell weight per litre of fermentation broth and amount of target protein in fermentation broth (grams/litre) as determined by HPLC. (b) Scanning electron microscope image of inclusion bodies bound to phenyl-A cryogel. Experimental conditions: (a) Two samples (1 ml each) were collected at 2, 4, 5 and 7 h after incubation of fermentation process, sonicated, centrifuged twice at 9400 g for 2 min and suspended in phosphate buffer saline (PBS) to an OD470 of 2.0. Anti-B17 or anti-A15 IgG (40 μl of 1 mg/ml) was added to 500 μl of inclusion body suspension, incubated in ice for 15 min, centrifuged and resuspended in 500 μl PBS. A 250-μl aliquot of the suspension was applied to protein-A cryogel columns, incubated for 15 min and unbound material was washed with PBS. Absorbance of the flow through fractions was measured at 470 nm. The difference between amount of applied and unbound material gives the amount of bound inclusion bodies. (b) Inclusion bodies were suspended in PBS to OD600 of 2.3 and 150 μl of this suspension was applied to phenyl-cryogels and incubated at 4°C overnight. Unbound inclusion bodies were washed with 1 ml PBS and 3 ml PBS-Tween and absorbance of the flow through fractions were measured at 600 nm. The difference between the amount of applied and unbound inclusion bodies gave the amount of bound inclusion bodies.

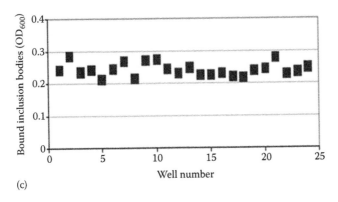

(c)

**FIGURE 13.5 (CONTINUED)** (c) Column-to-column variation in the binding of inclusion bodies to phenyl-cryogels. (Reprinted from *Anal. Biochem.* 354, Ahlqvist, J., Dainiak, M.B., Kumar, A., Hörnsten, E.G., Galaev, I.Y., and Mattiasson, B., Monitoring the Production of Inclusion Bodies during Fermentation and Enzyme-Linked Immunosorbent Assay Analysis of Intact Inclusion Bodies Using Cryogel Minicolumn Plates, 229–237. Copyright 2006, with permission from Elsevier.)

screening are also gaining importance, which eliminates initial extraction, cleaning and purification of samples from crude extracts, therefore speeding up the whole process.

### 13.4.1 Cell Surface Analysis

The large and interconnected pores of cryogels are not clogged and allow for easy passage of particulate matters (Galaev et al., 2005). Therefore, the cryogels make excellent adsorbents for integrated approaches for separation of analytes from a crude sample. In combination with open-ended 96-well plates, the cryogel minicolumn creates a platform or separation of desired cells with quantification of bound cells, therefore allowing fast detection of cells from a large number of samples. Parallel adsorption tests were carried out using different affinity cryogels, namely, Cu(II), Ni(II) and Zn(II)-IDA-cryogels for cell surface analysis of wild type *E. coli*, recombinant *E. coli* (presented with (His)$_6$-tags on cell surface) and *Bacillus halodurans* (*B. halodurans*) cells. The results of the screening tests revealed that the affinity of these three cells to the different affinity cryogels is highly dependent on cell type and growth phase. Both the wild type and recombinant *E. coli* cells showed affinity to the cryogels in the following order: Cu(II) > Ni(II) > Zn(II). *B. halodurans* did not show any affinity to these cryogels (Figure 13.6). The results are concurrent with the surfaces of the cells. *E. coli* cell surfaces contain histidine, arginine, lysine and tryptophan residues that promote their binding to the affinity cryogels whereas surfaces of *B. halodurans* are rich in acidic amino acids that do not bind with Cu(II), Ni(II) and Zn(II) (Arnold, 1991; Sulkowski, 1989; Trotsenko et al., 2002). The decrease in affinity of wild type *E. coli* cells to the cryogels with increase in the cell cultivation time can be attributed to the repelling effects of cell-released extracellular polymeric

# Cryogels in High Throughput Processes

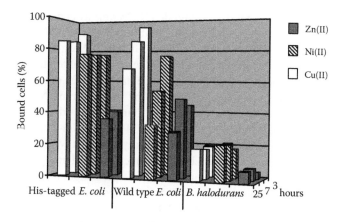

**FIGURE 13.6** Binding of recombinant *E. coli*, wild type *E. coli* and *B. halodurans* cells to Zn(II)-, Ni(II)- and Cu(II)-IDA-cryogel minicolumns as a function of cell fermentation time. Experimental conditions: Cells were suspended in binding buffer (20 mM HEPES, 200 mM NaCl, pH 7.0) to $OD_{450}$ of 0.9–1.0 and 100 µl aliquots were applied to 0.5 ml of Zn (II)-, Ni (II)- and Cu(II)-IDA-cryogel minicolumns. After incubation for 5 min, washing was done with 1.5 ml of binding buffer and the absorbance of the flowthrough was measured at 450 nm. The difference between the amounts of applied and unbound cells gives the amount of bound cells. (Reprinted from *Enzym. Microb. Technol.* 40, Dainiak, M.B., Galaev, I.Y., and Mattiasson, B., Macroporous Monolithic Hydrogels in a 96-Minicolumn Plate Format for Cell Surface-Analysis and Integrated Binding/Quantification of Cells, 688–695. Copyright 2007, with permission from Elsevier.)

substances (Boonaert et al., 2001; Tsuneda et al., 2003; Veenstra et al., 1996). The results of this study show that cell surface analysis using cryogels in 96-well plate format is a versatile process that allows for high throughput isolation, quantification and elution of viable cells (Dainiak et al., 2007b).

## 13.4.2 Cell-Based Assays

A majority of high throughput cytotoxicity assays employ two-dimensional (2D) monolayer culture systems that may produce erroneous results as 2D systems do not mimic the native environment of the cells. Of various three-dimensional (3D) cell culture systems, scaffolds generated by cryogelation techniques can provide 3D environment and can be easily modified for varied surface chemistries, elasticity and so on. These properties can account for providing the cells an environment similar to *in vivo* conditions thus producing results that are more appropriate. Cryogels with different functionalized surfaces have been used as cell culture scaffolds in the 96-well plate format to study proliferation, adhesion and chemosensitivity of three cell types, namely, HTC116 human colon cancer and human embryonic fibroblast cells (adherent cells) and KG-1 human acute myeloid leukemia cells (nonadherent cells) (Dainiak et al., 2008). The different functionalized surfaces are as follows: plain polyacrylamide (pAAm) cryogels (nonadhesive surface for cells and proteins); pAAm surfaces immobilized with type I collagen (cell adhesive protein);

and pAAm gels with N-isopropylacrylamide (NIPAAm) as comonomer for varied hydrophobicity containing agmatine-based RGD mimetics (abRGDm) that mimic cell adhesive peptide RGD (abRGDm- and abRGDm-NIPAAm-cryogels). Both the adherent cell, that is, HTC116 and fibroblasts adhered and proliferated well on collagen and abRGDm-cryogels as they mimicked the natural ECM of the cells. In addition, increasing the hydrophobicity of the gels by incorporating NIPAAm further enhanced proliferation of the adhered cells due to increase in adsorption of proteins on the surface of cryogels (Allen et al., 2006). HCT116 and KG-1 cells formed multicellular aggregates on collagen- and abRGDm-cryogels that were similar to the *in vivo* tumours. Therefore, cells cultured on 3D functionalized cryogels can produce more appropriate patterns of drug sensitivity. The presumption was recapitulated when HTC116 cells grown on different functionalized cryogels were treated with anticancer drug cisplatin. In the case of 2D cultures, less than 20% of cells were viable after 50 h of incubation with 70 μM cisplatin, while in the case of 3D cultures on both plain and functionalized cryogels, more cells survived even after 100 h of incubation (Figure 13.7).

Similarly, when KG-1 cells were treated with therapeutic agent cytosine 1-β-D-arabinofuranoside (AraC), there was a 30% decrease in cell viability in the case of plain cryogels after 18 h of incubation while the cells were less sensitive toward AraC when they were cultured on collagen-, abRGDm- and abRGDm-NIPAAm-cryogels (Figure 13.8). The results of this study show that many different viability assays can be performed and heterologous 3D models with multiple cell types can be created on different functionalized cryogel scaffolds in 96-well plate format that can be used for high throughput screening purposes (Dainiak et al., 2008).

### 13.4.3 Panning of Phage Display Library

In a phage display library, a foreign peptide is expressed and displayed on coat protein of M13 bacteriophage (Smith, 1985). To screen such a library, high affinity ligands are required. Isolation of such ligands is time-consuming and laborious as it requires several panning rounds. To isolate these high affinity ligands, they are immobilized on a solid phase and their affinity toward the displayed peptide is checked. Since the phage particles are highly diffusive owing to their small size, the immobilized ligands are practically unavailable for binding to the phages in case of conventional columns. Cryogels have many properties that may help in panning of phage display library. Cryogels can be synthesized in a manner so that they contain active groups on their surface upon which desired ligands can be immobilized. Such material can be chosen for the synthesis of cryogel so that the column produced is hydrophilic and results in minimal nonspecific binding. When chromatopanning procedure using cryogels was compared to conventional biopanning procedure, the former was found to be advantageous in many terms (Noppe et al., 2009). Chromatopanning using cryogels required approximately seven times less time than conventional biopanning as only one round of panning was enough to obtain high affinity targets for the displayed peptide (Table 13.4). Also, a mild elution solvent is required to elute the infected *E. coli* cells and phages due to which the viability of the cells is not affected and the ligands bound to the column are also not denatured. This also makes the

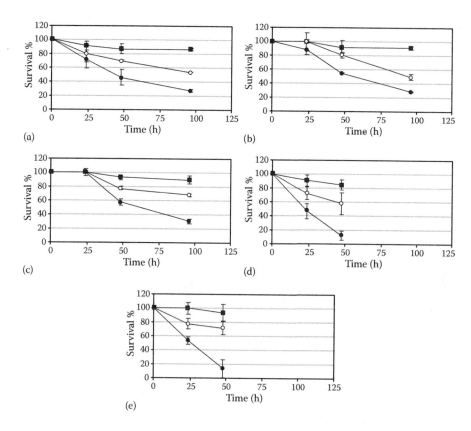

**FIGURE 13.7** Cytotoxic analysis of 5 μM (■), 35 μM (○) and 70 μM (●) cisplatin treatment on HCT116 cells cultured on (a) plain-, (b) collagen-, (c) abRGDm-NIPAAm-cryogel scaffolds and (d) plain, (e) collagen-coated polystyrene tissue culture plates as a function of incubation time with the drug. Experimental conditions: Different types of cryogel scaffolds were seeded with 1 ml (2–4 × $10^5$ cells/ml) suspension of HTC116 cells in McCoy's medium. Cells were treated with different concentration of drugs 4 days postseeding. After incubation with the drug for different time periods, viability of the cells was determined by XTT assay. (Dainiak, M.B., Savina, I.N., Musolino, I., Kumar, A., Mattiasson, B., and Galaev, I.Y.: Biomimetic Macroporous Hydrogel Scaffolds in a High-Throughput Format for Cell-Based Assays. *Biotechnol. Prog.* 2008. 24. 1373–1383. Copyright Wiley-VCH Verlag GmbH & Co. KGaA. Reproduced with permission.)

column reusable; therefore, minimizing the overall cost of panning (Noppe et al., 2009). The whole procedure can be further miniaturized by inserting the cryogel minicolumn in open-ended 96-well plates thereby allowing for parallel screening and increasing the throughput.

### 13.4.4 Pollutant Analysis

Cryogel minicolumns have been used along with 96-wells plate to form high throughput analysis systems for benzo[a]pyrene (BAP). BAP acts as an indicator of

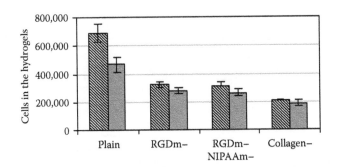

**FIGURE 13.8** Cytotoxic analysis of 20-μM Ara-C treatment on KG-1 cells grown on plain-, abRGDm-, abRGDm-NIPAAm- and collagen-cryogel scaffolds. Cells not treated with the drug were taken as control. Untreated (▨), treated (▪). Experimental conditions: Different types of cryogel scaffolds were seeded with 1 ml (2–4 × $10^5$ cells/ml) suspension of KG1 cells in RPMI medium. Cells were treated with 20-μM Ara-C during last 18 h of cultivation. Viability of the cells was determined by XTT assay. (Dainiak, M.B., Savina, I.N., Musolino, I., Kumar, A., Mattiasson, B., and Galaev, I.Y.: Biomimetic Macroporous Hydrogel Scaffolds in a High-Throughput Format for Cell-Based Assays. *Biotechnol. Prog.* 2008. 24. 1373–1383. Copyright Wiley-VCH Verlag GmbH & Co. KGaA. Reproduced with permission.)

**TABLE 13.4**
**Comparative Time Frame of Chromato-Panning Protocol and Tube Panning Protocol**

| Chromato-Panning | | Tube Panning | |
|---|---|---|---|
| Round I (h) | | Round I (h) | |
| Column equilibration | 0.25 | Target coating | 18 |
| On column panning | 1 | Wash-postcoat | 2.25 |
| On column infection | 0.5 | Phage incubation | 20 |
| Elution | 0.25 | Washing/elution | 0.5 |
| Plating/amplification | 18 | Infection | 0.5 |
| Phage recovery | 3 | Plating/amplification | 18 |
| | | Phage recovery | 3 |
| Time | ~23 h | Time | ~62 h |
| No additional panning rounds are required | | 2 additional panning rounds are performed (phage incubation R2/3) | 2.5 |
| Total time | ~23 h | | ~152 h |

*Source:* Noppe, W., Plieva, F., Galaev, I.Y., Pottel, H., Deckmyn, H., and Mattiasson B.(2009). *BMC Biotechnol.* 9.

contamination of poly aromatic hydrocarbon (PAH) (Gupta et al., 2013; Radecki et al., 1980). BAP is among the 16 priority pollutants included by the Environmental Protection Agency (EPA) of the United States as it is highly carcinogenic and abundant. Partial combustion of organic matters, wood and charcoal and so on leads to generation of PAHs (Baggiani et al., 2007; Krupadam et al., 2010; Wise et al., 1993; Zhu et al., 2005). The system developed is of significant importance as it can be used for simultaneous screening of a large number of samples having BAP contamination. Hydrophobic matrix composed of poly (4-vinyl pyridine-*co*-divinyl benzene) was synthesized, which helped in capture of BAP molecules directly on the column. The cryogel synthesized had large and interconnected pores; pore size ranged from 10–100 μm, and had porosity of 87.42% as measured by Archimedes principles. The presence of π–π interaction between BAP and polymeric matrix was responsible for binding of analytes on the cryogel column (Baggiani et al., 2007). The amount of BAP captured on the matrix in each well was approximately equal when the amount of analyte applied was the same. There was less significant column-to-column variation in the amount of BAP bound on each matrix. In all the wells, more than 70% of the BAP in the sample was captured (Figure 13.9a). The binding capacity of the 0.5 ml cryogel columns inserted in the well was 1 μg (Figure 13.9b). The columns were equilibrated using a mixture of acetonitrile and water in a ratio of 20:80. An equal amount of sample was applied on each well and flow through was collected. Captured amount of BAP was eluted by increasing the amount of acetonitrile in the mobile phase. Although acetonitrile is a polar solvent, it has sufficient dispersive properties to break interaction between BAP and polymer matrix (Hodgkinson et al., 1981). Further, this high throughput system was also used to screen for the presence of BAP in the spiked samples containing BAP and anthracene. In addition, it was found that BAP can still be analysed even in the presence of anthracene without affecting the binding capacity of the matrix. Furthermore, this synthesized cryogel matrix was used to capture BAP from the sample having only BAP and from a spiked sample having BAP along with anthracence. High recovery of 93.62% was achieved with yield of 79.2% when BAP alone was applied (Figure 13.9c). When a spiked mixture was applied, binding of BAP was 87.73% and anthracene binding was 95.6% (Figure 13.9d).

## 13.5 CONCLUSIONS

Cryogels have proved their potential in miniaturized platform of 96-well plate for screening/analysis of cell separation, cell-based assays, libraries, proteins and so on. These processes can also be easily scaled up. Therefore, it can be expected that cryogels will gain importance in the near-future for high throughput processes. In the future, cryogels can be explored for other high throughput purposes like drug screening and metabolism and automated processes will be implemented for much faster analysis and further reduction in time.

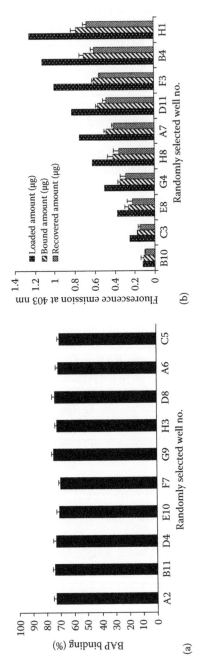

**FIGURE 13.9** (a) Minicolumn to minicolumn deviation of percentage of benzo[a]pyrene captured on poly (4-vinyl pyridine-co-divinyl benzene) cryogels. (b) Amount of benzo[a]pyrene loaded, bound on and recovered from poly (4-vinyl pyridine-co-divinyl benzene) minicolumns. Experimental conditions: Cryogel minicolumns (500 μl) were loaded with 250 μl of 1 μg/ml BAP (a) and different concentrations of BAP (b). After 15 min of incubation (only in case of minicolumns), washing was done with acetonitrile:water (20:80) and bound analytes were eluted with acetonitrile:water (90:10). Effluents collected were analysed in a spectrofluorimeter (excitation and emission wavelengths of 294 nm and 403 nm, respectively, for BAP and 241 nm and 401 nm, respectively, for anthracene).

*(Continued)*

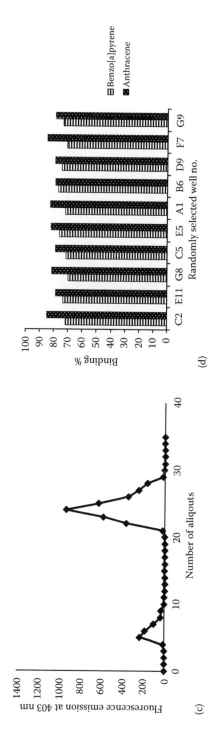

**FIGURE 13.9 (CONTINUED)** (c) Column to column deviation of percentage of benzo[a]pyrene captured from spiked sample containing mixture of benzo[a]pyrene and anthracene. (d) Chromatogram showing elution of BAP captured on poly (4-vinyl pyridine-co-divinyl benzene) cryogels matrix. Experimental conditions: Spiked sample of 0.357 ppm BAP and 0.091 ppm anthracene (c) and 2 ml cryogel column was loaded with 1 ml of 1 µg/ml BAP at flow rate of 1 ml/min (d). After 15 min of incubation (only in case of minicolumns), washing was done with acetonitrile:water (20:80) and bound analytes were eluted with acetonitrile:water (90:10). Effluents collected were analysed in a spectrofluorimeter (excitation and emission wavelengths of 294 nm and 403 nm, respectively, for BAP and 241 nm and 401 nm, respectively, for anthracene). (Reprinted from *J. Chromatogr. A.* 1278, Gupta, A., Sarkar, J., and Kumar, A., High Throughput Analysis and Capture of Benzo[a]pyrene using Supermacroporous Poly(4-vinyl pyridine-co-divinyl benzene) Cryogel Matrix, 16–21. Copyright 2013, with permission from Elsevier.)

## ACKNOWLEDGEMENTS

J.S. and A.G. are thankful to Council of Scientific and Industrial Research, India and Indian Institute of Technology Kanpur, India, respectively for providing their research fellowships.

## LIST OF ABBREVIATIONS

| | |
|---|---|
| 2D | Two dimensional |
| 3D | Three dimensional |
| abRGDm | Agmatine based RGD mimetics |
| AGE | Allyl glycidyl ether |
| AraC | Cytosine 1-β-D-arabinofuranoside |
| BAP | Benzo[a]pyrene |
| Con A | Concanavalin A |
| CV | Column volume |
| DNA | Deoxyribonucleic acid |
| EDTA | Ethylenediaminetetracetate |
| ELISA | Enzyme-linked immunosorbent assay |
| GFP | Green fluorescent protein |
| $His_6$-scFv | $His_6$-tagged single chain Fv-antibody fragment |
| IDA | Iminodiacetate |
| IMAC | Immobilized metal affinity chromatography |
| MBAAm | N, N'-Methylenebisacrylamide |
| NIPAAm | N- Isopropylacrylamide |
| pAAm | Polyacrylamide |
| PBS | Phosphate buffer saline |
| pDMAA | Polydimethylacrylamide |

## REFERENCES

Ahlqvist, J., Dainiak, M.B., Kumar, A. et al. (2006). Monitoring the production of inclusion bodies during fermentation and enzyme-linked immunosorbent assay analysis of intact inclusion bodies using cryogel minicolumn plates. *Anal. Biochem.* 354: 229–237.

Allen, L.T., Tosetto, M., Miller, I.S. et al. (2006). Surface-induced changes in protein adsorption and implications for cellular phenotypic responses to surface interaction. *Biomater.* 27: 3096–3108.

Arnold, F.N. (1991). Metal-affinity separations: A new dimension in protein processing. *Biotechnol.* 9: 150–5.

Arvidsson, P., Plieva, F.M., Lozinsky, V.I., Galaev, I.Y., and Mattiasson, B. (2003). Direct chromatographic capture of enzyme from crude homogenate using immobilized metal affinity chromatography on a continuous supermacroporous adsorbent. *J. Chromatogr. A.* 986: 275–290.

Baggiani, C., Anfossi, L., Baravalle, P., Giovannoli, C., and Giraudi, G. (2007). Molecular recognition of polycyclic aromatic hydrocarbons by pyrene-imprinted microspheres. *Anal. Bioanal. Chem.* 389: 413–422.

Boonaert, C.J.P., Dufrene, Y.F., and Derclaye, S.R. (2001). Adhesion of *Lactococcus lactis* to model substrata: Direct study of the interface. *Colloids Surf. B: Biointerf.* 22: 171–82.

Bulyk, M.L. (2006). DNA microarray technologies for measuring protein-DNA interactions. *Curr. Opin. Biotechnol.* 17: 422–430.

Chhatre, S., and Titchener-Hooker, N.J. (2009). Microscale methods for high-throughput chromatography development in the pharmaceutical industry. *J. Chem. Technol. Biotechnol.* 84: 927–940.

Cozens-Roberts, C., Quinn, J.A., and Lauffenburger, D.A. (1990). Receptor-mediated cell attachment and detachment kinetics. II. Experimental model studies with the radial-flow detachment assay. *Biophys. J.* 58:857–872.

Dainiak, M.B., Galaev, I.Y., Kumar, A., Plieva, F.M., and Mattiasson, B. (2007a). Chromatography of living cells using supermacroporous hydrogels, cryogels, In: Kumar, A., Galaev, I.Y., and Mattiason, B. (Eds.), *Cell Separation*. Springer, Berlin, Heidelberg 106: 101–127.

Dainiak, M.B., Galaev, I.Y., and Mattiasson, B. (2006a). Affinity cryogel monoliths for screening for optimal separation conditions and chromatographic separation of cells. *J. Chromatogr. A.* 1123: 145–150.

Dainiak, M.B., Galaev, I.Y., and Mattiasson, B. (2007b). Macroporous monolithic hydrogels in a 96-minicolumn plate format for cell surface-analysis and integrated binding/quantification of cells. *Enzym. Microb. Technol.* 40: 688–695.

Dainiak, M.B., Kumar, A., Galaev, I.Y., and Mattiasson, B. (2006b). Detachment of affinity-captured bioparticles by elastic deformation of a macroporous hydrogel. *Proc. Natl. Acad. Sci.* 103: 849–854.

Dainiak, M.B., Plieva, F.M., Galaev, I.Y., Hatti-Kaul, R., and Mattiasson, B. (2005). Cell chromatography: Separation of different microbial cells using IMAC supermacroporous monolithic columns. *Biotechnol. Prog.* 21: 644–649.

Dainiak, M.B., Savina, I.N., Musolino, I. et al. (2008). Biomimetic macroporous hydrogel scaffolds in a high-throughput format for cell-based assays. *Biotechnol. Prog.* 24: 1373–1383.

Desmet, C., Blum, L.J., and Marquette, C.A. (2012). High-throughput multiplexed competitive immunoassay for pollutants sensing in water. *Anal. Chem.* 84: 10267–10276.

Galaev, I.Y., Dainiak, M.B., Plieva, F.M., Hatti-Kaul, R., and Mattiasson, B. (2005). High throughput processing of particulate-containing samples using supermacroporous elastic monoliths in microtiter (multiwell) plate format. *J. Chromatogr. A.* 1065: 169–175.

Galaev, I.Y., Dainiak, M.B., Plieva, F., and Mattiasson, B. (2007). Effect of matrix elasticity on affinity binding and release of bioparticles. elution of bound cells by temperature-induced shrinkage of the smart macroporous hydrogel. *Langmuir.* 23: 35–40.

Gupta, A., Sarkar, J., and Kumar, A. (2013). High throughput analysis and capture of benzo[a]pyrene using supermacroporous poly(4-vinyl pyridine-co-divinyl benzene) cryogel matrix. *J. Chromatogr. A.* 1278: 16–21.

Halim, N.D., Joseph, A.W., and Lipska, B.K. (2005). A novel ELISA using PVDF microplates, *J. Neurosci. Methods.* 143: 163–168.

Hanora, A., Bernaudat, F., Plieva, F.M. et al. (2005). Screening of peptide affinity tags using immobilised metal affinity chromatography in 96-well plate format. *J. Chromatogr. A.* 1087: 38–44.

Hanora, A., Savina, I., Plieva, F.M. et al. (2006). Direct capture of plasmid DNA from non-clarified bacterial lysate using polycation-grafted monoliths. *J. Biotechnol.* 123: 343–355.

Hodgkinson, S.C., and Lowry, P.J. (1981). Hydrophobic-interaction chromatography and anion-exchange chromatography in the presence of acetonitrile. A two-step purification method for human prolactin. *Biochem. J.* 199: 619–627.

Huang, J.X., Mehrens, D., Wiese, R. et al. (2001). High-throughput genomic and proteomic analysis using microarray technology. *Clin. Chem.* 47: 1912–1916.

Jain, E., and Kumar, A. (2013). Disposable polymeric cryogel bioreactor matrix for therapeutic protein production. *Nat. Protoc.* 8: 821–835.

Jain, E., Srivastava, A., and Kumar, A. (2009). Macroporous interpenetrating cryogel network of poly(acrylonitrile) and gelatin for biomedical applications. *J. Mater. Sci. Mater. Med.* 20: 173–179.

Kemeny, D.M., and Challacombe, S.J. (1982). An introduction to ELISA, In: *ELISA and Other Solid Phase Immunoassays.* Kemeny, D.M. and Challacombe, S.J. (Eds.), John Wiley, New York, 1–29.

Kircher, M., and Kelso, J. (2010). High-throughput DNA sequencing-concepts and limitations. *Bioessays.* 32: 524–536.

Krupadam, R.J., Bhagat, B., and Khan, M.S. (2010). Highly sensitive determination of polycyclic aromatic hydrocarbons in ambient air dust by gas chromatography-mass spectrometry after molecularly imprinted polymer extraction. *Anal. Bioanal. Chem.* 397: 3097–3106.

Kumar, A., Plieva, F.M., Galaev, I.Y., and Mattiasson, B. (2003). Affinity fractionation of lymphocytes using a monolithic cryogel. *J. Immunol. Methods.* 283: 185–194.

Kumar, A., and Srivastava, A. (2010). Cell separation using cryogel-based affinity chromatography. *Nat. Protoc.* 5: 1737–1747.

Lozinsky, V.I., Galaev, I.Y., Plieva, F.M., Savina, I.N., Jungvid, H., and Mattiasson B. (2003). Polymeric cryogels as promising materials of biotechnological interest. *Trends Biotechnol.* 21: 445–451.

Martis, E.A., Radhakrishnan, R., and Badve, R.R. (2011). High-throughput screening: The hits and leads of drug discovery—An overview. *J. Appl. Pharm. Sci.* 01: 02–10.

Mayr, L.M., and Bojanic, D. (2009). Novel trends in high-throughput screening. *Curr. Opin. Pharmacol.* 5: 580–588.

Ming, F., Whish, W. J. D., and Hubble, J. (1998). Estimation of cell surface interactions: Maximum binding force and detachment constant. *Enz. Microb. Technol.* 22: 94–99.

Noppe, W., Plieva, F., Galaev, I.Y., Pottel, H., Deckmyn, H., and Mattiasson B. (2009). Chromato-panning: An efficient new mode of identifying suitable ligands from phage display libraries. *BMC Biotechnol.* 9.

Radecki, A., Lamparczyk, H., Grzybowski, J., Halkiewicz, J., and Fresenius, Z. (1980). Gas-chromatographic determination of benzo(a)pyrene in petroleum products used for the manufacture of drugs and cosmetics. *Anal.Chem.* 303: 397–400.

Sasakura, Y., Kanda, K., Yoshimura-Suzuki, T. et al. (2004). Protein microarray system for detecting protein-protein interactions using an anti-his-tag antibody and fluorescence scanning: Effects of the heme redox state on protein-protein interactions of heme-regulated phosphodiesterase from *Escherichia coli*. *Anal. Chem.* 76: 6521–6527.

Smith, G.P. (1985). Filamentous fusion phage: Novel expression vectors that display cloned antigens on the virion surface. *Science* 228: 1315–1316.

Stenuit, B., Eyers, L., Schuler, L., Agathos, S.N., and George, I. (2008). Emerging high-throughput approaches to analyze bioremediation of sites contaminated with hazardous and/or recalcitrant wastes. *Biotechnol. Adv.* 26: 561–575.

Sulkowski, E. (1989). The saga of IMAC and MIT. *Bioessays.* 10: 170–175.

Szymański, P., Markowicz, M., and Mikiciuk-Olasik, E. (2012). Adaptation of high-throughput screening in drug discovery-toxicological screening tests. *Int. J. Mol. Sci.* 13: 427–452.

Teilum, M., Hansson, M.J., Dainiak, M.B. et al. (2006). Binding mitochondria to cryogel monoliths allows detection of proteins specifically released following permeability transition. *Anal. Biochem.* 348: 209–221.

Treci, Ö., Luxemburger, U., Heinen, H. et al. (2004). ELISA: A fast and robust enzyme-linked immunosorbent assay bypassing the need for purification of recombinant protein. *J. Immunol. Methods.* 289: 191–199.

Trotsenko, I.A., and Khelenina, V.N. (2002).The biology and osmoadaptation of haloalkali-philic methanotrophs. *Mikrobiologiia.* 71: 149–159.

Tsuneda, S., Jung, J., Hayashi, H. et al. (2003). Influence of extracellular polymers on electrokinetic properties of heterotrophic bacterial cells examined by soft particle electrophoresis theory. *Colloids Surf. B: Biointerf.* 29: 181–188.

Villaverde, A., and Carrió, M. (2003). Protein aggregation in recombinant bacteria: Biological role of inclusion bodies. *Biotechnol. Lett.* 25: 1385–1395.

Veenstra, G.J.C., Cremers, F.F.M., Dijk, H., and Fleer, A. (1996). Ultrastructural organization and regulation of a biomaterial adhesion of *Staphylococcus epidermidis*. *J. Bacteriol.* 178: 537–41.

Wållberg, F. (2004). Flow Cytometry for Bioprecess Control. Licentiate thesis, KTH Biotechnology, Stockholm, Sweden.

Williams, S.L., Eccleston, M.E., and Slater, N.K. (2005). Affinity capture of a biotinylated retrovirus on macroporous monolithic adsorbents: Towards a rapid single-step purification process. *Biotechnol Bioeng.* 89: 783–787.

Wise, S.A., Sander, L.C., and May, W.E. (1993). Determination of polycyclic aromatic hydrocarbons by liquid chromatography. *J. Chromatogr. A.* 642: 329–349.

Wittrup, K.D., Mann, M.B., Fenton, D.M., Tasi, L.B., and Baily, J.E. (1988). Single cell light scatter as a probe of refractile body formation in recombinant *Escherichia coli*. *Biotechnol.* 6: 423–426.

White, R.E. (2000). High-throughput screening in drug metabolism and pharmacokinetic support of drug discovery. *Annu. Rev. Pharmacol. Toxicol.* 40: 133–157.

Zhu, L.Z., Cai, X.F., and Wang, J. (2005). PAHs in aquatic sediment in Hangzhou, China: Analytical methods, pollution pattern, risk assessment and sources. *J. Environ. Sci.* 17: 748–755.

# 14 Cryogels: Applications in Extracorporeal Affinity Therapy

*Handan Yavuz, Nilay Bereli, Gözde Baydemir, Müge Andaç, Deniz Türkmen and Adil Denizli**

## CONTENTS

14.1 Introduction .................................................................................................. 388
14.2 Affinity Therapy in Systemic Lupus Erythematosus ................................... 389
14.3 Affinity Therapy in Hyperbilirubinemia ...................................................... 393
14.4 Affinity Therapy in Rheumatoid Arthritis ................................................... 397
14.5 Affinity Therapy in Thalassemia .................................................................. 399
14.6 Affinity Therapy in Hypercholesterolemia .................................................. 403
14.7 Other Applications ....................................................................................... 409
14.8 Concluding Remarks ................................................................................... 411
List of Abbreviations .............................................................................................. 411
References .............................................................................................................. 411

### ABSTRACT

Extracorporeal **affinity therapies** are directed at the removal of potential toxic substances including bilirubin, pathogenic antibodies, various circulating immune-complexes, cholesterol and so on from human plasma. The conventional extracorporeal therapies including plasma exchange, hemodialysis, hemofiltration and hemoperfusion are nonselective techniques. In addition, the requirement for plasma substitutes such as albumin is very high. Moreover, the dangers of hepatitis or immune reactions accompany these therapies while using plasma for plasma products. Today, one of the most promising procedures for extracorporeal therapy is specific **affinity adsorption**. Affinity adsorbents including nanoparticles, microspheres, membranes, monoliths, fibres and hollow fibres may be used in a hemoperfusion system, where blood is directly perfused through the column. Among these adsorbents, **cryogels** are considered a reasonable option for the removal of toxic substances from plasma. This type of application is effective, simple and inexpensive. In this chapter, some interesting cryogel applications on extracorporeal affinity therapies for the removal of toxic

---

* Corresponding author: E-mail: denizli@hacettepe.edu.tr.

substances from human plasma are briefly reviewed. For more applications of cryogels, the reader can find some interesting biomedical uses of cryogels.

## KEYWORDS

Affinity therapy, cryogels, affinity adsorbents, extracorporeal treatment

## 14.1 INTRODUCTION

Conventional hemoperfusion is an **extracorporeal treatment** method in which the blood from the patient is circulated through an adsorption column in order to remove toxic substances from the blood (Figure 14.1) (Pişkin and Hoffman, 1986). The blood taken from the patient must be anticoagulated and generally separated into cells and plasma. The treated plasma is then combined with the previously separated cells and returned to the patient. Hemoperfusion is frequently utilized for acute blood purification as in the case of drug overdose, in which case the adsorbent is charcoal covered with a biocompatible coating layer (Xia et al., 2010). The charcoal binds medium to high molecular weight blood components due to the nonselectivity of conventional treatment protocols. It would be most desirable to selectively remove any undesired substances from human blood. For this purpose, affinity adsorption was suggested as an alternative to conventional treatment protocols for removing unwanted substances from the plasma of patients. Plasma exchange is the most commonly used treatment for extracorporeal affinity therapy. Currently designed affinity adsorbents made of synthetic polymers generally consider long-term contact biocompatibility and they allow blood purification protocols with optimal separation property as well

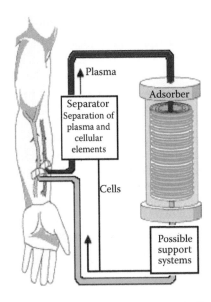

**FIGURE 14.1** Schematic representation of an extracorporeal affinity therapy.

(Tijink et al., 2013). The following lists are several diseases potentially suitable for treatment by extracorporeal affinity therapy.

- Hyperbilirubinemia
- Autoimmune diseases such as systemic lupus erythematosus, rheumatoid arthritis
- Haemophilia, complicated by antibodies to clotting factors VIII or IX
- Thrombocytopenia
- Certain neurological diseases (myasthenia gravis, multiple sclerosis)
- Familial hypercholesterolemia

Many adsorbents including nanoparticles, microspheres, membranes, monoliths, fibres and hollow fibres have been applied to remove toxic substances directly from human plasma (Wang et al., 1997; Yavuz et al., 2001; Şenel et al., 2002; Rad et al., 2003; Aşır et al., 2005; Shi et al., 2005; Zuwei et al., 2005; Demirçelik et al., 2009; Han and Zhang, 2009; Altıntaş et al., 2011). Among these adsorbents, cryogels are considered a reasonable option for the removal of toxic substances from human plasma. Specifically, the megaporous structure of cryogels makes them appropriate candidates as the basis for such processes (Kumar et al., 2001; Lozinsky et al., 2003; Dainiak et al., 2006). Owing to an interconnected megapore network, such a chromatographic matrix has a very low flow resistance (Babac et al., 2006; Bereli et al., 2008; Alkan et al., 2009; Yilmaz et al., 2009; Ertürk et al., 2013). In addition, the cryogels have many advantages such as quick swelling kinetics, short diffusion path and short residence time during adsorption-elution processes. This chapter reviews several recent cryogel applications on extracorporeal affinity therapies for removal of toxic substances from human plasma. The chapter ends by outlining some other biomedical applications of cryogels.

## 14.2 AFFINITY THERAPY IN SYSTEMIC LUPUS ERYTHEMATOSUS

Systemic lupus erythematosus (SLE) is a multisystem, autoimmune, convective-tissue disorder with a broad range of clinical presentations (Peeva et al., 2004). Many different autoantibodies are found in patients with SLE. The most frequent are anti-nuclear, particularly anti-dsDNA antibodies, and others include antibodies against ribonucleo-proteins and histones. Anti-dsDNA antibodies were first described in the sera of patients with SLE more than 40 years ago (Scofield, 2004). Since then, anti-dsDNA antibodies have emerged as a central focus in the investigation of the pathogenesis of SLE and of autoimmunity in general. Antibodies against DNA serve as markers of diagnostic and prognostic significance in SLE, and there is compelling evidence for an association between anti-dsDNA antibodies and tissue damage (Gilliam et al., 1980). The level of anti-dsDNA antibodies correlates well with the disease activity and organ involvements, such as nephritis and cerabritis.

Extracorporeal therapy with affinity adsorbents has become increasingly utilized as a therapeutic modality to remove autoantibodies from plasma of patients. Recently, Özgür and co-workers reported *in vitro* removal of anti-dsDNA antibodies from SLE plasma using DNA-immobilized poly(2-hydroxyethyl methacrylate) (PHEMA) cryogel (Özgür et al., 2011). PHEMA-based adsorbents are known to be

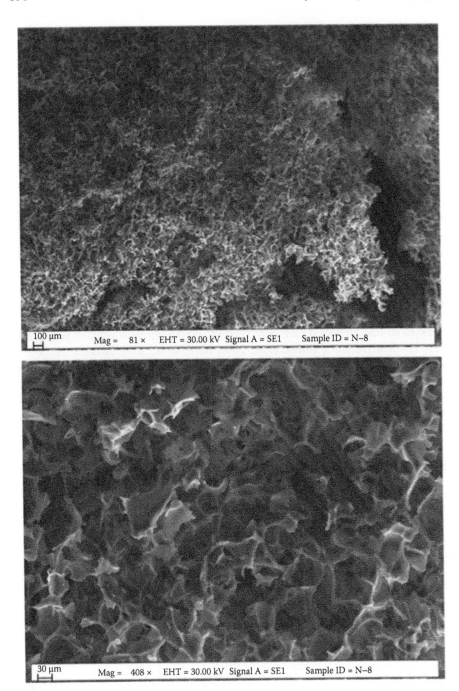

**FIGURE 14.2** SEM images of the PHEMA cryogel. (Reprinted from *Materials Science and Engineering C* 31, Özgür, E., Bereli, N., Türkmen, D., Ünal, S., Denizli, A., PHEMA cryogel for in-vitro removal of anti-dsDNA antibodies from SLE plasma, 915–920. Copyright 2011, with permission from Elsevier.)

biocompatible and frequently used as an affinity adsorbent. When the biocompatibility is a main concern, PHEMA is a good alternative due to its hydrophilicity. It can be used in direct extracorporeal therapy (Denizli and Pişkin, 1995b; Zhang et al., 2013). SEM images are shown in Figure 14.2. The PHEMA cryogel have nonporous walls and large continuous interconnected pores (10–200 μm in diameter) that provide flow channels for the mobile phase. The pressure drop needed to drive the liquid through any system should be as low as possible. Pressure drop studies through the PHEMA cryogel were performed in water as equilibration medium, and at linear flow rates from 38 to 382 cm/h (Figure 14.3). The water was passed through the column for 1 min at different flow rates. Due to the presence of large and interconnected megapores, the PHEMA cryogel column has very low flow resistance (Alkan et al., 2009).

The most important properties of adsorbents are high capacity and specificity of adsorbent. Özgür et al. noted that negligible amounts of about 90 IU/g of anti-dsDNA antibody adsorbed on the PHEMA cryogel. DNA binding significantly increased the anti-dsDNA antibody adsorption amount of the PHEMA-DNA up to $70 \times 10^3$ IU/g. With increasing anti-dsDNA antibody concentration, the amount of anti-dsDNA antibody adsorbed per unit mass first increased sharply (Figure 14.4), and then reached saturation level when the anti-dsDNA antibody concentration was greater than 200 IU/ml.

Traditional columns have a major limitation: incapability of processing whole blood. Blood cells are trapped between the particles of the column resulting in

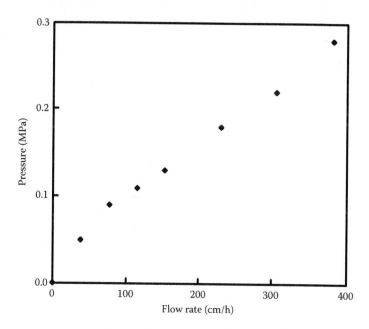

**FIGURE 14.3** Pressure drop at different flow rates. (Reprinted from *Biochemical Engineering Journal* 45, Alkan, H., Bereli, N., Baysal, Z., Denizli, A., Antibody purification with protein A attached supermacroporous PHEMA cryogel, 201–208. Copyright 2009, with permission from Elsevier.)

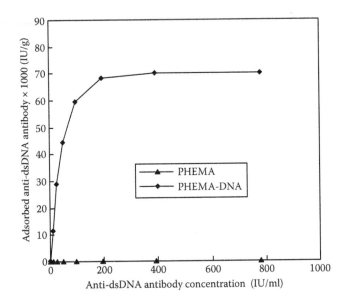

**FIGURE 14.4** Effect of anti-dsDNA antibody concentration on adsorption capacity; DNA loading: 53.4 mg/g; flow rate: 0.5 ml/min; T: 25°C. (Reprinted from *Materials Science and Engineering C* 31, Özgür, E., Bereli, N., Türkmen, D., Ünal, S., Denizli, A., PHEMA cryogel for in-vitro removal of anti-dsDNA antibodies from SLE plasma, 915–920. Copyright 2011, with permission from Elsevier.)

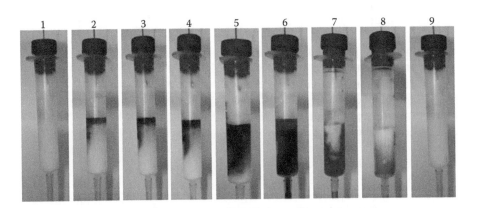

**FIGURE 14.5** Flow pattern of whole blood through a cryogel column. One millilitre of blood was applied to the cryogel column at a flow rate of 0.5 mL/min in isotonic buffer solution. Column 1: column before application; columns 2–8: columns during processing; column 9: column after the flow of blood sample. (Reprinted from *Materials Science and Engineering C* 31, Özgür, E., Bereli, N., Türkmen, D., Ünal, S., Denizli, A., PHEMA cryogel for in-vitro removal of anti-dsDNA antibodies from SLE plasma, 915–920. Copyright 2011, with permission from Elsevier Ltd.)

increased flow resistance and complete blockage of the flow. An expanded-bed chromatographic setup overcomes the problem of handling viscous solutions. However, the high shear stresses occurring in an expanded bed chromatographic setup could be detrimental for the integrity of blood cells. It is attractive to have a chromatographic adsorbent with pores large enough to accommodate blood cells without being blocked (Figure 14.5).

## 14.3 AFFINITY THERAPY IN HYPERBILIRUBINEMIA

Bilirubin is a negatively charged pigment formed in the normal metabolism of heme proteins in senescent red blood cells. High concentration of free bilirubin can evoke hepatic or biliary tract dysfunction and permanent brain damage or death in more severe cases (Lee et al., 2002). Neurological dysfunctions such as kernicterus or bilirubin encephalopathy may develop if the bilirubin concentration in the plasma rises above 15 mg/dL. Disorders in the metabolism of bilirubin may cause a yellow discoloration of the skin and other tissues.

Many bilirubin removal methods have been reported in literature. There are various kinds of methods that have been applied clinically for the treatment of hyperbilirubinemia (Takenaka, 1998). Phototherapy is one of the most commonly used treatments for mild cases. However, the effectiveness of phototherapy is limited by the fact that the light could only penetrate a few millimetres of skin and not reach a large proportion of the total bilirubin pool. It is also found that phototherapy may induce DNA damage (Tiribelli and Ostrow, 2005). Treatment with plasma exchange, on the other hand, requires large volumes of fresh frozen plasma, which is expensive and difficult to obtain. Alternatively, in exchange transfusions, an infant's blood is replaced with bilirubin-free adult blood. However, this procedure has been under concern due to its close relation to hypoglycaemia, hypocalcaemia, acidosis and more importantly the transmission of infectious diseases like hepatitis or acquired immune deficiency syndrome (Zhu et al., 1990). A haemodialysis relay upon the diffusive transport, therefore, is a rather slow process. This method is not specific as well. It should be noted that these systems are still complex and expensive. Hemoperfusion, that is, circulation of the blood through an extracorporeal column containing affinity adsorbents for bilirubin removal, has become the most promising technique (Yu et al., 2000; Avramescu et al., 2004; Ahmad et al., 2006).

Highly specific adsorbents underline various affinity-based separation techniques. For example, antibodies are routinely used as analytical reagents in clinical and research laboratories (Sellegren, 2001). For many practical reasons, attempts have been made to replace antibodies with more stable counterparts. One method that is being increasingly adopted for the generation of biomimetic antibodies is molecular imprinting of polymers (Özcan et al., 2006). Given the advantage of easy preparation and chemical stability, molecularly imprinted polymers (MIPs) possess a high potential for use in a variety of applications such as chromatographic stationary phases, immunoassay-type analyses and sensor development (Tatemichi et al., 2007; Ge and Turner, 2008; Haginaka, 2008). Generally, molecular imprinting is a synthetic strategy that is used to assemble a molecular receptor via template-guided synthesis. To prepare MIP, a template molecule is used to guide the assembly of functional monomers. A polymerization reaction is then employed to fix the preassembled binding groups around the template molecule. Following removal of the template molecule, the polymer revealed retains specific binding sites that can selectively rebind the original template molecule. Depending on the interactions between the template molecule and the functional monomers/groups involved at the imprinting and rebinding step, molecular imprinting has two different approaches: noncovalent and covalent (Büyüktiryaki et al., 2005; Takeuchi and Hishiya, 2008). In the noncovalent approach, various noncovalent interactions such as hydrogen bond, ionic interactions and hydrophobic effects are utilized. Given the fact that noncovalent interactions are prevalent in the biological world, exploitation of these binding forces, as it has turned out, has proven to be the most efficient and preferred method for generating robust, biomimetic materials (Baydemir et al., 2007; Takeda et al., 2009).

Conventional extracorporeal therapies used to remove bilirubin perform poorly due to low accessibility, insufficient adsorption capacities, low efficiency and economic limitations. Denizli and his co-workers proposed a cryogel column containing molecularly imprinted cryogel to remove bilirubin from human plasma. They reported research on the use of selective bilirubin adsorption using bilirubin-imprinted poly(hydroxyethyl methacrylate-N-methacryloyl-(L)-tyrosinemethylester) [poly(HEMA-MATyr)] [MIP-bilirubin] cryogel with large continuous interconnected pores (10–100 μm in diameter) that provide channels for the mobile phase to flow through (Figure 14.6) (Baydemir et al., 2009b). They reached a maximum bilirubin adsorption amount of 3.6 mg/g MIP-bilirubin cryogel. The relative selectivity coefficients of the MIP-bilirubin cryogel for bilirubin/cholesterol and bilirubin/testosterone mixtures were reported as 7.3 and 3.2 times greater than nonimprinted poly(HEMA-MATyr) [NIP] cryogel, respectively (Figure 14.7). In addition, they reported, MIP-bilirubin cryogel has potential as a clinical hemoperfusion material. For this purpose, they performed an experiment on whole blood flowing through the MIP-bilirubin cryogel column in order to show the superiority of cryogel column for direct affinity capture. The megaporous structure of cryogels allows for direct processing of whole blood containing blood cells (Kumar and Srivastava, 2010). They

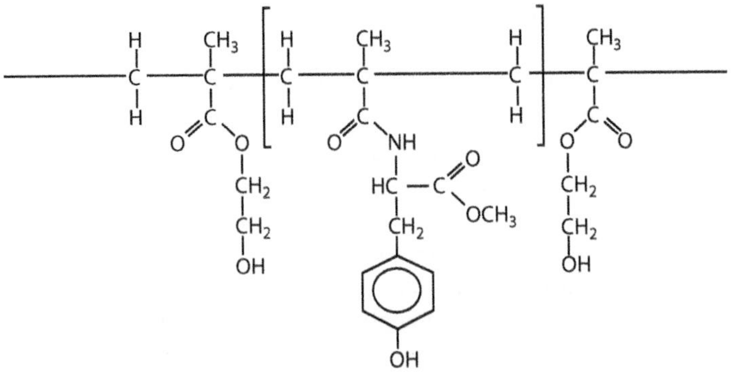

**FIGURE 14.6** Molecular structure of the poly(HEMA-MATyr) cryogel.

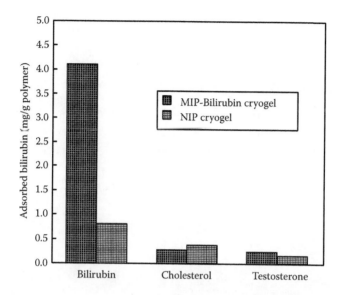

**FIGURE 14.7** Adsorbed template and competitive molecules in both MIP-Bilirubin and NIP cryogel 0.5 mL/min, 0.8 mg/dL, 30 mL solution, 125 mg polymer, T: 25°C. (Reprinted from *Reactive and Functional Polymers* 69, Baydemir, G., Andaç, M., Bereli, N., Galaev, I.Y., Say, R., Denizli, A., Supermacroporous PHEMA based cryogel with embedded bilirubin imprinted particles, 36–42. Copyright 2009, with permission from Elsevier.)

applied the pulse of blood on an MIP-bilirubin cryogel column under isotonic conditions, which passes through the column as a homogeneous plug without substantial tailing.

In another study, Baydemir and co-workers studied the use of selective bilirubin removal with PHEMA cryogel with embedded bilirubin-imprinted poly(HEMA-MATyr) particles [PHEMA/MIP composite cryogel] to improve the binding capacity of megaporous cryogel (Figure 14.8) (Baydemir et al., 2009a). Although the cryogels have many advantages (quick swelling kinetics, low-pressure drop, short diffusion path, short residence time during adsorption-elution processes, etc.), they have low surface area and low binding capacity due to the interconnected large-pores in their 3D structures. To overcome this problem, Baydemir et al. applied a new approach to increase the surface area and active recognition sites of the cryogel for improving the binding capacity of the cryogel column. The bilirubin adsorption capacity of the PHEMA/MIP composite cryogel (10.3 mg/g polymer) was improved significantly due to the embedded MIP particles into the polymeric matrix, which results in increased surface area (21.7 $m^2/g$ for 200 mg of embedded MIP particles) (Table 14.1; Figure 14.9). The relative selectivity coefficients of PHEMA/MIP composite cryogel for bilirubin/cholesterol and bilirubin/testosterone were calculated as 8.6 and 4.1 times greater than the PHEMA cryogel, respectively.

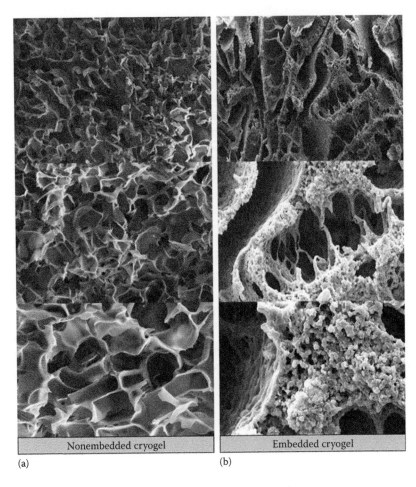

**FIGURE 14.8** SEM images of (a) PHEMA cryogel and (b) PHEMA/MIP composite cryogel. (Reprinted from *Reactive and Functional Polymers* 69, Baydemir, G., Andaç, M., Bereli, N., Galaev, I.Y., Say, R., Denizli, A., Supermacroporous PHEMA based cryogel with embedded bilirubin imprinted particles, 36–42. Copyright 2009, with permission from Elsevier Ltd.)

$N$-methacryloyl-L-tryptophan methyl ester (MATrp) containing PHEMA cryogel discs were prepared for removal of bilirubin out of human plasma (Figure 14.10) (Perçin et al., 2013). In the study, it was aimed to interact MATrp and the bilirubin molecule by hydrophobic interactions through the indole functional group of MATrp and pyrrole-like rings of bilirubin. They performed bilirubin adsorption studies in a batch system and reached the maximum bilirubin adsorption capacity of 22.2 mg/g cryogel disc without any significant decrease in adsorption capacity even after the adsorption–desorption cycle was repeated 10 times using the same group discs in a batch experimental setup.

### TABLE 14.1
### Surface Area of PHEMA/MIP Composite Cryogels

| Embedded Amount (mg) | Surface Area (m²/g) |
|---|---|
| – | 11.7 |
| 50 | 14.2 |
| 100 | 18.1 |
| 200 | 21.3 |
| 300 | 25.6 |

*Source:* Redrawn from Baydemir, G., Andaç, M., Bereli, N., Galaev, I.Y., Say, R., Denizli, A. (2009). *Reactive and Functional Polymers* 69, 36–42. Copyright 2009.

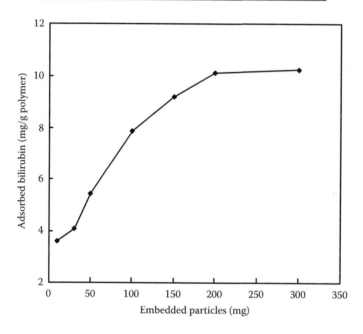

**FIGURE 14.9** Effect of embedded particles amount onto the adsorption amount: Bilirubin concentration: 1.0 mg/dL, flow rate: 0.5 ml/min; T: 25°C. (Reprinted from *Reactive and Functional Polymers* 69, Baydemir, G., Andaç, M., Bereli, N., Galaev, I.Y., Say, R., Denizli, A., Supermacroporous PHEMA based cryogel with embedded bilirubin imprinted particles, 36–42. Copyright 2009, with permission from Elsevier Ltd.)

## 14.4 AFFINITY THERAPY IN RHEUMATOID ARTHRITIS

Rheumatoid arthritis is a chronic, progressive, debilitating autoimmune disease that is characterized by chronic polyarthritis and destruction of multiple joints. Rheumatoid arthritis affects approximately 0.5–1.0% of the worldwide population (Strand et al., 2001). Plasma exchange has been applied to remove autoantibodies in

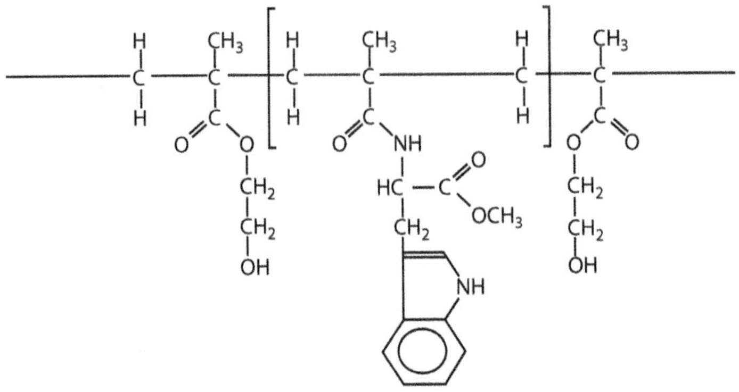

**FIGURE 14.10** Molecular structure of the PHEMATrp cryogel discs.

the treatment of severe forms of various autoimmune diseases (Poullin et al., 2005). In patients with rheumatoid arthritis, plasma exchange has shown no significant clinical benefits (American Medical Association Panel, 1985). Plasma exchange is a nonselective method and the requirement for plasma substitutes such as albumin is very high. Moreover, the dangers of hepatitis or an immune reaction accompany this therapy. It would be most desirable to remove selectively any pathogenic substances while using plasma products. For this purpose, adsorption was suggested as an alternative to plasma exchange for removing pathogenic substances from the plasma of patients with autoimmune diseases refractory to conventional treatments (Burnouf et al., 1998). Extracorporeal affinity adsorption has also been used for other conditions such as treatment of cancer, paraneoplastic syndromes and Guillain-Barre syndrome. Extracorporeal therapy is more selective than plasma exchange, as it does not remove plasma proteins such as albumin and clotting factors. In addition, adsorption does not require administration of plasma substitutes and therefore does not expose patients to the potential side effects of these compounds (Yilmaz et al., 2008).

IgM-antibody removal from human plasma was carried out with megaporous poly(hydroxyethyl methacrylate) [PHEMA] cryogel carrying protein A (Alkan et al., 2010). The PHEMA cryogel was prepared by bulk polymerization, which proceeds in an aqueous solution of monomer frozen inside a plastic syringe. After thawing, the PHEMA cryogel contained a continuous matrix, with interconnected pores in the size range of 10–200 μm. Pore volume in the PHEMA cryogel was 71.6%. Protein A molecules were covalently immobilized onto the PHEMA cryogel via cyanogen bromide activation. The PHEMA cryogel was contacted with blood in an *in vitro* system for determination of blood compatibility. Coagulation times in rheumatoid arthritis patients' plasma are summarized in Table 14.2. The megaporous structure of the PHEMA cryogel made it possible to process blood cells without blocking the cryogel column, as this is one of the main advantages of the megaporous cryogels (Kumar and Srivastava, 2010). IgM-antibody adsorption capacity decreased significantly with the increase of the plasma flow rate. An increase in the flow rate reduced the plasma volume treated efficiently until the breakthrough point and therefore decreased the retention time of the PHEMA cryogel column (Figure 14.11).

## TABLE 14.2
## Coagulation Times in Rheumatoid Arthritis Patient Plasma (Reported in Seconds)

| Experiments | APTT | PT | Fibrinogen Time |
| --- | --- | --- | --- |
| Control plasma | 27.8 | 12.2 | 14.1 |
| PHEMA | 29.3 | 12.4 | 18.3 |
| PHEMA-protein A | 26.2 | 12.5 | 12.1 |

*Source:* Reprinted from *Biochemical Engineering Journal* 51, Alkan, H., Bereli, N., Baysal Z., Denizli, A., 153–159. Copyright 2010, with permission from Elsevier.

*Note:* Each result is the average of six parallel studies.

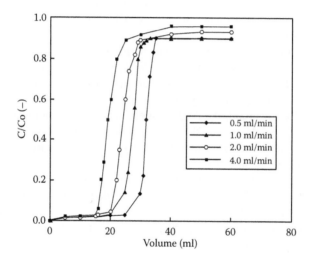

**FIGURE 14.11** Effect of flow-rate on the breakthrough curves. Protein A loading: 56 mg/g; IgM-antibody concentration: 2.98 mg/ml; T: 25°C. (Reprinted from *Biochemical Engineering Journal* 51, Alkan, H., Bereli, N., Baysal Z., Denizli, A., Selective removal of the autoantibodies from rheumatoid arthritis patient plasma using protein A carrying affinity cryogels, 153–159. Copyright 2010, with permission from Elsevier Ltd.)

The maximum IgM-antibody adsorption amount was 42.7 mg/g. IgM-antibody molecules could be repeatedly adsorbed and eluted without noticeable loss in the IgM-antibody adsorption amount.

## 14.5 AFFINITY THERAPY IN THALASSEMIA

Iron is an essential trace element for almost all organisms for a broad spectrum of biological processes, which include electron transfer, transport, storage and activation of oxygen, nitrogen fixation and DNA synthesis (Crichton, 1991). The toxic effects of iron overload are well known especially since the human body has no physiological route for the elimination of excess iron (Arena, 1970). Chronic iron overload may occur in a variety of diseases where the administration of parental iron is necessary (e.g., thalassemia, aplastic anemia). Acute iron intoxication is also a frequent, sometimes life-threatening form of poisoning, especially among young children. The toxicity of iron is related to its ability to induce oxidative stress in cells (Martin, 1986). In an occupational setting, inhalation exposure to iron oxide may cause siderosis. In the nonoccupational population, ingestion of large quantities of iron salts may cause nausea, vomiting and intestinal bleeding. There is accumulating evidence suggesting that an increase in iron storage may be associated with an increasing risk of developing cancer (Litovitz et al., 1987). Studies have demonstrated that there is an increased risk for developing colorectal carcinoma following ingestion of high amounts of iron (Mahoney et al., 1989). There is also an increase in hepatocellular carcinoma in patients with hereditary hemochromatosis, an inherited disorder in which there is hyperabsorption of iron from the intestinal tract and in lung cancer from exposure to asbestos fibres, which contain approximately 30% iron by weight (Hershko 1992).

The only available supportive treatment is chelation therapy and the only available clinical drug for transfusional iron overload and acute iron poisoning is desferrioxamine B (DFO), a linear hydroxamate and natural siderophore (Mahoney et al., 1989). The use of DFO has already been shown to result in prolonged life expectancy, reduced liver iron and the establishment of negative iron balance. However, the major limitation to the use of DFO is its lack of effectiveness when administered orally, the short half-life time in plasma and its potential toxicity when present in high concentrations (Jacobs, 1979). DFO is highly expensive as well. For these reasons, a number of orally active iron chelators are being tested but none of them is satisfactory (Denizli et al., 1998). To overcome the drawbacks of soluble iron chelators in the treatment of iron overload, binding of iron chelating ligands has been studied (Yavuz et al., 2005). Comparing to soluble iron chelators, iron-chelating resins might have advantages in stability, reusability and minimal damage to biological substances. Developing such a resin usually involves preparing selective and efficient adsorbents for the elimination of excess iron to safe levels. These requirements are mostly fulfilled by iron-imprinted polymers, in which a specific ligand is oriented around the template through coordination bonds that leads an adsorbent with iron ion cavities that is complementary in terms of coordination geometry, size and charge. Several studies show the efficiency of this method (Yavuz et al., 2006; Özkara et al., 2008).

A related study was introduced by Aslıyüce et al. (2010). They prepared iron imprinted PHEMA cryogel in order to remove excess $Fe^{3+}$ ions from beta-thalassemia patient plasma. For this purpose, they complexed $Fe^{3+}$ ions with N-methacryloyl-L-cysteine (MAC) monomer and the resulting MAC-$Fe^{3+}$ complex was cryo-polymerized in the presence of HEMA monomer. Their maximum $Fe^{3+}$ adsorption capacity onto $Fe^{3+}$ imprinted PHEMA-MAC cryogel was found as 75 μg/g while nonspecific adsorption was 11.2 μg/g. The $Fe^{3+}$ imprinted PHEMA-MAC adsorbent was also shown to be selective for the $Fe^{3+}$ ions. The relative selectivity was calculated as 10.5 and 2.3 for $Fe^{3+}/Cd^{2+}$ and $Fe^{3+}/Ni^{2+}$, respectively (Figure 14.12a,b).

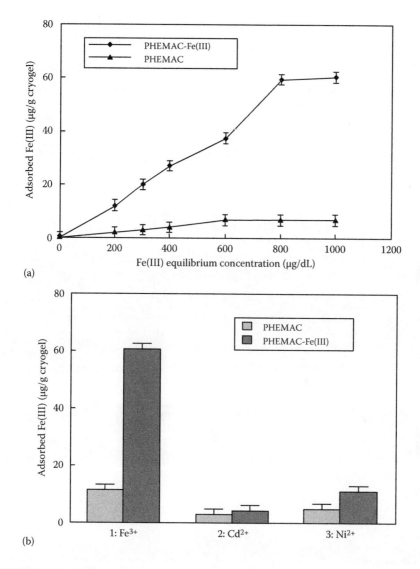

**FIGURE 14.12** (a) An adsorption isotherm for a PHEMAC-$Fe^{3+}$ column showing its adsorption levels at different $Fe^{3+}$ equilibrium concentrations in media, where the flow-rate was kept at 1.0 ml/min at T: 25°C. (b) Template ions adsorbed on PHEMAC-$Fe^{3+}$ and PHEMAC; flow-rate: 1.0 ml/min; ion concentration: 800 µg/dL; T: 25°C. (Reprinted from *Separation Purification Technology* 73, Aslıyüce, S., Bereli, N., Uzun, L., Onur, M.A., Say, R., Denizli, A., Ion-imprinted supermacroporous cryogel for in-vitro removal of iron out of human plasma with beta thalassemia, 243–249. Copyright 2010, with permission from Elsevier.)

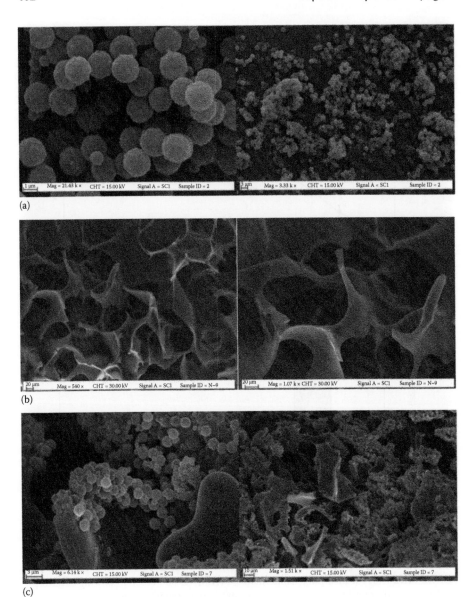

**FIGURE 14.13** Scanning electron photographs of (a) MIP beads, (b) PHEMA and (c) PHEMA-MIP cryogels. (Ergün, B., Baydemir, G., Andaç, M., Yavuz, H., Denizli, A.: Ion imprinted beads embedded cryogels or in-vitro removal of iron from β-thalassemic human plasma. *Journal of Applied Polymer Science*, 2012, 125, 254–262. Copyright Wiley-VCH Verlag GmbH & Co. KGaA. Reproduced with permission.)

As indicated in the previous sections, the main disadvantage of cryogel adsorbents is their low adsorption capacity because of their small surface area. In another study, the selective removal of $Fe^{3+}$ ions from thalassemia patient plasma was aimed by Ergün et al. to solve this problem with $Fe^{3+}$ imprinted poly(GMA-MAC) beads (MIP beads) embedded PHEMA composite cryogels (Ergün et al., 2012) (Figure 14.13). The embedding of MIP beads resulted in a sevenfold (76.8 m$^2$/g) increase in the surface area of the nonembedded PHEMA cryogels. In the study, the maximum adsorption amount of $Fe^{3+}$ ions was 2.23 mg/g. The relative selectivity of the composite cryogel toward $Fe^{3+}$ ions was 135.0, 61.4 and 57.0 times greater than that of nonimprinted composite cryogel with respect to $Ni^{2+}$, $Zn^{2+}$ and $Fe^{2+}$ ions, respectively (Figure 14.14a,b).

## 14.6 AFFINITY THERAPY IN HYPERCHOLESTEROLEMIA

Coronary heart disease is a leading cause of mortality with 2 million deaths in the European Union and 4.3 million deaths in all of Europe. It is also the cause of lifetime disability. The annual economic cost is estimated to be $192 billion while substantially increasing (Harland, 2012). Together with high blood pressure and lifestyle factors such as smoking, diet, and lack of physical activity, LDL cholesterol is widely recognized as one of the major risk factors in the development of coronary heart diseases because of its ability to build up in the lining of arteries to form atheroma and fatty acid deposits (Bruckert, 1999; Nobutaka, 2000; Thompson, 2003). The WHO estimates that excess total blood cholesterol is responsible for over 60% of coronary heart disease and 40% of ischemic stroke in developed countries. It has been reported that every 39 mg/dL decrease in LDL-cholesterol is associated with a 22% reduction in the risk of coronary heart disease. Thus, the reduction of elevated LDL-cholesterol is important (Denizli, 2002). In most cases, reduced dietary intake and drug therapy can control plasma cholesterol levels. However, in severe cases, more aggressive treatment is necessary. Conventional and most frequently used techniques are summarized in Table 14.3.

Immuno-affinity adsorbents, which use monoclonal antibodies against the main protein component of LDL, apoB100, as a ligand, offer most selective removal of LDL without decreasing useful HDL and other plasma proteins (Kane et al., 1975; Stoffel et al., 1981; Liu et al., 1996; Yavuz and Denizli, 2003, 2005). However, during the preparation of immune-affinity adsorbents, antibody-binding procedures usually result in a decrease in the antigen binding capacity because in the random immobilization, antibodies can bind to the matrix surface at many different points that finally prevents the accessibility of the antigen binding sites ($F_{ab}$) (Figure 14.15) (Dugas et al., 2010; Turkova, 1999).

To eliminate these drawbacks, several approaches have been developed for oriented antibody immobilization. In one scheme, antibodies were bound on solid supports through $F_c$ binding proteins such as protein A and protein G, leaving the antigen-specific sites free (Figure 14.16a) (Dugas et al., 2010). Another strategy for

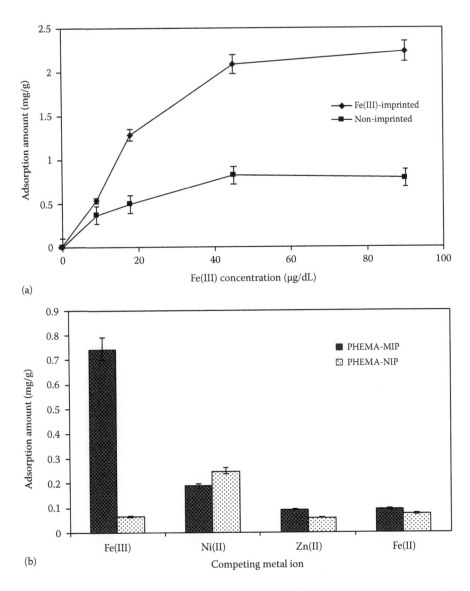

**FIGURE 14.14** (a) Adsorption isotherm for PHEMA-MIP and PHEMA-NIP composite cryogel at different $Fe^{3+}$ initial concentrations. Flow rate: 0.5 mL/min; T: 25°C. (b) Adsorption of $Fe^{3+}$ ions in the presence of competitor ions $Zn^{2+}$, $Ni^{2+}$ and $Fe^{2+}$. (Ergün, B., Baydemir, G., Andaç, M., Yavuz, H., Denizli, A.: Ion imprinted beads embedded cryogels or in-vitro removal of iron from β-thalassemic human plasma. *Journal of Applied Polymer Science*, 2012, 125, 254–262. Copyright Wiley-VCH Verlag GmbH & Co. KGaA. Reproduced with permission.)

## TABLE 14.3
### Extracorporeal Elimination Methods for LDL

| Method | Special Remarks |
| --- | --- |
| Plasmapheresis | - Virtually all plasma proteins removed<br>- Nonspecific and nonselective<br>- Frequent use is not desirable |
| Cascade filtration | - The size of the filter pores adapted for LDL (mean pore diameter 20 nm)<br>- Nonselective method |
| Heparin-induced extracorporeal LDL precipitation (HELP) | - Insoluble heparin-LDL complex develops at pH 5.1–5.2<br>- Decrease of fibrinogen<br>- Blood coagulation is influenced<br>- Need of water supply<br>- Need for intensive care<br>- Risk of infection by dialysis procedure |
| Thermofiltration | - Plasma is heated to 40°C, LDL and VLDL filtered |
| Dextran-induced LDL precipitation | - Selective adsorption through ApoB100 onto immobilized Dextran<br>- Enormous amount of waste<br>- Need for intensive care |
| Direct adsorption of lipoproteins (DALI) | - Adsorption through ApoB100 on polyacrylic acid |
| Immunoadsorption | - Adsorption onto polyclonal sheep antibodies against ApoB100<br>- Specific and selective |

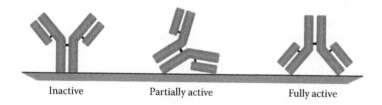

Inactive        Partially active        Fully active

**FIGURE 14.15** Antibody immobilization by random binding procedure.

oriented binding is the immobilization of antibody using its carbohydrate moieties that is generally linked to the $F_c$ part with covalent hydrazide bond after chemical or enzymatic oxidation or through the specific carbohydrate binding molecules such as lectins and boronate (Figure 14.16b) (Dugas et al., 2010; Aslıyüce et al., 2013). A third method is to utilize the sulfhydryl group of the $F_{ab}$ fragment (Figure 14.16c). This can be achieved by pepsin digestion, followed by a reduction of the sulphide bond between the monovalent $F_{ab}$ fragments, or by using artificial $F_{ab}$ fragments produced by molecular genetic techniques (Dugas et al., 2010; Aslıyüce et al., 2013). The sulfhydryl group in the C-terminal region of the fragment can be used to couple the fragment to an insoluble support leaving its antigen binding site available.

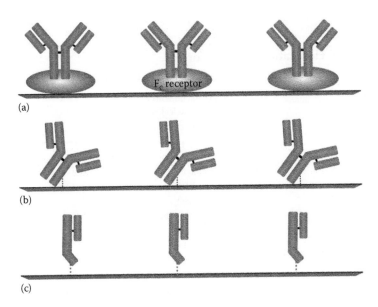

**FIGURE 14.16** Representation of oriented antibody immobilization methods. (a) Antibody binds to $F_c$ receptors on adsorbents, (b) antibody is immobilized via an oxidized carbohydrate moiety on $F_c$ fragment, (c) monovalent $F_{ab}$ fragment is bound via sulfhydril group in the C-terminal region.

Bereli and co-workers prepared immune-affinity cryogel discs in order to reach safe plasma LDL levels (Bereli et al., 2011). They immobilized antihuman β-lipoprotein antibody (anti-LDL antibody) molecules directly to the PHEMA cryogel and through protein A bound PHEMA cryogel to achieve orientation. The PHEMA cryogel is commented as blood compatible because all the clotting times were increased when compared with the control plasma. The maximum anti-LDL antibody immobilization amount was 63.2 mg/g in the case of random and 19.6 mg/g in the case of oriented antibody immobilization (protein A loading was 57.0 mg/g). Random and oriented anti-LDL antibody immobilized PHEMA cryogels adsorbed 111 and 129 mg LDL/g cryogel from hypercholesterolemia human plasma, respectively (Figure 14.17) with no significant adsorption of other lipoproteins, especially useful HDL. They reported that the oriented anti-LDL antibody immobilized PHEMA cryogels are more effective in LDL removal although they contain one third of the randomly antibody immobilized cryogels. This shows the mode of oriented immobilization makes the binding site of antibody always located on the top of its $F_{ab}$ variable regions, well accessible for interaction with LDL molecules. In their study, they also showed that up to 80% of the adsorbed LDL can be desorbed and reused many times (Bereli et al., 2011).

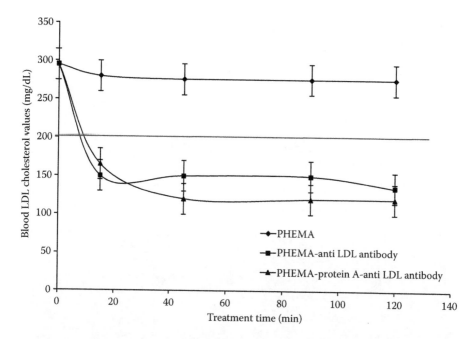

**FIGURE 14.17** Cholesterol removal from hypercholesterolemia human plasma with PHEMA, PHEMA-anti-LDL antibody and PHEMA protein A-anti-LDL antibody cryogels. Total cholesterol concentration: 431 mg/dL; LDL concentration: 295 mg/dL; anti-LDL antibody loading: 63.2 mg/g (random), 19.6 mg/g (oriented); T: 20°C; flow rate: 1.0 mL/min. (Reprinted from *Materials Science and Engineering C*, 31, Bereli, N., Şener, G., Yavuz, H., Denizli, A., Oriented immobilized anti-LDL antibody carrying PHEMA cryogel for cholesterol removal from human plasma, 1078–1083. Copyright 2011, with permission from Elsevier.)

In a recent study, cholesterol imprinted monosize microsphere-embedded cryogels were prepared for cholesterol removals from various samples (Çaktü et al., 2014). The monosize (2–3 µm) poly(glycidyl methacrylate-N-methacryloyl-(L)-tyrosine methylester) (PGMAT) microspheres were used, which were embedded into the poly(hydroxyethyl methacrylate) (PHEMA) cryogels (Figure 14.18). Specific surface area of the PHEMA cryogel was increased from 13 to 72.7 m$^2$/g by embedding of microspheres. By using these composites, the maximum adsorption capacity was found as 42.7 mg/g for intestinal mimicking solution. The relative selectivity coefficients of cholesterol imprinted PGMAT/PHEMA composite cryogel for cholesterol/estradiol and cholesterol/stigmasterol were 181.6 and 169.93 ($k'$) greater than the nonimprinted PGMAT/PHEMA matrix, respectively (Table 14.4). They removed 80% of the cholesterol from homogenized milk using the PGMAT/PHEMA composite cryogel (Table 14.5).

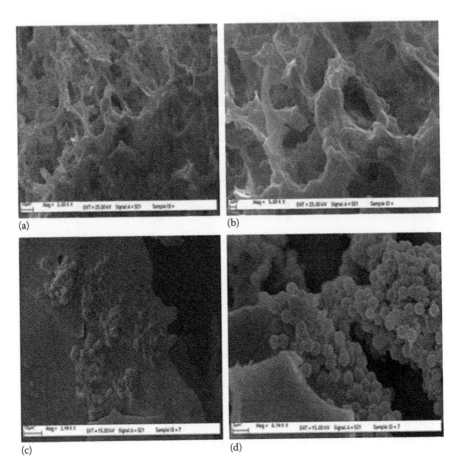

**FIGURE 14.18** SEM images of PHEMA cryogels (a, b); PGMAT/PHEMA composite cryogels (c, d) (macropore radius: ~100 µm). (Reproduced from Çaktü, K., Baydemir, G., Ergün, B., Yavuz, H., *Artificial Cells, Nanomedicine, and Biotechnology* 42(6), 365–375. Copyright 2014, Informa Healthcare.)

**TABLE 14.4**
$k$, $k_d$ and $k'$ Values of Estradiol and Stigmasterol with Respect to Cholesterol

| | Polymer | | | | |
|---|---|---|---|---|---|
| | nipPGMAT/PHEMA | | mipPGMAT/PHEMA | | |
| Molecule | $k_d$ | $k$ | $k_d$ | $k$ | $k'$ |
| Cholesterol | 3.01 | – | 343.3 | – | – |
| Estradiol | 22.7 | 0.13 | 14.26 | 24.07 | 181.6 |
| Stigmasterol | 59.01 | 0.05 | 39.57 | 8.6 | 169.93 |

*Source:* Redrawn from Çaktü, K., Baydemir, G., Ergün, B., Yavuz, H. (2014). *Artificial Cells, Nanomedicine, and Biotechnology* 42(6), 365–375. Copyright (2014), Informa Healthcare. With permission.

## TABLE 14.5
### Cholesterol Removal from Homogenized Cow's Milk

| Sample | Reference Value of Cholesterol Concentration in Milk mg/100g | Cholesterol Concentration in Milk Sample mg/100 g | Removal, % |
|---|---|---|---|
| Cow's milk | 13.4 | 12.3 | 80 |

*Source:* Redrawn from Çaktü, K., Baydemir, G., Ergün, B., Yavuz, H. (2014). *Artificial Cells, Nanomedicine, and Biotechnology* 42(6), 365–375. Copyright (2014), Informa Healthcare. With permission.

## 14.7 OTHER APPLICATIONS

Kumar and his group present the state-of-the-art use of supermacroporous cryogels for various applications in bioengineering (Kumar, 2008). By using a protein A affinity monolithic cryogel column, a generic approach was developed for specifically separating different cell types like lymphocytes and CD34+ stem cells from cord blood. A novel approach of releasing the cells from the cryogel matrix was designed by mechanically squeezing the cryogels. Another interesting application of the macroporous matrices has been the cultivation of the mammalian cells on the gelatin modified cryogels bioreactors. The cells grow, proliferate and secrete the protein therapeutics continuously in the circulating medium when allowed to culture on the cryogel matrices. These reactors also showed potential when used as an extracorporeal device for cryogel scaffolds from agarose, alginate and chitosan with gelatin. Cryogel showed good properties for cartilage tissue engineering.

In another study, Eichhorn and co-workers prepared composite monolithic columns by incorporating neutral polystyrene divinylbenzene (PS-DVB) microparticles into macroporous poly(vinyl alcohol) (PVA) or agarose-based cryogels (Eichhorn et al., 2013). They studied the adsorption characteristics of prepared composite cryogels for cytokines (TNF-$\alpha$, IL-6) and hydrophobic metabolites related to liver failure (bilirubin, cholic acid). They reported that composite cryogels have excellent flow-through properties; the flow-through properties were reported at a pressure of 15 kPa as 4.2 mL/min and 6.7 mL/min for 125-mg particle embedded PVA-based composite cryogels and for 125-mg particle embedded agarose-based composite cryogels, respectively. The bilirubin adsorption amounts were given as 2.1 ± 0.7 and 5.0 ± 0.5 µmol/g PS-DVB for PVA and agarose-based composite cryogels, respectively. They reported cholic acid adsorption amounts as 9.6 ± 0.5 µmol/g PS-DVB for PVA and 9.4 ± 0.5 µmol/g PS-DVB for agarose-based composite cryogels. The TNF-$\alpha$ adsorption amounts was reported as 0.10 ± 0.03, 0.11 ± 0.001 pmol/g PS-DVB respectively and IL6 adsorption amounts were reported as 0.06 ± 0.04, 0.11 ± 0.05 pmol/g PS-DVB, respectively, for PVA and agarose-based composite cryogels. They studied all adsorption experiments using plain PS-DVB microparticles and PS-DVB microparticles loaded PVA and agarose-based composite cryogels. They reported that the plain PS-DVB microparticles adsorbed all the target molecules in higher amounts than that of composite cryogels. They further reported that the embedding procedure affects the adsorption characteristics of PS-DVB microparticles due to the formation of a polymer layer on the microparticles' surface, which diminished the accessibility of the pores of microparticles. The effect of microparticle loading amount was also discussed in this study and they noted that the specific adsorption amounts were found to be higher for composite cryogels containing 125 mg of particles as compared to those containing 250 mg of microparticle. Hence, suggesting that it is very important to find the optimal amount of particles embedded within the prepared composite cryogels. The composite cryogels exhibited good blood compatibility, as global coagulation parameters (prothrombin time and activated partial thromboplastin time) remained stable after incubation of the cryogels with plasma. They showed that the produced PS-DVB containing composite cryogel is suitable for depletion of toxic metabolites from plasma.

Incorporation of activated carbon particles into poly(vinyl alcohol) (PVA) based cryogel resulted in macroporous monolithic composite materials (AC-PVA) exhibiting good flow-through properties. Ivanov and researchers prepared and characterized the AC-PVA composite cryogels with low and high carbon content for high abundant protein adsorption and for clearance of myoglobin (Mb) from blood in extracorporeal treatment of acute rhabdomyolysis, which results from increasing the amount of Mb in blood caused by traumas (Ivanov et al., 2012). They observed the highest initial rate of Mb adsorption for AC having the largest specific surface area (1939 $m^2 \cdot g^{-1}$) and pore volume (1.82 $cm^3 \cdot g^{-1}$). They found that adsorbed amounts of Mb were the highest among the proteins studied (up to 700 $mg \cdot g^{-1}$ carbon), which was attributed to the higher fraction of pores accessible for Mb. They also showed that adsorption of Mb on AC–PVA took place even in the presence of 500-fold higher concentration of BSA, which indicated a possibility of Mb clearance from blood plasma using the PVA–carbon monoliths.

Biocompatible and degradable PHEMA-based polymers could also be suitable for many different biomedical applications. Zhang and co-workers prepared degradable and less toxic PHEMA with ester linkages in the backbone, which could be successfully made by radical copolymerization with cyclic ketene acetal (Zhang et al., 2012). The resulting polymers were significantly less toxic with cell viabilities of more than 80% even for very high polymer concentrations (100 $mg \cdot mL^{-1}$). The authors showed that these polymers were hydrolytically degradable under basic conditions in addition to surface and bulk degradation using macrophages.

Apart from biomedical applications, cryogels have recently gained significant interest as three-dimensional extracellular matrixes (ECM) mimicking scaffolds for tissue engineering and *in vitro* cell culture applications due, to their inherent interconnected macroporous structure and ease of formation in comparison to other macropore forming techniques. Henderson et al. have significantly reviewed the recent advances in cryogelation techniques such as the monomers/precursors that can be utilized to synthesize biocompatible and biologically relevant cryogels as well as the various physicochemical characterization techniques used for these materials. They also discussed emerging trends in the application of cryogels, particularly as three-dimensional ECM mimicking scaffolds for cell culture and tissue engineering (Henderson et al., 2013). In a recent study, Berezhna et al. prepared macroporous monolithic composites with embedded divinylbenzene-styrene (DVB-ST) polymeric particles by cryogelation techniques using poly(vinyl alcohol) or agarose solutions for investigation of structure and biocompatibility in blood plasma (Berezhna et al., 2013). Biocompatibility of the composites was assessed by estimation of the C5a fragment of complement in the blood serum and the concentration of fibrinogen in the blood plasma. They presented that a time-dependent generation of C5a fragment indicated weak activation of the complement system. At the same time, an insignificant difference in fibrinogen concentration (one of the most important proteins in the coagulation system of the blood), which contacted the composites, between the pristine blood plasma and the plasma, which was circulated through the monolithic columns, was observed.

## 14.8 CONCLUDING REMARKS

**Extracorporeal treatment** is a process in which the blood is passed over an adsorbent in a packed bed column. Extracorporeal therapy is frequently used for acute blood purification as in the case of drug overdose, wherein the adsorbent is biocompatible polymer coated charcoal. The charcoal binds medium- to high-molecular weight blood components. Charcoal adsorption is nonspecific in nature, removing many desirable blood components as well as toxic substances. Consequently, a number of scientists have examined specific extracorporeal removal of toxic substances including bilirubin, pathogenic antibodies and so on. In this review, we outlined the developments in the extracorporeal affinity therapies. It has been demonstrated that cryogel-based affinity adsorbents can be used for the removal of toxic substances from human blood.

## LIST OF ABBREVIATIONS

| | |
|---|---|
| **AC** | Activated carbon |
| **ECM** | Extracellular matrix |
| **HDL** | High-density lipoprotein |
| **LDL** | Low-density lipoprotein |
| **MAC** | N-methacryloyl-L-cysteine |
| **Mb** | Myoglobin |
| **MIP** | Molecularly imprinted polymer |
| **PGMAT** | poly(glycidyl methacrylate-N-methacryloyl-(L)-tyrosine methylester) |
| **PHEMA** | poly(2-hydroxyethyl methacrylate) |
| **PS-DVB** | polystyrene divinylbenzene |
| **PVA** | poly(vinyl alcohol) |
| **SEM** | Scanning electron microscope |
| **SLE** | Systemic lupus erythematosus |

## REFERENCES

Ahmad, N., Arif, K., Faisal, S.M., Neyaz, M.K., Tayyab, S., Owais, M., (2006). PGLA-microsphere mediated clearance of bilirubin in temporarily hyperbilirubinemic rats, *Biochimica Biophysica Acta*, 1760(2), 227–232.

Alkan, H., Bereli, N., Baysal Z., Denizli, A. (2010). Selective removal of the autoantibodies from rheumatoid arthritis patient plasma using protein A carrying affinity cryogels. *Biochemical Engineering Journal*, 51, 153–159.

Alkan, H., Bereli, N., Baysal, Z., Denizli, A. (2009). Antibody purification with protein A attached supermacroporous PHEMA cryogel. *Biochemical Engineering Journal*, 45, 201–208.

Altıntaş, E.B., Türkmen, D., Karakoç, V., Denizli, A. (2011). Efficient removal of bilirubin from human serum by monosize dye-affinity beads. *Journal Biomaterials Science Polymer Edition*, 22, 957–971.

American Medical Association Panel. (1985). *JAMA*, 253, 819–825.

Arena, J.M. (1970). *Poisoning.*, 4th ed., Charles C Thomas, Springfield, IL, pp. 538–607.

Aşır, S., Uzun, L., Türkmen, D., Say, R., Denizli, A. (2005). Ion-selective imprinted superporous monolith for cadmium removal from human plasma, *Separation Science Technology*, 40, 3167–3185.

Aslıyüce, S., Bereli, N., Uzun, L., Onur, M.A., Say, R., Denizli, A. (2010). Ion-imprinted supermacroporous cryogel for in-vitro removal of iron out of human plasma with beta thalassemi. *Separation Purification Technology*, 73, 243–249.

Aslıyüce, S., Uzun, L., Say, R., Denizli, A. (2013). Immunoglobulin G recognition with $F_{ab}$ fragments imprinted monolithic cryogels: Evaluation of the effects of metal-ion assisted-coordination of template molecule. *Reactive Functional Polymers* 73, 813–820.

Avramescu, M.E., Sager, W.F.C., Borneman, Z., Wessling, M. (2004). Adsorptive membranes for bilirubin removal. *Journal of Chromatography B*, 803(2), 215–223.

Babac, C., Yavuz, H., Galaev, I.Y., Piskin, E., Denizli, A. (2006). Binding of antibodies to concanavalin A-modified monolithic cryogel. *Reactive and Functional Polymers*, 66, 1263–1271.

Baydemir, G., Andaç, M., Bereli, N., Galaev, I.Y., Say, R., Denizli, A. (2009a). Supermacroporous PHEMA based cryogel with embedded bilirubin imprinted particles. *Reactive and Functional Polymers*, 69, 36–42.

Baydemir, G., Andaç, M., Bereli, N., Say, R., Denizli, A. (2007). Selective removal of bilirubin from human plasma with bilirubin imprinted particles. *Industrial and Engineering Chemistry Research*, 46, 2843–2852.

Baydemir, G., Andaç, M., Bereli, N., Say, R., Galaev, I.Y., Denizli, A. (2009b). Bilirubin recognition via molecularly imprinted supermacroporous cryogels. *Colloids and Surfaces B: Biointerfaces*, 68(1), 33–38.

Behm, E., Ivanovich, P., Klinkmann, H. (1989). Selective and specific adsorbents for medical therapy. *International Journal of Artificial Organs*, 12, 1–10.

Bereli, N., Andac, M., Baydemir, G., Say, R., Galaev, I.Y., Denizli, A. (2008). Protein recognition via ion-coordinated molecularly imprinted supermacroporous cryogels. *Journal of Chromatography A*, 1190,18–26.

Bereli, N., Şener, G., Yavuz, H., Denizli, A. (2011). Oriented immobilized anti-LDL antibody carrying PHEMA cryogel for cholesterol removal from human plasma. *Materials Science and Engineering C*. 31, 1078–1083.

Berezhna, L. G., Ivanov, A.E., Leistner, A., Lehmann, A., Viloria-Cols, M., Jungvid, H. (2013). Structure and biocompatibility of poly(vinyl alcohol)-based and agarose-based monolithic composites with embedded divinylbenzene-styrene polymeric particles. *Progress in Biomaterials*, 2, 4–12.

Bruckert, E. (1999). LDL-apheresis: Questions for the future. *Transfusion Science*, 20, 43–47.

Burnouf, T., Goubran, H., Radosevich, M. (1998). Application of bioaffinity technology in therapeutic extracorporeal plasmapheresis and large-scale fractionation of human plasma. *Journal of Chromatography B: Biomedical Science Applications* 16, 65–80.

Büyüktiryaki, S., Say, R., Ersöz, A., Birlik, E., Denizli, A. (2005). Selective preconcentration of thorium in the presence of UO(2)(2+), Ce(3+) and La(3+) using Th(IV)-imprinted polymer. *Talanta*, 67(3), 640–645.

Çaktü, K., Baydemir, G., Ergün, B., Yavuz, H. (2014). Cholesterol removal from various samples by cholesterol-imprinted monosize microsphere-embedded cryogels. *Artificial Cells, Nanomedicine, and Biotechnology*, 42(6), 365–375.

Crichton, R.R. (1991). The importance of iron for biological systems. In: *Inorganic Biochemistry of Iron Metabolism from Molecular Mechanism to Clinical Consequences*, Wiley, UK, pp. 17–48.

Dainiak, M.B., Galaev, I.Y., Mattiasson, B. (2006). Affinity cryogel monoliths for screening for optimal separation conditions and chromatographic separation of cells. *Journal of Chromatography A*, 1123, 145–150.

Demirçelik, A.H., Andaç, M., Andaç, C.A., Say, R., Denizli, A. (2009). Molecular recognition based detoxification of aluminum in human plasma. *Journal of Biomaterials Science Polymer Edition*, 20, 1235–1258.

Denizli, A., Pişkin, E. (1995a). DNA immobilized PHEMA microbeads for affinity sorption of human IgG and anti-DNA antibodies. *Journal of Chromatography B*, 666, 215–222.

Denizli, A., Pişkin, E. (1995b). Protein A immobilized PHEMA beads for affinity sorption of human IgG. *Journal of Chromatography B*, 668, 13–19.

Denizli, A., Salih, B., Pişkin, E. (1998). New chelate-forming polymer microspheres carrying dyes as chelators for iron overload. *Journal Biomaterials Science, Polymer Edition*, 9, 175–187.

Denizli, A. (2002). Preparation of immuno-affinity membranes for cholesterol removal from human plasma. *Journal of Chromatography B*, 772, 357–367.

Dugas, V., Elaissari, A., Chevalier, Y. (2010). Surface sensitization techniques and recognition receptors immobilization on biosensors and microarrays. *Recognition Receptors in Biosensors*, SE(2), 47–134.

Eichhorn, T., Ivanov, A.E., Dainiak, M.B., Leistner A., Linsberger, Mikhalovsky, Weber V. (2013). Macroporous composite cryogels with embedded polystyrene divinylbenzene microparticles for the adsorption of toxic metabolites from blood. *Journal of Chemistry*, http://dx.doi.org/10.1155/2013/348412.

Ergün, B., Baydemir, G., Andaç, M., Yavuz, H., Denizli, A. (2012). Ion imprinted beads embedded cryogels or in-vitro removal of iron from β-thalassemic human plasma. *Journal of Applied Polymer Science*, 125, 254–262.

Ertürk, G., Bereli, N., Tümer, A., Say, R., Denizli, A. (2013). Molecularly imprinted cryogels for human interferon purification from human gingival fibroblast culture. *Journal of Molecular Recognition*, 26, 633–642.

Ge Y. and Turner, A.P.F. (2008). Too large to fit? Recent developments in macromolecular imprinting. *Trends in Biotechnology*, 26(4) 218–224.

Gilliam, A.C., Lang, D., LoSpalluto, J.J. (1980). Antibodies to double-stranded DNA: Purification and characterization of binding specificities. *Journal of Immunology*, 125, 874–885.

Haginaka. J. (2008). Monodispersed, molecularly imprinted polymers as affinity-based chromatography media. *Journal of Chromatography B*, 866, 3–13.

Han, X. and Zhang, Z. (2009). Preparation of grafted polytetrafluoroethylene fibers and adsorption of bilirubin. *Polymer International*, 58, 1126–1133.

Harland, Janice I. (2012). Food combinations for cholesterol lowering. *Nutrition Research Reviews*, 25, 249–266.

Henderson, T.M.A., Ladewig, K., Haylock, D.N., McLean, K.M., O'Connor, A.J. (2013). Cryogels for biomedical applications. *Journal of Materials Chemistry B*, 1, 2682–2695.

Hershko, C. (1992). Iron chelators in medicine. *Molecular Aspects of Medicine*, 13, 113–165.

Ivanov, A.E., Kozynchenko, O.P., Mikhalovska, L.I., Tennison, S.R., Jungvid, H., Gun'ko, V.M., Mikhalovsky, S.V. (2012). Activated carbons and carbon-containing poly(vinyl alcohol) cryogels: Characterization, protein adsorption and possibility of myoglobin clearance. *Physical Chemistry Chemical Physics*, 14, 16267–16278.

Jacobs, A. (1979). Annotation iron chelation theraphy for iron loaded patients. *British Journal of Haematology*, 43, 1–5.

Kane, J.P., Sata, T., Hamilton, R.L., Havel, R.J. (1975). Apoprotein composition of very low density lipoproteins of human serum. *Journal of Clinical Investigation*, 56, 1622–1632.

Kumar, A. (2008). Designing new supermacroporous cryogel materials for bioengineering applications. *European Cells and Materials*, 16, 11.

Kumar, A., Kamihira, M. Galaev, I.Y., Mattiasson, B., Iijima, S. (2001). Type-specific separation of animal cells in aqueous two-phase systems using antibody conjugates with temperature sensitive polymers. *Biotechnology and Bioengineering*, 75, 570–580.

Kumar, A., and Srivastava, A. (2010). Cell separation using cryogel-based affinity chromatography. *Nature Protocols*, 5(11) 1737–1747.

Lee, K-H., Wendon, J., Lee, M., Da Costa, M., Lim, SG., Tan, KC., (2002). Predicting the decrease of conjugated bilirubin with extracorporeal albumin dialysis MARS using the predialysis molar ratio of conjugated bilirubin to albumin. *Liver Transplantation*, 8(7), 591–593.

Litovitz, T.L., Schmitz, B.F., Matyunas, N.J., Martin, T.G. (1987). Annual report of the American Association of Poison Control Centers National Data Collection System. *American Journal of Emergency Medicine*, 6, 479–515.

Liu, B., Smyth, M.R., O'Kennedy, R. (1996). Orientated immobilisation of antibodies and its applications in immunoassays and immunosensors. *Analyst*, 121, 29–32.

Lozinsky, V.I., Galaev, I.Y., Plieva, F.M., Savina, I.N., Jungvid, H., Mattiasson, B. (2003). Polymeric cryogels as promising materials of biotechnological interest. *Trends in Biotechnology*, 21, 445–451.

Mahoney, John R., Hedlund, B.E., Eaton, J.W. (1989). Acute iron poisoning. *American Society for Clinical Investigation*, 84, 1362–1366.

Martin, R.B. (1986). The chemistry of aluminum as related to biology and medicine. *Clinical Chemistry*, 32(10), 1797–1806.

Nobutaka, T. (2000). Development of selective low-density lipoprotein (LDL) apheresis system: Immobilized polyanion as LDL-specific adsorption for LDL apheresis system. *Therapeutic Apheresis and Dialysis*, 4, 135–141.

Özcan, A.A., Say, R., Denizli, A., Ersöz, A. (2006). L-histidine imprinted synthetic receptor for biochromatography applications. *Analytical Chemistry*, 78(20), 7253–7258.

Özgür E., Bereli, N., Türkmen, D., Ünal, S., Denizli, A. (2011). PHEMA cryogel for in-vitro removal of anti-dsDNA antibodies from SLE plasma. *Materials Science and Engineering C*, 31, 915–920.

Özkara, S., Say, R., Andaç, C.A., Denizli, A. (2008). An ion-imprinted monolith for in-vitro removal of iron out of human plasma with beta thalassemia. *Industrial Engineering Chemistry Research*, 47, 7849–7856.

Peeva, E., and Diamond, B. (2004). Anti-DNA Antibodies. In R.G. Lahita, Ed. *Systemic Lupus Erythematosus*, Elsevier Inc., Oxford, pp. 283–314.

Perçin, I., Baydemir, G., Ergün, B., Denizli, A. (2013). Macroporous PHEMA-based cryogel discs for bilirubin removal. *Artificial Cells, Blood Substitutes, and Biotechnology*, 33, 172–177.

Pişkin, E., Hoffman, A.S. (1986). *Polymeric Biomaterials*, NATO ASI Series, Series E: Applied Sciences, Martinus Nijhoff Publishers, Dordrecht, p. 106.

Poullin P., Announ, N., Mugnier, B., Guis S., Roudier J., Lefevre P. (2005). Protein A-immunoadsorption (Prosorba® column) in the treatment of rheumatoid arthritis. *Joint Bone Spine*, 72, 101–103.

Rad, A.Y., Yavuz, H. Kocakulak, M., Denizli, A. (2003). Bilirubin removal from human plasma with albumin immobilized magnetic poly(hydroxyethyl methacrylate) beads. *Macromolecular Bioscience*, 3, 471–476.

Scofield, H.R. (2004). Autoantibodies as predictors of disease, *Lancet*, 363, 1544–1546.

Sellergren, B. (2001). *Molecularly Imprinted Polymers: Man-Made Mimics of Antibodies and Their Application in Analytical Chemistry.* Elsevier, Amsterdam, the Netherlands.

Şenel, S., Denizli, F., Yavuz, H., Denizli, A. (2002). Bilirubin removal from human plasma by dye affinity microporous hollow fibers. *Separation Science and Technology*, 37, 1989–2006.

Shi, W., Zhang, F., Zhang, G. (2005). Adsorption of bilirubin polylysine carrying chitosan-coated nylon affinity membranes. *Journal of Chromatogr. B.*, 819, 301–306.

Stoffel, W.H., Borberg, H., Greve, V. (1981). Application of specific extracorporeal removal of low density lipoprotein in familial hypercholesterolaemia. *Lancet* 2, 1005–1007.

Strand, V., Boers, M., Idzerda, L., Kirwan, J.R., Kvien, T.K., Tugwell, P.S., Dougados, M. (2001). It's good to feel better but it's better to feel good and even better to feel good as soon as possible for as long as possible. Response criteria the importance of change at OMERACT 10. *Journal of Rheumatology*, 38, 1720–1727.

Takeda, K., Kuwahara, A., Ohmori, K. Takeuchi, T. (2009). Molecularly imprinted tunable binding sites based on conjugated prosthetic groups and ion-paired cofactors. *Journal of American Chemical Society*, 131(25), 8833–8838.

Takenaka, Y. (1998). Bilirubin adsorbent column for plasma perfusion. *Therapeutic Apheresis*, 2, 129–133. doi: 10.1111/j.1744-9987.1998.tb00090.x.

Takeuchi, T., Hishiya, T. (2008). Molecular imprinting of proteins emerging as a tool for protein recognition. *Organic and Biomolecular Chemistry*, 6, 2459–2467.

Tatemichi, M., Sakamoto, M., Mizuhata, M., Deki, S., Takeuchi, T. (2007). Protein templated organic/inorganic hybrid materials prepared by liquid-phase deposition. *Journal of American Chemical Society*, 129(25) 10906–10910.

Thompson, G.R. (2003). LDL apheresis. *Atherosclerosis*, 167, 1–13.

Tijink, M.R., Janssen, J., Timmer, M., Austen, J., Aldenhoff, Y., Kooman J., Koole L., Damoiseaux, J., van Oerle, R., Henskens, Y., Stamatialis, D. (2013). Development of novel membranes for blood purification therapies based on copolymers of N-vinylpyrrolidone and n-butylmethacrylate. *Journal of Materials Chemistry B*, 1, 6066–6077.

Tiribelli, C. and Ostrow, J.D. (2005). The molecular basis of bilirubin encephalopathy and toxicity. *Journal of Hepatology*, 43, 156–166.

Turkova, J. (1999). Oriented immobilization of biologically active proteins as a tool for revealing protein interactions and function. *Journal of Chromatography B*, 722, 11–31.

Wang, H., Ma, J., Zhang, Y., He, B. (1997). Adsorption of bilirubin on the polymeric β-cyclodextrin supported by partially aminated polyacrylamide gel. *Reactive and Functional Polymers*, 32, 1–7.

Xia, S., Hodge, N., Laski, M., Wiesner, T.F. (2010). Middle molecule uremic toxin removal via hemodialysis augmented with an immunosorbent packed bed. *Industrial and Engineering Chemistry Research*, 49 (3), 1359–1369.

Yavuz, H., Andaç, M., Uzun, L., Say, R., Denizli, A. (2006). Iron removal from human plasma based on molecular recognition using ion-imprinted membranes. *The International Journal of Artificial Organs*, 29, 900–911.

Yavuz, H., Arıca, Y., Denizli, A. (2001). Therapeutic affinity adsorption of iron (III) with dye and ferritin. *Journal of Applied Polymer Science*, 82, 186–194.

Yavuz, H. and Denizli, A. (2005). Immunoadsorption of cholesterol on Protein A oriented beads. *Macromolecular Bioscience*, 5, 39–48.

Yavuz, H. and Denizli, A. (2003). Immunoaffinity beads for selective removal of cholesterol from human plasma. *Journal of Biomaterials Science Polymer Edition*, 14, 395–409.

Yavuz, H., Say, R., Denizli, A. (2005). Iron removal from human plasma based on molecular recognition using imprinted beads. *Materials Science and Engineering C*, 25, 521–528.

Yilmaz, E., Uzun, L., Rad, A.Y., Kalyoncu U., Ünal S., Denizli A. (2008). Specific adsorption of the autoantibodies from rheumatoid arthritis patient plasma using histidine-containing affinity beads. *Journal of Biomaterials Science Polymer Edition*, 19, 875–892.

Yilmaz, F., Bereli, N., Yavuz, H., Denizli, A. (2009). Supermacroporous hydrophobic affinity cryogels for protein chromatography. *Biochemical Engineering Journal*, 43, 272–279.

Yu, Y., He, B., Gu, H. (2000). Adsorption of bilirubin by amine-containing cross-linked chitosan resins. *Artificial Cells Blood Substitutes and Biotechnology*, 28(4), 307–320.

Zhang, Y., Chu, D., Zheng, M., Kissel, T., Agarwal, S. (2013). Biocompatible and degradable PHEMA based polymers for biomedical applications. *Polymer Chemistry*, 3, 2752–2759.

Zhang, Y., Chu, D., Zheng, M., Kissel, T., Agarwal, S. (2012). Biocompatible and degradable poly(2-hydroxyethyl methacrylate) based polymers for biomedical applications. *Polymer Chemistry*, 3, 2752–2759.

Zhu, X.X., Brown, G.R., St-Pierre, L.E. (1990). Adsorption of bilirubin with polypeptide-coated resin. *Biomaterials Artificial Cells and Artificial Organs*, 18, 75–93.

Zuwei, M., Kotaki, M., Ramakrishna, S. (2005). Electrospun cellulose nanofiber as affinity membrane. *Journal of Membrane Science*, 265, 115–123.

# 15 Cryogel Bioreactor for Therapeutic Protein Production

*Era Jain and Ashok Kumar**

## CONTENTS

15.1 Introduction .................................................................................................. 418
    15.1.1 Cryogel Bioreactors: Network and Physical Characteristics ........... 420
15.2 Cryogel Bioreactor for Protein Production: Setup and Advantages ............. 421
    15.2.1 Cryogel Bioreactor Matrix: Fabrication Method and Physical
            and Biological Characterization ...................................................... 422
    15.2.2 Urokinase Production Using Cryogel Bioreactor in an Integrated
            Setup ................................................................................................ 425
    15.2.3 Monoclonal Antibody Production in Cryogel Bioreactor ................ 430
            15.2.3.1 Monoclonal Antibody Production in PAAm-Gelatin
                    Cryogel ............................................................................. 430
            15.2.3.2 Monoclonal Antibody Production—Polyacrylamide-
                    Chitosan Cryogel, Non-Animal Derived Polymers .......... 434
15.3 Cryogel as Cell Support Matrix in an Extracorporeal Bioartificial Device.... 435
15.4 Conclusion .................................................................................................... 438
List of Abbreviations.............................................................................................. 439
References.............................................................................................................. 439

## ABSTRACT

Disposable bioreactors are becoming popular for small-scale therapeutic development and initial clinical trials due to their low cost and high efficiency. Our group has developed disposable cryogel bioreactor matrices for therapeutic protein production. A distinguishing feature of these cryogel bioreactor matrices is their low cost, a high surface-to-volume ratio and effective nutrient transport. Polyacrylamide coupled with gelatin or as a semi-interpenetrating network with chitosan (polyacrylamide–chitosan) has been used to formulate these cryogel bioreactor matrices. Both matrices are synthesized by free radical polymerization at −12°C in various formats such as monolith, beads and discs. Both adherent and nonadherent cells can be immobilized over a cryogel surface in high density for therapeutics production. The packed-bed, closed, continuous

---

* Contributing author: E-mail: ashokkum@iitk.ac.in; Tel. +91-512-2594051.

bioreactor configuration has been used to produce therapeutically important proteins like urokinase or monoclonal antibodies (mAb). The bioreactor can also be run as an integrated setup where protein purification can be done online with production using an affinity cryogel. The efficiency of these bioreactors has been found to be four times that of T-flask. Furthermore, the cryogels have also been used successfully to culture primary hepatocytes and hepatocarcinoma cell lines to be used in a bioreactor matrix in bioartificial liver devices (BAL).

## KEYWORDS

Bioprocessing, bioreactor, monoclonal antibody, polyacrylamide-chitosan cryogel, therapeutic protein production

## 15.1 INTRODUCTION

Production of recombinant proteins in host animal cells has become a method of choice for therapeutic protein production, and is considered a contemporary technique for monoclonal antibody manufacture (mAb) (Chu and Robinson, 2001; Warnock and Al-Rubeai, 2006). In order to achieve high productivity, a highly specialized, sophisticated and stringently regulated culture milieu is required, pertaining to the sensitive nature of mammalian cells and bioproducts produced by them. **Bioreactors** are culture devices that provide the necessary framework for culturing animal cells in a controlled and monitored environment. Low-density high-volume bioreactors (suspension cultures) are popularly used for industrial scale production of therapeutic proteins. On the other hand, high-density low-volume bioreactors (perfusion-based cultures) are used for small-scale and development purposes. A range of different perfusion bioreactors are available for serving a variety of needs and specialty applications (Chu and Robinson, 2001; Warnock and Al-Rubeai, 2006; Jain and Kumar, 2008).

An upcoming trend in the arena of bioreactor development is a disposable perfusion bioreactor (Eibl et al., 2010). This is because of the many advantages of culturing cells in a perfusion mode bioreactor. Usually, in a perfusion bioreactor, a high cell density of $10^7$–$10^8$ cells ml$^{-1}$ can be achieved, which is 10–30 times a stirred tank bioreactor. Moreover, cells are cultured over a matrix (most likely 3D matrix), which makes the microenvironment more similar to a cell's native environment as opposed to being suspended in media as in stirred tank bioreactors. This not only ensures a prolonged productive lifetime for cells, but also decreases the complexity of protein purification by concentrating it in a smaller volume. Additionally, proteins obtained by this process have a glycosylation pattern resembling natural proteins. Overall, this makes the process much more efficient and less labour intensive (Meuwly et al., 2007; Langer, 2011).

Packed-bed and hollow fibre bioreactors are widely used configurations in perfusion bioreactors (Meuwly et al., 2007). Immobilization matrices for packed-bed bioreactors come in different forms such as microcarrier beads, disks, foams and so on. Microcarrier beads or spheres for packed-beds composed of natural polymers such as collagen, alginate, dextran (Cytodex), agarose, gelatin and cellulose, or synthetic

# Cryogel Bioreactor for Therapeutic Protein Production

materials like plastic or glass are commercially available. SIRAN® glass spheres and Fibra-Cel® polyester-polypropylene disks have been successful as porous carriers for packed-bed bioreactors (Meuwly et al., 2007).

A hollow fibre bioreactor (HFBR) consists of a bundle of micron-sized fibres having selective permeability (Figure 15.1). The cells are situated in the extracapillary space while the media is circulated through the fibres. It is popularly used as a disposable perfusion bioreactor for the production of monoclonal antibodies in quantities ranging from 100 mg/ml to 1 g/ml. HFBR provides high surface-to-volume ratio permitting high cell density of up to $10^9$ cells ml$^{-1}$. This has several advantages like easy adaptation to serum-free media and concentration of product in a small volume, thus making product purification easier (Cadwell, 2005).

The benefits of a disposable bioreactor include easy validation and scale up, fewer chances of cross-contamination and cost efficiency. This makes them effective tools for initial start-ups, inoculum preparation, viral vaccine and diagnostics production. Due to the growing popularity and advantages of perfusion-based disposable bioreactors, quite a few prevailing disposable bioreactors have been adapted to run in perfusion mode and many have designs similar to HFBR. Other systems have a filtration unit to hold cells inside the bioreactor (Eibl et al., 2010; Poles-Lahille et al., 2011).

However, use of perfusion culture is restricted mostly because of complex operation and scale-up difficulties. Perfusion bioreactors are usually cost-intensive and have nonhomogenous culture conditions (Meuwly et al., 2007). Furthermore, HFBR often have cartridge clogging in addition to other limitations of a perfusion bioreactor (Cadwell, 2005). Thus, a simple and easy to adapt disposable perfusion bioreactor/matrix is required. Cryogel-based matrices can serve the purpose of making a disposable perfusion bioreactor by combining the advantages of a three-dimensional porous hydrogel matrix and a packed-bed perfusion bioreactor. A cryogel bioreactor

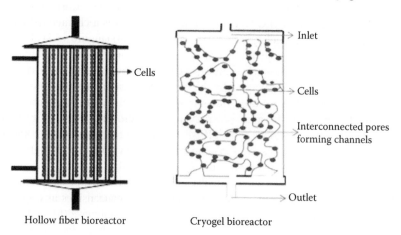

**FIGURE 15.1** Diagrammatic representation of cryogel bioreactor matrix in comparison to hollow fibre bioreactor. The interconnected pores in cryogel behave like fine capillaries similar to an HFBR. (Reprinted from *Biotechnology Advances* 26(1), Jain, E. and Kumar, A., Upstream processes in antibody production: Evaluation of critical parameters. 46–72. Copyright 2008, with permission from Elsevier.)

matrix is a versatile, cost-efficient, disposable, high-density culture system useful for adherent and nonadherent cells (Kumar et al., 2006; Nilsang et al., 2007, 2008; Jain and Kumar, 2013).

### 15.1.1 Cryogel Bioreactors: Network and Physical Characteristics

Cryogel bioreactor matrices possess some optimum features of an ideal bioreactor matrix for therapeutics production using mammalian cells (Jain and Kumar, 2008). As has been described earlier in previous chapters, cryogels are polymeric matrices synthesized at a temperature whereby most of the solvent used for their synthesis freezes to form ice crystals. Defrosting the frozen mixture after the required incubation time results in an interconnected porous structure. This imparts to cryogel some of the properties desired in a bioreactor matrix such as enough porosity and pore size to allow for cell immobilization and congestion-free media flow (Lozinsky et al., 2001; Kumar et al., 2005). This implicates ease of nutrient availability to the immobilized cells. Uniform and efficient nutrient distribution is one of the essential requirements for constructing a bioreactor matrix. In the case of cryogels, nutrient transport is facilitated both by fast convection and free diffusion of nutrients within the matrix as opposed to only diffusion transport in corresponding hydrogel-based matrices (Lozinsky et al., 2001; Kumar et al., 2005). Rapid swelling time (usually below 2 min) is one of the hallmark features of cryogels (Srivastava et al., 2007). This not only exemplifies an interconnected pore network within cryogels but also reduces pre-equilibration time with media and facilitates homogenous cell seeding/distribution (Plieva et al., 2005). Using a cryogel bioreactor matrix, cells can be directly seeded over a dried matrix without any pre-equilibration with media and a nearly 100% seeding efficiency can be achieved.

The interconnected porous network of a cryogel can be viewed as forming fine capillaries similar to an HFBR. Diameter of the capillaries is a characteristic of the pore dimension of the cryogel system (Persson et al., 2004) (Figure 15.1). The thick pore wall framework in cryogel provides a high surface area for cell attachment and growth. An estimation for the surface area of a cryogel bioreactor matrix lies between 4.3055 $m^2g^{-1}$ and 9 $m^2g^{-1}$, as dictated by the type of polymer and concentration used for cryogel synthesis. This is seven times more than the surface area provided by a commercially available 0.2-mm polyacrylamide microcarrier (0.6 $m^2g^{-1}$) for a packed-bed bioreactor and close to an HFBR (10 $m^2g^{-1}$) (Persson et al., 2004; Jain and Kumar, 2009). This is to say, cryogel matrices are a very compact high-density system with a high surface-to-volume ratio. This is particularly advantageous for culture of suspension-based cells such as hybridoma. The high surface-to-volume ratio facilitates the culture of cells in a small volume at high cell densities as opposed to a high volume system at low cell densities, which is commonly used for cells grown in suspension. In a laboratory setup, T-flasks are stacked to obtain increased surface area, thus occupying a large space. Comparatively, a small 10-ml column of cryogel matrix can be used in a much smaller space with a high density of cells making them potential alternatives (Kumar et al., 2006; Nilsang et al., 2007; Jain et al., 2011).

The simple process of making cryogel bioreactor matrices allows for the flexibility of choosing various different hydrophilic/natural/synthetic polymers to construct the bioreactor matrix. Numerous polymers and polymer combinations have been used to make up the cryogel bioreactor matrices. These cryogel bioreactor matrices have been used successfully to culture microbial as well as mammalian cells. Cryogel beads made from polyvinyl alcohol, polyacrylamide and dextran have ascertained to be extraordinarily worthy immobilization matrices for microbial cells including yeast. They make up a highly stable hydrophilic structure, which is mechanically robust even in the presence of or reinforced in the presence of carriers like bacterial, yeast and fungal spores. Furthermore, the gel strength remains unaffected by the extended presence of common low molecular weight solutes (salts and sugars) composing the nutrient broth of the immobilized carriers (Williams et al., 2005; Babac et al., 2006; Deraz et al., 2007). In addition to this, a cryogel constituted by a variety of polymers has efficiently been used as 3D scaffolds to culture for different mammalian cells like cartilage (Bhat et al., 2013), pancreatic cells (Bloch et al., 2005), bone (Mishra et al., 2014) and other tissues.

## 15.2 CRYOGEL BIOREACTOR FOR PROTEIN PRODUCTION: SETUP AND ADVANTAGES

Owing to their ease of production, flexibility of using various polymer combinations for fabrication and favourable cell supporting properties, cryogels have been applied successfully as bioreactor matrices for protein production. Moreover, cryogel matrices are cost-effective to be used as single use or disposable bioreactors.

The setup to run the bioreactor in a closed continuous mode can be done with little effort in a laboratory environment. Setting up a cryogel bioreactor in a packed-bed closed/continuous format involves fabricating the reactor in an appropriate mould/format. This can be either a monolith of varying size, beads or discs. It has been observed that the size and shape of the reactor matrices influence the productivity and cell behaviour in bioreactors (Jain et al., 2011). Thus, the flexibility of moulding the reactor matrix into a desired shape is an important feature. The reactor matrix can then be sterilized either by steam or by using alcohol washes and dried and stored for an extended period of time until use. For running the bioreactor, the matrix can be placed in a tightly sealed container having an inlet and an outlet. The cells can be seeded either manually or using a peristaltic pump on a pre-equilibrated or a dried matrix. Generally, a dried matrix captures cells effectively and a cell seeding efficiency of 90% to 100% can be achieved. Depending on the cell line, an incubation time of 6 to 24 h is required for proper cell attachment. Postincubation, the media reservoir can be connected in closed circuit to allow for circulation of nutrients within the cryogel bioreactor. The whole setup can be made to fit and run inside a $CO_2$ incubator. To monitor the system, samples of spent media can be collected from the outlet point and analysed for cell metabolism parameters and protein concentration of the protein of interest. Examples of protein production using cryogel matrices in a continuous bioreactor setup include monoclonal antibodies (mAb) and urokinase (Kumar et al., 2006a,b; Nilsang et al., 2008; Jain et al., 2011).

The continuous bioreactor setup for cryogel matrices is also capable of being used in an integrated setup combining production with simultaneous capture of the primary protein (Kumar et al., 2006b). In this setup, a captured cryogel column can be introduced inline with the cryogel bioreactor matrix. The capture column can facilitate the removal of the protein of interest from the reservoir media. This allows for concentrating the media broth and reducing the step in downstream processing. In particular, this is advantageous in increasing productivity of the proteins whereby the mammalian cells secrete proteins that inhibit their own production beyond a threshold concentration or other protein that inhibits the production of the protein of interest. This has been accomplished in the production of urokinase using polyacrylamide cryogel coupled to gelatin as a bioreactor matrix. The metal affinity cryogel column has been modified to isolate other protein types or cells. Examples of such ligands include concavalin A (for monoclonal antibody separation) (Babac et al., 2006), phenyl/octyl hydrobhobic ligands (for bacteriocin) (Deraz et al., 2007), streptavidin affinity interaction (for Moloney Murine leukaemia virus) (Williams et al., 2005), boronate (separation adherent and nonadherent cells) (Srivastava et al., 2012), protein A (Kumar and Srivastava, 2010) and so on. Most of these ligands have been immobilized on a polyacrylamide-based cryogel. A detailed protocol for designing a protein A column has been published earlier (Kumar and Srivastava, 2010). These affinity columns hold potential to be used in an integrated setup. Furthermore, a cryogel composed of different polymeric materials has also been used to culture various tissue cells. Thus, a variety of combinations of cell type and affinity ligands is possible while using cryogel as the support matrix.

### 15.2.1 Cryogel Bioreactor Matrix: Fabrication Method and Physical and Biological Characterization

Two types of polymer combinations have been used and characterized extensively to formulate cryogel bioreactor matrices for mammalian cell culture and protein production (Kumar et al., 2006; Jain and Kumar, 2009). The first is a gelatin containing cryogel bioreactor matrix and the second is a synthetic/nonanimal derived polymer containing a cryogel matrix. A gelatin containing cryogel matrix has been fabricated by covalent attachment of gelatin over a polyacrylamide (AAm) cryogel using N,N′methylene bis acrylamide (MBAAm) as a cross-linker. The presence of a gelatin-like polymer presents a cell binding motif, which facilitates cell attachment to the matrix (Nilsang et al., 2007, 2008).

The second cryogel matrix is a semi-interpenetrating network of polyacrylamide and chitosan (PAAC). The cryogel was fabricated by free radical polymerization of acrylamide and MBAAm in the presence of chitosan, which is physically entrapped in the network by noncovalent interactions. This results in a highly hydrophilic matrix with a positive surface charge due to chitosan, thus attracting a negatively charged cell surface (Jain and Kumar, 2009, 2013). The two types of polymer combinations selected represent two distinct types of strategies employed to assist cell adherence to the surface, thus demonstrating the diversity of the system. The presence of gelatin in a bioreactor matrix raises ethical concerns. Use of PAAC cryogel matrix provides an alternative to matrices containing animal-derived proteins like gelatin.

The process for synthesizing both matrices is simple and involves free radical polymerization at $-12°C$. A well-synthesized matrix of PAAm-gelatin and PAAC cryogel is characterized by an interconnected pore network with a pore size range of 10–100 µm and an average pore diameter of 70 µm (Figure 15.2). The amount of gelatin per millilitre of PAAm-gelatin cryogel is estimated to be 2.06 mg by this method. PAAC cryogels have a flow rate of 5 ml/min while PAAm gelatin has a flow rate of 8 ml/min or greater. The cryogels also consist of pores in the size range of 30 to 100 nm, which contributes 77.77% of the total surface area. The micropores present in this range inside the pore walls help in the diffusion of very low molecular weight solutes and molecules like oxygen through the gel walls. PAAC cryogels exhibit free diffusion and rapid convective transport as has been found by the water equivalent rate of diffusion of bovine serum albumin in cryogels and convective transport efficiency of $4 \times 10^{-12}$ $m^4N^{-1}s^{-1}$. The cryogels are thermally and mechanically stable and can be sterilized by autoclaving for the purpose of a bioreactor.

*In vitro* cell culture studies in a static culture have shown PAAC and other similar matrices to be suitable scaffolds for growing a number of hybridoma cell lines. The antibody production in a static 3D culture over cryogel is more efficient than the 2D

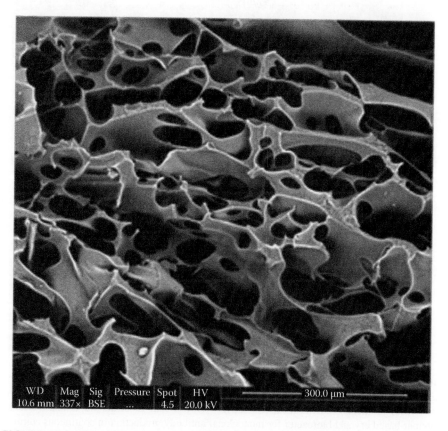

**FIGURE 15.2** Scanning electron microscope image of PAAC cryogel demonstrating the open and interconnected macroporous network.

culture in cell culture well plates. Culturing cells for a long period of time shows extensive proliferation and formation of a monolayer of cells (Figure 15.3).

Though these two cryogel matrices have been used for bioreactor application, there is a huge flexibility and potential to use other polymeric combinations for making cryogel matrices and use them as a bioreactor. We have also tested and characterized several other polymer combinations for their usage as bioreactor matrices for hybridoma culture and antibody production. These include poly(acrylonitrile) (PAN), poly(N-isoproppylacrylamide) (PNiPAAm) and poly(N-isoproppylacrylamide)–chitosan (PNiPAAm–chitosan). The polymer combinations were selected based on their hydrophobicity and hydrophilicity balance and ability to promote cell immobilization. The physical properties of these are summarized in Table 15.1. All the matrices have shown favourable growth of hybridoma cell lines and antibody production in the static culture (Jain et al., 2011) (Figure 15.4).

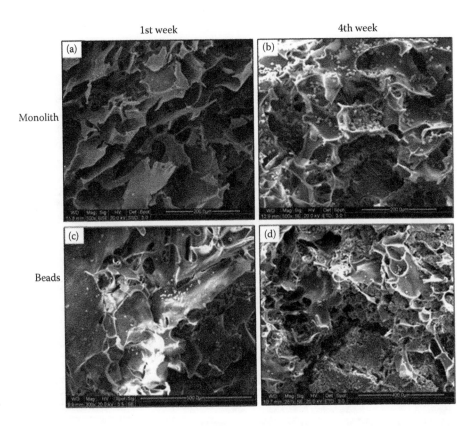

**FIGURE 15.3** Growth of 6A4D7 hybridoma cell line on PAAC cryogel matrices under static culture over a period of 4 weeks: (a) monolith first week, (b) monolith fourth week, (c) beads first week, (d) beads fourth week, respectively. (From Jain E. et al.: Supermacroporous polymer-based cryogel bioreactor for monoclonal antibody production in continuous culture using hybridoma cells. *Biotechnology Progress*. 2011. 27(1). 170–180. Copyright Wiley-VCH Verlag GmbH & Co. KGaA. Reproduced with permission.)

## TABLE 15.1
### Summary of Biological and Physical Properties of Various Cryogel Bioreactor Matrices

| Property | PAN | PAAC | PNiPAAm-Chitosan | PNiPAAm |
|---|---|---|---|---|
| Porosity | 92% | 95% | 87.6% | 95% |
| Resistance to flow | 2 ml/min | 5 ml/min | 10 ml/min | 4.7 ml/min |
| Pore size | 20–50 µm | 70–100 µm | Up to 70 µm | 30–100 µm |
| Surface property | Moderately hydrophobic 50°, uncharged | Hydrophilic, positively charged | Hydrophobic, positively charged | Hydrophobic 71° |
| Swelling ratio | 0.90 | 0.95 | 0.91 | 0.95 |
| FBS protein absorption | 1.64 mg/ml | 1.43 mg/ml | 1.07 mg/ml | – |

*Source:* Jain E. et al.: Supermacroporous polymer-based cryogel bioreactor for monoclonal antibody production in continuous culture using hybridoma cells. *Biotechnology Progress.* 2011. 27(1). 170–180. Copyright Wiley-VCH Verlag GmbH & Co. KGaA. Reproduced with permission.

### 15.2.2 Urokinase Production Using Cryogel Bioreactor in an Integrated Setup

Urokinase [EC 3.4.99.26] is a naturally occurring plasminogen activator and a protein of great therapeutic potential. It is a single polypeptide of about 55 kDa. It was first found in urine although it is also present in blood and other body fluids. Urokinase converts plasminogen to plasmin by cleaving the Arg-Val linkage in the Pro-Gly-Arg-Val sequence of plasminogen. The resulting plasmin dissolves the clots of fibrin. Due to its plasminogen activator activity, it is used for treating thromboembolic diseases. Urokinase is actively secreted by several mammalian cell lines as part of their physiological processes. Production of urokinase by immobilization of such mammalian cell lines in a solid microcarrier-based perfusion bioreactor has resulted in the formation of large cellular aggregates and low process productivity (Roychoudhury et al., 2006). Producing urokinase using tissue culture presents some typical challenges. An important factor regulating urokinase secretion by immobilized cells is the urokinase concentration itself. When the concentration of urokinase reaches a threshold value, it inhibits its own production. Urokinase activity can also be inhibited by complex formation between urokinase and plasminogen activator inhibitor (PAI), which may be present in the broth or by autocatalytic degradation of urokinase (Stump et al., 1986). The separation process of urokinase is furthermore complicated by the strong aggregating character of urokinase and the presence of urokinase-related proteins like tissue plasminogen activator (tPA), which competes with urokinase during the separation stage (Lewis, 1979). Many of these issues can be circumvented if the enzyme is continuously removed from the circulating media. This helps maintain the production at a constant rate for an extended period of time.

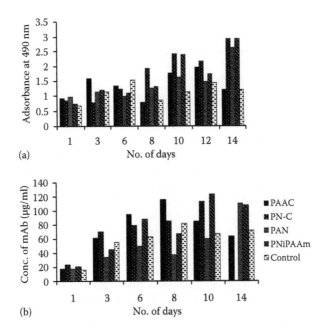

**FIGURE 15.4** Growth of hybridoma cell line 6A4D7 on different cryogel matrices for 15 days under static culture. (a) Proliferation of 6A4D7 on different cryogel scaffolds as measured by MTT assay. (b) Production of monoclonal antibody by 6A4D7 on different cryogel scaffolds as measured by ELISA. (PAAC: polyacrylamide–chitosan cryogel; PN-C: Poly(*N*-isopropyl-acrylamide)–chitosan cryogel; PAN: Polyacrylonitrile cryogel; PNiPAAm: Poly(*N*-isopropylacrylamide) cryogel; Control: 2D well plate culture). (From Jain E. et al.: Supermacroporous polymer-based cryogel bioreactor for monoclonal antibody production in continuous culture using hybridoma cells. *Biotechnology Progress*. 2011. 27(1). 170–180. Copyright Wiley-VCH Verlag GmbH & Co. KGaA. Reproduced with permission.)

An integrated setup for urokinase production and purification has been developed by our group, using a polyacrylamide cryogel coupled with gelatin as bioreactor matrices and a copper-coupled cryogel metal affinity matrix for separation (Figure 15.5). This integrated setup was useful in resolving several issues related to urokinase production.

Human fibrosarcoma HT1080 and human colon cancer HCT116 were cultured over the cryogel matrix, which attached to the matrix within 4–6 h postseeding and grew as a tissue sheet inside the cryogel matrix (Figure 15.6a,b). Continuous urokinase secretion into the circulating medium was monitored as a parameter of growth and viability of cells inside the bioreactor. No phenotypical changes were seen in the cells eluted from the gelatin-cryogel support and recultured in tissue culture flasks (Bansal et al., 2006; Kumar et al., 2006a,b).

The gelatin-cryogel bioreactor was further connected to a pAAm cryogel column carrying Cu(II)-iminodiacetic acid (Cu(II)-IDA)-ligands, which had been optimized for the capture of urokinase from the conditioned medium of the cell lines. Such affinity cryogel columns provide the possibility to process unclarified crude feeds

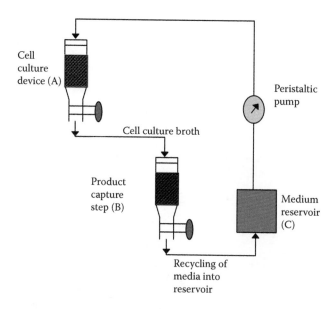

**FIGURE 15.5** Diagrammatic representation of integrated setup of cryogel bioreactor inline with a cryogel copper metal affinity column for simultaneous urokinase production and capture. (Kumar A. et al.: Integrated bioprocess for the production and isolation of urokinase from animal cell culture using supermacroporous cryogel matrices. *Biotechnology and Bioengineering.* 2006. 93(4). 636–646. Copyright Wiley-VCH Verlag GmbH & Co. KGaA. Reproduced with permission.)

in a chromatographic mode. Thus, an automated system was built, which combined the characteristics of a hollow fibre reactor with a chromatographic protein capture system. The urokinase was constantly isolated by the Cu(II)-IDA-cryogel column and periodically recovered through elution cycles (Kumar et al., 2006a). The urokinase activity increased from 250 PU/mg in the culture media to 2310 PU/mg postretrieval from the metal affinity column, which gave about ninefold purification of the enzyme. Increased output was attained by functioning an integrated bioreactor system nonstop for 32 days under product inhibition free conditions during which no backpressure or culture contamination was observed (Figure 15.6c,d). A total 152,600 Plough units of urokinase activity was recovered from 500 ml culture medium using 38 capture columns over a period of 32 days (Kumar et al., 2006a).

A comparative study for the practical applicability of Cu(II)-IDA cryogel and that of Cu(II)-IDA Sepharose columns for the direct capture of urokinase from culture broth of HT1080 cell line in an integrated setup as well as in offline capture revealed that the cryogel capture column showed better operational efficiency and selectivity compared to that of the Sepharose column (Table 15.2). In the integrated setup, the Sepharose-based column was clogged in time by the cells and cell debris released from the cell culture device, whereas in the cryogel matrix these cells passed easily through the capture column and presented no problem (Kumar et al., 2006b). During offline capture, the cryogel affinity column showed higher selectivity for urokinase

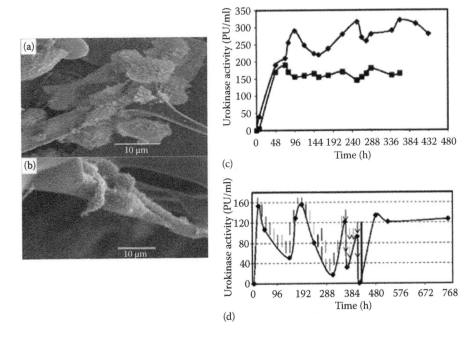

**FIGURE 15.6** Production of urokinase in PAAm-gelatin cryogel bioreactor using HT1080 and HCT116 cells. (a, b) Scanning electron images depicting the HT1080 cells growing as a confluent layer inside the cryogel matrix. (c) Secretion of urokinase by HT1080 human kidney cell line (filled diamonds) and HCT116 human colon cancer cell line (filled square) cultivated on gelatin-pAAm cryogel matrix (bed volume 5 mL) for 18 and 15 days, respectively. (d) Production and capture of urokinase secreted by human kidney cell line HT1080. The cells were grown on a gelatin–pAAm cryogel matrix for 32 days. The medium was circulated through the matrix with cultivated cells at a flow rate of 0.2 mL min$^{-1}$ and samples were withdrawn at regular intervals for determining the urokinase activity and any cells detached from the matrix scaffold. In the integrated setup, the urokinase produced was captured on Cu(II)–pAAm cryogel matrix (bed volume 5 mL). Short vertical lines indicate the replacement of the capture column with a fresh one. Arrows indicate connecting two capture columns (each with bed volume of 5 mL) in series. On day 18, the medium reservoir was changed with new medium (indicated by thick vertical line). (From Kumar A. et al.: Integrated bioprocess for the production and isolation of urokinase from animal cell culture using supermacroporous cryogel matrices. *Biotechnology and Bioengineering.* 2006. 93(4). 636–646. Copyright Wiley-VCH Verlag GmbH & Co. KGaA. Reproduced with permission.)

in comparison to the Sepharose columns. Although the yield was higher for the Sepharose affinity cryogel columns, enzyme purified via cryogel affinity column had a high increase in the fold purification of 26.6 as compared to 1.0 as obtained via Sepharose affinity column (Table 15.2).

Thus, cryogels provided a scaffold for the growth and proliferation of anchorage dependent cell lines and a suitable support matrix for preparing an affinity chromatographic column that could efficiently capture urokinase without developing backpressure. The bioreactor could be run for periods as long as 32 days without

## TABLE 15.2
## Comparison of Functional Efficiency of Sepharose and Cryogel Immobilized Metal Affinity Chromatography Columns for Separation of Urokinase Produced by HT1080

| Steps | Volume (ml) | Total Protein (mg) | Total Urokinase Activity (CPU) | Specific Activity (PU/mg Protein) | Fold Purification | Yield (%) |
|---|---|---|---|---|---|---|
| | | Cu(II)-IDA-Sepharose Chromatography | | | | |
| Cell culture broth | 10 | 43.9 | 1341.5 | 30.6 | 1.0 | 100 |
| Breakthrough + wash | 60 | 34.2 | 308.5 | 9.0 | 0.3 | 23 |
| Eluate | 8 | 7.9 | 1273.2 | 161.2 | 5.3 | 95 |
| | | Cu(II)-IDA-Polyacrylamide Cryogel Chromatography | | | | |
| Cell culture broth | 10 | 43.9 | 1341.5 | 30.6 | 1.0 | 100 |
| Breakthrough + wash | 34 | 38.6 | 440.0 | 11.4 | 0.4 | 33 |
| Eluate | 6 | 1.3 | 1058.5 | 814.2 | 26.6 | 79 |

*Source:* Reprinted from *Journal of Chromatography A* 1103(1), Kumar A. et al., 35–42. Copyright 2006, with permission from Elsevier.

*Note:* Cell culture broth (10 ml) was applied to the Cu(II)-IDA-Sepharose column (3-ml bed volume) and Cu(II)-polyacrylamide cryogel monolithic column (5-ml bed volume) at a flow rate of 1 ml/min. After washing with 20 mM HEPES, pH 7.0 containing 0.2 M NaCl the bound proteins were eluted with 0.2 M imidazole buffer, pH 7.4. Sepharose column was additionally washed with 2 mM imidazole containing above washing buffer before start of elution with 0.2 M imodazole buffer, pH 7.4.

contamination or other operational problems. This development is particularly significant with regard to the production of urokinase, an enzyme of immense therapeutic significance and, hence, is a pursuit central to the modern biotechnology and pharmaceutical industry. The system developed for urokinase is unique in the way that it integrates the production of urokinase with the simultaneous removal of the product online. This is a significant process upgradation in regard to using mammalian cells as the host for therapeutic protein production where feedback inhibition interferes with continuous production. This necessitates continuous online monitoring and removal of the inhibitory components (which can be the protein of interest) continually. Furthermore, removal of cell debris and other contaminants during the process can make the protein production more efficient. Thus, a system combining both production and preliminary protein purification in a single step can overcome such inhibitory mechanisms. This will also reduce the number of steps involved in subsequent protein purification and improve the operational economy (integrating the primary capture of the product from the cell suspension with the initial purification). Such integrated downstream processing implies also the application of the separation technologies capable of processing particulate-containing solutions.

### 15.2.3 Monoclonal Antibody Production in Cryogel Bioreactor

Cryogel bioreactors also have been applied for monoclonal antibody production using nonadherent hybridoma cells as host systems using PAAm-gelatin and PAAC cryogel matrix (Nilsang et al., 2007, 2008; Jain et al., 2011). The cryogel matrices were formatted as monolith, beads and discs for the purpose of study and to evaluate the effect of format on cell functioning (Figure 15.7a–c). Monoclonal antibody production using immobilized hybridoma cells in cryogel bioreactors shows similar patterns of cell metabolism irrespective of the shape, constitution of the bioreactor and the type of cells used for mAb production. The hybridoma cells usually take 36–48 h to completely attach to the matrix. The hybridoma cells grew in two stages: Stage 1: The lag phase characterized by slow glucose consumption and low mAb productivity; Stage 2: The log phase marked by a sudden change in glucose consumption and steady concentration thereafter accompanied by a high mAb productivity. The duration of the lag phase varies with the format of the bioreactor. The reactors are capable of running continuously for a period of 50 to 60 days. This requires a change of media reservoir as the medium is exhausted after a certain time depending on cell metabolic rate. A decrease in mAb productivity by 40–50% in all the bioreactors is seen after 28 to 35 days of run, which coincides with the first reservoir media exhaustion. The mAb productivity in each of the different cryogel bioreactors ranged between 110 mg/ml and 180 mg/ml, which is about 3–4 times the conventional T-flask and is dependent on the cell line and culture conditions used (Jain and Kumar, 2013). A cryogel bioreactor setup for monolith and beads format is shown in Figure 15.7d and e.

#### 15.2.3.1 Monoclonal Antibody Production in PAAm-Gelatin Cryogel

Cryogel matrix utility as a bioreactor has also been tested for production of an antibody against type II collagen using hybridoma cell line M2139. The matrix was

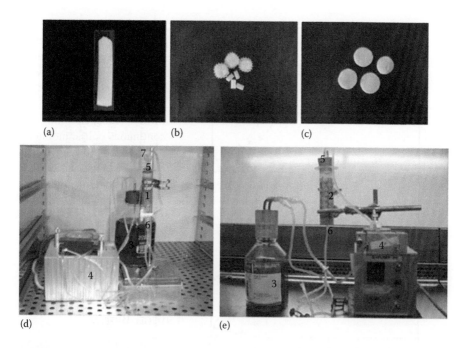

**FIGURE 15.7** Different formats and setup of cryogel bioreactor. Dried polyacrylamide-chitosan cryogels in three different formats. (a) Monolith column 5 ml, (b) beads, (c) discs, (d) monolith of polyacrylamide-chitosan bioreactor, (e) beads of polyacrylamide-gelatin bioreactor for monoclonal antibody production. (From Jain E. et al.: Supermacroporous polymer-based cryogel bioreactor for monoclonal antibody production in continuous culture using hybridoma cells. *Biotechnology Progress*. 2011. 27(1). 170–180. Copyright Wiley-VCH Verlag GmbH & Co. KGaA. Reproduced with permission; Reprinted by permission from Macmillan Publishers Ltd. *Nat Protoc*, Jain, E. and Kumar, A., 8(5): 821–835, copyright 2011.)

made in two different formats: monolith and beads. Hybridoma clone M2139 was immobilized over the porous bed matrix of polyacrylamide-gelatin cryogel (10-ml bed volume). The presence of gelatin in a cryogel bioreactor facilitated cell attachment within 48 h and cells grew as a confluent sheet inside the matrix (Figure 15.8a). A lag period of 15 days was observed in the monolith bioreactor during which cells secreted mAb into the circulating medium. During the entire run, the cell metabolism was monitored by glucose consumption and lactic acid metabolism, which gave an indication of cell viability and nutrient consumption rate. The immobilized hybridoma cells produced 6.5 µg mL$^{-1}$ day$^{-1}$ mAb during the exponential phase amounting to a mAb concentration of 130 µg mL$^{-1}$ post-nonstop run of the cryogel bioreactor for 36 days. The yield of the mAb after purification was 67.5 mg L$^{-1}$. The cells were active after change of medium reservoir and functioned continuously for 55 days (Nilsang et al., 2007) (Figure 15.8b).

When the same hybridoma cell line was cultured in a bead cryogel bioreactor, the lag phase lasted 5 days as indicated by the change in glucose consumption

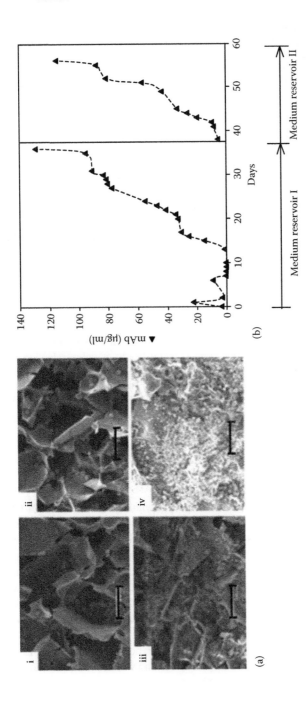

**FIGURE 15.8** Monoclonal antibody production using polyacrylamide-gelatin monolith cryogel bioreactor. (a) Hybridoma cells M2139 growing on polyacrylamide-gelatin bioreactor matrix i–iv: Day 1, Day 7, Day 36 and Day 55. (b) Production rate of monoclonal antibody against collagen II during the bioreactor run in monolith polyacrylamide-gelatin bioreactor. Scale bar is 50 μm. (From Nilsang S. et al.: Monoclonal antibody production using a new supermacroporous cryogel bioreactor. *Biotechnology Progress.* 2007. 23(4). 932–939. Copyright Wiley-VCH Verlag GmbH & Co. KGaA. Reproduced with permission.)

and increase in lactic acid and ammonia production (Figure 15.9a) (Nilsang et al., 2008). The mAb concentration in bead cryogel bioreactor increased gradually and reached 170 µg mL$^{-1}$ within 20 days with a rate of 8.11 µg mL$^{-1}$ day$^{-1}$ (Figure 15.9b). Monoclonal antibodies secreted from the bioreactor were found to be functionally similar to antibodies purified from the cells grown in commercial CL-1000 culture

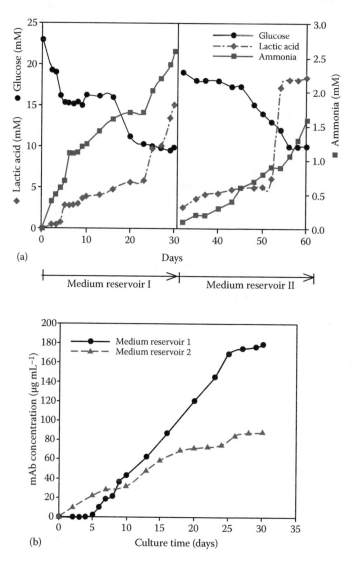

**FIGURE 15.9** Monoclonal antibody production using polyacrylamide–gelatin beads bioreactor with M2139 hybridoma cells immobilized inside the reactor: (a) Monitoring cell growth and viability by changes in glucose consumption and production of lactic acid and ammonia. Glucose concentration dropped steeply as the cells entered the log phase. (b) Rate of monoclonal antibody production during the reactor run. The medium reservoir was changed on day 30 owing to depletion of essential nutrients like glucose as seen in (a). *(Continued)*

(c)

**FIGURE 15.9 (CONTINUED)** (c) Arthritis development in B10.Q mice after antibody (produced in cryogel bioreactor) transfer and LPS injection (left, normal paw; right, arthritis paw). (From Nilsang S. et al.: Monoclonal antibody production using a new supermacroporous cryogel bioreactor. *Biotechnology Progress*. 2007. 23(4). 932–939. Copyright Wiley-VCH Verlag GmbH & Co. KGaA. Reproduced with permission; Nilsang S. et al.: Three-dimensional culture for monoclonal antibody production by hybridoma cells immobilized in macroporous gel particles. *Biotechnology Progress*. 2008. 24(5). 1122–1131. Copyright Wiley-VCH Verlag GmbH & Co. KGaA. Reproduced with permission; Jain E. et al.: Supermacroporous polymer-based cryogel bioreactor for monoclonal antibody production in continuous culture using hybridoma cells. *Biotechnology Progress*. 2011. 27(1). 170–180. Copyright Wiley-VCH Verlag GmbH & Co. KGaA. Reproduced with permission.)

flasks. Cells were found to be viable inside the porous matrix of the cryogel during the study period and secreted antibodies continuously. The mAb obtained after the cryogel reactor were biologically active *in vivo*, as shown by their ability to induce arthritis in mice (Figure 15.9c). The yield of the mAb postpurification was 67.5 mg $L^{-1}$, which was threefold higher than mAb yield obtained from T-flask batch cultivation.

### 15.2.3.2 Monoclonal Antibody Production—Polyacrylamide-Chitosan Cryogel, Non-Animal Derived Polymers

PAAC cryogels were used for mAb production using hybridoma cell line 6A4D7. PAAC cryogels, although hydrophilic and devoid of any animal-based polymer like gelatin, could be used successfully for hybridoma cell immobilization. Similar to PAAm-gelatin cryogel bioreactor the hybridoma cells took about 48 h to completely attach to PAAC cryogel bioreactor matrix. Three different cryogel bioreactor setups were made using monolith, beads and discs format. The hydridoma cells displayed a brief lag period of 6 days in the monolith and discs formats of the bioreactor matrix, while 2–3 days in the beads bioreactor. The cell activity inside the bioreactors was continuously monitored indirectly by measuring metabolic changes such as uptake of glucose and glutamine and production of lactic acid and ammonia. Comparing the glucose metabolism rate across the three reactor formats gave an indication of cell growth, viability and antibody production efficiency as hybridoma uses glucose/glutamine as energy-deriving substrates. Monolith and beads bioreactors had comparable metabolic activity of the cells with the beads bioreactor having the lowest ammonia production rate (Table 15.3). The discs bioreactor had the slowest glucose consumption and lactic acid production rate of 0.27 mM $day^{-1}$ and 0.642 mM $day^{-1}$,

### TABLE 15.3
### Summary of Reactor Running Conditions, Metabolic Changes and mAb Productivity for Different Cryogel Bioreactors

| Cryogel Reactor | PAAm-Gel (M) | PAAm-Gel (B) | PAAC (M) | PAAC (B) | PAAC (D) |
|---|---|---|---|---|---|
| Cell line | M2139 | M2139 | 6A4D7 | 6A4D7 | 6A4D7 |
| Incubation period (h) | 48 | 36 | 48 | 48 | 48 |
| Cells retained in bioreactor | $5 \times 10^6$ | — | $1.01 \times 10^7$ | $1.18 \times 10^7$ | $1.0 \times 10^7$ |
| Reactor size (ml) | 10 | 50 | 10 | 20 | 10 |
| Lag/log phase | 15/20 d | 5/20 d | 6/20 d | 4/20 d | 6/13 |
| Experiment days[a] | 36/55 | 30/60 | 30/60 | 30/60 | 30/60 |
| Glucose consumption (mM day$^{-1}$) | 0.75 | 0.55 | 0.56 | 0.53 | 0.27 |
| Lactic acid production (mM day$^{-1}$) | 2.48 | 0.18 | 1.1 | 1.2 | 0.6422 |
| Glutamine consumption (mM day$^{-1}$) | NA | NA | 0.08 | 0.075 | 0.056 |
| Ammonia production (mM day$^{-1}$) | NA | 0.042 | 0.026 | 0.021 | 0.054 |
| Maximal mAb productivity (µg ml$^{-1}$) | 130 | 180 | 115 | 120 | 62.5 |
| Total mAb produced (mg) | 67.5 | 90 | 57.5 | 60 | 31.25 |
| Overall mAb productivity (µg ml-1 day$^{-1}$) | 6.5 | 6 | 6 | 4.25 | 2.08 |

*Source:* Jain E. et al.: Supermacroporous polymer-based cryogel bioreactor for monoclonal antibody production in continuous culture using hybridoma cells. *Biotechnology Progress.* 2011. 27(1). 170–180. Copyright Wiley-VCH Verlag GmbH & Co. KGaA. Reproduced with permission; Reprinted by permission from Macmillan Publishers Ltd. *Nat Protoc,* Jain, E. and Kumar, A., 8(5): 821–835, copyright 2011.

[a] Experiment days are expressed as the number of days the bioreactor was run in the first and second change of media.

respectively. Interestingly, the ammonia production rate was highest for the discs bioreactor amongst the three, although it had a similar glutamine consumption rate as that of the other two bioreactors. Additionally, the disc bioreactor reached the stationary phase within 13 days of run beyond which glucose consumption was negligible, while lactate production continued. This indicates a low viability and oxygen-poor conditions inside the bioreactor whereby glutamine consumption is favoured leading to higher production of lactic acid and ammonia (Ozturk and Palsson, 1991).

After a continuous run of 30 days, the glucose concentration in the beads and monolith bioreactors dropped to 6 mM from 25 mM. Simultaneously, during this period a total of 32 mM of lactate and 0.8 mM of ammonia was produced (Figure 15.10a). Although the ammonia levels were much below the toxic levels (2 mM), the levels of lactic acid were toxic. This indicates complete depletion of glucose from the medium thereby compelling change in reservoir media. The media reservoir was changed post 30 days of run and the reactor was run for another 30 days. During the entire run, the monolith bioreactor produced a total of 57.5 mg of antibody at a concentration of 115 mg/l, which is fourfold higher than the T-flask culture. The mAb productivity rate during the log phase was 6.0 µg ml$^{-1}$ day$^{-1}$. The productivity decreased to 3.6 µg ml$^{-1}$ day$^{-1}$ and thereafter became constant for the entire period of the reactor run. The beads' bioreactor gave a productivity of 4.2 µg ml$^{-1}$ day$^{-1}$. Further, the productivity decreased to 2 µg ml$^{-1}$ day$^{-1}$ and remained constant for the entire run. Following the 60-day run, the bioreactor matrix was stable and showed minimal signs of degradation. The SEM images after 60 days showed that the hybridoma cells were still attached to the matrix in a high number (Figure 15.10c).

## 15.3 CRYOGEL AS CELL SUPPORT MATRIX IN AN EXTRACORPOREAL BIOARTIFICIAL DEVICE

There is a growing interest in developing hybrid artificial devices and still they are in their early stage of development. These bioartificial devices combine biological forms such as cells with some nonbiological forms such as biomaterial, which can maintain their activity outside the body. These then can be used as substitution or as an assist device for the damaged organ system. One of the most active areas of investigation involves hepatic assist devices. Currently, there are nine bioartificial liver devices (BAL) that are being tested clinically and most of them utilize a hollow fibre bioreactor (HFBR) (Rozga, 2006). By nature, a cryogel matrix is structurally similar to a hollow fibre reactor, which has been discussed earlier. Thus, a cryogel can potentially be used as a cell culture device. By appropriately modifying the polymeric composition and organization, it can be applied to culture hepatic cells in a BAL device, with fewer complications as associated with HFBR.

Thus, as a potential cell culture matrix, cryogels are being tested for hepatic cell immobilization/growth. The objective is to adapt the hepatic cell containing cryogel matrix as a biological component of a BAL. For this purpose, cryogel bioreactors made up of a combination of synthetic and natural polymers—copolymer of acrylonitrile and N-vinyl-2-pyrrolidone (poly(AN-co-NVP))—or interpenetrated polymer networks (IPN) of chitosan and poly (N-isopropylacrylamide), (poly(NiPAAm)-chitosan) were screened and proved to be favourable for growth of hepatocytes.

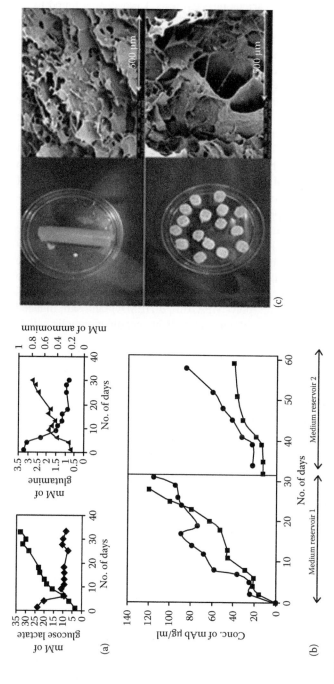

**FIGURE 15.10** Production of monoclonal antibody in PAAC cryogel matrix. (a) Hybridoma metabolism inside monolith bioreactor for 30 days of reactor run; glucose (filled diamonds), lactate (filled squares), glutamine (filled circles), ammonia (filled triangles). (From Jain E. et al.: Supermacroporous polymer-based cryogel bioreactor for monoclonal antibody production in continuous culture using hybridoma cells. *Biotechnology Progress*. 2011. 27(1). 170–180. Copyright Wiley-VCH Verlag GmbH & Co. KGaA. Reproduced with permission.) (b) Monoclonal antibody production in monolith bioreactor (filled circles) and beads bioreactor (filled squares) over a period of 60 days. (c) Digital image of monolith bioreactor and SEM image of attached hybridoma cells to the matrix after 60 days of run (upper panel); digital image of beads bioreactor and SEM image of attached hybridoma cells to the matrix after 60 days of run (lower panel). (Reprinted by permission from Macmillan Publishers Ltd. *Nat Protoc*, Jain, E. and Kumar, A., 8(5): 821–835, copyright 2011.)

The cryogels possessed interconnected macroporous networks with pore size lying between 20 and 100 µm. These were characterized to have efficient convective and diffusion transport as analysed by various physical tests. Cryogel was found to be biocompatible and hemocompatible in *in vivo* studies in rat and *in vitro* albumin absorption and platelet adherence tests. COS-7 and HepG2 cells proliferated inside the cryogel surface. HepG2 cells formed spheroids within the poly(NiPAAm)-chitosan cryogel and maintained ureagenesis. Further, primary hepatocytes grown on poly(NiPAAm)-chitosan cryogel formed small aggregates and performed the vital functions of the liver. *In vitro* clinical evaluation of the cryogel bioreactor loaded with primary hepatocytes showed cells' ability to detoxify the circulating plasma obtained from a liver patient.

Cryogel-based bioreactors for hepatocytes present an attractive device for cell immobilization in high numbers. These porous materials with large surface-to-volume ratios facilitate large quantities of hepatocytes to be cultured in small volumes of matrix. Highly porous structures could not only improve mass transfer rate of oxygen and nutrients into the inner pores, but also efficiently remove metabolic products. The BALs are based on the assumption to provide critical liver function in case of fulminant hepatic failure (FHF). Thus, such BAL devices should be able to carry out the liver's synthetic and metabolic functions, detoxification and excretion. The development of strategies that would improve the survival as well as the maintenance of the metabolic capacities of the artificial organ is of clinical importance. In this study, poly(NiPAAm)-chitosan and poly(NiPAAm-*co*-HEMA)-chitosan induced the formation of cellular aggregates with enhancing liver-specific metabolic activities. This demonstrates that PN-C based cryogel matrix presents a suitable environment for the growth of the liver cells. Furthermore, clinical testing of the devices under *in vitro* conditions shows their ability to purify toxic components of plasma from a liver patient efficiently. In addition to this, the PN-C cryogels are biocompatible as established by *in vivo* studies in mice. As a blood-contacting device, biocompatibility with body fluids is an important issue that is resolved in this regard. Altogether, cryogel-based bioreactors hold the potential to be developed into fully functional BAL devices. As future work, further modification of the matrix, *in vivo* testing in acute liver failure model of rats and higher animals, improvement in the oxygenation of the bioreactor and addition of a bile excretory compartment will help in improvising the present device. This will lead to the establishment of an efficient and economic BAL device.

## 15.4 CONCLUSION

The reported studies clearly demonstrate the great potential of cryogel matrices in two major areas of biotechnology and biomedicine: cell culture and integrated protein production setups. A combination of unique macroporous structure, mechanical strength and elasticity and the versatility of the cryogelation technology enables successful application of cryogels as adsorbents for cell chromatography and as scaffolds and supports for cells of various types in a continuous packed-bed bioreactor. Cryogels by their basic nature of fabrication present some ideal features such as open pores, easy convection and diffusion transport, which support cell growth. This makes them ideal candidates for a bioreactor matrix. In terms of efficiency and their high surface-to-volume ratio and economic cost, these can be accommodated in a small space, thus making them valuable bioreactor devices in the labaoratory setting. For their applications in biomedicine as BAL, various issues such as antigenicity, biocompatibility, rate of degradability and so on can be addressed by modulating the cryogel composition, concentration of polymeric precursors, freezing conditions and surface modification of cryogels. Further *in vitro* and *in vivo* studies of cryogel scaffolds are needed and are being carried out in order to facilitate cryogels entering the clinical trials phase.

## LIST OF ABBREVIATIONS

| | |
|---|---|
| AAm | Polyacrylamide |
| BAL | Bioartificial liver |
| Cu(II)-IDA | Cu(II)-iminodiacetic acid |
| FHF | Fulminant hepatic failure |
| HFBR | Hollow fibre bioreactor |
| IPN | Interpenetrated polymer networks |
| mAb | Monoclonal antibody |
| MBAAm | N,N'-methylene bis-acrylamide |
| PAAC | Polyacrylamide-chitosan |
| PAAm-gelatin | Polyacrylamide-gelatin |
| PAI | Plasminogen activator inhibitor |
| PAN | Poly(acrylonitrile) |
| PNiPAAm | Poly(N-isoproppylacrylamide) |
| PNiPAAm–chitosan | Poly(N-isoproppylacrylamide)–chitosan |
| poly(AN-co-NVP) | Poly (acrylonitrile-co-N-vinyl-2-pyrrolidone) |
| tPA | Tissue plasminogen activator |

## REFERENCES

Babac, C., Yavuz, H., Galaev, I. Y., Piskin, E. and Denizli, A. (2006). Binding of antibodies to concanavalin A-modified monolithic cryogel. *Reactive and Functional Polymers* **66**(11): 1263–1271.

Bansal, V., Roychoudhury, P. K., Mattiasson, B. and Kumar, A. (2006). Recovery of urokinase from integrated mammalian cell culture cryogel bioreactor and purification of the enzyme using p-aminobenzamidine affinity chromatography. *Journal of Molecular Recognition* **19**(4): 332–339.

Bhat, S., Lidgren, L. and Kumar, A. (2013). *In vitro* neo-cartilage formation on a three-dimensional composite polymeric cryogel matrix. *Macromol Biosci* **13**(7): 827–837.

Bloch, K., Lozinsky, V. I., Galaev, I. Y., Yavriyanz, K., Vorobeychik, M., Azarov, D., Damshkaln, L. G., Mattiasson, B. and Vardi, P. (2005). Functional activity of insulinoma cells (INS-1E) and pancreatic islets cultured in agarose cryogel sponges. *J Biomed Mater Res A* **75**(4): 802–809.

Cadwell, J. J. S. (2005). New developments in hollow-fiber cell culture. *Current Pharmaceutical Biotechnology* **6**: 397–403.

Chu, L. and Robinson, D. K. (2001). Industrial choices for protein production by large-scale cell culture. *Current Opinion in Biotechnology* **12**: 180–187.

Deraz, S., Plieva, F. M., Galaev, I. Y., Karlsson, E. N. and Mattiasson, B. (2007). Capture of bacteriocins directly from non-clarified fermentation broth using macroporous monolithic cryogels with phenyl ligands. *Enzyme and Microbial Technology* **40**(4): 786–793.

Eibl, R., Kaiser, S., Lombriser, R. and Eibl, D. (2010). Disposable bioreactors: The current state-of-the-art and recommended applications in biotechnology. *Applied Microbiology and Biotechnology* **86**(1): 41–49.

Jain, E., Karande, A. A. and Kumar, A. (2011). Supermacroporous polymer-based cryogel bioreactor for monoclonal antibody production in continuous culture using hybridoma cells. *Biotechnology Progress* **27**(1): 170–180.

Jain, E. and Kumar, A. (2008). Upstream processes in antibody production: Evaluation of critical parameters. *Biotechnology Advances* **26**(1): 46–72.

Jain, E. and Kumar, A. (2009). Designing supermacroporous cryogels based on polyacrylonitrile and a polyacrylamide-chitosan semi-interpenetrating network. *Journal of Biomaterials Science, Polymer Edition* **20**(7–8): 877–902.

Jain, E. and Kumar, A. (2013). Disposable polymeric cryogel bioreactor matrix for therapeutic protein production. *Nat Protoc* **8**(5): 821–835.

Kumar, A., Bansal, V., Andersson, J., Roychoudhury, P. K. and Mattiasson, B. (2006a). Supermacroporous cryogel matrix for integrated protein isolation–Immobilized metal affinity chromatographic purification of urokinase from cell culture broth of a human kidney cell line. *Journal of Chromatography A* **1103**(1): 35–42.

Kumar, A., Bansal, V., Nandakumar, K. S., Galaev, I. Y., Roychoudhury, P. K., Holmdahl, R. and Mattiasson, B. (2006b). Integrated bioprocess for the production and isolation of urokinase from animal cell culture using supermacroporous cryogel matrices. *Biotechnology and Bioengineering* **93**(4): 636–646.

Kumar, A., Rodriguez-Caballero, A., Plieva, F. M., Galaev, I. Y., Nandakumar, K. S., Kamihira, M., Holmdahl, R., Orfao, A. and Mattiasson, B. (2005). Affinity binding of cells to cryogel adsorbents with immobilized specific ligands: Effect of ligand coupling and matrix architecture. *Journal of Molecular Recognition* **18**(1): 84–93.

Kumar, A. and Srivastava, A. (2010). Cell separation using cryogel-based affinity chromatography. *Nat Protoc* **5**(11): 1737–1747.

Langer, E. S. (2011). Trends in perfusion bioreactors: Next revolution in bioprocessing. *BioProcess International* **9**(10): 18–22.

Lewis, L. J. (1979). Plasminogen activator (urokinase) from cultured cells. *Thrombosis and Haemostasis* **42**(3): 895–900.

Lozinsky, V. I., Plieva, F. M., Galaev, I. Y. and Mattiasson, B. (2001). The potential of polymeric cryogels in bioseparation. *Bioseparation* **10**(4–5): 163–188.

Meuwly, F., Ruffieux, P.-A., Kadouri, A. and von Stockar, U. (2007). Packed-bed bioreactors for mammalian cell culture: Bioprocess and biomedical applications. *Biotechnology Advances* **25**(1): 45–56.

Mishra, R., Goel, S. K., Gupta, K. C. and Kumar, A. (2014). Biocomposite cryogels as tissue-engineered biomaterials for regeneration of critical-sized cranial bone defects. *Tissue Eng Part A* **20**(3–4): 751–762.

Nilsang, S., Nandakumar, K. S., Galaev, I. Y., Rakshit, S. K., Holmdahl, R., Mattiasson, B. and Kumar, A. (2007). Monoclonal antibody production using a new supermacroporous cryogel bioreactor. *Biotechnology Progress* **23**(4): 932–939.

Nilsang, S., Nehru, V., Plieva, F. M., Nandakumar, K. S., Rakshit, S. K., Holmdahl, R., Mattiasson, B. and Kumar, A. (2008). Three-dimensional culture for monoclonal antibody production by hybridoma cells immobilized in macroporous gel particles. *Biotechnology Progress* **24**(5): 1122–1131.

Ozturk, S. S. and Palsson, B. O. (1991). Growth, metabolic, and antibody production kinetics of hybridoma cell culture: 2. Effects of serum concentration, dissolved oxygen concentration, and medium pH in a batch reactor. *Biotechnology Progress* **7**(6): 481–494.

Persson, P., Baybak, O., Plieva, F., Galaev, I. Y., Mattiasson, B., Nilsson, B. and Axelsson, A. (2004). Characterization of a continuous supermacroporous monolithic matrix for chromatographic separation of large biopraticles. *Biotechnology and Bioengineering* **88**(2): 224–236.

Plieva, F. M., Karlsson, M., Aguilar, M. R., Gomez, D., Mikhalovsky, S. and Galaev', I. Y. (2005). Pore structure in supermacroporous polyacrylamide based cryogels. *Soft Matter* **1**(4): 303–309.

Poles-Lahille, A., Richard, C., Fisch, S., Pedelaborde, D., Gerby, S., Kadi, N., Perrier, V., Trieau, R., Balbuena, D., Valognes, L. and Peyret, D. (2011). Disposable bioreactors: From process development to production. *BMC Proc* **5** Suppl **8**: P2.

Roychoudhury, P. K., Khaparde, S. S., Mattiasson, B. and Kumar, A. (2006). Synthesis, regulation and production of urokinase using mammalian cell culture: A comprehensive review. *Biotechnology Advances* **24**(5): 514–528.

Rozga, J. (2006). Liver support technology – An update. *Xenotransplantation* **13**(5): 380–389.

Srivastava, A., Jain, E. and Kumar, A. (2007). The physical characterization of supermacroporous poly(N-isopropylacrylamide) cryogel: Mechanical strength and swelling/deswelling kinetics. *Materials Science and Engineering A* **464**(1–2): 93–100.

Srivastava, A., Shakya, A. K. and Kumar, A. (2012). Boronate affinity chromatography of cells and biomacromolecules using cryogel matrices. *Enzyme and Microbial Technology* **51**(6–7): 373–381.

Stump, D. C., Thienpont, M. and Collen, D. (1986). Urokinase-related proteins in human-urine–isolation and characterization of single-chain urokinase (prourokinase) and urokinase-inhibitor complex. *Journal of Biological Chemistry* **261**(3): 1267–1273.

Warnock, J. N. and Al-Rubeai, M. (2006). Bioreactor systems for the production of biopharmaceuticals from animal cells. *Biotechnology and Applied Biochemistry* **45**(Pt 1): 1–12.

Williams, S. L., Eccleston, M. E. and Slater, N. K. H. (2005). Affinity capture of a biotinylated retrovirus on macroporous monolithic adsorbents: Towards a rapid single-step purification process. *Biotechnology and Bioengineering* **89**(7): 783–787.

# 16 Supermacroporous Functional Cryogel Stationary Matrices for Efficient Cell Separation

*Akshay Srivastava, Samarjeet Singh and Ashok Kumar**

## CONTENTS

16.1 Introduction ................................................................................................ 444
    16.1.1 Importance of Cell Separation Technology .................................... 444
    16.1.2 Current Techniques and Demand in Cell Separation Technology ... 445
16.2 Cryogel-Based Cell Separation Technology ............................................... 445
    16.2.1 Cryogel Features Ideal for Cell Separation Stationary Matrices ...... 446
        16.2.1.1 Flow Rate ........................................................................ 446
        16.2.1.2 Enabling Movement of Large Particles .......................... 447
        16.2.1.3 Functional Matrix ........................................................... 448
        16.2.1.4 Versatile Cell Recovery Properties ................................. 450
16.3 Application of Cryogel Chromatography Column ..................................... 451
    16.3.1 Separation of Prokaryotic Cells ....................................................... 451
    16.3.2 Separation of Mammalian Cells for Biomedical Applications ......... 454
        16.3.2.1 Fractionation of Immune Cell Types .............................. 454
        16.3.2.2 Separation of Stem Cells ................................................ 454
    16.3.3 Boronate Affinity Chromatography ................................................. 458
    16.3.4 Filtration of Leukocytes ................................................................... 458
List of Abbreviations .......................................................................................... 460
References .......................................................................................................... 460

## ABSTRACT

Cell separation is an essential tool in biological applications with increasing usage in cell therapy and tissue engineering applications. The available systems do not provide closed and sterile preparative cell separation processes for cell-based

---

* Corresponding author: E-mail: ashokkum@iitk.ac.in; Tel: +91-512-2594051.

therapy. These limitations could be overcome using recently developed cryogel-based cell separation technology. **Cryogels** are polymeric stationary matrices for the separation of cells and microorganisms in chromatographic mode. The interconnected porous network of cryogel enables the convective movement of large particles ≥20 μm, such as mammalian cells. Researchers have thoroughly investigated the physical properties of cryogel over the last decade and the separation of mammalian cells and microorganisms using cryogels gas been reported. Recently, cryogels have become an interesting matrix for the separation of therapeutic cells (e.g., stem cells) for various cell therapy applications. In this chapter, we discuss the key physical and functional features of cryogel and key examples of its applications in the separation of mammalian cells and microorganisms.

## KEYWORDS

Cryogel, interconnected pores, cell chromatography, mechanical cell recovery, stem cell separation

## 16.1 INTRODUCTION

### 16.1.1 IMPORTANCE OF CELL SEPARATION TECHNOLOGY

A highly purified single cell type population is a foremost requirement for diagnostic, biotechnological and biomedical applications. The importance of the area of cell separation could be realized by the growth in the global cell separation market, which is expected to reach over $5 billion by 2019 as reported by a Research and Markets report (Cell Isolation/Cell Separation Market by Product, Cell Type, Technique, Application & by End User–Forecast to 2019. N.p., n.d. Web. 21 Oct. 2015). Increase in demand of purified cell types in cancer research, tissue repair and regeneration, prevalence of infectious diseases and rising aging population are major factors in propelling the growth of this market. The global cell separation market can largely be categorized based on cell type, purification technique and biomedical application. The area of human cell isolation is expected to record the highest growth in the cell separation market. This growth is mainly attributed to increasing applications in human stem cell research and its usage for the treatment of various diseases. There are numerous research articles describing cell separation techniques and their applications; these were nicely reviewed in multiple books (Fisher et al., 1998; Kumar et al., 2007). These applications may include isolation of purified cell population for cell-based therapy, removal of malignant cells from normal cells, fractionation of immunocompetent and nonimmunocompetent cells and so on (Beaujean, 1997; Geifman-Holtzman et al., 2000; Racila et al., 1998; Siewert et al., 2001). The techniques used for specific cell separation are based on physicochemical (size, volume, density, light scattering properties, membrane potential, pH, electrical impedance and charge), functional (adherence, affinity or growth characteristics) and immunological properties (Beaujean, 1997; Nadali et al., 1995). Density gradient centrifugation is a widely employed technique based on the density and size of the cells. It is a time-consuming process and type-specific separation of particular cells is difficult to achieve. However, the method is mainly used for pretreating cells in the cell separation process. The development of monoclonal antibodies directed against specific cellular epitopes has led to the application of immunological methods for the classification and isolation of cell subsets. Affinity separation methods for the isolation of specific cell populations are generally based on specific antibodies against differentially expressed cell-surface antigens. The concept of affinity separation can be used either for negative selection or for positive selection of cells. In negative selection, unwanted cells adhere to the affinity ligand on the support matrix and the target cells pass through the matrix unretained (Berenson et al., 1986a). On the other hand, positive selection of target cells involves the binding of specific target cells to the affinity adsorbent and, after washing, the bound cells are released from the matrix in a purified form (Berenson et al., 1986b; Kumar et al., 2003).

### 16.1.2 Current Techniques and Demand in Cell Separation Technology

Fluorescent activated cell sorting (FACS) is a well-established technique in clinical diagnostics and biomedical research. Although the FACS technology provides impressive results regarding purity of separated cells, it is limited by the time required for separation, which can be several hours per sample. Being a serial sorting device (cells are analysed and sorted one by one), the capacity of FACS is limited by the speed of analysis and sorting, which is about $3 \times 10^5$ cells/min or when performing 'high speed' separations, $5 \times 10^8$ cells in 6 h. On the other hand, the introduction of colloidal magnetic particles of less than 100 nm in diameter, conjugated to specific ligands or antibodies, can be used to separate specific cell types by applying a magnetic field in magnetic affinity cell separation (MACS). MACS technology has become the standard method for cell separation. Numerous publications have proven its versatility for multiple applications: cell separations with consistent high quality results from lab bench to clinical applications, from small-scale to larg-scale separations, from frequently occurring cells to rare cells and complex subsets. However, the major limitation associated with both techniques is large-scale processing and the requirement of a special device. Additionally, these techniques are more utilized as analytical rather than as a preparative tool for cell separation. Both techniques are quite expensive; moreover, the available systems do not provide a closed and sterile preparative cell separation process for cell-based therapy. In addition, an added limitation is that both methods modify cell membranes and so neither FACS nor MACS is preferred for subsequent analysis or recultivation of the sorted cells (Neurauter et al., 2007; Recktenwald and Radbruch, 1997). The viability and functionality of the isolated specific cell type using the above methods is still under investigation and changes are required in purification procedures to achieve viable cells after complete separation.

## 16.2 CRYOGEL-BASED CELL SEPARATION TECHNOLOGY

Cryogels have provided new and promising cell separation matrices and chromatographic methods. The potential of cryogel-based cell chromatography methods was compared with other established cell separation technologies by Kumar and Bhardwaj (2008). In this chapter, we further focus on features of cryogels that make them suitable cell chromatography matrices and discuss their potential applications demonstrated to date.

### 16.2.1 Cryogel Features Ideal for Cell Separation Stationary Matrices

#### 16.2.1.1 Flow Rate

An important feature of the matrix material is the volume of the pores. Dry polymeric cryogels constitute only 3–4% of the total weight of saturated swollen cryogel. Approximately 4–5% of the swollen weight of cryogels is contributed by water tightly bound by the polymer (adsorbed from water vapour). Thus, the remaining 90% of the swollen wet cryogel weight represents the amount of water within the pores. The presence of interconnected pores in cryogels allows the convective flow through the porous channels called capillaries. The aqueous flow rate through the cryogel column is reported around 700–250 cm/h depending upon the cryogelation temperature. Such high flow rates, that is, 750 cm/h, with minimal pressure drop, are an important feature of supermacroporous structure of cryogels, which allows them to pack in the column format for chromatography purposes. The columns packed with poly(acrylamide-allyl glycidyl ether) cryogel or ion-exchange cryogel have not shown any deviation from flow linearity up to flow rates of 14 ml/min when compared with the columns of the similar geometry packed with Sephadex G-100 and Sephacryl S-1000 SF (Plieva et al., 2004a). Cryogels have received much attention as monolithic columns for chromatographic separation of large bioparticles such as bacteria, viruses and mammalian cell types (Arvidsson et al., 2002, 2003; Plieva et al., 2004b; Yao et al., 2006, 2007). In all these chromatographic applications, the column backpressure and column efficiency at different flow rates have been investigated. Due to large interconnected pores, cryogels exhibit low hydraulic flow resistance at high flow rates of up to 10 ml/min. This hydraulic flow resistance is calculated by measuring the flow velocities of water through the cryogel column under high hydrostatic pressure (0.01 MPa) (Arvidsson et al., 2003; Plieva et al., 2004b). Another way of calculating the hydraulic flow resistance is by calculating the permeability of the cryogels at different hydrostatic pressures (Yao et al., 2006, 2007). The permeability of the cryogels was found to be in the range of $3.43 \times 10^{-11}$ to $8.27 \times 10^{-13}$ $m^2$ (Persson et al., 2004; Yao et al., 2006, 2007; Yun et al., 2007). This range is considered higher than that for other organic polymeric monoliths ($1.11$–$7.2 \times 10^{-14}$ $m^2$) (Du et al., 2007; Mihelic et al., 2005). Yun et al. (2011) have presented a modified capillary model for describing the characteristics of microstructure and fluid hydrodynamics of proteins in cryogel columns (Figure 16.1). Based on the model predictions, it was found that the effective pore sizes of the pHEMA cryogels are in the range of 10–90 µm for a cryogel column with a smaller bed diameter and 10–110 µm for a cryogel with a larger bed diameter. The mean pore diameters for these two cryogels were found to be around 51 and 46 µm, and both these mathematical predictions were close to the values of the pore diameter observed through SEM. These models could also be applied for larger particles with further modification. Overall, cryogel columns can adopt greater range of flow rates for biomedical applications, for example, cell separation.

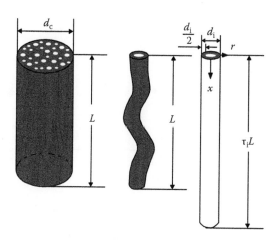

**FIGURE 16.1** Schematic diagram of a cryogel made up of tortuous capillaries. It displays schematically a cryogel structure made up of capillaries with different length ($L$) and tortuosities ($\tau$). Li is the capillary length and L is the length of the cryogel. (Reprinted from *J. Chromatogr. A* 1218, Yun, J., Jespersen, G.R., Kirsebom, H., Gustavsson, P., Mattiasson, B., and Galaev, I.Y., An improved capillary model for describing the microstructure characteristics, fluid hydrodynamics and breakthrough performance of proteins in cryogel beds, 5487–5497. Copyright 2011, with permission from Elsevier.)

### 16.2.1.2 Enabling Movement of Large Particles

Cryogel-packed columns show interesting chromatographic properties as a result of their low flow resistance due to inherent large interconnected pores. The transport of solutes inside the monolithic columns occurs primarily due to convection as opposed to diffusion and micron-sized solutes are transported with the same efficiency as soluble substances. The height equivalent to theoretical plate (HETP) values for the monolithic column were shown to be independent of the solute/particle size for acetone (MW 58 Da), BSA (MW 69 kDa) and Blue Dextran (MW 2000 kDa) at a wide range of flow rates 0.2–10 ml/min (Figure 16.2). Even the objects of micrometre-large size like *E. coli* cells were transported inside large pores of cryogel monolith with the same efficiency as soluble substances (Figure 16.2, open squares) (Plieva et al., 2004a).

The interconnected pore size in monolithic cryogels allows unhindered movement of microbial cells without any blocking of the cryogel. *E. coli* cells were not retained inside the plain cryogel columns. More than 90% of *E. coli* cells appeared in a flow-through fraction when the cell suspension was applied onto the top of the cryogel columns. However, large size yeast cells were partially retained by the cryogel columns but around 60–70% of yeast cells passed freely through the cryogel columns. The pore diameter ≥100 μm also enables the movement of much larger particle sizes (between 1 and 15 μm) without being mechanically entrapped inside the column. It has been shown that a pulse of unprocessed whole blood convectively transported through the monolithic cryogel column without any physical entrapment of blood cells (Figure 16.3). Convectively transported cells do not experience large shear forces because of laminar flow of liquid in the interconnected pores in the monolithic cryogel.

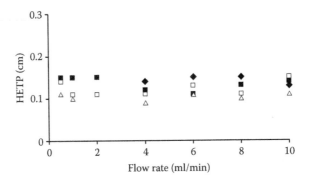

**FIGURE 16.2** The dependence of HETP on flow rate for acetone (MW 58 Da, open triangles), BSA (MW 69 kDa, closed squares), Blue Dextran (MW 2000 kDa, closed rhombus) and *E. coli* cells (open squares). Cryogel column was prepared at −12°C. All markers were applied in 20-mM Tris–HCl buffer, pH 7.0 containing 0.5 M NaCl. Chromatographic peaks were recorded at 280 nm after injecting 50 µl marker solution (or cell suspension, OD600 0.42). (Reprinted from *J. Chromatogr. B* 807, Plieva, F.M., Savina, I.N., Deraz, S., Andersson, J., Galaev, I.Y., and Mattiasson, B., Characterization of supermacroporous monolithic polyacrylamide based matrices designed for chromatography of bioparticles, 129–137. Copyright 2004, with permission from Elsevier.)

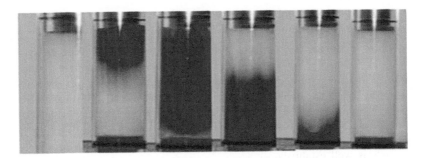

**FIGURE 16.3** Human blood is passed through a cryogel column showing the convective movement of large particles such as mammalian cells.

### 16.2.1.3 Functional Matrix

Inert cryogel chromatography matrices have been prepared from hydrophilic gel-forming polymers. These matrices were prepared either by polymerization of the appropriate monomers (e.g., polyacrylamide and poly(dimethylacrylamide)) or by physical gelation (e.g., agarose). It is well known that hydrophilic surfaces are less prone to nonspecific protein and cell attachment. Consequently, all the reported literature related to cryogel applications in cell separation shows the use of cryogels developed using hydrophilic polymers. These hydrophilic polymers can be modified later with specific groups to allow for cell capture. Cryogel matrices have been most commonly functionalized with protein A for the affinity capture of antibody-labelled

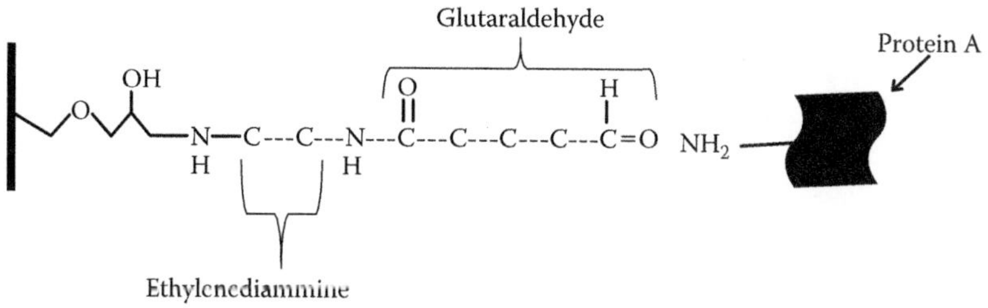

**FIGURE 16.4** Cross-linking of protein A ligand with epoxy containing cryogel matrix via 7-carbon linker including ethylenediamine and glutaraldehyde.

specific cell types. The choice of protein A as a ligand is attributed to its affinity toward the Fc portion of IgG antibody. In one study, the protein A ligand was coupled with a 7-carbon spacer arm generated on epoxy-activated poly(AAm-co-AGE) and poly(NiPAAm-co-AGE) cryogels using ethylenediamine and glutaraldehyde coupling chemistry. The 7-carbon spacers present on the epoxy surface appropriately protrude the ligand into the porous area of the matrix to efficiently capture the passing particles (Figure 16.4). Previous studies showed that 11-carbon spacer arms cause some nonspecific interactions with other cell types and carbon linkers less than 5 decreased the overall capacity of the columns since the ligand was not properly exposed in these cases (Kumar et al., 2003).

Grafting on macroporous cryogel materials provides functional materials for a chromatography purpose. In comparison to other hydrogel materials, cryogels provide a unique physical morphology ideal for grafting polymers within their porous network. As described earlier, cryogels retain their physical shape even after drying and can quickly absorb a large amount of liquid. This provides an easy approach for grafting of polymers, wherein the solution of polymer and initiator (initiates grafting over an activated polymer surface) quickly absorbs into the porous area. Beside protein, ligands and chemical ligands such as boronate containing groups have also been exploited for separation of cells. Boronate interaction with cis-diol group-containing compounds is well established and involves a reversible ester formation between boronate and the cis-diol group depending on the pH of the reaction medium. It is demonstrated that grafting of boronate majorly occurs inside the pores within the pore wall of the cryogel with high grafting efficiency and yield. It was shown that the degree of grafting was 40% with 35% grafting yield (Srivastava and Shakhya et al., 2012). The functionality of grafted boronate was further estimated by quantification of glucose-binding in grafted cryogel column. It was found that 0.8 mg glucose/ml cryogel monolith was absorbed via boronate interaction within the cryogel chromatography column. The grafting of boronate-containing copolymer was also confirmed by FTIR spectrum (Figure 16.5) where a peak at 1344 $cm^{-1}$ corresponds to a B–O bond and at 2354 $cm^{-1}$ corresponds to isopropyl group of PNiPAAm.

Similarly, it is also possible to couple other functional groups within the cryogel matrices for their application of capturing specific cell types. It has also been shown that cryogel matrices can be prepared with functional polymers directly without further functionalization. In one example, a unique combination of hydrophilic/

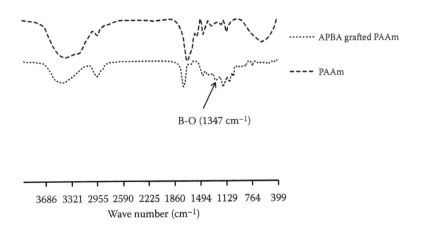

**FIGURE 16.5** FTIR spectrum of poly(NiPAAm-*co*-AAPBA) polymer grafted on PAAm cryogel. (Reprinted from *Enzyme Microb. Technol.* 10, Srivastava, A., Sakhya, A.K., and Kumar, A., Boronate affinity chromatography of cells and biomacromolecules using cryogel matrices, 373–381. Copyright 2012, with permission from Elsevier.)

hydrophobic polymers was selected to prepare cryogel matrices for the separation of leukocytes (unpublished work). These matrices were directly applied without adding any further functionality.

### 16.2.1.4 Versatile Cell Recovery Properties

The sponginess and the presence of functional groups in cryogel matrices are well adapted to two main types of cell recovery properties. After attachment of specific cells within cryogel matrices, they can be eluted out using two different approaches:

*Treatment with immunoglobulin (Ig) solution*: Bound cells can be specifically eluted by immunoglobulin solution through competitive binding on protein A. Fresh solution of immunoglobulin solution is applied on the top of the cryogel column and allowed to pass through the affinity cryogel chromatography matrix bound with specific cells. This causes desorption of the bound cells due to displacement with a higher-affinity immunoglobulin (Kumar et al., 2003).

*Mechanical deformation of cryogel matrix*: A breakthrough study has shown that cryogel matrices can be used for releasing cells by squeezing the cryogels when the cells are affinity bound on them (Dainiak et al., 2007, 2006a,b). Initial results have shown the tremendous potential of such a cell separation strategy applicable to different types of cell populations particularly stem cells and other medically relevant cell types. The method provides a very convenient and elegant way to release bound cells from the matrices, which otherwise poses as a major bottleneck in positive cell selection and separation. Later, it was again verified by Kumar and Srivastava that the sponginess of cryogel matrices allows the release of bound cell via length-wise compression (Kumar and Srivastava, 2010). The captured cells can be recovered by the compression of the affinity cryogel matrix to up to ~50% of its original length and the extracted liquid collected from the capillaries of the cryogel (Figure 16.6). Thereafter, the compressed affinity matrix is reswollen with elution buffer to regain

# Supermacroporous Functional Cryogel Stationary Matrices

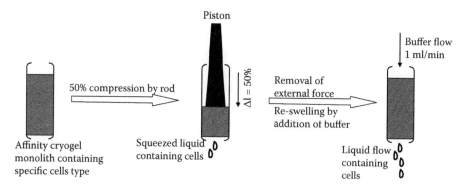

**FIGURE 16.6** Mechanical squeezing of cryogel matrix for the recovery of bound cells. Affinity bound cells on cryogel were removed by compressing the cryogel column by external piston up to 50% of its original length; it dispenses liquid that contains cells. In the next step, the buffer was added to reswell the affinity cryogel matrix and then the column was flushed with fresh cold buffer. (Reprinted by permission from Macmillan Publishers Ltd. *Nat. Protoc.* Kumar, A., and Srivastava, A., 5:1737–1747, copyright 2010.)

its original column length, followed by immediate flushing with buffer at a flow rate of 2 ml min$^{-1}$. Uniaxial compression of pAAm cryogels is well demonstrated by analysing the compressive strength at various uniaxial compressions (Figure 16.7). The shape of the pore walls in the pAAm MG monoliths changes upon mechanical compression. In addition, the volume of the internal porous area also changes and causes the forced expulsion of liquid present in the capillaries of the cryogels (Figure 16.7a). The compression curve for pAAm cryogel at different degrees of compression shows typical features of elastic materials with small- and large-deformation regions (Figure 16.7b). Up to 30–40% compression, the strength of pAAm cryogels was increased linearly followed by the nonlinear region at compression above 60% (Plieva et al., 2009).

## 16.3 APPLICATION OF CRYOGEL CHROMATOGRAPHY COLUMN

### 16.3.1 Separation of Prokaryotic Cells

Separation of microbial cells by chromatographic method is generally very difficult due to many factors such as large size of the cells, low diffusivity, complex cell surface chemistry and multipoint attachment to a functional surface. More importantly, the absence of an appropriate matrix with high porosity does not allow the processing of cells by chromatography (Plieva et al., 2008). However, pAAm cryogel chromatography columns with various functionalities (e.g., ion-exchange, affinity and hydrophobic) have been successfully utilized for the separation of microbial cells (Dainiak et al., 2005, 2006a; Plieva et al., 2004). It has been reported that *E. coli* cells can be bound to an ion-exchange cryogel column and eluted with 70–80% recovery at 0.35–0.4 M NaCl. Similarly, *E.coli* cells can also bind to a cryogel column bearing Cu(II)-loaded iminodiacetate (Cu(II)-IDA) ligand and eluted with 80%

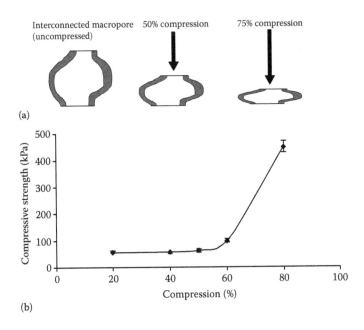

**FIGURE 16.7** (a) Changing porous structure of pAAm cryogel due to uniaxial compression. (b) Compressive strength at different strains for the pAAm cryogel. (Reprinted from *Sep. Purif. Technol.* 65, Plieva, F.M., Seta, E.D., Galaev, I.Y., and Mattiasson, B., Macroporous elastic polyacrylamide monolith columns: processing under compression and scale-up, 110–116. Copyright 2009, with permission from Elsevier.)

recovery using either 10 mM imidazole or 20 mM ethylenediaminetetraacetic acid (EDTA) (Arvidsson et al., 2002). It is well known that different microbial cells have different cell surface properties with specific chemical groups being exposed to the medium. Hence, it is possible to separate specific microbial cells from a mixed population by exploiting these differences. This was demonstrated on two models where mixtures of wild-type *E. coli* and recombinant *E. coli* cells, displaying poly-His peptides (His-tagged *E. coli*), and of wild-type *E. coli* and *Bacillus halodurans* cells, were separated from each other (Dainiak et al., 2005). Further, this approach was also tested with concanavalin A (ConA) ligand immobilized on the cryogel monolith used for the separation of yeast and *E. coli* cells (Dainiak et al., 2006a). A nearly baseline chromatographic separation was achieved under optimized conditions. For instance, an almost complete separation of *E. coli* and yeast cells was achieved under optimized conditions upon application of a mixture of these cells onto a Concavalin A-cryogel (ConA-cryogel) column (Dainiak et al., 2006a; Persson et al., 2004). The chromatogram (Figure 16.8) of separation of *S. cerevisiae* cells from *E. coli* cells on ConA-cryogel column includes loading and washing of unbound cells at a flow rate of 21 cm/h (peak I), further washing at a higher flow rate of 430 cm/h (peak II), conventional elution using 0.3 M methyl α-d-manno-pyranoside at a flow rate of 430 cm/h (peak III), and finally elution was done by compressing the column (peak IV) (Dainiak et al., 2006a).

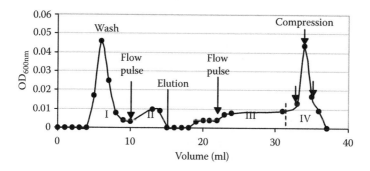

**FIGURE 16.8** Chromatogram of a mixture of *E. coli* and *S. cerevisiae* cells (0.23 ml; $OD_{600}$ 1.3 containing equal amounts of cells of both types) on ConA-cryogel monolithic column (112.8 mm × 7.1 mm diameter). Running buffer: 0.95 mM $CaCl_2$, 5.56 mM KCl, 137 mM NaCl, 0.8 mM $KH_2PO_4$, 0.41 mM $NaHCO_3$, 0.01 mM Tris–HCl, pH 7.4. Flow rate: 21 cm/h. Elevated flow rate (430 cm/h; the starting points of flow pulses are marked with arrows) was applied prior to elution (marked with a vertical line indicating introduction of 0.3 M methyl α-d-manno-pyranoside into the running buffer) and during the elution step prior to compression, respectively. Compressions were carried out repeatedly (indicated with arrows) as described in the text. (Reprinted from *J. Chromatogr. A* 1123, Dainiak, M.B., Galaev, I.Y., Mattiasson, B., 145–150. Copyright 2006, with permission from Elsevier.)

Apart from microbial cells, other large bioparticles, for example, inclusion bodies (Ahlqvist et al., 2006), mitochondria (Teilum et al., 2006) and viruses (Williams et al., 2005), were also captured and separated using monolithic cryogel columns. A cryogel monolith was developed and modified with streptavidin to enable the capture of biotinylated Moloney Murine Leukaemia Virus (MoMuLV) from crude, unclarified cell culture supernatant. Functional polyacrylamide cryogel columns were prepared (Arvidsson et al., 2003) and streptavidin was immobilized to the epoxy functionalized monoliths. Biotinylated MoMuLV containing cell culture supernatant was directly passed through the monolith without any preclarification and the adsorption capacities of $2 \times 10^5$ cfu/ml of adsorbent were demonstrated (cf. Fractogel streptavidin, at $3.9 \times 10^5$ cfu/ml of adsorbent). It was found that a specific titre of the collected fraction was increased by 425-fold. However, less than 8% recovery of particulate was achieved (Williams et al., 2005). In another example, selected phage clones expressing a binding peptide specific to recombinant human lactoferrin or von Willebrand factor (vWF) were covalently attached to macroporous poly(dimethylacrylamide) monolithic column. Large pore size (10–100 μm) of macroporous poly(dimethylacrylamide) cryogel enables the coupling of long (1 μm) phage particles as ligands without column blocking. The phage macroporous monolithic columns were successfully used for the direct affinity capture of target proteins from particulate containing feeds like milk containing casein micelles and fat globules (1–10 m in size) or even whole blood containing blood cells (up to 20 m in size) (Noppe et al., 2007). Recently, cryogel monoliths with boronate as a ligand have demonstrated the capture of yeast cells (Srivastava et al., 2012). Boronate grafted polyacrylamide (PAAm) cryogel is able to bind yeast cells at basic pH and the captured cells were eluted by the treatment of 0.1 M fructose in PBS. It was shown that more than 90% cells were recovered from the boronate grafted cryogel column. This again verified the ability of cryogel monoliths for separation of large molecules such as yeast cells.

### 16.3.2 Separation of Mammalian Cells for Biomedical Applications

#### 16.3.2.1 Fractionation of Immune Cell Types

The first example of cell separation using cryogel was reported by Kumar et al. (2003) for the separation of immune cell types. The interaction of protein A with cells bearing IgG antibodies on their surfaces has been successfully used for the fractionation of blood lymphocytes in a chromatographic format. B lymphocytes are Ig-positive cells and express IgG on the cell surface. When loaded with 1 ml of treated cells ($3.0 \times 10^7$ cells per ml), ~91% of B cells were retained in the 2-ml cryogel column. With ~11.6% of B cells in the total mononuclear cells, the B-cell–binding capacity was determined to be ~$1.6 \times 10^6$ cells per millilitre of adsorbent. The cells in the breakthrough fraction were enriched in T cells (81%). The viability of cells investigated in the breakthrough fraction was >90%, as measured by Trypan blue staining of cells. The electron micrograph image of bound cells on the protein A–cryogel matrix clearly demonstrated the specific binding of the cells. The cells were attached to the pore walls rather than mechanically entrapped in the dead-flow zones. In all, 70% of the total bound cells were recovered by the treatment of Ig solution, and contain ~70% of B cells (Figure 16.9). The overall recovery of B cells from the lymphocyte fraction can further be increased if the sample is passed through the column in recycle mode. Moreover, the capacity of the column can be amplified by increasing the size of the affinity column to achieve >90–95% cell recovery. The recovery percentage could be increased by mechanical squeezing of gel. However, the recovered cells also contained ~20% of T cells. Probably, most of the T cells were retained in the column through some weak interactions of cell receptors with protein A or through some cells aggregated with B cells. To achieve higher purification of the recovered B cells, a pulse gradient of human Ig (10, 30 and 50 mg ml$^{-1}$) was applied. At the highest concentration of IgG, 60% B-cell recovery was achieved with <10% T cells. Furthermore, the cells that were directly applied to protein A–cryogel without labelling with goat antihuman IgG were completely recovered by the flow of buffer (Figure 16.9). Affinity chromatography of human blood mononuclear cells on supermacroporous cryogel–protein A was found in this study to result in about a twofold enrichment of T cells by negative selection. The very low B-cell contamination (~1%) showed significant selectivity of the separation process. With mechanical squeezing of the cryogel matrix, ~95% of bound cells were recovered, which contained ~70% of B cells. The viability of the recovered cells was ~90%.

#### 16.3.2.2 Separation of Stem Cells

The application of protein A cryogel matrices was further extended to capture CD34$^+$ human acute myeloid leukaemia KG-1 cells (Kumar et al., 2005). The CD34 surface antigen is a well-recognized surface marker for hematopoietic stem cells. Anti-CD34 monoclonal antibody labelled cells were applied on affinity cryogel matrices and

# Supermacroporous Functional Cryogel Stationary Matrices

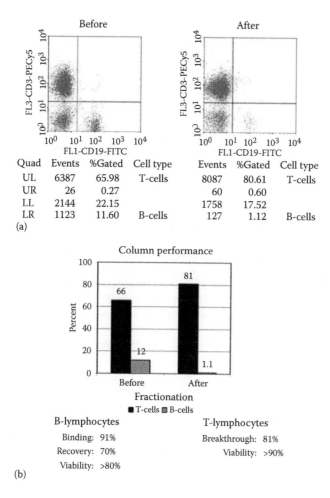

**FIGURE 16.9** Composition of human peripheral blood lymphocytes before and after passage through the supermacroporous monolithic cryogel-protein A column. Lymphocytes (1 ml, $3.0 \times 10^7$ cells/ml) were treated with goat antihuman IgG(H + L) and applied to a 2-ml cryogel-protein A column. Flow cytometric analysis (a). Column performance (b). The scatter gates were set on the lymphocyte fraction. UL (upper left); UR (upper right); LL (lower left); and LR (lower right) quadrants. The cells bound in the column were released with 2 ml of dog IgG (30 mg/ml). (Reprinted from *J. Immunol. Methods* 283, Kumar, A., Plieva, F.M., Galeav, I.Y., and Mattiasson, B., Affinity fractionation of lymphocytes using a monolithic cryogel, 185–194. Copyright 2003, with permission from Elsevier.)

affinity cryogel beads. About 95% positively labelled cells were retained within the protein A–cryogel column. The cell capture efficiency was lower in the case of protein A–cryogel beads, where approximately 76% cell binding was achieved. In a parallel strategy, direct immobilization of anti-CD34 antibodies on the protein-A cryogel column before passing the pool of cells was conducted; however, lower cell binding (50%) was observed. This may be attributed to the poor orientation and

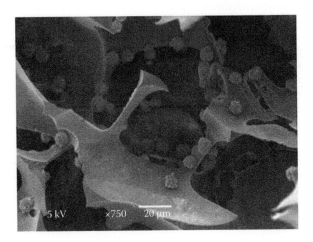

**FIGURE 16.10** Scanning electron micrograph of the bound CD34+ human acute myeloid leukaemia KG-1 cells in the inner part of supermacroporous protein A–cryogel monolithic matrix. Magnification ×750. (From Kumar, A., Rodriguez-Caballero, A., Plieva, F.M., Galaev, I.Y., Nandakumar, K.S., Kamihira, M., Holmdahl, R., Orfao, A., and Mattiasson, B.: Affinity binding of cells to cryogel adsorbents with immobilized specific ligands: effect of ligand coupling and matrix architecture. *J. Mol. Recognit.* 2005. 18. 84–93. Copyright Wiley-VCH Verlag GmbH & Co. KGaA. Reproduced with permission.)

inactivation of the antibody molecules when coupled directly to the matrix. The scanning electron microscopic images (Figure 16.10) of the bound CD34$^+$ cells on the protein A-cryogel matrix clearly demonstrated the specific binding of these cells. The cells were also attached specifically to the bead surface. In the case of monolithic cryogel, the cells were specifically attached to the pore walls rather than being mechanically entrapped in the dead-flow zones as shown in SEM micrograph (Figure 16.10).

Thus, this system could be a good model for the separation of CD34$^+$ hematopoietic stem cells from bone marrow, peripheral blood or umbilical cord blood (UCB). The study was extended to separate CD34$^+$ cells from UCB (Kumar and Srivastava, 2010). The antibody labelled UCB and RBC sedimented UCB cells (0.8 ml) was applied on the top of the column in two separate experiments and 0.5 ml flow through was collected before cells were allowed to run completely through protein-A cryogel column. Approximately $8 \times 10^6$ nucleated cells were passed through the column during the washing step. The affinity bound CD34$^+$ cells were recovered by mechanical squeezing of cryogel. Where the cells were released by 50% column compression, followed by buffer wash, to elute loosen CD34$^+$ cells. The FACS analysis showed more than 95% cell recovery of bound cells. The study has optimized the procedure for cell recovery through mechanical squeezing by lengthwise compression of up to 50–60%, followed by 10 ml of flushing by PBS. Finally, it was concluded that a pure population of $1 \times 10^5$ CD34$^+$ hematopoietic stem cells could be purified from 0.8 ml of labelled UCB using cryogel matrix (Figure 16.11). However, higher cell binding and recovery were observed when RBC-sedimented UCB was applied on the column and purified. Around $7 \times 10^5$ cells per ml of cryogel column could be recovered from RBC sedimented UCB. Overall, 20% CD34+ cell recovery from UCB and 90% cell recovery from RBC-sedimented UCB is possible using the 1 ml cryogel column (height × diameter: $1.0 \times 0.9$ cm$^2$). The cell binding capacity and recovery of the cryogel column can be altered by changing the physical dimensions to process large volumes of cell mixture. Affinity bound CD34+ cells can be concentrated (50% compression followed by 1 ml buffer wash) in 1.5 ml of an elution aliquot containing > 90% of bound cells by mechanical squeezing cell recovery method. Later, isolated cells were grown and tested for CD34+ expression by immunostaining. It demonstrated that isolated stem cells could maintain proliferation capacity and retained CD34+ cell-surface markers.

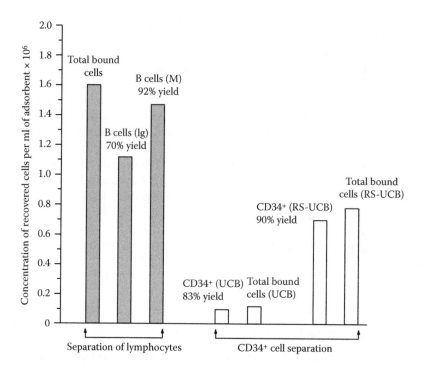

**FIGURE 16.11** The cell separation efficiency on an affinity-based cryogel chromatographic column (per millilitre adsorbent). Separation of lymphocytes (gray bars) showing B-cell recovery after incubation with immunoglobulin (B cells [Ig]) with 70% yield, and recovery by mechanical squeezing (B cells [M]) with 92% yield from total bound cells on the cryogel adsorbent. CD34+ cell separation (white bars) showing recovery of CD34+ cells (CD34+ [UCB]) with 83% yield directly from UCB and CD34+ cell recovery (CD34+ [RS-UCB]) with 90% yield from RBC-sedimented UCB from total CD34+ cells bound on cryogel adsorbent when direct UCB and RBC-sedimented UCB (RS-UCB) were applied, respectively. In both types of control experiments, that is, when labelled cells were passed through a plain cryogel column and when unlabelled cells were passed through an affinity cryogel column, no cell binding was observed. (Reprinted by permission from Macmillan Publishers Ltd. *Nat. Protoc.* Kumar, A., and Srivastava, A., 5:1737–1747, copyright 2010.)

### 16.3.3 BORONATE AFFINITY CHROMATOGRAPHY

The immobilization of boronate ligands provides a unique chromatography matrix for cell and biomolecule separation (Srivastava et al., 2012). The separation of adherent and nonadherent cells was demonstrated and the bound cells were recovered by mechanical squeezing, which is a generic approach for cell release. It was found that a low concentration of fructose can be used to elute nonadherent cells and high concentration of fructose as eluent can be used for adherent cell recovery from boronate grafted cryogel matrix. This may provide a new approach and matrix for the fractionation of adherent and nonadherent cells from the mixture. The cell recovery was done three different ways to compare the recovery yield while maintaining high viability. The mechanical squeezing was done to demonstrate that cells could be recovered irrespective of their dissociation constant with the ligand. In all three different experiments using CC9C10, D9D4 and Cos-7 cells, 75–80% of cells were recovered through mechanical squeezing. It provides a generic approach to separate cells from affinity cryogel matrices. The results demonstrated that both the nonadherent cells were recovered with 0.1 M fructose concentration and adherent cell (Cos-7) could be recovered at high fructose concentration of 0.5 M in PBS. At 0.1 M fructose concentration in PBS (0.1 M, pH 7.2) around 75–80% nonadherent cells were recovered but only 5% adherent cells could be released from boronate grafted cryogel matrices. Moreover, 80–90% bound adherent cells (Cos-7) could be recovered at 0.5 M fructose concentration. It may be explained based on glycan content and its difference in composition on adherent and nonadherent cells. It is anticipated that there will be differences in cell surface carbohydrates, laminin and fibronectin synthesis, which has already been shown between two Ehrlich ascites tumour (EAT) cell lines, that is, the adherent and nonadherent EAT cells (Song et al., 1993). In another work, differential expression of integrin subunits on adherent and nonadherent cells was also studied and demonstrated that the difference in glycoprotein content has an effect on cell adherence property (Grodzki et al., 2003). This might affect the binding affinity of different cell types. These experiments provide a new approach to fractionate the mixture of adherent and nonadherent cells.

### 16.3.4 FILTRATION OF LEUKOCYTES

In a recent application of cryogel (unpublished work), a cryogel filter has been developed that can allow cells like RBCs and platelets to pass through unhindered while capturing large leukocytes in the inner porous surface area of cryogel. Promising results have been obtained using these cryogel-based leukocyte depletion filters. The cryogel-based filter is easy to prepare, is mechanically stable, and can be made in different shapes and sizes. The developed filter is made of polyacrylate cryogels having a cotton mesh as a template. Briefly, interpenetrating network (IPN) of cryogel was made from a combination of poly(dimethylaminoethyl methacrylate) poly(DMAEMA), hydroxyethylmethacrylate (HEMA), butylmethacrylate (BMA) and methylmethacrylate (MMA) in various weight ratios. The choice of polymers provides a nice balance between hydrophilicity and hydrophobicity along with large number of cross-junctions. A developed cryogel sheet (9 cm dia.) contains 60%:40% ratio of hydrophilicity and hydrophobicity monomers. It has been verified earlier that a filter that has an appropriate hydrophilic/hydrophobic balance can be used as an effective leukocyte depletion filter (Kova et al., 1997). The interpenetrating network cryogel was prepared to provide a rough surface with a number of cross-junctions (a net-like structure within pores) to retain large and rough cell surface leukocytes. The smooth-walled RBCs can easily travel through the network. The prototype cryogel filter for leukocyte depletion was fabricated by optimizing the shape and size of filtration device based on the desired flow properties through the filter, pressure drop across the filtration unit, flow dynamics and finally depletion of leukocytes on the cryogel filter. The percentage of platelets and RBCs passing through the cryogel filter was optimized in such a way that a maximum number of these cells pass through the cryogel filter whereas a maximum depletion of the leukocytes binding can be achieved. After complete cytotoxicity testing using laboratory-based methods, whole blood and component blood filtration was performed with the cryogel filter sheets (Figure 16.12). These sheets were tightly packed in plastic

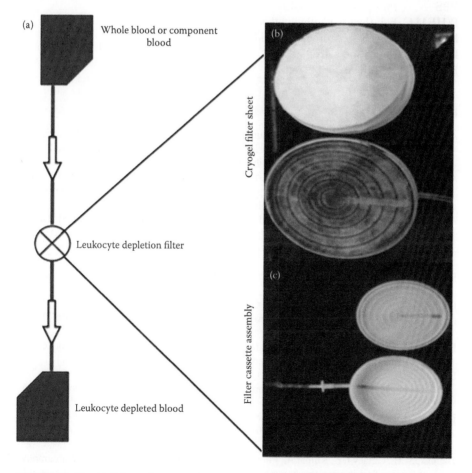

**FIGURE 16.12** (a) Schematic representation of passage of blood from developed cryogel filter; (b) digital image of macroporous cryogel matrices for filter; (c) digital image of cassette for filter.

filter mould and saline solution (0.9%, w/v) was passed at a flow rate of 100 ml/min. After saline, one unit whole blood or component blood was passed through these cryogel filters for leukocyte depletion. It was found that more than 95% leukocyte was depleted with minimal RBCs and platelet retention.

## LIST OF ABBREVIATIONS

| | |
|---|---|
| BSA | Bovine serum albumin |
| ConA | Concanavalin A |
| FACS | Fluorescence activated cell sorting |
| FTIR | Fourier transform infra-red spectroscopy |
| HETP | Height equivalent to a theoretical plate |
| MACS | Magnetic-activated cell sorting |
| pAAm | Poly(acrylamide) |
| PBS | Phosphate buffer saline |
| P(HEMA) | Poly(hydroxyl ethylmethacyrlate) |
| PNiPAAm | Poly(N-isopropylacrylamide) |
| RBCs | Red blood cells |
| UCB | Umbilical cord blood |

## REFERENCES

Ahlqvist, J., Kumar, A., Sundstrom, H., Ledung, E., Hornsten, E.G., Enfors, S.O., and Mattiasson, B. (2006). Affinity binding of inclusion bodies on supermacroporous monolithic cryogels using labeling with specific antibodies. *J. Biotechnol.* 122: 216–225.

Arvidsson, P., Plieva, F.M., Savina, I.N., Lozinsky, V.I., Fexby, S., Bulow, L., Galaev, I.Y., and Mattiasson, B. (2002). Chromatography of microbial cells using continuous supermacroporous affinity and ion-exchange columns. *J. Chromatogr. A* 977: 27–38.

Arvidsson, P., Plieva, F.M., Lozinsky, V.I., Galaev, I.Y., and Mattiasson, B. (2003). Direct chromatographic capture of enzyme from crude homogenate using immobilized metal affinity chromatography on a continuous supermacroporous adsorbent. *J. Chromatogr. A* 986: 275–290.

Beaujean, F. (1997). Methods of CD34+ cell separation: Comparative analysis. *Transfus. Sci.* 18: 251–261.

Berenson, R.J., Bensinger, W.I., Kalamasz, D., and Martin, P. (1986a). Elimination of daudi lymphoblasts from human-bone marrow using avidin-biotin immunoadsorption. *Blood* 67: 509–515.

Berenson, R.J., Bensinger, W.I., and Kalamasz, D. (1986b). Positive selection of viable cell populations using avidin-biotin immunoadsorption. *J. Immunol. Methods* 91: 11–19.

Dainiak, M.B., Galaev, I.Y., Mattiasson, B. (2006a). Affinity cryogel monoliths for screening for optimal separation conditions and chromatographic separation of cells. *J. Chromatogr. A* 1123: 145–150.

Dainiak, M.B., Kumar, A., Galaev, I.Y., and Mattiasson, B. (2006b). Detachment of affinity-captured bioparticles by elastic deformation of a macroporous hydrogel. *Proc. Natl. Acad. Sci. U. S. A.* 103: 849–854.

Dainiak, M.B., Galaev I.Y., Kumar A., Plieva F.M., and Mattiasson B. (2007). Chromatography of living cells using supermacroporous hydrogels, cryogels. *Adv. Biochem. Eng. Biotechnol.* 106: 101–127.

Dainiak, M.B., Plieva F.M., Galaev I.Y., Hatti-Kaul R., and Mattiasson B. (2005). Cell chromatography: Separation of different microbial cells using IMAC supermacroporous monolithic columns. *Biotechnol. Prog.* 21: 644–649.

Du, K-F., Yang, D., and Sun, Y. (2007). Fabrication of high-permeability and high-capacity monolith for protein chromatography. *J. Chromatogr. A* 1163: 212–218.

Fisher, D., Francis, G.E., and Rickwood, D. (1998). *Cell Separation: A Practical Approach.* New York: Oxford University Press.

Geifman-Holtzman, O., Makhlouf, F., Kaufman, L., Gonchoroff, N.J., and Holtzman, E.J. (2000). The clinical utility of fetal cell sorting to determine prenatally fetal E/e or e/e Rh genotype from peripheral maternal blood. *Am. J. Obstet. Gynecol.* 183: 462–468.

Grodzki, A.C.G., Pastor, M.V.D., Sousa, J.F., Oliver, C., and Jamur, M.C. (2003). Differential expression of integrin subunits on adherent and nonadherent mast cells. *Braz. J. Med. Biol. Res.* 36: 1101–1109.

Kumar, A., Galaev, I.Y., and Mattiasson, B. (2007). *Cell Separation: Fundamentals, Analytical and Preparative Methods.* Berlin: Springer-Verlag.

Kumar, A., Plieva, F.M., Galeav, I.Y., and Mattiasson, B. (2003). Affinity fractionation of lymphocytes using a monolithic cryogel. *J. Immunol. Methods* 283: 185–194.

Kumar, A., and Srivastava, A. (2010). Cell separation using cryogel-based affinity chromatography. *Nat. Protoc.* 5: 1737–1747.

Kumar, A., Rodriguez-Caballero, A., Plieva, F.M., Galaev, I.Y., Nandakumar, K.S., Kamihira, M., Holmdahl, R., Orfao, A., and Mattiasson, B. (2005). Affinity binding of cells to cryogel adsorbents with immobilized specific ligands: Effect of ligand coupling and matrix architecture. *J. Mol. Recognit.* 18: 84–93.

Kova, M. J., Bartunkova, J., Smetana, K., Lukas, J., Vacik, J., and Dyr, J. E. (1997). Comparative study of human monocyte and platelet adhesion to hydrogels *in vitro*: Effect of polymer structure, *J. Mater. Sci. Mater. Med.* 8: 19–23.

Mihelic, I., Nemec, D. A., Podgornik, A., Koloini, T. (2005). Pressure drop in CIM disk monolithic columns. *J. Chromatogr. A* 1065: 59–67.

Nadali, G., de Wynter, E.A., and Testa, N.G. (1995). CD34 cell separation: From basic research to clinical applications. *Int. J. Clin. Lab. Res.* 25: 121–127.

Neurauter, A.A., Bonyhadi, M., Lien, E., Nokleby, L., Ruud, E., Camacho, S., and Aravak, T. (2007). Cell isolation and expansion using dyna beads. *Adv. Biochem. Eng. Biotechnol.* A. Kumar, I.Y. Galaev, and B. Mattiasson, Eds. Berlin: Springer, pp. 41–73.

Noppe, W., Plieva, F.M., Vanhoorelbeke, K., Deckmyn, H., Tuncel, M., Tuncel, A., Galaev, I.Y., and Mattiasson, B. (2007). Macroporous monolithic gels, cryogels, with immobilized phages from phage-display library as a new platform for fast development of affinity adsorbent capable of target capture from crude feeds. *J. Biotech.* 131: 293–299.

Plieva, F.M., Savina, I.N., Deraz, S., Andersson, J., Galaev, I.Y., and Mattiasson, B. (2004a). Characterization of supermacroporous monolithic polyacrylamide based matrices designed for chromatography of bioparticles. *J. Chromatogr. B* 807: 129–137.

Plieva, F.M., Andersson, J., Galaev, I.Y., Mattiasson, B. (2004b). Characterization of polyacrylamide based monolithic columns. *J. Sep. Sci.* 27: 828–836.

Plieva, F.M., Seta, E.D., Galaev, I.Y., and Mattiasson, B. (2009). Macroporous elastic polyacrylamide monolith columns: Processing under compression and scale-up. *Sep. Purif. Technol.* 65: 110–116.

Plieva, F.M., Galaev, I.Y., Noppe, W., and Mattiasson B. (2008). Cryogel applications in microbiology. *Trends Microbiol.* 16: 543–551.

Persson, P., Baybak, O., Plieva, F., Galaev, I.Y., Mattiasson, B., Nilsson, B., and Axelsson, A. (2004). Characterization of a continuous supermacroporous monolithic matrix for chromatographic separation of large bioparticles. *Biotechnol. Bioeng.* 88: 224–236.

Racila, E., Euhus, D., Weiss, A.J., Rao, C., McConnell, J., Terstappen, L.W., and Uhr, J.W. (1998). Detection and characterization of carcinoma cells in the blood. *Proc. Natl. Acad. Sci. U.S.A.* 95: 4589–4594.

Recktenwald, D., and Radbruch, A. (1997). *Cell Separation Methods and Applications.* New York: Dekker.

Siewert, C., Herber, M., Hunzelmann, N., Fodstad, O., Miltenyi, S., Assenmacher, M., and Schmitz, J. (2001). Rapid enrichment and detection of melanoma cells from peripheral blood mononuclear cells by a new assay combining immunomagnetic cell sorting and immunocytochemical staining. *Recent Results Cancer Res.* 158: 51–60.

Srivastava, A., Sakhya, A.K., and Kumar, A. (2012). Boronate affinity chromatography of cells and biomacromolecules using cryogel matrices. *Enzyme Microb. Technol.* 10: 373–81.

Song, Z., Varani, J., and Goldstein, I.J. (1993). Differences in cell surface carbohydrates, and in laminin and fibronectin synthesis, between adherent and non-adherent Ehrlich ascites tumor cells. *Int. J. Cancer* 55: 1029–1035.

Teilum, M., M.J. Hansson, M.B. Dainiak, R. Mansson, S. Surve, E. Elmer, P. Onnerfjord, and G. Mattiasson. (2006). Binding mitochondria to cryogel monoliths allows detection of proteins specifically released following permeability transition. *Anal. Biochem.* 348: 209–221.

Williams, S.L., M.E. Eccleston, and N.K. Slater. (2005). Affinity capture of a biotinylated retrovirus on macroporous monolithic adsorbents: Towards a rapid single-step purification process. *Biotechnol. Bioeng.* 89: 783–787.

Yao, K., Yun, J., Shen, S., Wang, L., He, X., and Yu, X. (2006). Characterization of a novel continuous supermacroporous monolithic cryogel embedded with nanoparticles for protein chromatography. *J. Chromatogr. A* 1109: 103–110.

Yao, K., Yun, J., Shen, S., and Chen, F. (2007). In-situ graft-polymerization preparation of cation-exchange supermacroporous cryogel with sulfo groups in glass columns. *J. Chromatogr. A* 1–2, 246–251.

Yun, J., Shen, S., Chen, F., and Yao, K. One-step isolation of adenosine triphosphate from crude fermentation broth of *Saccharomyces cerevisiae* by anion-exchange chromatography using supermacroporous cryogel. (2007). *J. Chromatogr. B* 860: 57–62.

Yun, J., Jespersen, G.R., Kirsebom, H., Gustavsson, P., Mattiasson, B., and Galaev, I.Y. (2011). An improved capillary model for describing the microstructure characteristics, fluid hydrodynamics and breakthrough performance of proteins in cryogel beds. *J. Chromatogr. A* 1218: 5487–5497.

# Index

Page numbers followed by f and t indicate figures and tables, respectively.

## A

Acetone, 97, 447, 448f
Acetonitrile, in mobile phase, 379
Acetoxycoumarins, 318
Acetylcholine, sensor for, 26–27
Acrylamide (AAm)
    cryogel, 48t, 53, 104–108, 105f, 106f, 107f
    leakage, 337, 338
Acrylic acid, 47t
Actifuse, 228t
Actin, 201
Addition polymerization, 44, 45–53
    free radical polymerization
        cryogelation by, 44, 45, 46t–48t, 48, 49t
        mechanism of, 48, 49–51
    initiation
        photochemical, 52
        radiation, 52–53
        redox, 52
        thermal, 51–52
Adipic acid hydrazide, cross-linking with, 55
Adipose derived stem cells (ADSCs), growth of, 201, 202t, 203, 209
Adsorbents
    cryogel-based, 333–335
        functionalized, 333–334
        protected, 334–335, 335f, 336f
    immuno-affinity, 403
    ligand-based and particle-based cryogel composites, 342t–343t
    MIPs, 335, 336, 341, 344–345, 344f
    pollutant source, 345–349
        industrial and municipal wastewater, 345–346, 346f
        metal removal, in biogas production, 347–349, 348f
        microbial cryogel systems, 349
        oil removal, advantage of cryogel pore space, 347
        POU filters, 346–347
Adsorption
    equilibrium equation, 132
    method, nitrogen, 67
    "tentacle," of protein, 107–108
    wastewater treatment, 333
Affine binding, protein immobilization by, 320–325
    antigen-antibody binding, 321

    lectin, 322
    metal ion chelating binding, 322–325, 323t
Affinity chromatography, 454
Affinity separation methods, 445
Affinity therapies, extracorporeal, *see* Extracorporeal affinity therapy
Agarose (AGS)
    for cartilage tissue engineering, 22
    chitosan-agarose-gelatin, 71, 206, 207f, 209, 240, 243
    gelatin and, 238, 239f
    gel-based TMMs, 154
    with polymers, cross-linkers/initiators, and mechanism, 17t
    properties of MSCs within, 291, 292–293, 292f
    regeneration of liver and pancreas, 189–190
Agmatine-based RGD mimetics (abRGDm), 376
Alamar blue fluorescence, 292
Albumin absorption, 438
Aldehydes, cross-linking with, 44, 44f, 55
Alginate
    based macroporous cryogel scaffolds, 289–294, 290f, 292f, 293f
        MSCs seeded, cryopreservation of, 294–295
    intraperitoneal implantation, 279
Allografts, 217, 225, 229, 257
Allylglycidyl ether (AGE), 16
Alpha-ketoglutarate (α-KG), 264, 266
Ammonium persulphate, 49t
Anthracene, 379
Antibody
    anti-dsDNA, concentration on adsorption capacity, 391, 392f
    antigen-antibody binding, 321
    immobilization of, 403, 405, 405f, 406f
    PHEMA-anti-LDL antibody, 406, 407f
Anti-dsDNA antibody, concentration on adsorption capacity, 391, 392f
Antigen-antibody binding, protein immobilization via, 321
Anti-inflammatory effects, of stem cells, 264
Anti-LDL antibody, PHEMA and, 406, 407f
Antimicrobials, in wound dressing, 186
Apaceram Bone Graft Substitute, 228t
ApoB100, protein component of LDL, 403

**464**  Index

Applications, 19–27
  beads and composites, 135
  biotechnological, 13f
  in environmental biotechnology, 23, 24–26, 25f
  in immobilization, separation, and imprinting, 19–21, 20f
  others, 26–27
  paradigm shift in, 14f
  in tissue engineering and biomedical applications, 21–23, 24f
  in water and wastewater treatment, *see* Water and wastewater treatment
Aqueous drops, in microchannels
  freezing, in water-immiscible fluid, 119–120, 120f
  generation, 118–119
Aqueous solutions, freezing of, 92–93, 92f, 93f
Archimedes principle, 65
Arsenic
  environmental pollutant, 332
  removal of, 26, 345–346, 346f
  use of, 350
Arthritis development, 434f
Arthro-Kinetics, 226t
Articulating cartilage, 186–188
Atomic force microscopy (AFM), 199
Attenuated total reflection (ATR) method of FTIR, 73
Autografts, 217, 225, 229, 256–257
Axial dispersion coefficient, 132
1,1′-Azobis(cyclohexanecarbonitrile) (ABCN), 49t
2,2′-Azo-*bis*-isobutyronitrile (AIBN), 45, 49t

**B**

*Bacillus halodurans*, 374
Beads, cryogel
  applications of, 135
  characterization, 112–113, 129–134; *see also* Characterization
  composite monolithic cryogels, 124–128; *see also* Composite monolithic cryogels
  defined, 113
  fabrication, 112–124; *see also* Fabrication
Beads of polyacrylamide-gelatin bioreactor, 431f, 433f
Benzo[a]pyrene (BAP), 377, 379
  removal of, 26
Benzoyl peroxide, 49t
BiceraResorbable Bone Substitute, 228t
Bilirubin, affinity therapy, 393–397; *see also* Hyperbilirubinemia
Bioactive glass composite cryogels, for bone tissue engineering, 225, 228, 232–236, 233f, 234f, 235f, 236f, 237f

Bioartificial liver devices (BAL), 436
Bio-artificial pancreas (BAP)
  concept, 278
  as delivery of oxygen and nutrients, 280
  designs and configurations, 279
  prototypes of, 278–279
Bioceed-C contains chondrocytes, 217
Biocompatibility, cryogels
  defined, 196
  *in vitro*, 75–76, 200–205, 202t–203t, 203f, 204f
  *in vivo*, 205–209, 206t, 207f, 208f
  of macroporous cryogel materials, *see* Macroporous cryogels, materials
  material, factors affecting, 197–200, 198f
  properties, 18
Biogas production, metal removal in, 347–349, 348f
Biological properties, of cryogels, 74–76
  *in vitro* biocompatibility, 75–76
  protein adsorption studies, 74–75
Biomaterials, for musculoskeletal tissue engineering, 218–230; *see also* Musculoskeletal tissue engineering
  bone, 225, 228–229, 228t
    ceramic/bioactive glass, 225
    polymer-ceramic/bioactive glass composites, 225, 228
    polymeric, 225
  cartilage, 219–225, 220t–224t, 226t–227t
  ligament, 229–230
  muscle, 229
  tendon, 229
Biomedical applications, cryogels in, 21–23, 24f
Biomet, 227t
Biomolecules at injury site, delivery of, 258
Bioreactors, defined, 418
Bioseed C, 227t
Biosphere Bioactive Bone Graft Putty, 228t
Biotechnological applications of cryogels, 13f
  environmental, 23, 24–26, 25f
Biotinylated MoMuLV, 453
Bio Tissue Technologies GmbH, 227t
Blue Phantom™, 152–153
Bone
  neoplasia, treatment, 217
  regeneration of, 188–189
Bone marrow stromal cells (BMSCs), differentiation of, 231
Bone mineral, defined, 340
Bone tissue engineering
  biomaterials for, 225, 228–229, 228t
  cryogels for, 230–235
    gelatin-based, 231
    HEMA based, 232
    polyethylene glycol-based, 231–232
    polymeric, 231–232

Index

polymeric-ceramic/bioactive glass composite, 232–236, 233f, 234f, 235f, 236f, 237f
silk fibroin based, 232
Boronate affinity chromatography, 458; *see also* Cryogel chromatography column
Bovine serum albumin (BSA), HETP on flow rate for acetone, 447, 448f
Brain phantoms, 154–155, 154f
Breast phantoms, 152–154, 153f
Brunauer-Emmet-Teller (BET) surface area, 338, 340
Butylmethacrylate (BMA), 458

C

Cadmium
adsorption, in real seaweed leachate, 348, 348f
environmental pollutant, 332
use of, 350
CallosPromodelBone Void Filler, 228t
Capillary-based models, for cryogel-bead packed beds/monolithic cryogels, 129–132
adsorption equilibrium equation, 132
axial dispersion coefficient, 132
bed porosity and permeability, 130–131
mass balance equation, 131
mass transfer coefficient, 132
pore structure, 129f
pore tortuosity and size distributions, 130
Carbohydrate-based scaffolds, 217
7-carbon spacers, 449
Carboxymethyl cellulose (CMC), 40, 41–42, 42f
Carboxymethyl curdlan (CMc), 40
CaReS-1S, 226t
Cartilage repair and regeneration, cryogels for, 186–188
Cartilage tissue engineering
biomaterials for, 219–225, 220t–224t, 226t–227t
cryogels for, 236–243
natural polymers, fabricated using, 237–243, 239f, 241f, 242f
synthetic polymers, fabricated using, 243
Cartipatch, 226t
Cascade filtration, 405t
Cassette for filter, digital image, 459f
Cell-based assays, 375–376, 377f, 378f
Cell delivery, microcryogels for, 190
Cell proliferation assay, 201
Cell recovery properties, versatile
mechanical deformation of cryogel matrix, 450–451
treatment with immunoglobulin (Ig) solution, 450

Cell separation
efficiency, 457f
optimal conditions for, 364–366, 365t, 367f
Cell separation stationary matrices, cryogel features for; *see also* Supermacroporous functional cryogel stationary matrices
flow rate, 446
functional matrix, 448–450
movement of large particles, enabling, 447, 448f
versatile cell recovery properties, 450–451
Cell separation technology; *see also* Supermacroporous functional cryogel stationary matrices
cryogel-based, 445–451
current techniques/demand in, 445
importance of, 444–445
Cell support matrix in extracorporeal bioartificial device, 436, 438
Cell surface analysis, 374–375, 375f
Cell therapy, for neural regeneration, 254–255, 255f
Central nervous system (CNS), 253–254, 253f, 254f
Ceramic biomaterials, for bone tissue engineering, 225
Characterization, of cryogels, 59–76; *see also* Synthesis and characterization
biological properties, 74–76
*in vitro* biocompatibility, 75–76
protein adsorption studies, 74–75
chemical properties, 73–74
FTIR, 73
XPS and elemental analysis, 74
XRD, 73–74
cryogel phantoms, 160–166
mechanical properties, 160–166, 161f, 162f, 163f, 164f, 165f
wave propagation and MRI properties, 166
mechanical and thermal properties, 69–72
compression test, 69–70
fatigue test, 70–71
rheology studies, 71–72, 72f
modelling of fluid flow and mass transfer for, 129–134
capillary-based, 129–132; *see also* Capillary-based models
numerical solution of model, 132–134
morphological and textural features, 59–69
CLSM, 60, 60f, 61, 61f, 62f
hydrated polysaccharide-based cryogels, 60, 60f
μ-CT, 61–64, 62f, 63f, 64f, 65f, 65t
mercury porosimetry, 65–67, 67f
porosity, 64–69; *see also* Porosity

Chelatotropic cryogels, 38
Chemical cross-linking, 44, 44f, 45f, 45t
Chemical properties, of cryogels, 73–74
  FTIR, 73
  XPS and elemental analysis, 74
  XRD, 73–74
Chemical stability, cryogels, properties, 18
Chemotropic cryogels, 38
Chitosan cryogels
  for cartilage tissue engineering, 22
  chitosan-agarose-gelatin, 71, 206, 207f, 209, 240, 243
  cross-linked network and, 15
  derivative cryogel, 183
  fatigue test, 70–71
  gelatin and, 238, 239f
  porosity, 66
Chlorhexidine (CX), antimicrobial agent, 186
Chondriotin sulfate, 253
Chondrocytes
  in cartilage tissue engineering, 237–238
  goat, 240
Chondro-Gide, 227t
Chondux, 227t
Chromatopanning procedure, 376, 378t
α-Chymotrypsin, 310t, 312–313
Circular dichroism (CD) technique, 199
*Citrobacterium intermedius*, 10, 19
Collagen-nanohydroxyapatite biocomposite cryogels, 235, 236
Composite cryogels, 335, 336–345
  ligand-based and particle-based, characteristics, 342t–343t
  metal-oxide, 336–340, 337f, 339f
  with molecularly imprinted polymers, 341, 344–345, 344f
  other types, 340, 341, 342t–343t
Composite monolithic cryogels, 124–128
  applications of, 135
  overview, 112–113, 113f
  preparation, via cryo-polymerization, 124–128
    with embedded macroporous beads, 124–125, 125f
    grafting polymerization, 125–126, 126f
    properties, 126–128, 127f, 128f
Compression test, 69–70
Compressive strength, 452f
Computerized Imaging Reference Systems (CIRS) Inc., 152
Concanavalin A, 317, 322, 365–366, 365t
Concavalin A-cryogel (ConA-cryogel), 452
Concavlin A, 19
Conducting cryogels, for neural tissue engineering, 267–268, 267f
Conducting polymer, 261–262
CONDUIT® TCP Granules, 228t

Confocal laser scanning microscopy (CLSM), 60, 60f, 61, 61f, 62f, 180
Conventional gels *vs.* cryogels, 8–9
Conventional strategies, in neural tissue engineering, 255–258, 256f
Cord blood stem cells (CBSCs), 264
Coronary heart disease, development, 403
Costs, water and wastewater treatment, 352–353
Covalent immobilization, on cryogel-type carriers, 309–315, 310t, 314f
Cross-linking, cryogelation by functional group, 55–56
  adipic acid hydrazide and, 55
  aldehydes and, 55
  EDC-NHS and, 56, 56f
Cryoconcentration
  defined, 7, 8, 37
  NFLMP, 56
  phase diagram of NaCl, 93, 93f
Cryogel(s)
  applications, 19–27
    biomedical, 21–23, 24f
    in environmental biotechnology, 23, 24–26, 25f
    in extracorporeal affinity therapy, *see* Extracorporeal affinity therapy
    in immobilization, separation, and imprinting, 19–21, 20f
    others, 26–27
    in tissue engineering, 21–23, 24f
    in water and wastewater treatment, *see* Water and wastewater treatment
  beads
    applications of, 135
    characterization, 112–113, 129–134; *see also* Characterization
    composite monolithic cryogels, 124–128; *see also* Composite monolithic cryogels
    defined, 113
    fabrication, 112–124; *see also* Fabrication
  category, 38
  as cell support matrix in extracorporeal bioartificial device, 436, 438
  composite monolithic, *see* Composite monolithic cryogels
  conducting, 267–268, 267f
  conventional gels *vs.*, 8–9
  cryogelation, process, 7–8, 8f
  defined, 6, 36, 262, 444
  enzymatic biocatalysts, *see* Enzymatic biocatalysts
  gels and different gelation methods, 5, 6t
  in high throughput processes, *see* High throughput processes
  historical perspective, 9–11
  morphology, *see* Morphology

Index

musculoskeletal tissue engineering, biomaterials for, see Musculoskeletal tissue engineering
for neural tissue engineering, 262–266, 263f, 265f, 266f; see also Neural tissue engineering
origin, 4–5
overview, 6–15
pore space, advantage, 347
properties, 16, 17–18
  biocompatibility, 18
  chemical stability and modifiability, 18
  ease of fabrication, economical operations and longer shelf life, 18
  mechanical stability and viscoelasticity, 17–18
  rapid swelling mechanism, 16, 17
  supermacroporous architecture with interconnected pores, 16
in regenerative medicine, see Regenerative medicine
related research, 4–27
statistical perspective, 11–15, 11f, 12f, 12t, 13f, 14f
supermacroporous
  architecture with interconnected pores, 16
  as scaffolds for pancreatic islet transplantation, see Pancreatic islet transplantation
synthesis, 15–16, 17t; see also Synthesis and characterization
synthetic, see Synthetic cryogels
tissue phantoms with uniform elasticity for medical imaging, see Tissue phantoms
tortuous capillaries, 447f
Cryogelation
  for fabricating macroporous hydrogels, 53
  by free radical polymerization, 44, 45, 48, 53–56
    commonly used monomers, 46t–48t
    cross-linking with adipic acid hydrazide, 55
    cross-linking with aldehydes, 55
    cross-linking with EDC-NHS, 56, 56f
    functional group cross-linking, 55–56
    initiators, 49t
    using UV irradiation, 54–55, 54f
  overview, 36–37
  process, 7–8, 8f, 64
Cryogel-based cell separation technology, see Cell separation stationary matrices, cryogel features for
Cryogel-bead packed beds, capillary-based models for, 129–132
  adsorption equilibrium equation, 132
  axial dispersion coefficient, 132
  bed porosity and permeability, 130–131
  mass balance equation, 131
  mass transfer coefficient, 132
  pore structure, 129f
  pore tortuosity and size distributions, 130
Cryogel bioreactor
  matrix, 422–425
  network and physical characteristics of, 420–421
Cryogel chromatography column; see also Supermacroporous functional cryogel stationary matrices
  boronate affinity chromatography, 458
  filtration of leukocytes, 458–460
  mammalian cell separation for biomedical applications
    fractionation of immune cell types, 454
    separation of stem cell, 454–457
  prokaryotic cell separation, 451–454
Cryogel immobilized metal affinity chromatography columns, 429t
Cryogel phantoms, 155–160
  characterization and homogeneity, 160–166
    mechanical properties, 160–166, 161f, 162f, 163f, 164f, 165f
    wave propagation and MRI properties, 166
  defined, 155
  formation process, 155f
  other polymers, 158–160, 159t, 160f
  PVA homopolymer, 156–158, 156f, 157f, 158f
Cryopolymerization
  fabrication of cryogel beads by, 114–124
    aqueous drops and immiscible slug flow in microchannels, 118–119
    freezing of aqueous drops in water-immiscible fluid, 119–120, 120f
    microchannel liquid-flow focusing and, 114–116, 115f
    microchannels for fluid-focusing, 116–117, 117f
    overview, 114
    properties of cryogel beads, 120–124, 121f, 122f, 123f
  mechanism of, 94–97, 95f, 96f, 97f
  preparation of composite monolithic cryogels via, 124–128
    with embedded macroporous beads, 124–125, 125f
    grafting polymerization, 125–126, 126f
    properties, 126–128, 127f, 128f
  solutes on, 97–99, 98f, 100f
Cryoporometry, NMR, 67–68
Cryopreservation, of MSCs seeded macroporous alginate-based cryogel scaffolds, 294–295
Cryo-SEM techniques, 60

Cryo-structurization, principle of, 9–10
Cryotropic cryogels, 38
Crystallinity of materials, XRD and, 73–74

## D

Deep cartilage defect, 242f
Defrosting, 420
Delivery, of biomolecules at injury site, 258
Density gradient centrifugation, 444
Deposition, in water and wastewater treatment, 350–351
Depression, freezing point, 93, 94
Dermal tissue regeneration, cryogels for, 177–184, 179f, 180f, 181f, 182f, 183f
Desferrioxamine B (DFO), 400
Dextran
  induced LDL precipitation, 405t
  macroporous hydrogels of, 53
Diabetic environment, neovascularisation in, 280–282
Differential scanning calorimetry (DSC) method, 71
  temperature-dependent behaviour of water, 101–103
Diffusion tensor imaging (DTI), MRE and, 165
3-(4,5-Dimethylthiazol-2-yl)-2,5-diphenyltetrazolium bromide (MTT) dye, 201
1,4-Dioxane, as porogen, 57–58, 60f
Direct adsorption of lipoproteins (DALI), 405t
Direct contact assay, 200–201
Disposable bioreactor, benefits of, 419
Dried polyacrylamidechitosan cryogels, 431f
Dry polymeric cryogels, 446

## E

Economical operations, 18
Ehrlich ascites tumour (EAT), 458
Elastography
  phantoms, need for, 150–155
    brain, 154–155, 154f
    breast, 152–154, 153f
    liver, 151–152, 152f
    mechanical properties for TMMs, 151t
  techniques, for medical imaging, 147–150; see also Tissue phantoms
    MRE, 149–150
    OCE, 148–149
    USE, 147–148
Electrospun fibres, for neural regeneration, 258, 259f
Element analyser, 74
Embedded macroporous beads, preparation of cryogels with, 124–125, 125f
Engineered nanomaterials (ENMs), 351
Entrapment technique, protein immobilization by, 315–319, 316t

Environmental biotechnology, cryogels in, 23, 24–26, 25f
Environmental pollutants, 332, 362
Environmental scanning electron microscopy (ESEM), 60, 60f
Enzymatic biocatalysts, 307–325
  overview, 308–309, 309f
  protein immobilization
    affine binding, 320–325
    antigen-antibody binding, 321
    combined methods, 325
    covalent attachment to cryogel matrix, 309–315, 310t, 314f
    entrapment technique, 315–319, 316t
    ionic binding, 320
    lectin affinity binding, 322
    metal ion chelating binding, 322–325, 323t
Enzyme-linked immunosorbent assay (ELISA), 369, 371, 372
*Escherichia coli*, 313, 365–366, 365t, 369, 369t, 374, 376
  cells, 447
Esperase, 310t
Ethyl(dimethylaminopropyl) carbodiimide-$N$-hydroxysuccinimide (EDC-NHS), 56, 56f
Ethylenediaminetetraacetic acid (EDTA), 350, 366, 452
EXPRESS-ION™ Exchanger Q particles, 320
Extracellular matrix (ECM)
  in cartilage tissue engineering, 186–188, 219–225, 220t–224t
  collagen, 178
  production, 187–188
Extracorporeal affinity therapy, applications in, 387–411
  in hyperbilirubinemia, 393–397
    MATrp, 396, 398f
    MIP-bilirubin cryogel, 394, 395f
    PHEMA/MIP composite cryogel, 395, 396f, 397f, 397t
    poly(HEMA-MATyr), 394, 394f
  in hypercholesterolemia, 403, 405–409, 405f, 405t, 406f, 407f, 408f, 408t, 409t
  other applications, 409–411
  overview, 388–389, 388f
  in rheumatoid arthritis, 397, 398–399, 399f, 399t
  in SLE, 389–393
    anti-dsDNA antibody concentration, 391, 392f
    flow pattern of whole blood, 392f, 393
    PHEMA-based adsorbents, 389–391, 390f, 391f
  in thalassemia, 399–403, 401f, 402f, 404f
Extracorporeal bioartificial device, cell support matrix in, 436, 438

# Index

## F

Fabrication, of cryogels, 112–124
   beads by micro-flow focusing and cryo-polymerization, 114–124
      aqueous drops and immiscible slug flow in microchannels, 118–119
      freezing of aqueous drops in water-immiscible fluid, 119–120, 120f
      microchannel liquid-flow, 114–116, 115f
      microchannels for fluid-focusing, 116–117, 117f
      overview, 114
      properties of cryogel beads, 120–124, 121f, 122f, 123f
   for cartilage tissue engineering
      using natural polymers, 237–243, 239f, 241f, 242f
      using synthetic polymers, 243
   ease of, 18
   overview, 112–113, 113f
Fabrication, of scaffolds, 258–259
   delivery of biomolecules at injury site, 258
   electrospun fibres for neural regeneration, 258, 259f
   hydrogels for neural recovery, 258, 259, 260t
Fabrication method, 422–425
Fatigue test, 70–71
Fibra-Cel® polyester-polypropylene disks, 419
Fibroscan®, 161, 164, 165
Fidia Advanced Biopolymers, 226t
Fiji software, 99
Filters, POU, 346–347
Filtration of leukocytes, 458–460; *see also* Cryogel chromatography column
Flow-through method, 338, 339f
Fluid flow, modelling, for characterization of cryogel beds, 129–134
   capillary-based, 129–132; *see also* Capillary-based models
   numerical solution of model, 132–134
Fluorescent activated cell sorting (FACS), 445
Focusing, micro-flow
   fabrication of cryogel beads by, 114–124
      aqueous drops and immiscible slug flow in microchannels, 118–119
      freezing of aqueous drops in water-immiscible fluid, 119–120, 120f
      microchannel liquid-flow, 114–116, 115f
      microchannels for fluid-focusing, 116–117, 117f
      overview, 114
      properties of cryogel beads, 120–124, 121f, 122f, 123f
Formamide, as porogen, 57–58, 58f
Formation, gel, 38–53
   addition polymerization, 44, 45–53

   chemical cross-linking, 44, 44f, 45f, 45t
   cryogelation by free radical polymerization, 44, 45, 46t–48t, 48, 49t
   mechanism of free radical polymerization, 48, 49–51
   photochemical initiators, 52
   physical cross-linking, 38–43, 42f, 43f
   radiation initiation, 52–53
   redox initiation, 52
   thermal initiators, 51–52
Fourier transform infrared spectroscopy (FTIR), 73, 449, 450f
Fractionation of immune cell, 454
Free radical polymerization, 422
   cryogelation by, 44, 45, 48, 53–56
      commonly used monomers, 46t–48t
      cross-linking with adipic acid hydrazide, 55
      cross-linking with aldehydes, 55
      cross-linking with EDC-NHS, 56, 56f
      functional group cross-linking, 55–56
      initiators, 49t
      using UV irradiation, 54–55, 54f
   mechanism of, 48, 49–51
Freeze–thaw treatment
   cryogels/hydrogels, 147
      physically cross-linked, 40
      PVA, 38–39, 156–158, 156f, 157f, 158f
      starch, 41
      xanthan gum, 40–41
   PVA
      cryogels, 38–39
      homopolymer, 156–158, 156f, 157f, 158f
Freezing
   aqueous drops in water-immiscible fluid, 119–120, 120f
   aqueous solutions, 92–93, 92f, 93f
Freezing point
   defined, 93
   depression, 93, 94
Freezing rate, temperature and, 57
Fulminant hepatic failure (FHF), 438
Functional group cross-linking, cryogelation by, 55–56
   adipic acid hydrazide, 55
   aldehydes, 55
   EDC-NHS, 56, 56f
Functionalized cryogels, 333–334

## G

Geistlich Biomaterials, 227t
Gelatin cryogels
   agarose and, 238, 239f
   chitosan-agarose-gelatin, 71, 206, 207f, 209, 240, 243
   chitosan and, 238, 239f

cross-linked network and, 15
in dermal tissue regeneration, 178–183, 178f, 180f, 181f, 182f, 183f
fatigue test, 70–71
laminin cryogel, 263, 263f
macroporous, 55
porosity, 63f, 64, 64f, 65f, 66
for tissue engineering
bone, 231, 235, 236
cartilage, 22
Gelation
defined, 37–38
different methods, 5, 6t
process, 37–38
Gel point, defined, 38
Gels
conventional, cryogels vs., 8–9
defined, 5
formation, methods of inducing, 38–53
addition polymerization, 44, 45–53
chemical cross-linking, 44, 44f, 45f, 45t
cryogelation by free radical polymerization, 44, 45, 46t–48t, 48, 49t
mechanism of free radical polymerization, 48, 49–51
photochemical, 52
physical cross-linking, 38–43, 42f, 43f
radiation initiation, 52–53
redox initiation, 52
thermal initiators, 51–52
macroporous polymeric, porosity of, 99–104, 100f, 101f, 102f, 104f
Gibbs-Thomson equation, 103
Glucose oxidase, 310t
Glutaraldehyde
bacteriostatic effect, 186
for cross-linking synthetic and natural polymers, 55, 176
cryogel-type carrier activated by, 311–315
in dermal tissue regeneration, 178–183, 180f, 181f, 182f, 183f
wound dressings and, 186
Goat chondrocytes, 240
Grafting on macroporous cryogel, 449
Grafting polymerization
of composite cryogels, 125–126, 126f
on pore surface, 104–108, 105f, 106f, 107f
Graham, Thomas, 6
Gravimetric method, 101
Green fluorescent proteins (GFP), 369, 370f, 371t
Group cross-linking, cryogelation by functional, 55–56
adipic acid hydrazide, 55
aldehydes, 55
EDC-NHS, 56, 56f
Guillain-Barre syndrome, 398

## H

Haemodialysis, 393
Haemolytic activity, 200
Hagen-Poiseuille equation, 131
Height equivalent to theoretical plate (HETP), 447, 448f
Hemoperfusion, 388, 393
Heparin-induced extracorporeal LDL precipitation (HELP), 405t
HETP, see Height equivalent to theoretical plate (HETP)
High throughput processes, cryogels in, 361–379
analysis of proteins, 366–372, 368f, 369t, 370f, 371t, 373f–374f
other processes, 372–379
cell-based assays, 375–376, 377f, 378f
cell surface analysis, 374–375, 375f
panning of phage display library, 376, 377, 378t
pollutant analysis, 377, 379, 380f–381f
overview, 362–364, 363f
screening of optimal conditions for cell separation, 364–366, 365t, 367f
Histiogenics Corporation, 226t
Historical perspective, cryogels, 9–11
$^1$H NMR spectroscopy, low-temperature, 101, 103, 104f
Hollow fibre bioreactor (HFBR), 429, 429f, 436; see also Therapeutic protein production, cryogel bioreactor for
Homogeneity, cryogel phantoms, 160–166
mechanical properties, 160–166, 161f, 162f, 163f, 164f, 165f
wave propagation and MRI properties, 166
Homopolymer, PVA, 156–158, 156f, 157f, 158f
Horseradish peroxidase, 310t
Human colon cancer, 426, 428f
Human cord blood-derived stem cells, 189
Human fibrosarcoma, 426
Human mesenchymal stem cells (hMSCs), 178, 202t
Human peripheral blood lymphocytes, composition, 455f
Hyalograft C, 226t
Hyaluronic acid hydrogels, 42–43
Hybridoma clone, 431
Hybridoma metabolism, 437f
Hydrated polysaccharide-based cryogels, 60, 60f
Hydrogels
hyaluronic acid, 42–43
for neural recovery, 258, 259, 260t
water-swollen polymeric materials, 147
Hydrogen bonding induced gelation, 43, 43f
Hydrophilic polymers, 448
Hydrostatic pressures, 446
Hydroxyapatite cryogels, for bone tissue engineering, 235, 236

# Index

Hydroxyethyl methacrylate (HEMA), 48t, 458
Hydroxyethyl methacrylate (HEMA) cryogels for bone tissue engineering, 232
   morphology, 101, 101f, 102f
Hyperbilirubinemia, affinity therapy in, 393–397
   MATrp, 396, 398f
   MIP-bilirubin cryogel, 394, 395f
   PHEMA/MIP composite cryogel, 395, 396f, 397f, 397t
   poly(HEMA-MATyr), 394, 394f
Hypercholesterolemia, affinity therapy in, 403, 405–409, 405f, 405t, 406f, 407f, 408f, 408t, 409t

## I

Ice crystallization, 7, 59, 92–93, 94–97, 99
ImageJ software, 99, 101
Imaging, cryogel tissue phantoms with uniform elasticity, see Tissue phantoms
Iminodiacetate, 451
Iminodiacetic acid, 426
Immiscible fluids
   flow, in microchannels, 118–119
   water, freezing of aqueous drops in, 119–120, 120f
Immobilization
   cryogels in, 19–21, 20f
   matrices, 418
   protein, see Protein(s), immobilization
Immobilized metal affinity chromatography (IMAC), 10, 333, 366
Immune cell types, fractionation of, 454
Immunoadsorption, 405t
Immunoglobulin (Ig) solution, treatment with, 450
Implanted insulin-producing cells, 3D-house for, 282–286, 283f, 284f, 285f
Implanted pancreatic islet, cryogels as scaffolds for, 280–286
   neovascularisation in diabetic environment, 280–282
   3D-house for implanted insulin-producing cells, 282–286, 283f, 284f, 285f
Imprinting, cryogels in, 19–21, 20f
Induction, of neovascularisation in diabetic environment, 280–282
Industrial wastewater, 345–346, 346f
Inert cryogel chromatography, 448
Infuse, 228t
Initiators
   concentrations, effect of, 58–59
   free radical, see Free radical polymerization
   photochemical, 52
   radiation, 52–53
   redox, 52
   thermal, 51–52

Injury site, biomolecules at, 258
Insulin-producing cells, 3D-house for, 282–286, 283f, 284f, 285f
Integra®, 178, 179, 179f, 180f
Interconnectivity of pores
   in μ-CT, 64, 64f, 65f
   supermacroporous architecture with, 16
   swelling–deswelling kinetics analysis, 68–69
Interpenetrated polymer networks (IPN), 341, 436
   of cryogel, 458, 459
Inulinase, immobilizing, 322
*In vitro* biocompatibility, of cryogels, 75–76, 200–205, 202t–203t, 203f, 204f
*In vivo* biocompatibility, of cryogels, 205–209, 206t, 207f, 208f
Ion-exchange cryogel beads, 135
Ionic binding, protein immobilization by, 320
Ionotropic cryogels, 38
Iron
   PHEMA cryogel and, 400, 401f, 403
   toxic effects of, 399–400
Islet transplantation, pancreatic, see Pancreatic islet transplantation

## K

Kaldnes, 334, 336f
Kensey Nash Bone Void Filler XC, 228t

## L

Laccase, 310t
Lamellar columns, defined, 231–232
Laminin-gelatin cryogel, 263, 263f
Langmuir isotherm equilibrium adsorption, 132
Large particles, movement of, 447, 448f
Lectin affinity binding, protein immobilization via, 322
Leukocytes, filtration of, 458–460
Ligament tissue engineering, 229–230
Ligand-based cryogel composites, characteristics, 342t–343t
Lipase, 310t
Liver, cryogels for regeneration of, 189–190
Liver phantoms, 151–152, 152f
Localized toxicity, defined, 205
Low-density lipoprotein (LDL)
   coronary heart diseases, 403
   dextran-induced LDL precipitation, 405t
   extracorporeal elimination methods for, 405t
   PHEMA-anti-LDL antibody, 406, 407f
   protein component, 403
Lysozyme, 310t, 313

## M

Macrophages, on implant surface, 197
Macroporous cryogels
   beads, preparation with, 124–125, 125f
   materials, biocompatibility, 195–209
      *in vitro* biocompatibility, 200–205, 202t–203t, 203f, 204f
      *in vivo* biocompatibility, 205–209, 206t, 207f, 208f
      overview, 196–197
   matrices for filter, digital image, 459f
   MSCs seeded, alginate-based cryogel scaffolds, 294–295
Macroporous polymeric gels, porosity of, 99–104, 100f, 101f, 102f, 104f
Magnetic affinity cell separation (MACS), 445
Magnetic resonance elastography (MRE)
   mechanical properties of tissues, 27
   medical imaging, 146–147, 149–150, 165
Magnetic resonance imaging (MRI), properties, 149–150, 154f, 166, 167
Magnifuse bone graft, 228t
Maioregen, 217
Mammalian cell separation for biomedical applications; *see also* Cryogel chromatography column
   fractionation of immune cell types, 454
   separation of stem cell, 454–457
Mass balance equation, in mobile fluid phase, 131
Mass transfer, modelling
   for characterization of cryogel beds, 129–134
      capillary-based, 129–132; *see also* Capillary-based models
      numerical solution of model, 132–134
Mass transfer coefficient, 132
Material biocompatibility, factors affecting, 197–200, 198f
Mechanical deformation of cryogel matrix, 450–451, 451f
Mechanical properties, of cryogels, 69–72
   compression test, 69–70
   fatigue test, 70–71
   phantoms, 160–166, 161f, 162f, 163f, 164f, 165f
   physically cross-linked, 41
   rheology studies, 71–72, 72f
Mechanical stability, 17–18
Medical imaging
   cryogel tissue phantoms with uniform elasticity for, *see* Tissue phantoms
   elastography techniques for, 147–150
      MRE, 149–150
      OCE, 148–149
      USE, 147–148
Mercury porosimetry, 65–67, 67f
Mesenchymal stem cells (MSCs)
   cryo-preservation of, 231–232
   differentiation, 258
   immobilization of functional, 280
   immunomodulatory effect of, 287
   for neuroprotective and neurorestorative effect, 255, 255f
   for pancreatic islet support, 286–289, 289f
   role, protection of pancreatic islets, 280
   seeded macroporous alginate-based cryogel scaffolds, cryopreservation, 294–295
Metal affinity cryogel column, 422
Metal hydroxide coated macroporous polymers (MHCMPs), 338, 340, 340f
Metal ion chelating binding, protein immobilization via, 322–325, 323t
Metal-oxide composite cryogels, 336–340, 337f, 339f
Metal removal, in biogas production, 347–349, 348f
Methacrylic acid, 47t
$N$-Methacryloyl-L-tryptophan methyl ester (MATrp), 396, 398f
Methanol, 97
$N,N$-Methylene-bis-acrylamide (MBAAm), 15, 16, 48t
Methylmethacrylate (MMA), 47t, 458
Michael addition, cross-linking by, 44, 45f
Microbial cryogel systems, 349
Microcarrier beads, 418
Microchannel(s)
   for fluid-focusing, fabrication of, 116–117, 117f
   generation of aqueous drops and immiscible slug flow in, 118–119
   liquid-flow focusing, cryopolymerization and, 114–116, 115f
Micro-computed tomography (μ-CT) method
   SEM *vs.*, 64, 65t
   textural features of cryogels, 61–64, 62f, 63f, 64f, 65f, 65t
Microcontact printing, 200
Microcryogels, for cell delivery, 190
Micro-cutting technique, 116–117
Micro-flow focusing method
   fabrication of cryogel beads, microchannels, 114–124
      aqueous drops and immiscible slug flow, 118–119
      for fluid-focusing, 116–117, 117f
      freezing of aqueous drops in water-immiscible fluid, 119–120, 120f
      liquid-flow, 114–116, 115f
      overview, 114
      properties of cryogel beads, 120–124, 121f, 122f, 123f
      preparation of cryogels, 124–125, 125f
Micromachining, techniques for, 116–117

# Index

Micropollutants, defined, 341
Mobile fluid phase, mass balance equation in, 131
Modelling, of fluid flow and mass transfer, 129–134
   capillary-based models, 129–132
      adsorption equilibrium equation, 132
      axial dispersion coefficient, 132
      bed porosity and permeability, 130–131
      mass balance equation, 131
      mass transfer coefficient, 132
      pore structure, 129f
      pore tortuosity and size distributions, 130
   numerical solution, 132–134
Modifiability, 18
Modulus of elasticity, 70
Molecularly imprinted polymers (MIPs)
   adsorbent materials, 341
   composite cryogels with, 335, 336, 341, 344–345, 344f
   MIP-bilirubin cryogel, 393–395
   PHEMA/MIP composite cryogel, 395, 396f, 397t
   PHEMA-MIP cryogels, 402f, 403
Moloney Murine Leukaemia Virus (MoMuLV), 453
MoMuLV, *see* Moloney Murine Leukaemia Virus
Monoclonal antibody (mAb)
   manufacture, 418
   in PAAC cryogel matrix, 437f
   productivity, 435t
   and urokinase, 421
Monoclonal antibody production; *see also* Therapeutic protein production, cryogel bioreactor for
   in PAAm-gelatin cryogel, 430–434
   polyacrylamide and chitosan (PAAC) cryogel, 434–436, 435t
Monolithic cryogels, composite, 124–128
   capillary-based models for, 129–132
      adsorption equilibrium equation, 132
      axial dispersion coefficient, 132
      bed porosity and permeability, 130–131
      mass balance equation, 131
      mass transfer coefficient, 132
      pore structure, 129f
      pore tortuosity and size distributions, 130
   overview, 112–113, 113f
   preparation, via cryo-polymerization, 124–128
      with embedded macroporous beads, 124–125, 125f
      grafting polymerization, 125–126, 126f
      properties, 126–128, 127f, 128f
Monolith of polyacrylamide-chitosan bioreactor, 431f, 432f
Monomer, effect of, 58–59

Morphology, of cryogels, 59–69
   CLSM, 60, 60f, 61, 61f, 62f
   hydrated polysaccharide-based cryogels, 60, 60f
   μ-CT, 61–64, 62f, 63f, 64f, 65f, 65t
   porosity, 64–69
      chitosan cryogels, 66
      gelatin cryogels, 63f, 64, 64f, 65f, 66
      interconnectivity of pores, 64–69, 64f, 65f
      mercury intrusion porosimetry, 65–67, 67f
      nitrogen adsorption method, 67
      NMR cryoporometry, 67–68
      SEM *vs.* micro-CT, 65t
      swelling–deswelling kinetics analysis, 68–69
      total pore surface area, 66
      total pore volume, 66
      Washburn equation, 66
Mouse model
   biocompatibility of cryogels
      *in vitro*, 200–204, 202t
      *in vivo*, 205–209, 206t, 207f, 208f
   chitosan-agarose-gelatin cryogel matrices, 189
   cranial defect model, 233, 235f
   PAAm cryogel matrix, 204, 204f
   pancreatic islets, 190
   PVA cryogel, for culturing mouse fibroblasts, 39
Mouthwashes, antibacterial, 190
MTF new bone void filler, 228t
Multiphoton microscopy (MPM), 61
Municipal wastewater, 345–346, 346f
Muscle neoplasia, treatment, 217
Muscle tissue engineering, 229
Musculoskeletal tissue engineering, 215–243
   biomaterials for, 218–230
      bone, 225, 228–229, 228t
      cartilage, 219–225, 220t–224t, 226t–227t
      ligament, 229–230
      muscle, 229
      tendon, 229
   conditions, prevalence of, 218
   cryogels for bone tissue engineering, 230–235
      gelatin-based, 231
      HEMA based, 232
      polyethylene glycol–based, 231–232
      polymeric, 231–232
      polymeric-ceramic/bioactive glass composite, 232–236, 233f, 234f, 235f, 236f, 237f
      silk fibroin based, 232
   cryogels for cartilage tissue engineering, 236–243
      natural polymers, fabricated using, 237–243, 239f, 241f, 242f
      synthetic polymers, fabricated using, 243
   overview, 216–218
   significance of cryogel technology, 230
Myocardial tissue, regeneration of, 189

## N

Nanocomposites, defined, 336
Nanoparticles (NPs), composite cryogels, 335, 336–337
Naringinase, 318
Natural polymers, cryogel fabricated using, 237–243, 239f, 241f, 242f
NeoCart, 226t
Neo-cartilage generation, 240, 242f
Neovascularisation, in diabetic environment, 280–282
Nerve conduits, 259, 260–261, 260f
Nervous system, 253–254, 253f, 254f
Neural recovery, hydrogels for, 258, 259, 260t
Neural regeneration
    cell therapy for, 254–255, 255f
    electrospun fibres for, 258, 259f
    tissue, 189
Neural tissue engineering, 251–268
    conducting polymer, 261–262
    conventional strategies, 255–258, 256f
    cryogels, 262–266, 263f, 265f, 266f
        conducting, 267–268, 267f
    fabrication of scaffolds, 258–259
        delivery of biomolecules at injury site, 258
        electrospun fibres for neural regeneration, 258, 259f
        hydrogels for neural recovery, 258, 259, 260t
        nerve conduits, 259, 260–261, 260f
        nervous system, 253–254, 253f, 254f
    overview, 252–253
    regeneration, cell therapy for, 254–255, 255f
Neutrophils, in material degradation, 197
$N$-isopropyl acrylamide (NIPAAm), 48t, 376
Nitrogen adsorption method, 67
Nonadsorption conditions, mass balance equation in mobile fluid phase under, 131
Non-animal derived polymers, 434–436
Nonfrozen liquid microphase (NFLMP)
    cryoconcentration, 56
    polymer-rich region, generation, 41
NovaBoneMacroPor-Si$^+$—Bioactive Bone Graft, 228t
Novocart 3D, 217
Nuclear magnetic resonance (NMR) cryoporometry, 67–68
Nucleation
    crystallization of water and, 92
    temperature, defined, 92
Nutrient transport, 420
$N$-Vinyl pyrrolidone, 47t

## O

Oil removal, 347
OP-1, 228t
Optical coherence elastography (OCE), 146, 148–149
Optimal conditions, for cell separation, 364–366, 365t, 367f
Organophosphate hydrolase (OPH)
    covalent immobilization on cryogel-type carriers, 310t
    immobilization of, 323
Origin, cryogels, 4–5
Osseofit, 217
OsteoSelect demineralized bone matrix putty, 228t

## P

PAAm, *see* Polyacrylamide (PAAm)
Packed-bed bioreactors, 418–419
Palpation, defined, 146
Pancreas, cryogels for regeneration of, 189–190
Pancreatic islet transplantation, 277–295
    cryogels as scaffolds, 280–286
        neovascularisation in diabetic environment, 280–282
        3D-house for implanted insulin-producing cells, 282–286, 283f, 284f, 285f
    overview, 278–280, 279f
    ready-to-use 3D cryogel/stem cell, 286–295
        alginate-based macroporous cryogel scaffolds, 289–294, 290f, 292f, 293f
        cryopreservation of MSCs seeded macroporous alginate-based cryogel scaffolds, 294–295
        MSCs, for support, 286–289, 289f
Panning, of phage display library, 376, 377, 378t
Paraneoplastic syndromes, 398
Particle-based cryogel composites, characteristics, 342t–343t
Particle induced X-ray emission (PIXE), 351
Perfusion bioreactors, 419
Peripheral nervous system (PNS), 253–254, 253f, 254f
Permeability, cryogel bed, 130–131
Phage display library, panning of, 376, 377, 378t
Phantoms
    cryogel, 155–160
        characterization and homogeneity, 160–166; *see also* Cryogel phantoms
        defined, 155
        formation process, 155f
        other polymers, 158–160, 159t, 160f
        PVA homopolymer, 156–158, 156f, 157f, 158f
    elastography, need for, 150–155
        brain, 154–155, 154f
        breast, 152–154, 153f
        liver, 151–152, 152f
        mechanical properties for TMMs, 151t
    tissue, *see* Tissue phantoms

# Index

Phosphate buffer saline (PBS), 456, 458
Photochemical initiators, 52
Phototherapy, effectiveness of, 393
Physically cross-linked cryogels, 38–43, 42f, 43f
Plasma exchange, rheumatoid arthritis and, 398
Plasmapheresis, 405t
Plasminogen activator inhibitor (PAI), 425
Point-of-use (POU) filters, 346–347
Pollutant analysis, 377, 379, 380f–381f
Pollutants, environmental, 332, 362
Pollutant source, as adsorbents, 345–349
    industrial and municipal wastewater, 345–346, 346f
    metal removal, in biogas production, 347–349, 348f
    microbial cryogel systems, 349
    oil removal, advantage of cryogel pore space, 347
    POU filters, 346–347
Polyacrylamide and chitosan (PAAC) cryogel, 434–436, 435t; *see also* Monoclonal antibody production
    biological and physical properties, 425t
    growth of 6A4D7 hybridoma cell line on, 424f, 426f
    scanning electron microscope image of, 423f
Polyacrylamide (pAAm) cryogels, 453
    flow-focusing and cryo-polymerization method, 119–120, 120f
    growth of mouse hybridoma cells, 204, 204f
    with polymers, cross-linkers/initiators, and mechanism, 17t
    pore architecture in, 58f
    properties, 120–124, 121f, 122f, 123f
Polyacrylamide (pAAm)-gelatin cryogel, 430–434; *see also* Monoclonal antibody production
Poly(acrylonitrile) (PAN), 15, 16, 46t, 74–75, 424, 425t
Poly-aromatic hydrocarbons (PAHs)
    contamination, 377, 379
    removal, 25–26
Poly(dimethylaminoethyl methacrylate) poly(DMAEMA), 458
Polyester (PES) cryogels, crystallinity of, 71
Poly(ethylene glycol) (PEG), 46t
    based cryogels, for bone tissue engineering, 231–232
    glucose, presence, 26–27
    molecular weight of, 318
Polyethylene glycol diacrylate (PEGDA), 231
Polyfuse—Neurobone, 228t
Poly-hydroxyethylmethacrylate (PHEMA) cryogels
    based adsorbents, 389–391, 390f, 391f
    freezing temperature, 119
    gelatin and, 16
    in immobilization, separation, and imprinting, 20, 21
    immobilization by ionic binding, 320
    iron imprinted, 400, 401f, 403
    MATrp containing, 396
    PHEMA-anti-LDL antibody, 406, 407f
    pHEMA-lactate-dextran cryogels, 22
    PHEMA-MIP cryogels, 402f, 403
    composite, 395, 396f, 397t
    properties, 121, 123
    with protein A, 398, 399f
    urease, immobilization, 319
Poly(hydroxyethyl methacrylate-N-methacryloyl-(L)-tyrosinemethylester) (poly(HEMA-MATyr)), 394, 394f
Poly isoprene, 47t
Polymeric biomaterials, for bone tissue engineering, 225
Polymeric-ceramic composite cryogels, for bone tissue engineering, 225, 228, 232–236, 233f, 234f, 235f, 236f, 237f
Polymeric cryogels, for bone tissue engineering, 231–232
    gelatin-based, 231
    HEMA based, 232
    polyethylene glycol-based, 231–232
    silk fibroin based, 232
Polymerization
    at cryo-conditions, 53
    cryopolymerization, *see* Cryopolymerization
    free radical
        cryogelation by, *see* Free radical polymerization, cryogelation by
        mechanism of, 48, 49–51
    grafting
        of composite cryogels, 125–126, 126f
        on pore surface, 104–108, 105f, 106f, 107f
Polymers
    with AGS, cross-linkers/initiators, and mechanism, 17t
    conducting, 261–262
    cryogel fabricated using
        natural, 237–243, 239f, 241f, 242f
        synthetic, 243
Poly(methylmethacrylate), 47t
Poly(N-isoproppylacrylamide) (PNiPAAm), 424
Poly(N-isoproppylacrylamide)–chitosan (PNiPAAm–chitosan), 424
Polysaccharide cryogels, physically cross-linked, 39–40
Polystyrene, 47t
Polystyrene divinylbenzene (PS-DVB)
    microparticles, 409–410
Polytetrafluoroethylene (PTFE), 46t
Poly(vinyl acetate), 46t

Poly-vinyl alcohol (PVA) cryogels
   cryo-structurization of polymeric systems, 9
   crystallinity of, 71–72, 73–74
   by freeze–thaw treatment, 38–39
   haemolytic activity of, 200
   homopolymer, 156–158, 156f, 157f, 158f
   monomers for addition polymerization, 46t
   physically cross-linked, 41–42, 42f
   study of physical gelation, 38
   thermal stability, 72, 72f
Polyvinyl alcohol-tetraethylorthosilicate-agarose-calcium chloride (PTAgC) biocomposite cryogels, 233, 236f
Polyvinyl alcohol-tetraethylorthosilicate-alginate-calcium oxide (PTAC), 188, 232–233, 233f, 234f
Poly(vinyl butyral), 46t
Polyvinylcaprolactam (pVCl), 205–209, 207f, 208f
Poly(vinyl chloride), 46t
Poly(vinylidene chloride), 46t
Polyvinyl pyrrolidone (PVP), 159, 159t, 164, 164f, 199
Pore space, cryogel, 347
Pore tortuosity and size distributions, capillaries, 130
Porogen, effect of, 57–58, 58f
Porosity, of cryogels, 64–69
   addition of methanol/acetone, 97–99, 98f, 100f
   capillary-based models, 130–131
   chitosan cryogels, 66
   in completely hydrated state, 99, 100f
   gelatin cryogels, 63f, 64, 64f, 65f, 66
   grafting polymers on pore surface, 104–108, 105f, 106f, 107f
   interconnectivity of pores, 64–69, 64f, 65f
   macroporous polymeric gels, methods for, 99–104, 100f, 101f, 102f, 104f
   mercury intrusion porosimetry, 65–67, 67f
   nitrogen adsorption method, 67
   NMR cryoporometry, 67–68
   SEM vs. micro-CT, 65t
   swelling–deswelling kinetics analysis, 68–69
   total pore surface area, 66
   total pore volume, 66
   Washburn equation, 66
Postincubation, 421
Potassium diperiodatocuprate, graft polymerization, 105, 106f
Potassium persulphate, 49t
Prevalence, of musculoskeletal conditions, 218
Properties, cryogels, 16, 17–18
   biocompatibility, 18
   biological, 74–76
      *in vitro* biocompatibility, 75–76
      protein adsorption studies, 74–75
   chemical, 73–74
      FTIR, 73
      XPS and elemental analysis, 74
      XRD, 73–74
   chemical stability and modifiability, 18
   composite cryogels, 126–128, 127f, 128f
   cryogel beads, 120–124, 121f, 122f, 123f
   ease of fabrication, economical operations and longer shelf life, 18
   mechanical and thermal, 69–72
      compression test, 69–70
      fatigue test, 70–71
      rheology studies, 71–72, 72f
   mechanical stability and viscoelasticity, 17–18
   physical, factors affecting, 56–59
      freezing rate and temperature, 57
      monomer and initiator concentrations, effect of, 58–59
      porogen and solvent, effect of, 57–58, 58f
   rapid swelling mechanism, 16, 17
   regenerative medicine, 176–177
   supermacroporous architecture with interconnected pores, 16
Prosthetics, polymeric/ceramic/metal, 225
Protected cryogels, 334–335, 335f, 336f
Protein(s)
   adsorption studies, 74–75
   high throughput analysis of, 366–372, 368f, 369f, 370f, 371t, 373f–374f
   immobilization
      affine binding, 320–325
      antigen-antibody binding, 321
      combined methods, 325
      covalent attachment to cryogel matrix, 309–315, 310t, 314f
      entrapment technique, 315–319, 316t
      ionic binding, 320
      lectin affinity binding, 322
      metal ion chelating binding, 322–325, 323t
Proteoglycans, 253
*Pseudomonas aeruginosa*, 186
Psychrotropic cryogels, 38

## R

Radiation initiation, 52–53
Rapid swelling mechanism, 16, 17
Ready-to-use 3D cryogel/stem cell constructs, 286–295
   alginate-based macroporous cryogel scaffolds, 289–294, 290f, 292f, 293f
   cryopreservation of MSCs seeded macroporous alginate-based cryogel scaffolds, 294–295
   MSCs, for support, 286–289, 289f

# Index

Red blood cells (RBC)-sedimented UCB (RS-UCB) cells, 456, 457, 457f
Redox initiation, 52
Regeneration
  neural
    cell therapy for, 254–255, 255f
    electrospun fibres for, 258, 259f
    tissue, 189
    system, in water and wastewater treatment, 349–350
Regenerative medicine, cryogels in, 175–190
  bone, regeneration, 188–189
  cartilage repair and regeneration, 186–188
  component of topical wound dressings, 184–186, 184f, 185f, 186f
  dermal tissue regeneration, 177–184, 179f, 180f, 181f, 182f, 183f
  liver and pancreas, regeneration, 189–190
  microcryogels for cell delivery, 190
  myocardial tissue, regeneration, 189
  neural tissue, regeneration, 189
  overview, 176
  properties, cryogel, 176–177
Repair, cartilage, 186–188
ReproBone, 228t
Rheology studies, 71–72, 72f
Rheumatoid arthritis, affinity therapy in, 397, 398–399, 399f, 399t
*Rhodococcal* cells, 24
Risk assessment, in water and wastewater treatment, 351–352, 352f

## S

*Saccharomyces cerevisiae*, 54, 365–366, 365t
Savinase, 310t
Scaffolds
  fabrication, 258–259
    delivery of biomolecules at injury site, 258
    electrospun fibres for neural regeneration, 258, 259f
    hydrogels for neural recovery, 258, 259, 260t
  supermacroporous cryogels, for pancreatic islet transplantation, *see* Pancreatic islet transplantation
Scanning electron microscopy (SEM)
  cryo-SEM techniques, 60
  ESEM, 60, 60f
  of human acute myeloid leukaemia cells, 456f
  image of PAAC cryogel, 423f
  μ-CT *vs.*, 64, 65t
Schiff-base formation, 55
Screening, high throughput, 364–366, 365t, 367f
Scrubs, antibacterial, 186

Seeded chondrocytes, in cartilage tissue engineering, 237–238
Seeded macroporous alginate-based cryogel scaffolds, MSCs, 294–295
Separation, cryogels in, 19–21, 20f
Sepharose adsorbents, 135
Sepharose column, 427
Sepharose immobilized metal affinity chromatography columns, 429t
Serial sorting device, 445
Shelf life, long, 18
Silk fibroin based cryogels, for bone tissue engineering, 232
SIRAN® glass spheres, 419
Smith & Nephew, 226t
Soft lithography, 200
Sol, defined, 37
Sol–gel transition, defined, 37–38
Solutes, on cryopolymerization, 97–99, 98f, 100f
Solvents, effect of, 57–58, 58f
Solvotropic cryogels, 38
Speckle tracking, 149
Stability, of cryogels
  properties
    chemical, 18
    mechanical, 17–18
    thermal, physically cross-linked cryogels, 41
*Staphylococcus aureus*, 186
Starch, cryogels of, 41
Statistical perspective, cryogels, 11–15, 11f, 12f, 12t, 13f, 14f
Stem cells
  ADSCs, 201, 202t, 203, 209
  anti-inflammatory effects, 264
  CBSCs, 264
  differentiation, 257
  hMSCs, 178, 202t
  human cord blood-derived, 189
  MSCs, *see* Mesenchymal stem cells (MSCs)
  neural progenitor, 262–263
  ready-to-use 3D cryogel/stem cell, 286–295
    alginate-based macroporous cryogel scaffolds, 289–294, 290f, 292f, 293f
    cryopreservation of MSCs seeded macroporous alginate-based cryogel scaffolds, 294–295
    MSCs, for support, 286–289, 289f
  seeded, 258
  separation of, 454–457; *see also* Mammalian cell separation for biomedical applications
Stoke's law, 338
Subtilisin, 310t, 311–312
Sulfamethazine, 325
Supermacroporous cryogels
  architecture with interconnected pores, 16
  matrices, 428f, 432f

as scaffolds for pancreatic islet
transplantation, *see* Pancreatic islet
transplantation
Supermacroporous functional cryogel stationary
matrices
for cell separation stationary matrices
flow rate, 446
functional matrix, 448–450
movement of large particles, enabling,
447, 448f
versatile cell recovery properties,
450–451
cell separation technology
current techniques/demand in, 445
importance of, 444–445
cryogel chromatography column, application
of
boronate affinity chromatography, 458
filtration of leukocytes, 458–460
mammalian cell separation for biomedical
applications, 454–457
prokaryotic cell separation, 451–454
Swelling–deswelling kinetics analysis, 68–69
Symphony® I/C graft chambers, 228t
Synthesis and characterization, of cryogels,
36–76
characterization, 59–76; *see also*
Characterization
biological properties, 74–76
chemical properties, 73–74
mechanical and thermal properties,
69–72
morphological features, *see* Morphology
textural features, *see* Textural features
cryogelation by free radical polymerization,
53–56
adipic acid hydrazide, cross-linking with,
55
aldehydes, cross-linking with, 55
EDC-NHS, cross-linking with, 56, 56f
functional group cross-linking, 55–56
using UV irradiation, 54–55, 54f
gelation, process, 37–38
gel formation, different methods of inducing,
38–53
addition polymerization, 44, 45–53
chemical cross-linking, 44, 44f, 45f, 45t
cryogelation by free radical
polymerization, 44, 45, 46t–48t, 48,
49t
mechanism of free radical
polymerization, 48, 49–51
photochemical initiators, 52
physical cross-linking, 38–43, 42f, 43f
radiation initiation, 52–53
redox initiation, 52
thermal initiators, 51–52

overview, 15–16, 17t, 36–37
physical properties and factors affecting
properties, 56–59
freezing rate and temperature, 57
monomer and initiator concentrations,
effect of, 58–59
porogen and solvent, effect of, 57–58, 58f
Synthetic cryogels, 91–108
cryopolymerization, mechanism, 94–97, 95f,
94f, 95f
freezing of aqueous solutions, 92–93, 92f, 93f
grafting polymers on pore surface, 104–108,
105f, 106f, 107f
methods, for porosity of macroporous
polymeric gels, 99–104, 100f, 101f,
102f, 104f
overview, 92
solutes on cryopolymerization, 97–99, 98f, 100f
Synthetic polymers, cryogel fabricated using, 243
Systemic lupus erythematosus (SLE), affinity
therapy in, 389–393
anti-dsDNA antibody concentration, 391,
392f
flow pattern of whole blood, 392f, 393
PHEMA-based adsorbents, 389–391, 390f, 391f
Systemic toxicities
category, 205
characteristics, 205
System regeneration, 349–350

## T

Taylor-Aris correlation, 132
Technology, cryogel, in musculoskeletal tissue
engineering, 230
Teflon, 46t
Temperature, freezing rate and, 57
Tendon tissue engineering, 229
"Tentacle" adsorption, of protein, 107–108
tert-butyl peroxide, 49t
Tetramethylethylenediamine (TEMED), 16, 53
Tetrapeptides, synthesis of, 311, 312
Textural features, of cryogels, 59–69
CLSM, 60, 60f, 61, 61f, 62f
hydrated polysaccharide-based cryogels, 60,
60f
μ-CT, 61–64, 62f, 63f, 64f, 65f, 65t
porosity, 64–69
chitosan cryogels, 66
gelatin cryogels, 63f, 64, 64f, 65f, 66
interconnectivity of pores, 64–69, 64f, 65f
mercury intrusion porosimetry, 65–67, 67f
nitrogen adsorption method, 67
NMR cryoporometry, 67–68
SEM *vs.* micro-CT, 65t
swelling–deswelling kinetics analysis,
68–69

# Index

total pore surface area, 66
total pore volume, 66
Washburn equation, 66
Thalassemia, affinity therapy in, 399–403, 401f, 402f, 404f
Therapeutic protein production, cryogel bioreactor for
  cell support matrix in extracorporeal bioartificial device, 436, 438
  hollow fibre bioreactor (HFBR), 419, 419f
  network and physical characteristics, cryogel bioreactor, 420–421
  overview, 418
  setup and advantages
    about, 421–422
    fabrication method/physical and biological characterization, 422–425, 423f, 425t
    monoclonal antibody production, 430–436, 431f–434f, 435t, 437f
    urokinase production in integrated setup, 425–430, 426f–428f, 429t
Thermal initiators, 51–52
Thermally stimulated depolarization (TSD), 101
Thermal properties, cryogels, 69–72
  compression test, 69–70
  fatigue test, 70–71
  rheology studies, 71–72, 72f
Thermal stability, of physically cross-linked cryogels, 41
Thermodynamic feasibility, monomers to polymers on, 50
Thermofiltration, 405t
Thermogravimetric analysis (TGA) method
  temperature-dependent behaviour of water, 101
  thermal properties of cryogel, 71
Thermolysin, 310t, 311
Thermotropic cryogels, 38
*Thiobacillus denitrificans*, 349
Three-dimensional (3D) cryogel/stem cell, ready-to-use, 286–295
  alginate-based macroporous cryogel scaffolds, 289–294, 290f, 292f, 293f
  cryopreservation of MSCs seeded macroporous alginate-based cryogel scaffolds, 294–295
  MSCs, for support, 286–289, 289f
Three-dimensional (3D)-house, for implanted insulin-producing cells, 282–286, 283f, 284f, 285f
Tissue Bank of France, 226t
Tissue engineering, cryogels in, 13, 21–23, 24f
  musculoskeletal, biomaterials for, *see* Musculoskeletal tissue engineering
  neural, *see* Neural tissue engineering

Tissue-mimicking materials (TMMs), as elastography phantoms, 146, 147, 150–151, 151t, 154
Tissue phantoms, 146–166
  cryogel, 155–160
    characterization and homogeneity, 160–166; *see also* Cryogel phantoms
    defined, 155
    formation process, 155f
    other polymers, 158–160, 159t, 160f
    PVA homopolymer, 156–158, 156f, 157f, 158f
  elastography phantoms, need for, 150–155
    brain, 154–155, 154f
    breast, 152–154, 153f
    liver, 151–152, 152f
    mechanical properties for TMMs, 151t
  elastography techniques, 147–150
    MRE, 149–150
    OCE, 148–149
    USE, 147–148
  overview, 146–147
Tissue plasminogen activator (tPA), 425
Tissue regeneration, cryogels for
  bone, 188–189
  dermal, 177–184, 179f, 180f, 181f, 182f, 183f
  myocardial, 189
  neural, 189
Titanate nanotubes (TNTs), 336, 338, 342t, 348
Topical wound dressings, cryogels as component, 184–186, 184f, 185f, 186f
Tortuosities, 447f
Toxicities
  iron, 399–400
  systemic and localized, 205
Trabecular bone, 188
Tracking, speckle, 149
Transplantation, of pancreatic islets, *see* Pancreatic islet transplantation
TRS Cranial Bone Void Filler (TRS C-BVF), 228t
TrueFit Plug, 226t
Trypan blue staining of cells, 454
Trypsin, 310t, 311
Tube panning protocol, 378t
Tubulin, 201

## U

UCB, *see* Umbilical cord blood (UCB)
Ultrasound elastography (USE), for medical imaging, 146, 147–148
Umbilical cord blood (UCB), 456–457
Unfrozen liquid microphase (ULMP), 7, 8
Uniform elasticity, cryogel tissue phantoms with, *see* Tissue phantoms
Urease, 316t, 319

Urokinase production in integrated setup, 425–430, 426f–428f, 429t; *see also* Therapeutic protein production, cryogel bioreactor for
US Environmental Protection Agency (USEPA), 350
UV irradiation, cryogelation by free radical polymerization using, 54–55, 54f

## V

Vascular endothelial growth factor (VEGF), 188–189
VeriCart, 226t
Vibration amplitude sonoelastography, 148
Vinyl acetate, 47t
Viscoelasticity, 17–18
Vitoss, 228t
Von Willebrand factor (vWF), 453

## W

Wallerian degeneration, 254, 254f
Washburn equation, 66
Wastewater
   industrial and municipal, 345–346, 346f
   treatment; *see also* Water and wastewater treatment
      adsorption, 333
      costs, 352–353
      deposition, 350–351
      risk assessment, 351–352, 352f
      system regeneration, 349–350
      WWTPs, 334
Wastewater treatment plants (WWTPs), 334
Water and wastewater treatment, 331–353
   adsorbents, cryogel-based, 333–335
      functionalized, 333–334
      protected, 334–335, 335f, 336f
   adsorbents, pollutant source, 345–349
      industrial and municipal wastewater, 345–346, 346f
      metal removal, in biogas production, 347–349, 348f
      microbial cryogel systems, 349
      oil removal, advantage of cryogel pore space, 347
      POU filters, 346–347
   composite cryogels, 335, 336–345
      ligand-based and particle-based, characteristics, 342t–343t
      metal-oxide, 336–340, 337f, 339f
      with molecularly imprinted polymers, 341, 344–345, 344f
      other types, 340, 341, 342t–343t
   future challenges, 349–353
      costs, 352–353
      deposition, 350–351
      risk assessment, 351–352, 352f
      system regeneration, 349–350
   overview, 332–333
   WWTPs, 334
Water-immiscible fluid, freezing of aqueous drops in, 119–120, 120f
Wave propagation, 166
Wound dressings, topical, 184–186, 184f, 185f, 186f

## X

Xanthan gum, cryogels of, 40
Xenografts, 217, 225, 257
X-ray diffraction (XRD), 73–74
X-ray photoelectron spectroscopy (XPS), 74

## Y

Young's modulus
   chitosan-gelatin cryogels and, 70
   defined, 70
   of PVA-C, 161–163, 162f

## Z

Zerdine™ gel, 152
Z-stacks, 61

For Product Safety Concerns and Information please contact our EU representative GPSR@taylorandfrancis.com Taylor & Francis Verlag GmbH, Kaufingerstraße 24, 80331 München, Germany

Printed and bound by CPI Group (UK) Ltd, Croydon, CR0 4YY

08/06/2025

01896985-0008